AF443070

Rare-Earths and Actinides in High Energy Spectroscopy

Progress in Theoretical Chemistry and Physics

VOLUME 28

Honorary Editors

Sir Harold W. Kroto (*Florida State University, Tallahassee, FL, USA*)
Prof. Yves Chauvin (*Institut Français du Pétrole, Tours, France*)

Editors-in-Chief

J. Maruani (*formerly Laboratoire de Chimie Physique, Paris, France*)
S. Wilson (*formerly Rutherford Appleton Laboratory, Oxford, UK*)

Editorial Board

E. Brändas (*University of Uppsala, Uppsala, Sweden*)
L. Cederbaum (*Physikalisch-Chemisches Institut, Heidelberg, Germany*)
G. Delgado-Barrio (*Instituto de Matemáticas y Física Fundamental, Madrid, Spain*)
E.K.U. Gross (*Freie Universität, Berlin, Germany*)
K. Hirao (*University of Tokyo, Tokyo, Japan*)
E. Kryachko (*Bogolyubov Institute for Theoretical Physics, Kiev, Ukraine*)
R. Lefebvre (*Université Pierre-et-Marie-Curie, Paris, France*)
R. Levine (*Hebrew University of Jerusalem, Jerusalem, Israel*)
K. Lindenberg (*University of California at San Diego, San Diego, CA, USA*)
A. Lund (*University of Linköping, Linköping, Sweden*)
R. McWeeny (*Università di Pisa, Pisa, Italy*)
M.A.C. Nascimento (*Instituto de Química, Rio de Janeiro, Brazil*)
P. Piecuch (*Michigan State University, East Lansing, MI, USA*)
M. Quack (*ETH Zürich, Zürich, Switzerland*)
S.D. Schwartz (*Yeshiva University, Bronx, NY, USA*)
A. Wang (*University of British Columbia, Vancouver, BC, Canada*)

Former Editors and Editorial Board Members

I. Prigogine (†)	W.F. van Gunsteren (*)
J. Rychlewski (†)	H. Hubač (*)
Y.G. Smeyers (†)	M.P. Levy (*)
R. Daudel (†)	G.L. Malli (*)
M. Mateev (†)	P.G. Mezey (*)
W.N. Lipscomb (†)	N. Rahman (*)
H. Ågren (*)	S. Suhai (*)
V. Aquilanti (*)	O. Tapia (*)
D. Avnir (*)	P.R. Taylor (*)
J. Cioslowski (*)	R.G. Woolley (*)

†: deceased; *: end of term

More information about this series at http://www.springer.com/series/6464

Christiane Bonnelle · Nissan Spector

Rare-Earths and Actinides in High Energy Spectroscopy

 Springer

Christiane Bonnelle
Laboratoire de Chimie Physique
  Matière et Rayonnement
Sorbonne Universités, UPMC 06
Paris
France

Nissan Spector
Sorbonne Universités, UPMC 06
Paris
France

and

Soreq Nuclear Research Center
Yavne
Israel

ISSN 1567-7354   ISSN 2215-0129   (electronic)
Progress in Theoretical Chemistry and Physics
ISBN 978-90-481-2878-5   ISBN 978-90-481-2879-2   (eBook)
DOI 10.1007/978-90-481-2879-2

Library of Congress Control Number: 2015954591

Springer Dordrecht Heidelberg New York London
© Springer Science+Business Media Dordrecht 2015

Springer Science+Business Media B.V. Dordrecht is part of Springer Science+Business Media
(www.springer.com)

PTCP Aim and Scope

Progress in Theoretical Chemistry and Physics

A series reporting advances in theoretical molecular and material sciences, including theoretical, mathematical and computational chemistry, physical chemistry and chemical physics and biophysics.

Aim and Scope

Science progresses by a symbiotic interaction between theory and experiment: theory is used to interpret experimental results and may suggest new experiments; experiment helps to test theoretical predictions and may lead to improved theories. Theoretical Chemistry (including Physical Chemistry and Chemical Physics) provides the conceptual and technical background and apparatus for the rationalization of phenomena in the chemical sciences. It is, therefore, a wide ranging subject, reflecting the diversity of molecular and related species and processes arising in chemical systems. The book series *Progress in Theoretical Chemistry and Physics* aims to report advances in methods and applications in this extended domain. It will comprise monographs as well as collections of papers on particular themes, which may arise from proceedings of symposia or invited papers on specific topics as well as from initiatives from authors or translations.

The basic theories of physics—classical mechanics and electromagnetism, relativity theory, quantum mechanics, statistical mechanics, quantum electrodynamics—support the theoretical apparatus which is used in molecular sciences. Quantum mechanics plays a particular role in theoretical chemistry, providing the basis for the valence theories, which allow to interpret the structure of molecules, and for the spectroscopic models, employed in the determination of structural information from spectral patterns. Indeed, Quantum Chemistry often appears synonymous with Theoretical Chemistry; it will, therefore, constitute a major part of this book series. However, the scope of the series will also include other areas of theoretical chemistry, such as mathematical chemistry (which involves the use of algebra and topology in the analysis of molecular structures and reactions); molecular mechanics, molecular dynamics, and chemical thermodynamics, which play an important role in rationalizing the geometric and electronic structures of molecular assemblies and polymers, clusters, and crystals; surface, interface, solvent, and solid state effects; excited-state dynamics, reactive collisions, and chemical reactions.

Recent decades have seen the emergence of a novel approach to scientific research, based on the exploitation of fast electronic digital computers. Computation provides a method of investigation which transcends the traditional division between theory and experiment. Computer-assisted simulation and design may afford a solution to complex problems which would otherwise be intractable to theoretical analysis, and may also provide a viable alternative to difficult or costly laboratory experiments. Though stemming from Theoretical Chemistry, Computational Chemistry is a field of research in its own right, which can help to test theoretical predictions and may also suggest improved theories.

The field of theoretical molecular sciences ranges from fundamental physical questions relevant to the molecular concept, through the statics and dynamics of isolated molecules, aggregates and materials, molecular properties and interactions, to the role of molecules in the biological sciences. Therefore, it involves the physical basis for geometric and electronic structure, states of aggregation, physical and chemical transformations, thermodynamic and kinetic properties, as well as unusual properties such as extreme flexibility or strong relativistic or quantum-field effects, extreme conditions such as intense radiation fields or interaction with the continuum, and the specificity of biochemical reactions.

Theoretical Chemistry has an applied branch (a part of molecular engineering), which involves the investigation of structure-property relationships aiming at the design, synthesis and application of molecules and materials endowed with specific functions, now in demand in such areas as molecular electronics, drug design or genetic engineering. Relevant properties include conductivity (normal, semi- and super-), magnetism (ferro- and ferri-), optoelectronic effects (involving nonlinear response), photochromism and photoreactivity, radiation and thermal resistance, molecular recognition and information processing, biological and pharmaceutical activities, as well as properties favouring self-assembling mechanisms and combination properties needed in multifunctional systems.

Progress in Theoretical Chemistry and Physics is made at different rates in these various research fields. The aim of this book series is to provide timely and in-depth coverage of selected topics and broad-ranging yet detailed analysis of contemporary theories and their applications. The series will be of primary interest to those whose research is directly concerned with the development and application of theoretical approaches in the chemical sciences. It will provide up-to-date reports on theoretical methods for the chemist, thermodynamician or spectroscopist, the atomic, molecular or cluster physicist, and the biochemist or molecular biologist who wish to employ techniques developed in theoretical, mathematical and computational chemistry in their research programs. It is also intended to provide the graduate student with a readily accessible documentation on various branches of theoretical chemistry, physical chemistry, and chemical physics.

Preface

According to the Roman poet Ovid, the Ages of Man are four: Gold, Silver, Bronze and Iron. In his book 'Metamorphoses', he tells about the myth of the four Ages. While the Golden Age and the Silver Age are symbolic and represent spirituality, justice and peace, in the Bronze Age men actually used bronze and during the Iron Age they used iron. These two metals played a cardinal role in the development of mankind. Their importance was evident upon their discovery and their use was immediate.

Saint Jerome gave the chronology of these Ages, placing the Bronze Age between the seventeenth and fifteenth centuries B.C. and the Iron Age from the fifteenth century B.C. to his days, around 400 A.D. In modern history, the Bronze Age started around 5000 B.C. and lasted until about 1000 B.C., when the Iron Age began. It lasted until 1 B.C.

Two millennia later, the twentieth century has witnessed a new "Metallic Age": the Rare-Earth Age. It is named after a group of metals that, unlike bronze and iron, had to wait 2000 years to be discovered. Unlike the two ancient metals, their importance was not evident upon their discovery and their use was not immediate. Had they been discovered a millennium ago they would probably have been doomed to oblivion. Even after their modern discovery they were considered as an oddity and, like another discovery of the twentieth century—the laser—one could say they were a solution waiting for a problem.

However, they had one big advantage over the ancient metals: They aroused curiosity. And curiosity is the driving force of mankind and of research. They presented a challenge: first, to geologists, who found their ores just by chance, on very rare regions on the earth—hence their name: rare-earths. Then to chemists, who laboured hard to separate them, because they are all so similar chemically. Then to physicists, and in particular to spectroscopists, who found their spectra so difficult to analyze that they called them "complex spectra".

But the twentieth century was ready for them. The scientists rose to the challenge and the mystery has started to lift. The realization of their importance—exactly due to their exotic characteristics—started to dawn both on the scientific community and on the industry. Their applications became a plethora. We are now in the middle of a fascinating period of an interplay between theoretical and applied research. The rare-earths attracted the interest of the industry that found them indispensably useful in many areas. Exactly this usefulness incited the researchers to delve even deeper into the structure of the rare-earths and explain the origin of their extraordinary properties. They have also extended their attention to the Actinides—the Rare-Earths heavier, man-made (except thorium and uranium) homologues in the Periodic Table. As a result there has recently been accumulated a vast amount of knowledge and insight into the physical mechanisms that are at the basis of these properties. This book tries to present an up-to-date review of the achievements in this subject, that may inspire future accomplishments.

Contents

Introduction

The rare-earths and the actinides form two series of elements characterized by the filling of the 4f and 5f sub shells. By reference to their place in the periodic table, the rare-earths are the 15 metallic elements from lanthanum ($Z = 57$) to lutetium ($Z = 71$). They are also known as the lanthanides. However, lanthanum metal has no 4f electron in the ground state. Moreover, ytterbium (70) is the first metal for which the 4f sub shell is full and it is generally considered as the last member of the series. Among them, promethium is radioactive and is not present in nature, but it might be prepared artificially. The rare-earths were discovered and isolated between 1796 and 1947. The first works were made in France by L.-N. Vauquelin around 1790. In 1907, G. Urbain realized the first separation method of these elements by fractionized crystallization. Their abundance is a controversial subject. They are naturally present in the form of compounds of several rare-earths. The elements with even Z are more abundant. Their treatment and their purification are difficult and that makes the pure metals very dear.

By analogy, the actinides could be considered as the 15 metallic elements from actinium ($Z = 89$) to lawrencium ($Z = 103$). However, actinium, discovered by A.L. Debierne in 1899, does not have 5f electrons and the element with equivalent properties to lanthanum is thorium (90). The series of the actinides is thus considered as having 14 elements from 90 to 103. Uranium (92), the first recognized element of the series, was discovered in 1789 by M.H. Klaproth in pitchblende. The extracted material was in reality the oxide UO_2; the pure metal was obtained by E.M. Péligot only half a century later. Thorium was identified by J. J. Berzelius in 1828. Protactinium was discovered by O. Hahn and L. Meitner in 1917. In the beginning of the twentieth century, uranium was yet considered as the heaviest of the elements. Only from 1940 neptunium, plutonium, then successively the other actinides were discovered. The last two members of the series, nobelium and lawrencium, were identified in 1959 and 1961.

The rare-earths are characterized by the filling of the 4f sub shell from one for cerium up to fourteen for ytterbium. They form two groups, the light ones, lanthanum-europium, and the heavy ones, gadolinium-lutecium. The electronic configurations of the rare-earths are generally not the same in the free atom and in the solid. The free atoms are divalent with the ground configuration $4f^m6s^2$ except for lanthanum, cerium, gadolinium and lutecium, which have the ground configuration $4f^{m-1}5d^16s^2$. In the solid, the rare-earths become trivalent, of electronic configuration $4f^{m-1}(5d6s)^3$, except for europium with a half-filled 4f sub shell and ytterbium with a completely full 4f shell. They remain divalent with configurations $4f^m(5d6s)^2$ in the metal but can become trivalent in various compounds. Exceptions exist also for cerium, samarium, terbium and thulium, which are tetravalent in some compounds. The valence 5d and 6s electrons form the bond and the metals exhibit some d-like character.

From a chemical point of view, the trivalent rare-earth metals are very similar due to an almost identical outer electron arrangement. Indeed, the number of the 5d and 6s valence electrons is practically constant. These electrons are strongly coupled to each other. They are the electrons that interact with the ligand electrons in the compounds and participate in the chemical bond formation. In contrast, the electrons in the partially occupied 4f shell have a small radial extension. They are shielded by the 5s and 5p electrons from interacting with the other valence electrons of the rare-earth and with the ligands. Therefore, they hardly participate in the chemical bond formation. However, because the occupation numbers of the 4f shell changes from 0 to 14 through the series, the rare-earth elements and their compounds have a wide range of electronic structures and magnetic properties.

The interest in the rare-earths is connected with their particular properties that have found numerous industrial applications. Their optic as well as magnetic properties are characteristic of the presence of very narrow energy levels among the distribution of the valence levels. One of their peculiarities is their optical spectra of very narrow and intense lines, leading to very well-defined luminescence emissions of energies measurable with precision, which can be used in numerous applications such as laser sources, television screens, electronic industries, solar cells. The rare-earth compounds are also magnetic materials, which can be used as magnets or for magneto-calorific applications. They are present in hybrid car engines and wind turbines. In the 1960s, the development in the study of rare-earth compounds was mainly motivated by a search for new ferromagnetic semiconductors. At present, one of the purposes is to obtain materials with novel properties, such as spintronic and multiferroic materials, superconductors at high temperature, materials with giant magnetoresistance. The use of rare-earth compounds is in such development that it has been asserted that "the life in the twenty first century would not be the same without rare-earths".

These particular properties result from the specific characteristics of the 4f electrons. The predominant role played by the 4f electrons in those particular physical properties of the rare-earths has been widely discussed. A large part of the important work made on the rare-earths has been to investigate the "character" of the 4f electrons in the various rare-earths and their influence on the environment and physico-chemical properties of the solid. The problem is to know what model must be used to treat the 4f electrons in the metals, what are the interactions of the 4f electrons with the other valence electrons and what changes occur when the rare-earth is in a compound of particular properties.

It appears clearly that the 4f electrons are relatively insensitive to external perturbations because, as already underlined, they were well shielded from the electronic distribution of the neighbouring ion, due to their small radial distribution and the presence of the outer 5s and 5p electrons. Moreover the f–f laser transitions are narrow as compared to transitions, which involve the $f^{m-1}5d$ configuration. The 4f electrons could then be considered as core electrons. However, in some cases, a strong connection exists between the 4f distribution and the crystalline structure revealing the presence of a partial delocalization of the 4f electrons.

Concerning the actinides, they hold a special position because of their technological importance due to their radioactive properties. All the actinides are radioactive, their half-life varying from a few thousand years for thorium and uranium to a few days for actinium down to several minutes or less for the elements beyond americium. The actinides are metals; they are soft, of silvery colour and have high plasticity. In air they tarnish and then spontaneously ignite. In addition they undergo eventual electronic changes due to the accumulation of damage by self-irradiation. As an example, helium atoms, vacancies and interstitials are introduced into the lattice due to the decay of plutonium by α-emission. The introduction of these defects produces an expansion of the lattice.

The number of the 5f electrons increases progressively from one for protactinium up to fourteen for nobelium. The electronic ground configuration of the actinides in the solid also differs from that of the free atoms. From plutonium on, all the atoms have the $5f^m 7s^2$ configuration, except for at the half-full shell where curium, as gadolinium, has the $5f^7 6d^1 7s^2$ configuration. The light actinides have the $5f^m 6d^1 7s^2$ or $5f^{m-1} 6d^2 7s^2$ ground configurations in the free atom. In solids, several oxidation states coexist and this denotes the presence of several possible configurations for the valence electrons. In the rare-earths, important characteristics are connected with the elements having almost half-filled 5f shell. Characteristics analogous to those of the rare-earths appear in the heavy actinides, from americium and beyond. For the light actinides from thorium to plutonium, it was initially supposed that the 5f and 6d electrons had similar characteristics. This had suggested that the actinide series could bridge a gap between transition metals and rare-earth metals. Indeed, the 5f electrons in the light actinides have peculiar characteristics, which are different from those of the other materials. Properties of actinides, as for example structural, electric and magnetic properties, unconventional superconductivity, raise interesting fundamental questions, which are all related to their particular electronic structures (Table 1).

The study of the characteristics of the nf electrons is needed to understand the properties of the lanthanides and actinides. Indeed, physical and chemical properties of the solids depend on the electron distribution of the elements present. For the rare-earths and the actinides, the number of the valence electrons intervening in the chemical bonds and the structure of the solid depends upon whether the f electrons are localized or mixed with the valence electrons. This valence electron distribution conditions all the physico-chemical properties and various parameters such as crystal structure, melting point, phase transition, response to pressure, cohesive energy, electrical resistance, magnetic susceptibility. Consequently, it is essential to

Table 1 Atomic number and ground configuration for rare-earth and actinide atoms

Element	Z	Config	Element	Z	Config
Lanthanum (La)	57	$5d^16s^2$	Actinium (Ac)	89	$6d^17s^2$
Cerium (Ce)	58	$4f^15d^16s^2$	Thorium (Th)	90	$6d^27s^2$
Praseodymium (Pr)	59	$4f^36s^2$	Protactinium (Pa)	91	$5f^26d^17s^2$ $5f^16d^27s^2$
Neodymium (Nd)	60	$4f^46s^2$	Uranium (U)	92	$5f^36d^17s^2$
Promethium (Pm)	61	$4f^56s^2$	Neptunium (Np)	93	$5f^46d^17s^2$ $5f^57s^2$
Samarium (Sm)	62	$4f^66s^2$	Plutonium (Pu)	94	$5f^67s^2$
Europium (Eu)	63	$4f^76s^2$	Americium (Am)	95	$5f^77s^2$
Gadolinium (Gd)	64	$4f^75d^16s^2$	Curium (Cm)	96	$5f^76d^17s^2$
Terbium (Tb)	65	$4f^96s^2$	Berkelium (Bk)	97	$5f^97s^2$ $5f^86d^17s^2$
Dysprosium (Dy)	66	$4f^{10}6s^2$	Californium (Cf)	98	$5f^{10}7s^2$
Holmium (Ho)	67	$4f^{11}6s^2$	Einsteinium(Es)	99	$5f^{11}7s^2$
Erbium (Er)	68	$4f^{12}6s^2$	Fermium (Fm)	100	$5f^{12}7s^2$
Thulium (Tm)	69	$4f^{13}6s^2$	Mendelevium (Md)	101	$5f^{13}7s^2$
Ytterbium (Yb)	70	$4f^{14}6s^2$	Nobelium (No)	102	$5f^{14}7s^2$
Lutetium (Lu)	71	$4f^{14}5d^16s^2$	Lawrencium (Lw)	103	$5f^{14}6d^17s^2$

study the electron interactions and spectroscopy is an indispensable tool to use in this aim.

Optical spectroscopy supplies a large number of analyzes of electronic transitions of the type $4f^m\text{-}4f^{m-1}5d^{m'+1}$ involving 4f electrons. These transitions have been observed both in atoms and in numerous inorganic compounds. Ground state, energy between the ground state and the first excited levels and crystal field splitting were deduced from these observations. However, these data can describe neither the valence electron distributions nor that of conduction levels while these characteristics can be deduced for any solid from high energy spectroscopy. These methods enable the measurements of both large and small energy transfers and this is an advantage over the thermodynamic and magnetic probes. Another advantage is the extremely short time characteristic of the measurement because this allows a direct observation of the states between which the electron system can fluctuate. Radiative and electronic spectroscopies are widely used as well in the X-ray as in X-UV ranges to determine the electronic configuration in the solids. Applications to rare-earth and actinide compounds are numerous. This type of information varies according to the method used and a good knowledge of the various methods is necessary to obtain the maximum of data on the studied material.

One important question is to know the characteristics of the nf electrons and the interplay between them and the valence electrons. A variety of phenomena such as the simultaneous presence of two different valences involving changes of the nf states make the problem complex and in spite of numerous studies describing the characteristics of the 5f electrons in the light actinides there remains actually a challenge. The same difficulties are also present for various rare-earth compounds. The electronic characteristics of these materials need to be well understood in order to obtain progress and future perspective. Since the first Conferences in the field (*First Rare-Earth Research Conference*, E.V. Kleber, ed., Macmillan, New York, 1961) (*The Actinides: Electronic Structure and Related Properties*, edited by A.J. Freeman and J.B. Darby, Jr., Academic Press, New York, 1974), numerous reviews have been published separately on the rare-earths and actinides or together. The Handbook *on the Physics and Chemistry of the Rare-Earths* has been published since 1978 and *on the Physics and Chemistry of the Actinides* since 1984. Numerous international conferences take place regularly on these subjects.

To realize these researches it is necessary to obtain impurity-free single crystals of the rare-earths and actinides. The same difficulties exist both for rare-earths and actinides compounds and this may be responsible for some of the long-standing controversies concerning their electronic structure and their transport and magnetic properties.

In Chap. 1 of this book, a survey of the electronic structure of the metal rare-earths and actinides is given along with a description of their crystal arrangements. In Chap. 2, the same is presented for various compounds among the more studied actually. Resonant processes characteristic of the rare-earth and actinide spectra are presented in Chap. 3 and widely discussed for lanthanum as an example. Remarkable results concerning the intensities of the radiative and non-radiative transitions in the X-ray range appear in these spectra and reveal unexpected characteristics of these resonant states. Description of the various spectroscopies among the more used is given. All the methods are presented for lanthanum because its excited configuration has only one 4f electron, which makes its calculations and the interpretation a good model for the rest. Chapters 4 and 5 describe the results obtained from these various spectroscopies for the rare-earths and the actinides and for several of their compounds among the more representative of the particularities of these materials. An important point is to discuss the localization of the f electrons and an eventual interplay between the delocalized valence electrons and the strongly localized f electrons, which are responsible for the particular properties of these materials. Indeed, novel phenomena and new understanding are yet to be found in the field.

Chapter 1
Electron Distributions and Crystalline Structures

Abstract Theoretical models to describe the electron distributions are mentioned with attention to the specific case of the nf distributions and their characteristics in relation with the physical properties of rare-earths and actinides. The crystalline structures of these metallic elements are discussed in connection with pressure effect.

Keywords Electron distribution · Valence states · Electron correlation · Heavy fermion · Pressure effect

1.1 Introduction

Valence electron distributions and spatial arrangements of atoms are two interdependent characteristics, which are responsible for the physical and chemical properties in the solids. The electron distributions determine the size of the atoms and their magnetic and optical properties. The atomic arrangements govern the interactions between atoms, their cohesion and the response to pressure or to temperature. The two electronic and atomic structures together are responsible for the bonds between atoms. All these characteristics depend on the mobility of the valence electrons, which, in turn, depends on the extension of their wave functions and the electron–electron interactions. Indeed, the characteristics of the valence electrons are responsible for the large diversity of metals and their partition into good metals, transition metals, rare-earths and actinides. After a brief review concerning the free atoms and the good and transition metals, the observed particularities resulting from the presence of 4f and 5f electrons are discussed along with the existing theoretical models for the rare-earths and the actinides.

© Springer Science+Business Media Dordrecht 2015
C. Bonnelle and N. Spector, *Rare-Earths and Actinides in High Energy Spectroscopy*, Progress in Theoretical Chemistry and Physics 28, DOI 10.1007/978-90-481-2879-2_1

1.2   The Electrons in Atoms

1.2.1   Energy Levels and Wavefunctions

Each electron in the atom, called a "bound" electron, moves in the central field of the positive nucleus and in the field of all the other negative electrons. The possible energetic positions of the electrons in the atom are called "energy levels". They are the eigenvalues of the Hamiltonian matrix that contains all the interactions within the atom. As such they have discrete values, that is, they are quantized. The level with the lowest energy in this system is called the "ground level" or the "ground state". The others—with higher energy—are named "excited levels". An electron can make transitions ("jump") between levels, following "selection rules" that govern the allowed transitions. Upon a transition from a higher level to a lower level, a photon is emitted whose energy is equal to the energy difference between the initial and final levels. The ensemble of the photons emitted in such transitions is the spectrum of the atom. "The electrons in the atom see each other in two ways: (a) through Coulomb's law, (b) through the Pauli principle" [1]. Coulomb's law controls the arrangement of the energy levels while the Pauli principle determines the grouping of the electrons in the atom.

The spatial arrangement of the electrons in the atom is determined by their eigenfunctions that are the solutions of the Schrodinger equation of the atomic system. The eigenfunctions themselves can be expressed as a product of a radial part and an angular part. The radial part of the equation cannot be solved analytically and therefore its solutions are expressed in terms of radial parameters called the Slater integrals or Slater parameters. The angular part, on the contrary, has an exact solution in terms of Legendre polynomials that are characterized mainly by two integers, n and l. n is the degree of the polynomial and is related to the total energy of the atom, which is a scalar. l is related to the quantified movement of the electron around the nucleus, i.e. to the orbital momentum of the electron which is a vector. Another orbital momentum connected with the electron is its spin s, related to the quantified movement of rotation of the electron around itself, also a vector.

1.2.2   Quantum Numbers and Configurations

In addition to its inherent spin, it is possible to assign unequivocally to each electron four numbers, n, l, m_l and m_s, called "quantum numbers". n is called the principal quantum number. It can take any integral value from 1 and up. l is called the orbital quantum number and can go from zero to $n - 1$. s is always equal to 1/2. m_l and m_s are the projections of l and s on an arbitrary z axis and are called magnetic moments or magnetic quantum numbers. For example, an electron may have the quartet of its quantum numbers as follows: 3, 2, 1, 1/2. For each electron, the vector sum of its orbital angular momentum l and its spin angular momentum s gives its total angular momentum j.

The electrons with the same n form a "shell". A shell can contain electrons with different l's. Those with the same l are called "equivalent electrons". They form a "sub shell". According to the Pauli principle, no two electrons can have the same four quantum numbers. This limits the number of electrons in a sub shell. The maximum number of electrons in a sub shell of a given l is $2(2l + 1)$. Thus, in a shell with $l = 3$ there can be 14 electrons. A shell with the maximum possible number of electrons in it allowed by the Pauli principle is called a "closed shell" or a "core". The electrons in closed shells are called "core electrons". They do not participate in chemical bonds. However, under certain conditions an electron can jump out of the core and leave behind a "core hole". In such cases both the jumping electron and the hole it left behind play an important role in various interactions both inside and outside the atom. The electrons of a sub shell that participate in chemical bonds are called "valence electrons".

Traditionally electrons were assigned "names" according to their l values. Those with $l = 0$ are called s-electrons. Those with $l = 1$ are called p electrons; with $l = 2$ they are called d electrons and with $l = 3$ f electrons. An electron whose l value is even is called an "even" electron. When its l value is odd it is called an "odd" electron. As mentioned above there can be only 14 f electrons. However, we have not mentioned the principal quantum number n yet. If $l = 3$ it means that n is equal to 4 or more. Therefore there can be 14 4f-electrons, 14 5f-electrons, etc. The ensemble of all the nl electrons in an atom is called a "configuration". The configuration that contains the ground level is called the "ground configuration". Since, depending on their excitation, the electrons can change their n and l values, there can be a number of configurations in the excited atom. They are called "excited configurations". A configuration is called "even" when the sum of all the l's of its electrons is even; similarly for an odd configuration. Thus the configuration ds and fp are even. This notion of parity is very important, since in the case of electric dipole radiation only transitions of electrons between even and odd configurations are allowed. Since the electrons that are in closed shells are not involved in optical transitions, nor do they participate in chemical bonds, the term "configuration" may be used to designate those partial groups of electrons that participate actively in the transitions or reactions. The term "subconfiguration" may also apply under such conditions.

1.2.3   Electrostatic and Spin–Orbit Interactions, Couplings

For a single electron with l and s, there can be only two possible j values: $l + 1/2$ and $l - 1/2$. For a many-electron configuration the resultant angular momentum J is the vector sum of all the j's. It should be noted that the total S, L, and J of closed sub shells are zero. Therefore they do not contribute to the total angular momentum. The number of energy levels in each configuration is equal to the number of possible values of the different J's obtainable from the above vector addition. However, the arrangement of the levels, and consequently the characteristics of the spectra, depend on two main interactions: the Coulomb interaction between the electrons, called the electrostatic interaction, which is a multielectron interaction represented by the sum over all the pairs ($e_i e_j / r_{ij}$), and the spin–orbit interaction which is a single-electron interaction represented by ($s_i \cdot l_i$) summed over all the single electrons. For an s electron the spin–orbit term vanishes because $l_i = 0$. The resultant J values can be obtained in two different ways: In the first case all the orbital angular momenta are added together (addition here is vectorially) to form a resultant L, all the spin angular momenta add to a resultant S and then L and S add to a resultant J. Alternatively, all the j's can add up to give the resultant J. The outcome, as far as number of levels and of different J values, is the same. The first addition is called Russell Saunders coupling and the second j–j coupling. In the Russell Saunders case a group of levels with the same S and L is called a "term". According to the previous example for orbital momenta l, the different L values of the terms are given traditionally the following designations: Terms whose $L = 0$ are called S terms. $L = 1$ are P terms. $L = 2$ are D, $L = 3$ are F, $L = 4$ are G, etc. The expression $2S + 1$ for each term is called its "multiplicity". Thus a term with $S = 0$ is called a singlet. $S = 1/2$ is a doublet. $S = 1$ is a triplet. $S = 3$ is a septet, etc. or, in general, a multiplet. The designation of a level is $^{(2S+1)}L_J$. When the spin–orbit interaction vanishes all levels of the same term have the same energy. A non-vanishing spin–orbit interaction "splits" the term into a multiplet. The order of the levels in a term is normally such that the lowest energy level has the lowest J. Sometimes it is the other way round and in this case the term is called "inverted".

The Hamiltonian is diagonal with respect to J meaning that there are no off diagonal matrix elements connecting different J's. Therefore J is called good quantum number. In Russell Saunders coupling S and L are also good quantum numbers. The part in the Hamiltonian of the electrostatic interaction is divided into two expressions called direct and exchange interactions, with radial direct and exchange parameters designated as F_k and G_k, respectively. For equivalent electrons only F_k are used. The spin–orbit part is represented, for each electron l, by a single parameter called zeta l. The spin–orbit interaction of the various electrons can be strong as opposed to a weak electrostatic interaction among them. Or those two can be of similar strength. The first case is called j–j coupling since, owing to the strong spin–orbit interaction the l's and s's of each electron couple together to give a total j and only then the different j's couple together to give a final resultant J. In this case the j's are good quantum numbers and can be used to designate the energy levels. In the second case, often named intermediate coupling, neither the j's nor the L's and S's are good quantum numbers and they cannot be clearly assigned to the energy levels. The latter can, therefore, be identified by their energy and their total J.

As already mentioned the lowest level of the set of the energy levels of an atom is called the ground level. The configuration to which it belongs is called the ground configuration. Given the ground configuration, it is interesting to know which, among all its levels, has the lowest energy. For this there is an empirical rule called "Hund's rule" that predicts this level. It is valid mostly in multielectron configurations, and in Russell Saunders coupling. According to Hund's rules: (1) among all the terms of the ground configuration the one with the highest multiplicity will be the lowest; (2) if there are several terms with the highest multiplicity the one with the highest L value will be the lowest; (3) among the levels of this term the one with the lowest J will have the lowest energy, unless the term is inverted; in this case the level with the highest J will have the lowest energy.

1.2.4   The 4f and 5f Shells-Lanthanides and Actinides

The 4f sub shell can have 14 electrons. This sub shell starts filling up at atomic number 57-Lanthanum. It becomes a closed sub shell 14 elements later—at atomic number 71-Ytterbium. The 14 elements of this group are called rare-earths or lanthanides despite the fact that Lanthanum does not have a 4f electron in its ground configuration. The 4f electrons start to show up in the ground configuration only in Cerium. Then their number increases by one for each successive element until, for Ytterbium, there are 14 4f electrons in the ground configuration. Only 7 4f electrons are in the ground configuration of Gadolinium (the eighth element in the series). As for the 5f electrons, the 5f sub shell starts filling up at atomic number 89-Actinium, and ends up being full at atomic number 102-Nobelium. The 14 elements of this group are called actinides despite the fact that Actinium does not have a 5f electron in its ground configuration. The filling of the 4f shell in the rare-earths results in an increasing number of terms followed by a growing complexity of their level structure and consequently of their spectra. In Table 1.1 [2], we give a list of all the possible terms for the $4f^n$ configurations. From this table it can be seen that while in the nd shell the maximum possible value for L is 6, this value reaches 12 for the nf shell. In addition, the 5d electrons can join the 4f core to form configurations of the type $4f^n5d$ to increase even more the total number N of levels. The number of allowed transitions increases as N^2 and a rare-earth spectrum in the optical as well as in the X-ray regions can easily consist of 50,000 lines.

1.2.5   "Collapse" of the 4f Shell Wave functions

All the above is related to the quantitative fact of the existence of up to 14 4-f electrons in a single shell. The most important qualitative characteristic of this shell is its "collapse" inside the more external shells of the rare-earth. One calls the "collapse" of the 4f shell the fact that the main node of the 4f radial wave function approaches the nucleus and is inside the centrifugal potential well associated with the high angular momentum l. It should be noted that the wave functions themselves are not observable and their shapes and nodes are derived from quantum theoretical calculations solving the Schrodinger equation.

It turns out that the radii of the 4f orbitals are small—even smaller than those of the 5s and 5p ones. The result is that the 4f electrons are shielded from outside influence by the valence electrons like 6s and 5d and they do not take part in

Table 1.1 Allowed LS terms of the s, p, d, f configurations

					Total number
s	^2S				1
s^2	^1S				1
p, p^5	^2P				1
p^2, p^4	$^1(SD)$	^3P			3
p^3	$^2(PD)$	^4S			3
d, d^9	^2D				1
d^2, d^8	$1(SDG)$	$^3(PF)$			5
d^3, d^7	$^2(PD_2FGH)$	$^4(PF)$			8
d^4, d^6	$^1(S_2D_2FG_2I)$	$^3(P_2DF_2GH)$	^5D		16
d^5	$^2(SPD_3F_2G_2HI)$	$^4(PDFG)$	^6S		16
f, f^{13}	^2F				1
f^2, f^{12}	$^1(SDGI)$	$^3(PFH)$			7
f^3, f^{11}	$^2(PD_2F_2G_2H_2IKL)$	$^4(SDFGI)$			17
f^4, f^{10}	$^1(S_2D_4FG_4H_2I_3KL_2N)$	$^3(P_3D_2F_4G_3H_4I_2K_2LM)$	$^5(SDFGI)$		47
f^5, f^9	$^2(P_4D_5F_7G_6H_7I_5K_5L_3M_2NO)$	$^4(SP_2D_3F_4G_4H_3I_3K_2LM)$	$^6(PFH)$		73
f^6, f^8	$^1(S_4PD_6F_4G_8H_4I_7K_3L_4M_2N_2Q)$	$^3(P_6D_5F_9G_7H_9I_6K_6L_3M_3NO)$	$^5(SPD_3F_2G_3H_2I_2KL)$	^7F	119
f^7	$^2(S_2P_5D_7F_{10}G_{10}H_9I_9K_7L_5M_4N_2OQ)$	$^4(S_2P_2D_6F_5G_7H_5I_5K_3L_3MN)$	$^6(PDFGHI)$	^8S	119

combinations with other atoms or ions. The small radius of the 4f shell orbitals decreases the distance between the 4f electrons leading to an increase of the Coulomb electrostatic interaction among them. As mentioned above the electrostatic interaction is at the basis of the arrangement of the energy levels in the atom. A strong electrostatic interaction, expressed by large values of the Slater direct and exchange parameters F_k and G_k, spreads out the configurations involving 4f electrons. Since the 4f shell is found inside the outer shells of 5d and 6s electrons, it screens them from the nucleus. The combined effect of this is twofold: on the one hand, it reduces the binding energy of the outer electrons thereby rendering their ionization easy. A typical value for the first ionization potential of a rare-earth is 6 eV. On the other hand, it increases the energy spread of the configurations of the type $4f^n5d$ up to twice this value. Thus, some of the levels of these configurations can be found in the first continuum of the atom.

In the case of the actinides, the collapse of the 5f wave functions is not as clear as in the case of the lanthanides. The 5f orbitals are more diffuse, the relativistic effects are stronger due to the higher Z. Consequently, the various theoretical calculations do not agree about the onset of the collapse. Most of them place it around element $Z = 90$, namely thorium. This means that actinium is not an actinide.

1.3   Electrons in Solids

In a solid, one considers with a good approximation that no overlap exists between the wave functions of the electrons of the closed shells, or *core electrons*, that belong to neighbouring ions; those are localized electrons. Consequently, they can be described by using the same wave functions of the electrons in the free atoms. However, the presence of the *crystal potential*, i.e. the static electric field produced by the charge distribution due to ions and electrons in a crystalline solid, modifies the energy levels with respect to those of the free atom. All the core levels of the same atom are reduced by approximately the same amount whose value can vary from a few electronvolts up to 10 eV according to the atomic number.

For the electrons of the open shells, or *valence electrons*, shifts of their energy levels and changes of their distribution occur simultaneously. These changes involve modifications of the charge state, level hybridizations, more generally modifications of the electronic interactions. The valence electrons ensure the cohesion between the atoms and contribute to chemical bonds. They are responsible for the electrical and thermal conductivity and the magnetic properties. They are considered as moving in the solid. The movement of each electron depends on the potential in the solid, i.e. on the arrangement of the atoms and on the movement of all the other electrons.

From a general point of view, the s, p, d valence electron distributions in solids are considered as characteristic of itinerant electrons more or less strongly correlated [3]. The first description of the valence electrons in solids was the nearly free electron gas approximation. In this approximation, used only for a few cases of s and p valence electrons, several basic definitions were introduced; they are given in the next paragraph. Numerous other theoretical models have been developed. Their complexities increase along with the magnitude of the electron–electron interactions that need to be taken into account. These models had initially for purpose to study the behaviour of the d electrons because these electrons are responsible for characteristic properties of the transition metal solids.

The 4f electrons of the rare-earths make only a small contribution to bonding and most of them may be considered as non-bonding. However, they have often been treated by the same models as the d electrons. An important question to be discussed is whether the 4f electrons in the lanthanides are well described as strongly correlated itinerant electrons or as localized electrons, given that localized valence electrons should have behaviour analogous to that of core electrons. The same question was posed for the 5f electrons in the actinides. This question necessitates a comparison between the theoretical electron distributions deduced from various models and the corresponding experimental data. Rapid glance of the main models used to treat the electron distributions will be presented.

1.3.1 Energy Distributions of the Valence Electrons

In a first approximation, the valence electrons are considered as delocalized in the entire solid and viewed as a uniform distribution of non-interacting charges moving in a zero external field. This model is the *free electron gas* approximation. Each valence electron is characterized by an energy ε, a wave number k and an associated wavelength λ. Its wave function is a plane wave $\psi_k(r) = e^{ik \cdot r}$ where $\mathbf{r}$ is the electron position, $\mathbf{k}$ the wave vector and $(\hbar)^2 k^2/2m$ the energy.

From a more general point of view, the valence electrons are considered as moving in a uniform potential of positive charges that represents the crystal ions. This is the *nearly free electron gas* approximation. The calculations are made in the Born–Oppenheimer approximation that neglects the movements of the nucleus and assumes the crystal to be a rigid lattice. The crystalline structure is considered perfect, infinite and at 0 K. The wave functions of the electrons are Bloch functions, $\psi_k(r) = u_k(r) \cdot e^{ik \cdot r}$, i.e. plane waves multiplied by a function $u_k(r)$ that has the periodicity of the crystal, $u_k(r) = u_k(r + R_n)$ where R_n is a vector of the lattice. The periodic potential energy $V(x)$ has small amplitude and is treated as a perturbation applied to a free electron gas. The wave functions are normalized in the length of the lattice; they are degenerate and $\psi_k(r) = \psi_{-k}(r)$. In this model, the movements of the electrons are treated as independent. It has been used to treat the solids having s and p valence electrons, as, for example, alkaline metals, magnesium and aluminium, called *good metals*. If s and p valence electrons are present simultaneously, their number in each sub shell is fractional.

In the nearly free electron approximation, the dispersion relation of the electronic waves, relating energy ε and wave vector $\mathbf{k}$, is $\varepsilon(\mathbf{k}) = \varepsilon(\mathbf{k} + 2\pi n/R)$ where R is the period in the considered direction and n is an integer. The wave vector and the energy are quantized and the interval between consecutive values of k in the considered space direction is $2\pi/N_A R$ where N_A is the number of atoms per unity volume. To each value of k corresponds an energy level $\varepsilon(k)$. For a macroscopic solid, the interval between the k values is negligible and the electron energy is a quasi-continuous periodic function of k.

The energy levels of the valence electrons are thus distributed in a continuous energy band, labelled *valence band*, of width W. This width increases with the number of the valence electrons in the atom; it varies from 3 eV for sodium to 10 eV for aluminium. The good metals are *wide-band* materials. The top of the valence band is the limit of the occupied levels; its energy position is the *Fermi level* E_F. The limit of the valence band is abrupt at 0 K. At temperature T, this limit is a function of the Fermi–Dirac distribution, which varies in a narrow region about $4\,kT$ wide around E_F; k is the Boltzmann constant. Since kT is equal to 0.026 eV at room temperature, the effect on the valence band is negligible at this temperature. The *valence* is defined as the number of electrons per atom available for band formation. The unoccupied energy levels are distributed continuously above the Fermi level. This distribution is named *conduction band*. The electrical and thermal conductivities depend on the electron density in the vicinity of the Fermi level.

The number of energy levels in the valence and conduction bands per unit interval of energy is named *density of valence, or conduction, states, $n(\varepsilon)$*. For free electrons, this density of states is uniform and follows a quadratic equation of the energy. The densities of states $n(\varepsilon)$ are represented in the *k-space*, which is the Fourier space of the real crystal. For each atom present in the unit cell of the crystal there is associated one electronic level of wave vector k in the unit cell of the *k*-space, named *first Brillouin zone* (cf. Sect. 1.4). The density of the states $n(\varepsilon)$ is the number of states between the surfaces of energy ε and $\varepsilon + d\varepsilon$ of the *k*-space. This is a continuous function of the energy ε given by

$$n(\varepsilon) = 1/8\pi^3 \int dV_k$$

The integral is calculated in the volume V_k of the *k*-space situated between the surfaces of energy ε and $\varepsilon + d\varepsilon$. For free electrons, the surfaces of constant energy are spheres. In a periodic lattice, ε is a discontinuous function of k and energy gaps are present at the Brillouin zone limits. Maxima of $n(\varepsilon)$ are formed when the energy surfaces reach the zone limits. The increase of $n(\varepsilon)$ is due to a deformation of the constant energy surfaces. The surface that corresponds to the Fermi energy E_F is named *Fermi surface*. Generally, the Fermi surfaces are not spherical. They can touch the zone in certain directions. The distortions of the Fermi surface are chiefly noticeable when they are near a plane of energy discontinuity. There is no exact analytical solution for a real solid. However, quite accurate results can be obtained for particular points of the k-space by numerical methods.

In a crystal, an electron cannot be displaced in the directions for which its wavelength λ satisfies the Bragg relation. Consequently, energy gaps are present in some directions depending on the crystalline structure. In such a k direction, the energy distribution is spread out in quasi-continuous bands separated by forbidden energy regions. In a three-dimensional space, the positions of the forbidden bands are different for each direction k and an overlap of the allowed bands is possible. This is verified in metals for which the bands of allowed energies overlap in space. The *k* vectors of equal value determine surfaces of equal energy, or constant energy surfaces, in all directions. These surfaces are the places of the corresponding electronic levels. In a non-conductor, no overlap exists between the allowed energy bands. The valence band is separated from the conduction band by a forbidden energy band, named *band gap* and the Fermi level of a perfect insulator crystal is positioned in the middle of the gap.

In the nearly free electron gas approximation, the valence electrons are considered as independent itinerant particles. A model assuming a random distribution of the non-interacting electrons is a *one-particle* picture. All the electrons are considered as moving in the same potential; they have thus the same energy band system. The band structure expected from this model is in agreement with that observed for the alkaline metals, named also *s* metals.

1.3.2   *Models for Weakly Correlated Electrons*

A significant characteristic of the electrons in a solid is the extended character of their wave functions. Among the calculation methods of the density of states, those using plane waves in a Hartree scheme are convenient if the interactions between the electrons, i.e. *electron correlations*, are weak. These methods are *one-electron approximations*, valid in treating the s and p valence electrons. The general equation, representing the movement of N_E electrons, is split into N_E equations, each representing the movement of a single electron. Each electron is in the same potential and has the same wave function $\psi_k(r)$ as the others, the total wave function being the product of the N_E one-electron wave functions. Each wave function $\psi_k(r)$ is, thus, a solution of a one-electron Schrödinger equation having as Hamiltonian

$$H_H = T + V$$

where H_H is the Hartree Hamiltonian, T is the kinetic energy term, $(\hbar)^2 k^2/2m$. V is the potential energy term and represents the average interaction of one electron with the crystal. It consists of two contributions, the potential energy of the electron moving in the nuclei field, V_n, and the *direct Coulomb energy* V_e, i.e. the repulsion energy acting on one electron in the average-field of the other $N - 1$ electrons, named Hartree potential. The Hamiltonian is of the form

$$H_H = T + V_n + V_e$$
$$= T + \sum_n Z_n e^2/(\mathbf{r_i} - \mathbf{R_n}) + 1/2 \sum_{j \neq i} Z_n e^2/(\mathbf{r_i} - \mathbf{r_j})$$

R_n designates the nucleus coordinates, r_i the coordinates of the i electron and Z_n the atomic number of nucleus n. Several representations using plane waves as wave functions have been developed to describe the valence electrons. Among the representations using plane waves in a Hartree scheme, the orthogonalized plane waves (OPW) method and the augmented plane waves (APW) method are the most widely used. The OPW method [4] uses a non-local potential. It is a Hartree self-consistent field method in which the valence electrons wave functions are plane waves orthogonal to the wave functions of the core electrons. A linear combination of atomic orbitals of the ion is added to represent the wave function in the vicinity of each ion and to make the convergence more rapid. The total wave function is orthogonalized with respect to core orbitals. This assumes that the core electrons are localized and their wave functions do not overlap, and that the variation of the plane wave in the vicinity of the ion can be neglected. The potential energy is the sum of the potentials due to the ions and to a homogeneous distribution of free electrons. The potential is the same for all the electronic states because these are all eigen-states of the same Hamiltonian.

The APW method is a local potential method [5–7]. In contrast to the previous method, the crystal potential is approximated by a *muffin-tin* potential. It has the spherical symmetry inside the sphere centred on each atom, named *atomic sphere*, of radius equal to the half-distance between nearest neighbours, and it is constant outside the sphere. Continuity of the potential is assumed at the surface of each sphere. The spheres are distinctive for each type of atoms and the potential at the interior is equal to the atomic potential. For a system of N interacting electrons, the wave function of the ground state is approximated by a Slater determinant of N mono-electronic functions. The wave equation can be solved exactly for each sphere; each wave function is the product of a radial function and spherical harmonics. To facilitate the convergence, the total wave function is taken as a linear combination of wave functions, each being a series of functions with the spherical symmetry inside the spheres and a plane wave outside the spheres. To obtain the continuity, the plane wave is developed in spherical coordinates.

Other one-electron methods have been developed, among them the *KKR*, or *Green function* method [8, 9]. As in the APW method, the potential is of muffin-tin type but the inverse process is used to obtain the wave functions. A function taking exactly into account the atom arrangement periodicity, i.e. the lattice periodicity, is determined and is constrained to satisfy the Schrödinger equation around the atoms. The wave function is thus expanded in a series of lattice harmonics, solutions of the radial part of the Schrödinger equation. This method is an application to electronic waves of the dynamical theory, which was developed by P.P. Ewald in order to treat the interferences of the electromagnetic waves with an atomic lattice. The complexity of the KKR equations is due, for a large part, to the energy dependence of the parameters characteristic of the structure. An approximate form of the KKR method using energy-independent structure parameters has been developed. Then the calculations become several orders of magnitude faster than by the standard KKR method while the deviations from the previous calculations are at most a few percent [10].

The analogy between the band structure of the s and p valence electrons materials and that expected from the approximation of the nearly free electrons had led to the introduction of the *pseudo-potential* model, where the interaction between the electrons is described by an average potential. As the potential depends on the occupied states and the occupied states depend on the potential, a self-consistent calculation is performed by successive iterations. Moreover, when a plane wave is orthogonalized to a core function, the spherical symmetry is lost. All these calculations are long and complex. In order to achieve simplification, the notion of pseudo-potentials was introduced to the OPW method [11]. The pseudo-potential is equal to a non-local repulsive potential V_R which adds on to the crystal attractive potential V and tends to cancel it. The wave equation of the pseudo-potential is

$$(T + V_R + V)\phi = E\phi.$$

The pseudo wave function ϕ is developable in plane waves and represents the quasi-monotone part of the total wave function ψ situated at the exterior of the cores. Since the pseudo-potential is small, ϕ varies only slightly and a small number of plane waves are sufficient to describe the s electron distributions. The convergence is slower for p electrons. Calculations, using not the OPW method but that of the quantum defect, have also been developed [12] see also [13].

The one-electron models result in correct energy distributions for metals with s and p valence electrons and semi-conductors such as Si and Ge, or generally, for systems with weak electron–electron interactions as, for example, the noble metals [14]. Indeed, for the s and p valence electrons, the Hartree models, which neglect all the correlations and consider the electrons to move independently of each other, seem more adequate than models that include only a part of the correlations.

The number of d electrons in a transition element in metallic state is fractional. However, the d electrons are not completely free to move in the material and they cannot be treated by a free or a quasi-free electron model. This is because, when two, or more, d electrons are present on the same site, a strong Coulomb repulsion exists between them. Calculations that do not treat the atomic potential as a perturbation have been developed in the *Hartree–Fock* approximation, i.e. by taking into account the exchange interactions. In this approximation, each electron is considered as surrounded by an "exchange hole", i.e. a sphere with no parallel spin electron density and a positive charge +e. Each hole is associated with a negative potential energy, the "exchange energy", which depends on the density of the electrons with parallel spins. The Hamiltonian is

$$H_{H-F} = T + V_n + V_e + V_x$$

where the Hartree-Fock exchange potential V_x is a non-local potential implying the anti-symmetrization of the wave function. The Hartree–Fock model is an independent particle scheme, ignoring the many-body nature of the problem. Among the treatments using this model, the *tight binding method* [15] was widely used. In this method, the periodic potential is the sum of the atomic potentials centred on each atomic site in the crystal. One determines Bloch sums. Each Bloch sum $\psi_{l,k}(r) = \sum_n \chi_l(r - R_n) \cdot e^{ik \cdot R_n}$ corresponds to the sum of the atomic orbital χ_l on all the n atoms at an equivalent position in the crystal and it fulfils the Bloch condition $\psi_k(r + R_n) = \psi_k(r) \cdot e^{ik \cdot R_n}$. There is a Bloch sum for each atomic orbital present in the unit cell of the crystal. The wave functions are linear combinations of the Bloch sums and, therefore, are linear combinations of atomic orbitals (LCAO), which are eigenfunctions of the atomic potentials. They are slightly different from the atomic functions in the vicinity of each ion. The exchange potential can be treated analytically; it is associated with all the ions of the lattice. The tight binding method was used for the transition metals [16]. It gives a suitable description of the d electronic distributions only if their interaction with other distributions is weak.

The previous model was simplified by introducing an effective mean potential that is an average on all the potentials of occupied states [15]. This potential is the same for all the states. It is obtained by using a model of interacting electron gas and by neglecting the influence of the periodic field. The exchange and correlation interactions are taken into account by introducing the Slater exchange potential, which is proportional to $n^{1/3}$, where n is the electronic density. Values of the splitting due to spin are in agreement with those obtained from more sophisticated methods. This one-electron picture, known as $X\alpha$ *method*, tends to average the effects due to the electron–electron interactions and it has given suitable results for several correlated systems.

Among the preceding methods, APW and KKR self-consistent calculations have been used with success for a wide class of materials but an important computational effort is needed and no simplifications are possible. In contrast, the OPW and LCAO methods have led, respectively, to the pseudo-potential and tight binding approximations. Combining some features of these various methods, in particular tight binding and APW, an efficient band structure method was developed [17]. It uses a muffin-tin potential and energy-independent muffin-tin orbitals. The use of energy-independent basis functions is advantageous because the secular equations are linear in energy. The basis functions are constructed from linear combinations of muffin-tin orbitals (LMTO); these are Bloch sums. The structural parameters, characterizing the crystal structure, do not depend on the energy or on the atomic volume. They can be tabulated once and for all for a given structure throughout the Brillouin zone. The LMTO method was simplified by using the atomic sphere approximation (ASA). This approximation supposes the potential to vanish in the interstitial region, located between the atomic polyhedron and the sphere. The introduced error is very small; it is proportional to the volume between the atomic spheres. In order to reduce this volume, overlapping spheres are generally used. The ASA-LMTO formalism establishes a compromise between accuracy and efficiency; it is computationally fast and particularly suited for closely packed structures. Relativistic effects may be included.

After the construction of appropriate one-electron potentials was achieved and band structure self-consistent calculations were widely developed, it became necessary to introduce systematically the exchange and correlation effects to compute in a realistic manner the solid properties. The exchange interactions are important in the treatment of atomic properties, such as the mechanical properties, but less so in treating electronic properties, such as the conductivity. Indeed, the Hartree–Fock model makes possible a good description of the core electrons and is well adapted to treat localized electrons. However, concerning the valence electrons, this model neglects the correlation effects, which may be as important as the exchange interactions. In atomic and molecular systems, configuration interaction corrections are introduced instead of correlation effects. In solids, unrealistic features can be present in the Hartree–Fock energy eigenvalues, such as vanishing density of states at the Fermi level in metals or unphysically large band gaps in insulators. The complexity of the corrections to Hartree-Fock model has incited the development of other theories.

1.3.3   The Electron Correlations

If the correlations between electrons are neglected, as in the Hartree approximation, the average electron density is obtained from the wave functions $\psi^* \psi$ and the electrostatic potential is computed from the charge distribution. In reality, the electrons interact among themselves via the Coulomb's law

$$V_{ee} = 1/2 \sum_i \sum_{j \neq 1} e^2 / \mathbf{r_i} - \mathbf{r_j}$$

and repel each other. Their movements are correlated. The Coulomb interaction is a long-range interaction. One names *correlation hole* the depletion of electron charge around each electron owing to the Coulomb repulsion, which is spin independent. Each electron moves with a correlation hole around it and the potential seen by the electron will be lower than in the absence of correlation. The average electron density varies with the position and so does the average potential arising from the correlations. The true potential is a function of the coordinates of all the electrons and of the spatial fluctuations in the electron density. Numerous formalisms have been developed to describe a gas of interacting electrons.

Electron correlations were computed for the first time by Bohm and Pines by considering the Coulomb interaction with screening between electron pairs and by taking into account the correlation and exchange holes moving with the electrons [18, 19]. The *random phase approximation* (RPA) was developed for describing the dynamic behaviour of electron systems. According to this approximation, the movement of the electrons includes a short-distance term, corresponding to weakly screened Coulomb interactions between low-energy electrons, and also, at long distance, one corresponding to the movement of the electron ensemble, giving rise to plasma oscillations.

Independently, a calculation of the correlation energy for an electron gas in a uniform background of positive charges has been performed in second-order perturbation theory and in the limit of high electron densities [20]. The electron density n is defined from the inter-electron spacing r_s by the relation

$$4\pi a_0^3 r_s^3 / 3 = n^{-1}$$

where a_0 is the electron radius. In the high-density limit, the Coulomb interaction perturbs only slightly the movement of the electrons, making the treatment of the electron gas simple. It has been remarked that also at very low densities ($r_s \gg 10\, a_0$), the Coulomb interaction has a little influence. The electrons may be expected to become quasi-localized [21] and the correlation energy can be expanded as a power series of $(1/r_s)^{1/2}$. For metals, the electron densities are in an intermediate domain and the kinetic and potential energies play comparable roles in their determination. As indicated above, the correlation energies were calculated [18, 19] and satisfactory agreement with experiment had been obtained for the *simple metals*. However, this model did not give exact results in the high-density and low-density regimes. The validity range of the calculations has then been estimated and all the results obtained in both extreme regimes widely discussed [22, 23]. Numerous studies have been made in order to describe better the physical behaviour of an electron gas and to calculate metallic properties such as conductivity, specific heats, etc. Correlation effects have been included in the treatment of the dielectric constant [12, 24, 25]. In parallel, calculations have been developed from models adapted to treat the case of the nearly free electrons. Among these, we mention the development of a scheme specifically designed for use with the OPW method [26].

A phenomenological theory of interacting particles obeying Fermi–Dirac statistics was proposed by Landau [27]. In this model, the system was described as a gas of *quasi-particles*, i.e. electrons with a large effective mass m^*, due to their interaction with the other electrons and inducing a change of their mobility. This system was designated as a *Fermi liquid*, probably because of Landau's interest in describing liquid He^3. Only short-distance forces were considered in this model. In the case of propagating electrons in interaction with their surrounding, each electron was "dressed", altering its mass and its dynamic properties. The ground state of the valence electrons in a normal metal at sufficiently low temperature can be described as a Fermi liquid. A Fermi liquid is in fact a quantum state of the matter observed at very low temperature for the majority of crystalline solids. A formalism based on the Landau model of the Fermi liquid was developed by considering simultaneously the short and long-distance forces [28]. The long-range Coulomb interactions have indeed an important role in the case of electrons in metals.

The correlation effects for d electrons are different from those of quasi-free electrons owing to the differences in their wave functions, and therefore their movements. Indeed, contrary to the Coulomb repulsion, which is at long range, the other electronic interactions are at very short range. The density of the d electrons is concentrated in the vicinity of the nucleus and is sparse between the atoms, making it possible, in principle, to attach a d electron to a specific atom. Consequently, each d electron interacts strongly with other d electrons present in the same atom and only weakly with the d electrons of the neighbouring atoms. It was shown experimentally that in the transition metals, the d electrons have simultaneously characteristics of the band model and of the quasi-atomic model. For example, the magnetic moments per atom are non-integral multiples of Bohr magneton while the spin-wave processes in the ferromagnetic metals are quasi-atomic. Consequently, the d electrons do not move freely in the entire solid. Exchange and correlation interactions must be taken into account to determine their movements and many-electron wave functions must be used. Several effects are to be considered: d–d intra-atomic correlations, d–d inter-atomic transitions and d–sp excitations.

Models taking into account the correlations between d electrons have been developed to describe the behaviour of the transition metals and compounds. A few among these models have been adapted to the case of very strongly correlated electrons such as the f electrons in the lanthanides and actinides. Among them, we consider in the first place the Hubbard model. Only the intra-atomic interactions were taken into account [29]. The Hubbard Hamiltonian is

$$H = H_0 + \sum U \, n_{i,\sigma} \cdot n_{i,-\sigma}$$

where H_0 is the Hamiltonian in the absence of correlations and $n_{i,\sigma}$ is the density operator of d electrons of spin σ on the atom i. U is the Coulomb interaction energy between two electrons present on the same atomic site, i.e. the energy needed to put two electrons at the same site; it is

$$U = (ii|1/r|ii)$$

where i is the wave function of a d electron present on the atomic site i. Terms of the type $(ij|1/r|ij)$ are neglected. In spite of these approximations, the Hubbard Hamiltonian cannot be solved analytically and, initially, exact solutions were only obtained in a one-dimensional space. More recently, exact solutions have been obtained for the case of infinite number of dimensions and it was shown that these results were very close to those for the three-dimensional case and, therefore, could be considered as a reliable approximation [30].

In the Hubbard model, the aim was to take into account the screening effects due to internal atomic sub shells. These internal sub shells reduce the range of the effective interactions among valence electrons. The model was equivalent to treating the electron interactions as quasi-atomic interactions. In this model, a potential term allows for the one-site interaction and a kinetic term represents the hopping of electrons between sites of the lattice. Indeed, the d electrons are described as being simultaneously atomic and quasi-free, meaning they can hop from one site to another. In order to account for the occurrence of a non-integral number of d electrons per atom, it was supposed that a transition metal could be a collection of atoms in different configurations, i.e. having different numbers of d electrons. The atoms then should change rapidly their configurations because of their interactions with each other [31]. This is a good approximation as long as the long-range electronic interactions can be ignored. It was suggested that this model of *hopping* could be applied to d bands in transition metals and their compounds.

In the early works, only the correlations of the intra-atomic type were taken into account. Improvements were attempted by considering solid-state effects, such as s–d hybridization and the Coulomb repulsion between electrons present on neighbouring atoms. The energy corresponding to the sum of the electron correlation terms was named U *parameter*, independently of the various terms intervening the calculation. A large number of works were done in the Hubbard model by using improved calculation methods, various wave functions like free-quasi-particle (Fermi gas) Green functions, various values of the interaction parameter U, etc. The more important ones will be mentioned in the course of this paragraph.

Another formalism, the *density functional theory* (DFT), was developed independently as an alternative approach to the treatment of many-body effects. The energy of an electron system was expressed, for the first time, as a function of electron density by Thomas and Fermi in the late 1920s. This concept has been widely developed after that Hohenberg and Kohn [32] had demonstrated that all the ground state properties of an electron system are completely defined in terms of its electron charge density $n(r)$. Consequently, there exists a functional relating the ground state energy of an electron gas to $n(r)$, i.e. a one to one correspondence between the energy and the density. This model includes all correlation effects; it is, in principle, exact for a constant or slowly varying electronic density. It makes possible the description of the perturbations due to a change in the charge distribution.

The development of the DFT by Kohn and Sham [33] had as aim to treat the case of interacting inhomogeneous electron gas. Since such system cannot be solved analytically, Kohn–Sham proposed to treat the interacting electrons as a virtual ensemble of independent electrons whose ground state density is that of the real system. The functional of the total energy is

$$E_{K-S}(n) = T_S + V_{K-S}$$

where T_S is the kinetic energy operator of the non-interacting electrons and V_{K-S} is defined as

$$V_{K-S} = V_n + V_e + V_{XC}$$

V_n is the external potential due to nuclei, V_e is the Hartree potential, V_{XC} is the sum of two terms, the exchange-correlation potential per electron of the interacting electron gas and the correction to the kinetic energy term accounting for the electron–electron interaction. Thus, the term V_{XC} contains all effects beyond the

Hartree model. $V_{\text{K-S}}$ is an effective mono-electron potential. In this formulation the Kohn–Sham mono-electronic equations

$$\{T_{\text{S}} + V_{\text{K-S}}\}\phi_i(r) = \varepsilon_i\phi_i(r)$$

can be resolved analytically. The N-particle problem was reduced to N one-particle problems and the electron density was given by

$$n(r) = \sum_{i=1\,\text{to}\,N} (\phi_i(r))^2$$

where N is the number of electrons. The Kohn–Sham DFT consists of replacing $V_{\text{K-S}}$ by

$$V = V_n + \int \{n(r)n(r')/r - r'\} \cdot d^3r' + V_{\text{XC}}(n(r))$$

The many-body effects are included in the potential V_{XC}, which is local, in contrast to the Hartree–Fock potential. The functional relating the density $n(r)$ and the energy is not known exactly because of the presence of the exchange-correlation term. It is necessary to select a functional as a model of approximation. A treatment of self-consistent field type makes, then, possible the determination of the wave functions and electron densities and that of the electron energy.

The quality of the DFT arises from characteristics of the exchange-correlation functional. Among the representations of the latter, the simplest and the most used has been the *local density approximation* (LDA). In this approximation, the electron density is that of a homogeneous gas, the exchange-correlation energies are treated together and the exchange-correlation energy at any point is assumed to depend only on the electron density at that point. However, LDA is an approximation even for the extended s and p orbitals. It improves with the increasing size of the system. In spite of its simplicity, LDA gives relatively good results for the d electrons because of the presence of a compensation effect: indeed, it underestimates the exchange energy and overestimates the correlation energy. It was extended to spin interaction in an electron gas [33]. Spin dependence was introduced into the functional in order to treat systems present in an external magnetic field, homogeneous spin-polarized Fermi liquids or systems with important relativistic effects. This approximation, named *local-spin-density approximation* (LSDA), was applied to magnetic transition materials [34]. LDA and LSDA contain no adjustable parameters.

The DFT differs from the other methods by the fact that the quantum mechanic effects enter in the electronic density rather than through the many-body wave functions [35, 36]. The calculations are more rapid because $n(r)$ depends only on three coordinates, independently of the number of electrons, while the wave function of a system with N electrons depends on three coordinates for each electron. As already underlined, this model is valid for a slowly varying density as well as for high densities. Owing to its simple applicability, the Kohn–Sham density functional formalism was very widely applied to valence electrons in solids in the ground state. Physical properties associated with crystalline stability of the ground state were calculated, such as cohesive energy, compressibility and elastic constants. These studies explained in a qualitative way the respective effects of exchange and correlation on each particular property. As an example, for the compressibility of transition elements compounds, the results with both exchange and correlation are very near those obtained from a Hartree model because the correlation effects practically compensate the exchange effects [37].

For over a decade, the DFT-LDA had been considered as the adequate method for realistic solid-state calculations. Its success was astonishing because, in the LDA, every electron r_i moves independently within the local potential $V_{LDA}(\rho(r_i))$ associated with the time-averaged density of the other electrons at point r_i, $\rho(r_i)$. The potential is a one-electron potential, orbital momentum independent and in which the Coulomb interactions between electrons are taken into account in an averaged way. The many-body problem is thus reduced to a single-electron calculation in a local potential and the method is reliable only for materials for which the Coulomb interactions remain smaller than the kinetic energy of the electrons, i.e. U is smaller than the bandwidth W. The LDA and LSDA failed, however, to describe materials with highly localized electrons. Moreover, these approximations have an intrinsic deficiency because the self-interaction term is not correctly taken into account. In the Hartree–Fock theory, it is completely cancelled while in LDA and LSDA, only a partial cancellation is achieved. A partial self-interaction remains and vanishes only for a one-electron system. It is negligible for orbitals delocalized over extended systems. In contrast, this term leads to systematic errors for finite systems exhibiting strong electron–electron interactions. The presence of this term can explain why the LDA gives large deviations from experiment for quasi-localized orbitals. A model of *self-interaction correction* (SIC) was proposed [38] to remove the interaction of each electron with itself that is present in LDA. In this model, a term representing the sum of the self-interaction of each orbital is subtracted from the $V_{XC}(n(r))$ exchange-correlation potential. The correction vanishes for extended orbitals; it improves the agreement between theory and experiment for a number of physical problems where the LDA approach produces systematic errors. We mention as an example the variation of the band gaps of the alkali halides: the self-interaction correction is larger for the localized orbitals than for the delocalized ones; consequently, it predicts for the gap energy a bigger value than the one obtained from LDA, in better agreement with the experiment. In the case of the rare-earths, the SIC was applied only to 4f electrons and not to delocalized s, p and d electrons.

Other density functionals are now available [39]. Their performances vary with the studied property. Among those, we mention the *generalized gradient approximation* (GGA) [40–42]. It was developed to treat spatially inhomogeneous systems, i.e. closer to the real systems. In this approximation, the exchange-correlation energies depend directly on the electronic density and its gradient. Other functionals have been proposed, the differences between them is of the order of the differences between GGA and LDA calculations. New approximations are now being developed for higher order density gradient.

The d electrons of the transition metals were often described in an orthonormal atomic-like basis, for example the basis of local orthonormal LMTOs of Andersen in their recent formulation [43], or other orbital basis sets, by including the correlation effects in LSDA [44]. The success of this formalism was due to the use of the SIC-LSDA scheme. The ability of this model to describe the characteristics of the transition metals was proved by the agreement between experimental and calculated data concerning, among others, the spin splittings of the Fe, Co and Ni ferromagnetic metals and the cohesive energies of the 3d and 4d metals [38]. In addition, self-consistent ab initio SIC-LSDA calculations had correctly predicted the metal transition oxides to be insulators [45]. This was due to a calculated downward shift of the valence levels with respect to the conduction levels. However, this shift implies a redistribution of valence orbitals in disagreement with the experimental results. Moreover, only the characteristics of the ground state are correctly described by this approximation while the predictions of the excited energy levels may be erroneous. It was thus difficult to determine the energy gaps in insulators from this method because the band gap is not a ground state property. It must also be recalled that the approach used in the various DFT schemes, including the SIC-LSDA, is an independent particle approximation that ignores the many-body nature of the wave functions. Despite their impressive successes, the LDA and LSDA failed to predict the insulating and magnetic properties of transition metal compounds and to describe the characteristics of the *heavy fermion* systems, i.e. of materials with open 4f or 5f sub shells. Indeed, LDA predicts the presence of f electrons at the Fermi level for the rare-earth metals. Owing to the highly localized nature of the f electrons, strong on-site Coulomb repulsions exist between them and calculations beyond LSDA are necessary.

This failure of the SIC-LSD and GG approximations of the density functional theory to describe solids characterized by strong correlation effects has stimulated further theoretical activity. It was suggested that the conventional homogeneous electron gas framework used in the LSDA was not a good starting point to treat the correlations of the entire system of the electronic levels in these compounds and that it was necessary to use an orbital-dependent potential. The electron gas model was, thus, abandoned as a means to describe strongly correlated electrons and these were represented in an atomic-like basis identifiable with the single-particle basis of orthonormal LMT orbitals. A Hubbard-like term was added to the LSDA functional, forming a scheme called the *LSDA + U method* [46]. The Hubbard term described the localized d or f electrons while the delocalized s or p electrons are described by the usual LSDA calculation. A correction term corresponding to correlations included in the LSDA term for the localized electrons is subtracted and the energy functional is

$$E_{\text{LSDA}+\text{U}} = E_{\text{LSDA}} - (1/2)UN(N-1) + (1/2)U\sum n_i n_j.$$

where $n_{i\ (\text{or}\ j)}$ is the localized orbital occupancy. The on-site Coulomb repulsion U splits the LDA bands into two sets of bands of width $W < U$ and only the lower band is occupied with an integral number of electrons as Mott insulator. If the average number of localized electrons is one per lattice site, U is the energy necessary to have two electrons on the same lattice site. The Coulomb interaction energy of localized electrons, identified with the Slater integral F^k, is calculated self-consistently but this energy is strongly reduced by the screening due to the other valence electrons and consequently U is difficult to determine theoretically. The essential assumption in this model is that the strongly correlated electrons are subject to on-site quasi-atomic interactions and, consequently, the number of electrons per site is an integer. This implicitly introduces a separation of the electron orbitals into localized and itinerant. The LSDA + U method was initially used to reproduce the ground state of 3d transition metal oxides and good values were obtained for the splitting between the occupied and unoccupied d bands and for the magnetic moments. As an example of application of the method to rare-earths, we cite the study of gadolinium nitride GdN [47]. However, this method cannot describe electronic transitions because it does not take into account many-body interactions.

Self-consistent electronic structure calculations of strongly correlated systems have also been developed by putting $U = \infty$ [48] or using the slave-boson method [49]. These works aim to study the high-T_c superconducting materials, designated as "extremely correlated Fermi liquids". In these systems, various excited particles, electrons and holes, are present, the latter considered as bosons carrying a charge $-e$. Coupling exists between the boson field and the fermion operators and the neutral fermion excitation spectrum was derived from a fermion-boson field theory [50, 51]. As an example, it was shown by the slave-boson method in the scheme of the LMTO-ASA [52] that the on-site Coulomb repulsion is responsible for the existence of Cu 3d levels at the top of the valence band of $YBa_2Cu_3O_7$, as seen experimentally, but in contrast with calculations not considering correlation effects.

Another approach to the problem of strong correlations was developed [53–56] by taking the LDA + U calculation as starting point and by using a quantum impurity Anderson model subject to a self-consistency condition. In this model, the lattice with many degrees of freedom was replaced by a single-site impurity having a small number of quantum degrees of freedom and embedded in an effective medium determined self-consistently. The Hamiltonian associated with the impurity model is [57]

$$H = H_{ov} + H_{ol} + H_{corr} + H_{vl}$$

H_{ov} and H_{ol} are the unperturbed energies of the quasi-free valence electron system and of the levels of the impurity atom, H_{corr} is the on-site Coulomb repulsion term of the Hubbard model associated with the localized levels of the impurity atom and H_{vl} is the valence-localized electron interaction term. This interaction is of the type $\Sigma V_{vl}(c_v * c_l + c_l * c_v)$, where V_{vl} is a hybridization term, allowing the localized electron l to become mobile despite the fact the electrons v and l are separated. The physical idea was that the dynamics at a given site might be determined by taking into account simultaneously the degrees of freedom at this site and the external milieu created by all other degrees of freedom on other sites. This impurity model gives a picture of the local dynamics of a quantum many-body system. A large number of techniques already developed for treating the single-impurity Anderson model could, then, be used to obtain a picture of the local dynamics of the quantum many-body systems, thus to study the electronic correlations in the solid. The calculations were performed in the frame of a mean field theory by replacing the many-body system by a one-body system in a suitably chosen mean external field, representative of the mean configuration of the system.[1] This approach, called the *dynamical mean-field theory* (DMFT), extends to the quantum many-body problems the mean-field theory used in classical statistical mechanics. The main difference from the classical case is that the on-site quantum problem remains a many-body problem. This non-perturbative model was also named *local impurity self-consistent approximation* (LISA) [48] and many calculation methods have been developed to solve the dynamical mean-field equations.

In the DMFT, applied to quantum many-body problems, only the spatial fluctuations are frozen, contrary to the Hartree–Fock approximation, where both spatial and temporal fluctuations are frozen. The DMFT is dynamical because it takes into account the Coulomb interactions, considered as local quantum fluctuations, i.e. the temporal fluctuations between quantum states at a given lattice site. They govern the frequency dependence of the *self-energy*, which is the operator representing the effects of exchange and correlation. The general importance of the frequency dependence of the Coulomb interaction, also for low-energy physics, has been emphasized [58]. Fluctuations around the mean value of the field represent fluctuations among configurations, the various parameters being determined from self-consistent calculations. This makes it possible to find the self-consistent changes in both charge densities and local Green functions associated with the removal of one electron. It becomes, thus, possible to treat the excitation processes and to describe the corresponding spectra.

The above-mentioned impurity approximation was, then, considered as a starting point to investigate the finite-dimensional strongly correlated electron systems. The potential applications of the DMFT have been widely used [48, 59–61] to study the physics of these systems, in particular their phase diagrams, thermodynamic properties and their electronic characteristics. The model was used with the help of a variety of analytical and numerical techniques, such LDA, quantum Monte Carlo or numerical renormalization group. In its association with LDA, the regime of large correlations is treated within the Hubbard approximation, i.e. by supplementing the one-particle LDA Hamiltonian with the local Coulomb repulsion U and Hund's rule exchange term. By taking into account the specific lattice structure and the density of states as obtained from LDA calculations, the LDA-DMFT describes strongly correlated metals as well as Mott insulators in a single framework. It has

---

[1]A mean field calculation is as follow: from the Schrödinger equation for a mean-field $V(r)$, one calculates a set of single-particle wave functions ψ. From ψ, one obtains the density of particles. After the density is computed, the potential is re-calculated. One stops when there is no more change from one iteration to the next.

been used to treat the dynamics of heavy fermion systems, i.e. systems with local electronic correlations between f electrons [62].

However, because they are dependent on the LDA, both LDA + U and LDA + DMFT add non-local potentials to certain localized configurations, leaving some ambiguity about how a localized configuration is to be defined. In the aim to remove the failures due to the LDA, the GWA model (G designs the Green function and W the screened Coulomb interaction) [63], was associated with the impurity model by introducing the GWA + DMFT approach [64, 65]. The GWA was developed with the aim to characterize the excited levels of stable solid systems, contrary to previous methods designed to obtain the ground state properties. GWA is a perturbative treatment based on the random phase approximation and on an electron self-energy, obtained by the Green's function method by calculating the lowest order diagram in the dynamically screened Coulomb interaction [66]. Such a fully self-consistent GW + DMFT scheme is a formidable task and a simplified implementation of the calculations is generally used. Progress has been made by employing LDA eigenfunctions to generate the self-energy. The quality of the GWA is then closely tied to the quality of LDA starting point. The GWA is successful for weakly correlated systems and improves the description of excitations with respect to the LDA but both confront the same difficulties concerning the treatment of the open f sub shells [67]. A new method, the *quasi-particle self-consistent GW* approximation, QSGW, has been developed in the aim to replace the conventional self-consistent GWA [68, 69]. The QSGW approximation is a self-consistent perturbation theory, parameter-free, independent of the basic set of eigenfunctions, where the self-consistency is used to optimize the effective Hamiltonian by minimizing the perturbation. The self-consistency is an essential requirement because it improves the agreement with the experimental results. The QSGW approximation has given accurate predictions of ground and excited state properties for a large number of weakly and moderately correlated spd materials, both metals and wide band gap insulators. The errors are larger for f solids; however, it has been found that the method can reasonably describe localized f electron systems [67]. The advantage of the QSGWA is to due to the fact that no special treatment of the f electrons is necessary; both valence and f electrons are treated in the same unified framework, without introduction of parameters dependent on the material.

In parallel to these various methods, another theoretical model, initially developed to treat the *Kondo effect*, has been used to interpret specific properties of inter-metallic heavy fermions compounds, especially those involving rare-earths and actinides. The Kondo effect corresponds to the unusual temperature-dependent behaviour of the thermal, electrical and magnetic properties of non-magnetic metals containing very small quantities of magnetic impurities. Initially, the Kondo effect was observed in a wide variety of magnetic alloys, named *Kondo alloys*. An example is the anomalous logarithmic increase of the electrical resistivity with decreasing temperature. Properties, such as heat capacity, magnetic susceptibility, thermoelectric power, also display this anomalous temperature dependence. At low temperature, the Kondo effect can induce the opening of a narrow band gap of a few hundredths of electronvolt, leading to the formation of a *Kondo insulator*. From a fundamental point of view, the Kondo effect corresponds to many-body scattering processes of the itinerant electrons by magnetic impurities in a metallic solid. A magnetic impurity having at least one unpaired electron is coupled to the itinerant electrons of the metal valence band and the strength of this coupling varies as a function of the temperature. At high temperatures, the coupling between the impurity magnetic moment and the magnetic moments of itinerant electrons is weak, resulting in a small anti-ferromagnetic correlation in the vicinity of the impurity. In contrast, at low temperatures, this coupling becomes strong. The scattering cross section of the itinerant electrons off the magnetic impurity diverges as the temperature approaches 0 K, while the impurity spin moment is gradually screened off by the valence electrons. The magnetic moments of the impurity and one itinerant electron couple very strongly to form a non-magnetic state. The system can be described as a Fermi liquid, i.e. in terms of quasi-particles. The density of states associated with these quasi-particles is expected to have a narrow peak at the Fermi level. This peak, which is a many-body effect, is known as *Abrikosov-Suhl resonance*, or *Kondo resonance*. Its presence is due to an unusual scattering mechanism of the valence electrons by the local moments associated with the impurities, inducing a resonant effect in the scattering amplitude. This resonance is observed at temperatures smaller than the *Kondo temperature* T_K and its width is equal to the Kondo energy kT_K, i.e. of the order of only a few meV. This anomalous scattering from the magnetic impurities leads to the enhanced contribution to the specific heat coefficient observed at low temperatures, i.e. at $T \ll T_K$. The same temperature dependence is observed for the resistivity. The Kondo temperature defines the limit of the validity of the Kondo results.

The Kondo effect concerns, in principle, an example of very simple magnetic system, a single magnetic atom or ion in a non-magnetic environment. The impurities are supposed to be so dilute that the interaction between them can be ignored. However, this effect requires very sophisticated mathematical techniques. The treatment of the Kondo effect uses renormalization field theory techniques. The exact solution permits a systematic calculation of all properties, resistivity, thermal conductivity, thermoelectric power, specific heat, magnetic susceptibility, and so forth, and provides a physical understanding of these properties. The theoretical works on the Kondo effect have been connected with a large variety of subjects, such as heavy fermion physics, quantum dots, quantum tunnelling in metals, metallic glasses.

In summary, to describe the valence electron behaviour in solids, three energy terms are to be taken into account, the kinetic energy, the lattice potential and the Coulomb interactions among the electrons. The treatments vary widely from the case where the electron–electron interaction is neglected to the case where this interaction is considered as strong. It was generally thought that the existence of a quasi-atomic behaviour of the valence electrons was due to the presence of strong correlations between them. This has led to a development of models, such as the Hubbard model, in which the movement of the electrons is described by Fermi gas wave functions, supplemented by the introduction in the Hamiltonian of a correlation term U that can take very large arbitrary values, enhancing the charge fluctuations. Various other models were used to approximate exchange and correlation, e.g. the Slater $n^{1/3}$ approximation, the Hedin–Lundquist potential. Developed in parallel, the Kohn–Sham formalism presents a double interest, the inclusion of all the correlation effects and the simplicity of an independent particle model, its basic equations being one-electron Schrödinger equations with local potentials, in analogy with the molecular orbital model. But the success of this formalism depends on the exchange-correlation functional. The most commonly used functional gives a good description of the weakly correlated systems but fails to describe the strongly correlated ones. Indeed, LDA can yield an uncorrelated metal instead of strongly correlated metals or insulators while LSDA + U method can result in an insulator even for metallic materials. Other functional representations like the Gutzwiller approach, the slave-boson formulation, gave equivalent results.

DMFT is a non-perturbative many-body model based on the representation of the lattice onto the Anderson impurity model, recently developed and specially designed to investigate systems strongly correlated with local Coulomb interactions. Associated with the conventional LDA or GWA, it has succeeded in describing the insulating states for which the self-energy effects are crucial. The LDA-DMFT and GWA-DMFT are linear response many-body methods whose advantage lies in their ability to reproduce the results in the limit of small and large U and to account for the main contribution of electron correlations. However, the use of these fully self-consistent schemes is cumbersome. Consequently, simplified implementations are employed. Moreover, although these models were successful in the case of several complex systems, they have their weak points. Firstly, the interacting orbitals are judicially selected, casting doubt on the determination of the Coulomb interaction. Secondly, DMFT does not take into account the non-local correlations, even though many important phenomena stem from long-range correlations. The fit obtained from these approximations is better for strongly localized orbitals than for moderately localized orbitals, making DMFT more suitable for dealing with strongly correlated electron systems. In the LDA + DMFT, the Coulomb interaction is used as a free parameter or chosen ad hoc, as opposed to ab initio calculations, which are parameter-free. In the case of GW + DMFT, the Coulomb interaction is inherently obtained from GW and the challenging problem should be calculations that take into account the interaction dependence on the excitation frequency because the prediction of the excitation spectra is important for the control of the theoretical models. Other specific models have been developed, each adapted to describe the physical properties of a specific system. However, it appears that at present none of the above-mentioned theoretical methods can be considered universal and parameter-free.

1.3.4   The s, p, d Densities of States

The distributions of the s and p valence electrons in solids are generally well described by assuming a random distribution of electrons moving in a self-consistent field. The electrons are represented by plane waves in a Hartree scheme and the extended character of the wave functions is considered as a significant property of the electrons in a solid. The densities of states associated with such quasi-free valence electrons are *wide energy bands* and good agreement exists between the observed band structures for the s and p materials and those calculated by the above method.

However, it must be mentioned that energy levels associated with localized s and p valence electrons can be present in solids. The first ones to be observed were due to the presence of atomic defects in non-conducting solids. These discrete "defect" levels are present in the band gap. The number of defects being small in a crystalline solid, no interaction exists between them and they are considered as localized. If the number of the defects increases sufficiently, electron transfer between the defect levels becomes possible and these levels are present within a narrow defect band. From a general point of view, localization of the valence electrons is possible if no overlap exists between their orbitals, i.e. if the atoms are widely spaced and do not interact with their neighbours. Consequently, if impurity atoms are present in a solid, metallic or not, localized levels are present. Their existence in the alloys in the presence of a small concentration of one of the constituents or of magnetic impurities [70, 71] has been widely discussed. The same effect of localization is observed in the case of defects present on the surface of a solid. It is also observed for aggregates with a number of atoms sufficiently small so that the valence orbitals are well separated.[2]

The mobility of the valence electrons decreases with the increasing strength of the electron–electron correlations, thus with the increasing number of the valence electrons per atom. This change of mobility can be taken into account by the introduction of an effective mass, which increases with the correlations. As a consequence of the reduced mobility, the valence band becomes narrower than that expected for a free electron gas and the observed density of the states at the Fermi level is clearly higher than that given by a quasi-free electron model. This is verified for the materials with d electrons. Let us note that for each atomic electron l there corresponds in a solid a number $(2l + 1)$ of energy bands. Since the d atomic levels are five times degenerate, one expects the presence of five d bands. These bands partially overlap in the solid, making the density of d states considerably higher than that of the s states. Moreover, the d electrons of each atom are intra-correlated

---

[2]In a metal aggregate with N atoms of valence v, the average interval between the energy levels of valence electrons is of the order of $2E_F/Nv$. A continuous energy band is present at room temperature only if this interval is smaller than $k_B T$, of the order of 0.025 eV. For a monovalent metal, this condition is satisfied for $N < 2E_F/k_B T \approx 240$ atoms, that is for spherical aggregates of diameter about 20 Å.

and only weakly inter-correlated with the d electrons of the neighbouring atoms in the solid. Owing to the d–d electronic correlations of the *intra-atomic,* or *intra-site,* type, the d electron density is concentrated mostly near the ions and is small elsewhere. These correlations increase with the number of the electrons present in the d sub shell. At the same time, the bandwidths decrease and the corresponding metals are designated as *narrow energy band* materials.

While a single-particle approach can be used for wide-band materials, a many-body description is necessary to treat the narrow band materials. According to the Hubbard model, the term of electrostatic repulsion between the d electrons present in the same atom, $(ii|1/r|ii)$, calculated by using the free atom wave functions, resulted in a value for U of the order of 15–20 eV. The terms of the type $(ij|1/r|ij)$ were about one order of magnitude smaller. This value of U is much larger than that expected for a solid. From $U/W \geq \sqrt{3} \cdot (1/2)$, with the same wave functions, the valence band should divide into sub-bands, the electrons would be localized, thus unable to carry current and the system should behave as an insulator. Moreover, if the separation between the sub-bands is large enough, i.e. U is sufficiently large, each atom can have an integral number of electrons.

For the transition metals, the number of d electrons is non-integral and the sub-bands overlap. The d electrons move from one atom to another as in a band model but their motions are correlated. Evaluations of the Coulomb repulsion energy U, including the intra- and inter-atomic interactions and the d–s interaction, indicate for these metals values of U of about 2–3 eV. The bandwidth W is then larger than U, except for the metals at the end of the first transition series, for which W is of the order of U. If, however, a transition element is present as an impurity in a solid, U becomes larger than W. Local perturbation is created and it induces the long-range charge oscillations [72], which can be described by the density functional theory including all correlation effects [33]. This underlines the totally different behaviour of an atom according to whether or not its interaction with its neighbouring atoms in the solid is taken into account. Numerous calculations have been performed to determine the energy distribution of the d levels by introducing as a parameter the electronic correlation energy U, which corresponds to the Coulomb repulsion energy between electrons localized on one site; U is then named *on-site Coulomb interaction energy* or *intra-atomic Coulomb term.* The five valence d orbitals are generally assumed to have the same interaction term U, making the correlations equal for all the d electrons.

As already underlined, the d–d correlations increase with the number of d electrons in the sub shell. They increase also with a decreasing of the sub shell radius. This radius is shorter for the 3d sub shell than for the other nd series. Thus, among the s, p, d valence electrons, the more strongly correlated are the d electrons in elements belonging to the end of the 3d transition series. For all the transition metals, the s and d wave functions partly overlap. This causes a mixing of the different sub-bands, resulting in a continuum of d levels. However, structures characteristic of the presence of the various occupied d sub-bands appear in some curves of density of states observed by high-resolution spectroscopy. The d–d correlations are large in the metals with an almost closed d shell, such as Fe, Co, Ni, which are ferromagnetic at ordinary temperature.[3] In contrast, all other transition metals have a Pauli paramagnetism. Let us recall that in metals, ferromagnetism is due almost entirely to the spin magnetic moment while the orbital magnetic moment, much weaker, is responsible of the Pauli paramagnetism.[4]

Theoretical methods described in the precedent paragraph, like APW, tight binding, have been used to determine the densities of d states in the metals obtaining satisfactory agreement between the band calculations and the spectro-scopic results. Alloys have also been described by these methods or with the help of the Fermi liquid model which considers, in the first approximation, the system as a free electron gas with an *effective electron mass m** heavier than the mass of a free electron.

When the wave functions of the valence electrons have an extended character, the number of electrons present in the valence band fluctuates continuously around an average, for all the cells of the solid. Conversely, if, for a transition group element, the 3d electron number is integral on each site, the d orbitals must be considered as being of atomic-like type. The 3d electrons present in these orbitals have a localized character by opposition to the delocalized character of the 3d electrons present in bands.

Initially the rare-earth metals were considered as consisting of a core of the type $4f^N$ and three 6s–5d valence electrons per atom, except for europium and ytterbium with only two 6s–5d valence electrons. The 6s–5d electrons were assumed to occupy free electron bands, their number in each one of the s and d sub shells being fractional. However, the rare-earth metal electronic properties indicated the pres-ence of a density of states at the Fermi surface clearly higher than predicted by the nearly free electron model. One of the first calculations of the energy bands in a rare-earth metal used the APW method [73]; it concerned gadolinium. This metal is a natural choice because of the simplifications resulting from its half-filled 4f shell and the amount of accurate experimental data available. It was shown that the valence band resembles that of the beginning of the shell transition metal owing to the mixed s–d character of the valence electrons, the small number of the d elec-trons and the absence of 4f electrons in the band. The presence of valence levels with d character at the Fermi level accounts for the high state density and the high electronic specific heat of the rare-earth metals. This first calculation was non-self-consistent, non-relativistic, non-spin-polarized. The general shape of the bands was hardly affected by improved approximations. However, relativistic effects must be taken into account for rare-earths and relativistic APW (RAPW) calculations have been developed [74, 75]. They take into account the spin–orbit coupling and induce a modification of the band crossings. These calculations indicate a shift of the relativistic bands from the non-relativistic ones, depending on the s or d symmetry. Thus, the energy of the d bands increased relatively to that of

---

[3]The Curie temperature is of the order of several hundreds of degrees Celsius for Fe, Co, Ni.

[4]In the transition metals, below the Curie temperature, the ferromagnetism results from the dif-ference between the number of electrons with spin parallel or antiparallel in the d shell.

the s–p band as one moves along the 4f series [76]. It has been shown that the energy bands of the metal rare-earths were similar to those of hexagonal transition metals such as yttrium and scandium. The 5d valence orbitals are widely spread out spatially. The presence of 5d electrons explains some of the rare-earth metals properties, such as their conductivity.

1.3.5 4f Distributions in Rare-Earths

As indicated above, the 4f electrons of the rare-earth metals were initially treated as core electrons, i.e. as localized on each ion. In this atomic-like model, the ions were considered as separate and without interaction between them. This treatment was equivalent to ignoring the influence of the 4f electrons on the solid electronic properties since these properties depend on the characteristics of extended electronic wave functions [77]. Indeed, a very narrow 4f band, 0.05 eV wide, was predicted about 10 eV below the bottom of the 6s–5d bands for gadolinium metal [73], indicating that the 4f electrons are highly localized. Consequently, a big difference was expected between the magnetic properties of the rare-earths and those of the transition elements. It was due to the presence of the strongly localized 4f electrons in an open shell, carrying large magnetic moments whose magnitude is approximated by atomic Russell-Saunders coupling (Table 1.1). The five light rare-earths from cerium to europium (radioactive Pm is not considered) are anti-ferromagnetic at very low temperatures.[5] The five heavy rare-earths from terbium to thulium are ferromagnetic at low temperatures; they become antiferromagnetic before getting to the paramagnetic phase at temperatures that decrease from 230 K for terbium to 60 K for thulium. Gadolinium alone is magnetic at the standard temperature with a $T_{curie} = 16$ °C.

In the rare-earth metals, the 4f electrons appeared then to be absent from the valence band and to be present in a narrow energy range, located well below the Fermi level E_F [78], whose width depends only on the direct f–f overlap. The unoccupied 4f levels are located well above E_F at the beginning of the 4f series. They approach it as Z increases. It was shown that, under standard temperature and pressure (STP) conditions, the trivalent $4f^n$ configuration was stable in all the rare-earths, except in divalent europium and ytterbium. Consequently, the gradual addition of 4f electrons along the series does not modify the valence electron distribution and the atomic volume varies a little, with two exceptions for the divalent metals, whose atomic volumes are clearly larger. The direct f–f interactions between neighbouring rare-earth atoms can be considered as negligible and the behaviour of the 4f electrons appears significantly different from that of the s, p, d electrons. A model of very narrow bands has, then, been used to describe the 4f electrons. In this model,

---

[5]Antiferromagnetism is described as the coupling between spin magnetic moments belonging to neighbouring ions in the solid.

the distance between 4f electrons present on a same lattice site was small and the electron correlations strong. Thus, two electronic sub-systems coexisted, one formed by the itinerant s, p, d electrons, the other by the localized f electrons. The levels near E_F consisted of wide bands dominated by s, p, d orbitals while very narrow "4f bands" were present far below. The single contribution of the 4f electrons was to polarize the spd sub-system through an indirect exchange mechanism.

Nevertheless, calculations treating the 4f electrons like the other valence electrons of the solid have been developed for the rare-earth metals and their compounds. These developments were justified by the fact that, in some compounds, the 4f bands were found to contribute to the cohesive energy. The LDA, or LSDA, of the density functional theory was used. In these calculations, all the electrons, 6s, 6p, 5d and 4f, were described by extended Bloch functions and the electron correlations were treated assuming a homogeneous electron gas; the electrostatic exchange interaction was, then, stronger than the spin–orbit interaction. A relativistic technique was developed that used the LSDA for exchange and correlation and included the spin–orbit interaction as a perturbation, thus keeping the spin as a good quantum number. The band structure of the metals was calculated from this model. Application to gadolinium was made [76, 79]. Disagreement was found between the experimental values of the cohesion factors and those obtained from self-consistent relativistic calculations using the LMTO method and the LSDA [80]. Too small equilibrium lattice constants and too large cohesive contributions were found, making it seem that the 4f electrons participated in the cohesion. In fact, these results showed that the LSDA was not adapted to describe systems having electrons of a truly localized nature.

The difficulties of the traditional band methods in treating the 4f electrons of the rare-earths were evident. The independent particle approximation was no longer valid and calculations based on the LSDA failed completely to describe the open 4f shell systems because it left 4f electrons at E_F. This inability of the LSA and LSDA to treat properly the correlations within the 4f shell was due, for a large part, to the presence of the self-interaction terms. As the self-interaction increases with electron localization, its contribution has a substantial effect on the 4f electrons while the effect on the 6s and 5d delocalized valence electrons is negligible. Consequently, the results depend on the considered sub shells. For cerium, for example, the unphysical interaction of the 4f electron with itself in the LDA would reduce the attractive potential, resulting in an extended 4f orbital and increasing the calculated hybridization. A big improvement was obtained by using the SIC-LSD approximation, introduced by Perdew [38]. Let us recall that in this model, a negative potential term is added, arising from the SIC contribution. Ab initio SIC-LSDA, and SIC-LDA, calculations of the density functional theory have been widely used to describe both the 4f electrons and the bonding 6s-5d electrons in the rare-earths [81]. The agreement between the SIC-LSDA or SIC-GGA calculations and experimental data obtained for some properties of the rare-earths justified the use of this theoretical treatment, initially developed to treat solid systems such as the transition compounds.

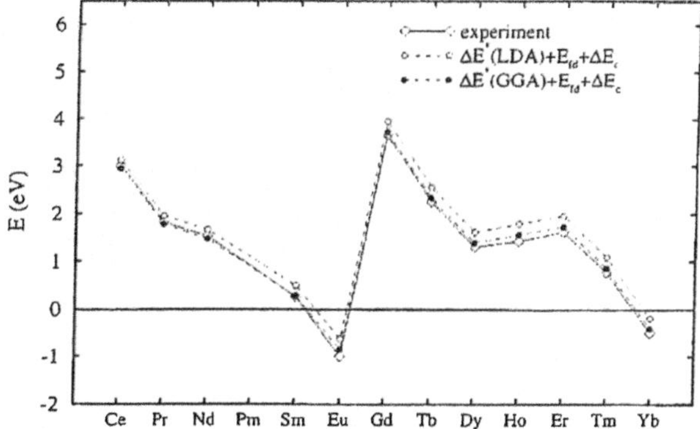

Fig. 1.1 Di-trivalent energy difference E at zero pressure for the elements 58–70, Pm excepted [82]. The experimental points are taken from Ref. [83]. Calculated results are obtained from the LDA and GGA approximations

As an example of calculations with the help of both SIC-LSD and SIC-GG approximations, the energy difference between divalent and trivalent configurations is presented in Fig. 1.1 [82].

Positive values were obtained for all the rare-earths except for europium and ytterbium, confirming the bivalence of these two elements in the metal and the trivalence for all the other rare-earths. The divalent-trivalent energy difference increases strongly between europium and gadolinium, revealing a large stability of the trivalent gadolinium; then it decreases progressively going to ytterbium. Lattice sizes were also calculated as a function of the atomic number Z. The SIC-LSDA, known to give different results depending on the presence of band-like or core-like electrons, has been used to determine the number of the localized 4f electrons. In this aim, the total energy over all possible configurations of the localized or itinerant 4f electrons was minimized in the considered solid [84]. These calculations confirmed the rare-earth valences indicated above. The SIC-LSDA has also provided an accurate description of the cohesive properties throughout the lanthanide series but these properties are known to depend essentially on the s, p, d electrons while their dependence on the 4f electrons remains weak. Moreover, it must be underlined that SIC-LSDA overestimated the separation between the occupied and unoccupied 4f levels and that the success of the calculations clearly depended on the type of studied properties.

The discrete or extended character of 4f levels in the rare-earth metals and compounds remained, therefore, an open problem and the question was to know whether to treat the 4f levels in a core-like model or in a strongly correlated band model. Another question was: to what extent does the behaviour of the 4f electrons depend on their interactions with the s–d hybridized valence electrons, or conversely, what was the influence of the 4f electrons on the valence bands, particularly near the Fermi level? These interactions depend on the overlap between the various orbitals. The radial densities of valence electrons are shown in Fig. 1.2 for gadolinium [85].

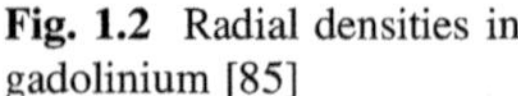

Fig. 1.2 Radial densities in gadolinium [85]

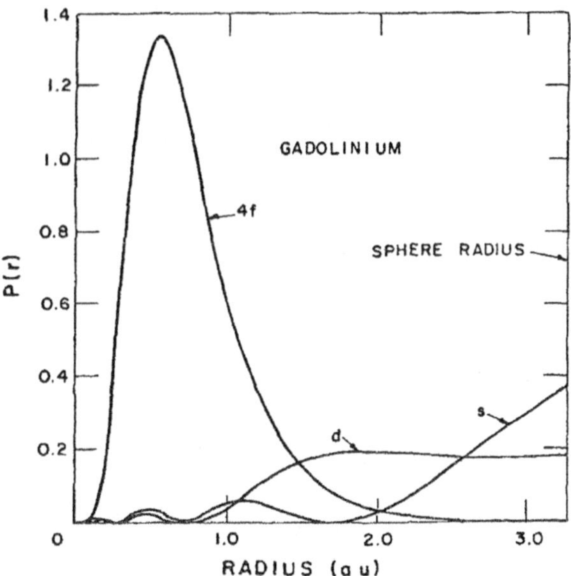

From this figure, both 4f–5d and 4f–6s overlaps are very small, making the 4f–5d6s interactions very weak. It has been suggested that the 4f–6s interaction was probably stronger than the 4f–5d interaction because the 4f sub shell is concentrated near the nucleus, in a region where the 5d wave functions are small. However, the 4f–5d exchange interaction is larger than the 4f–6s one.

The shape and the spatial extension of the electron wave functions have, then, been studied for the entire rare-earth series. They permitted to determine the electron characteristics and their degree of localization. The spatial extension of the valence 6s, 6p, 5d, 4f orbitals within the Wigner–Seitz atomic cell was calculated from the radial wave functions at the Fermi level. The calculations used the semi-relativistic Dirac equation for the ground state charge density [86]. The radial distances of the s, p, d and f orbitals at the ground state, normalized to the experimental Wigner–Seitz radius, are plotted in Fig. 1.3.

The s, p and d valence orbitals vary slightly with Z and are of the order of 0.7–0.8 times the Wigner–Seitz radius. The extension of the 4f orbital is much smaller and decreases rapidly between lanthanum and γ-cerium, then progressively along the series, being 0.2 times the Wigner–Seitz radius for lutecium. The small deviation for europium should be due to the clearly larger Wigner–Seitz radius associated with the bcc crystalline structure of this element (cf. Sect. 1.4). From these results, the overlap of the 4f distributions between neighbouring atoms in the solid is small. This is an evidence of the localization of the 4f orbitals and its increase with the increasing atomic number. Confirmation follows from the fact that the mean radii of the 4f orbitals are smaller than the radii of the 5p core orbitals. Consequently, the 4f electrons are well shielded from the effects of the surrounding atoms. The 4f wave functions are localized in the nuclear potential well and are

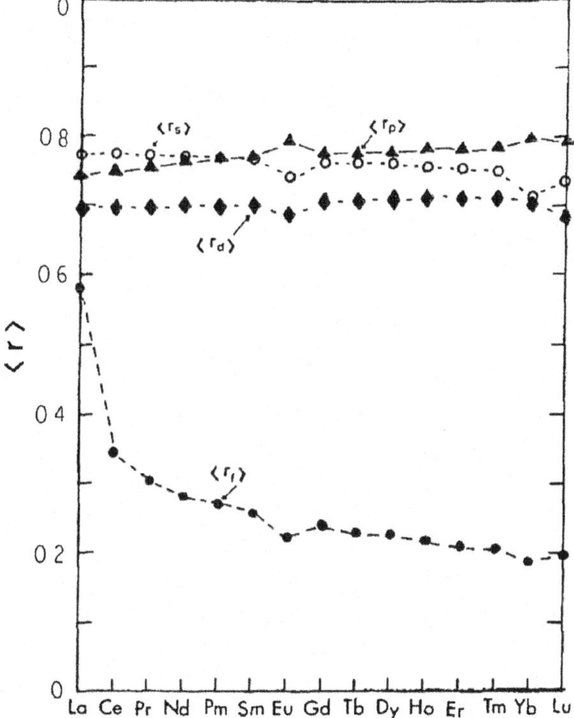

Fig. 1.3 Radial distances of the s, p, d and f orbitals at the ground state, normalized to the experimental Wigner–Seitz radius [86]

strongly dependent on the Coulomb attraction of the inner sub shells. Owing to the considerable overlap of the 4f and 4d radial wave functions of the rare-earths the direct and exchange Slater parameters of the electrostatic interaction between their sub shells are large and the 4f–4d interaction is strong. It is stronger than the 4f spin–orbit interaction in these elements. In contrast, it must be recalled that since the d wave functions of the transition elements are external the d valence electrons interact essentially with other d electrons.

The 4f electrons of the metal rare-earths in their ground state could not be treated by the same methods as the d electrons of the transition metals because of the differences in the shape and the extension of their wave functions. Since the 4f wave functions are concentrated in a range of strongly negative potential, the 4f electrons are expected to be quasi-localized on each rare-earth ion and to keep some atomic character in the solid. Unlike the d electrons, in the metallic state, the 4f electrons do not contribute to the conductivity and give rise to a well-defined localized magnetic moment. The open 4f shell should retain an integral occupation number and no 4f electron should contribute to the cohesion. There exists, however, an apparent anomaly between the limited spatial extent of the 4f electrons and the reduced value of their energy levels: these electrons are confined to the region close to the nuclei while their energy is in the vicinity of the energy of valence electrons, at about an energy of the order of U below the Fermi level, i.e. at only a few electronvolts. That explains why the electronic properties of the rare-earths are more strongly dependent on the energy level overlap than on the wave function hybridization. The f-level position relative to the valence band cannot be determined in a one-particle picture. It results from the total energy difference between the system with or without one f electron in the valence band. All the above differences between f and d electrons and the particularities of the f electrons required adapted theoretical models for treating the f electrons. A special technique taking into account the *l*-dependence of the potential seemed necessary in order to improve the agreement with the experimental results. This type of treatment, analogous to a Hartree–Fock-type one and similar to the relativistic core-like model, was used with success to describe the excited levels observed in the high energy spectra (cf. Chap. 2). But it did not allow the description of the 4f electrons for the ensemble of the rare-earth metals and compounds.

Indeed, it was known that in some particular cases, as for example in the tetravalent cerium compounds, a small number of 4f electrons could behave as band electrons [87], and not as core electrons, A model suggesting that the 4f electrons should have a *dual* character has then been proposed [84, 88, 89]. In this model, an f band should be present, split into two narrow sub-bands: one, occupied, lying well below E_F; the other, above it, is virtual and unoccupied. In the trivalent metals, the unoccupied f sub-band should lie just at the Fermi level. It should hybridize with the partially occupied s–d bands and should make the valence band acquire some f character, thus *"creating a different type of f electrons which can participate in electronic bonding"*. In this model, among the electrons occupying the valence band, a few should be f electrons having the characters of valence electrons, i.e. participating in chemical bonding and contributing to the Fermi surfaces. Their number should be fractional. It should be small for the first rare-earths, should increase up to about 0.6 for samarium; the same variation should occur between gadolinium and thulium. In this model, the virtual unoccupied f sub-band of the trivalent rare-earths is supposed to acquire electrons from the s–d valence band. This initially unoccupied f sub-band should, then, be shifted just at the Fermi level and occupied levels of f symmetry, widely separated in energy from the other occupied core-like f levels, should be present. In the divalent metals, in the unoccupied f sub-band located well above the Fermi level, no valence-like f electron could be present.

It appears thus that two types of 4f electrons, "pure" and hybridized with the other valence electrons, could be present in the solid. The pure 4f electrons are localized core-like electrons with an integral number present on each ion. The hybridized 4f electrons should be delocalized band-like electrons, mixed with the s–d electrons, participating in the bonding and contributing to the Fermi surfaces. These latter should be present only if the valence of the rare-earth is superior to the number of the electrons present in the external s, d sub shells of the free atom. Then, for the rare-earths of $4f^n 6s^2$ configuration in the atom and divalent in the metal, europium and ytterbium, the two 6s electrons of each free atom are in the valence band and each ion has the $4f^n$ configuration. All the f electrons are localized in the solid and there are the s–f exchange interactions between the ions and the valence electrons, which intervene to explain the physico-chemical properties. In contrast, for a rare-earth of $4f^n6s^2$ configuration in the free atom but trivalent in the metal, the ions have the $4f^{n-1}$ configuration and only $(n-1)4f$ electrons are localized. The trivalence of the metal induces a decrease in the number of the 4f-localized electrons by a transfer of a 4f electron into the valence band [87]. Two rare-earths, cerium and gadolinium, have the $4f^n5d6s^2$ configuration in the free atom and are trivalent in the metal. In both cases, the three s, d electrons form the valence band and the n 4f electrons do not participate in the bonding and are localized. Indeed, each gadolinium ion is known as having the $4f^7$ configuration in the metal and all its compounds, the seven 4f electrons being pure atomic-like electrons. Nevertheless, when submitted to particular conditions of temperature or pressure, cerium metal can undergo a valence change with a transfer of 4f electron into the valence band. In this particular case, it becomes possible to speak of 4f band. Under high pressure, an increase of valence has also been observed for ytterbium [90, 91] and europium [92] and both become trivalent rare-earths when they are sufficiently compressed.

An important point must then be considered: the behaviour of *delocalized* 4f electrons in the rare-earth metals. Their presence is due to a valence change with respect to that of the metal in the STP conditions, which can be accompanied by a structural transition. The most widely studied element is cerium, often considered as an emblematic representative of the rare-earth series because the metal undergoes a valence increase under a change of the pressure-temperature conditions. Studies concerning this metal are summarized in the next paragraph.

1.3.5.1 Cerium Metal

Cerium is a highly reactive metal, making its experimental research difficult. Its ground configuration in the free atom is $4f^15d6s^2$. Under the STP conditions, the metal is trivalent and the ion Ce^{3+} has the ground configuration $4f^1$ outside closed shells. From the spectroscopic data, this 4f electron is known to be of the atomic-type in the metal and the trivalent compounds (cf. Chap. 4) while the three hybridized 5d–6s electrons form the valence band. The Ce^{3+} trivalent ion may easily lose its f electron because of the instability of the almost empty shell. Cerium becomes then tetravalent. This valence change is known to take place in numerous compounds (cf. Chap. 2) [93]. The pressure–temperature phase diagram of cerium metal is complex (cf. Sect. 1.4). Only the $\gamma \Rightarrow \alpha$ iso-structural transition observed under high pressure is considered in this paragraph because it is accompanied by a widely discussed electronic change. Upon increased pressure, the face-centred cubic phase γ undergoes an unusually large volume contraction of about 15 %. This contraction is accompanied by a change of most of the physical parameters, inter-atomic distances, energies of the levels, charge density, magnetic properties.

Several models have been developed to explain the electronic change accompanying the first-order $\gamma \Rightarrow \alpha$ structural transition in metal cerium. These models are independent of the temperature because they considered the transition as entirely due to electronic processes. In the face-centred cubic phase γ, stable in STP conditions, large on-site Coulomb interaction exists, causing one 4f electron to be localized on each Ce^{3+} ion. This localization is in agreement with the presence of a well-defined magnetic moment close to the value expected for a singly occupied 4f shell, with $J = 5/2$ [94, 95]. The size decrease of the cerium ions, which accompanies the contraction of the atomic volume, induces a partial delocalization of the 4f electron, which, in turn, changes the cohesion. But this cohesion variation is too small to produce a discontinuous phase transition. The $\gamma \Rightarrow \alpha$ structural change is accompanies by a change of the valence and of the electronic properties and a diminishing of the intrinsic 4f magnetic moment. Thus, the γ phase has a temperature-dependent Curie–Weiss susceptibility while a temperature-independent paramagnetic susceptibility is present in the α phase. Various models have been developed to describe the 4f electron in the non-magnetic α phase. Initially, it had been proposed independently by W.H. Zachariasen and by L. Pauling that this change corresponded to the passage of the 4f electron into the 5d valence band, $4f^1$ trivalent cerium being transformed to $4f^0$ tetravalent cerium. In this model, the 4f electron moved from a localized orbital to a level of the valence band and the total magnetic moment was quenched during the transition [94]. The 4f electron lost its own orbital characters and acquired the orbital characteristics of the other valence electrons [96]. The mechanism leading to the $\gamma \Rightarrow \alpha$ phase transition could, then, be explained from the electron–electron interaction matrix element between 4f electrons and valence band electrons. But positron annihilation measurements [97] and various properties such as cohesion [98], Compton scattering [99, 100], can be explained only by the presence of 4f electrons in the α phase [93].

It has, thus, been considered that the average number of 4f electrons did not decrease significantly during the phase transition and remained close to one [101]. The localized 4f electrons of the γ phase were assumed to become *itinerant 4f electrons* in the α phase, in agreement with the fact that $\langle r_f \rangle$ is more extended for α-Ce than for γ-Ce and that the 4f electrons contribute to the cohesive energy in the α phase. It was suggested that this transition arose from a reduction of 4f electron correlations following the decrease of the lattice spacing. In fact, the correlation reduction follows the decrease of the ratio between Coulomb energy and kinetic energy under pressure. Interpretation of the γ-α transition with the help of a Mott metal-insulator transition-like scheme had been proposed [98, 102]. But initially, the contribution of the 4f band to the cohesive energy was overestimated. Based on this scheme, LDA calculations of the s, d, p, f electron distribution in the α phase and of the spd bands in the γ phase have been made by decoupling the localized 4f electron of γ-Ce and treating it as a core-electron [103]. The pressure–temperature phase diagram of the γ-α transition has been calculated and this model correctly described the linear variation of the transition temperature with the pressure and the location of the critical point [104].

Independently, the Kondo model has been adapted to treat the specific properties resulting from the scattering of itinerant electrons by localized electrons. The electron–electron coupling induced by the scattering process is considered as producing a delocalization of the localized electrons by hybridization with the itinerant electrons. The coupling strength is characterized by the Kondo temperature T_K and by the presence of a scattering resonance near the Fermi level at temperatures less than T_K. By applying this model to the rare-earths, the scattering between the valence electrons and the localized f electrons, which are immersed in the valence electron distribution, is predicted to lead to a hybridization of the f and valence electrons. For cerium, it was shown that the γ-α structural transition resulted from an increase of the hybridization between the f and valence electrons in the α phase [105, 106]. Strengthening of the bound character of the 4f electrons was thus expected [107, 108]. However, the change was predicted to be small and the 4f electron was considered as remaining rather localized in the α phase. The pressure-temperature phase diagram was calculated with the Kondo model [109]. Good agreement was obtained between theory and experiment for the critical point but the model predicted a nonlinear variation of the transition temperature with the pressure, in contrast with the experimental observations [93].

Contrary to what was expected for a localized electron, a weak peak was observed at the Fermi level in the valence state density of the α-cerium phase [110, 111]. This peak, observed by high-resolution photoemission spectroscopy (cf. Chap. 4), at temperatures between 10 and 15 K, was identified as the Kondo resonance expected in the valence level distribution around the Fermi level at temperatures inferior to the Kondo temperature; indeed, it is in agreement with the 4f spectral function calculated by LDA + DMFT with increasing atomic volume [112]. This peak is characteristic of a Fermi liquid behaviour [113]. It corresponds to the many-body singlet ground state formed at T close to 0 K by a gradual screening of the 4f magnetic moment. The presence of this peak showed that cerium presents the unconventional properties of the heavy fermion systems in its α phase. The Kondo model is suitable to describe the peculiar properties resulting from the interplay between localized f electrons and the ones that tend to delocalize by hybridization with the valence band [114]. However, the observation of the Kondo resonance has been a controversial subject. Another interpretation of the variation of the spectra with the temperature, based on the role of phonon broadening, has been proposed [115] based on experiments with resolution lower than that obtained by Garnier et al. [111], which appears to obtain better experimental results in the field. Indeed, the phonon contribution must be taken into account when an electron moves from a localized level to an extended level if the localized and extended levels are not mixed. Consequently, if the localized 4f levels are located below the valence band, phonons intervene in the scattering process. Other processes can take place like the excitation of levels split by the crystal field and the excitation of electron–hole pairs. These various processes are usually very weak but in some particular cases, they intervene and make the interpretations difficult. The observation of a Kondo resonance in α-cerium has been widely discussed [61, 116]. Several experimental aspects connected with these observations will be reported in Chap. 4.

LDA-DMFT model has also been used to describe the cerium γ-α transition. As previously underlined, the DMFT self-consistency condition agrees with the Anderson impurity model, which is known to be a local representation of the

Hubbard model, thus of the Mott transition model. By considering an spd band and an f band, the position of the f band and the Coulomb repulsion energy U were calculated along with the Kondo temperatures [117–121]. Temperatures of the order of 1000 and 30 K have been estimated for α- and γ-cerium, respectively. Values of U equal to 5.72 eV for α-Ce and 5.98 eV for γ-Ce were calculated in reference [119] while value of U $\approx$ 6 eV was generally introduced in the other calculations. The difference between these numerical values is due to differences in the calculation procedures used. The ratio of Coulomb interaction to the bandwidth energy was found to decrease from 3.8 to 2.5 across the γ-α transition. The large values of U are characteristic of the presence of strong correlations among the localized f electrons as well as between the f electrons and the delocalized spd electrons. The variation of the energy U between α- and γ-Ce is small and this indicates that the correlations remain the same for the two phases and that they are stronger than initially expected for the "delocalized" phase α-Ce. This result is in disagreement with the initial suggestion that the structural transition in cerium arose from a rapid change of f–f correlations of the Mott transition-type. In contrast, it is supported by the Kondo model, according to which the change remains small because of the weak f–s, d hybridization, that remains limited due to the small overlap of the f and s, d wave functions. Moreover, it is important to note that the calculated values of U are clearly larger than those estimated from spectroscopic data, which are around 4 eV. However, it was mentioned [119] that the γ-α transition could be explained by the Mott model with a reduced value of U.

Infrared and optical spectroscopy experiments showed that the γ-α transition in metal cerium was accompanied by an electronic structure change. Indeed, a peak was observed at 1 eV in the real part of the optical conductivity for the α phase and was absent in the γ phase [113]. It was concluded that the two phases had different electronic structures on an energy scale widely exceeding the Kondo energy, all the four fsd electrons forming the valence band in the α phase. However, it was not clear how differences in electronic properties could result from small variations in pressure. It was suggested that such a structural change required taking into account the lattice field. Research of an eventual modification in the thermodynamic properties at the γ-α transition was then undertaken. The dependence of the total energy on the volume was calculated from the LDA-DMFT model (cf. Fig. 1.4) [119].

At $T = 0.054$ eV ($\approx$600 K), the calculation revealed a thermodynamic instability region, indicating the presence of a first-order phase transition, whose minimum corresponds to the atomic volume of the α phase. The total energy is related to the volume. The volume reduction associated with the γ-α transition induces then a reduction of the total energy, which could be attributed to a drop in the entropy due to the phonons. This interesting result has hardly been exploited.

Actually, it is generally admitted that a localization-delocalization process of the 4f electron accompanies the γ-α volume collapse and makes observable the change of the electronic and physical properties. As has already been underlined, two alternative models are used to explain the variation in the hybridization between the f orbitals and the band-like s–d orbitals, the Mott transition model and the Kondo

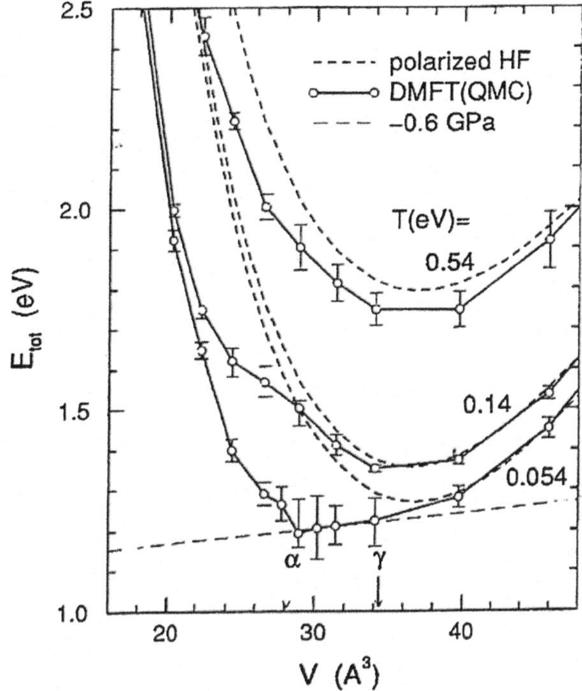

Fig. 1.4 Comparison of the total energy calculated in LDA-DMFT and polarized Hartree–Fock approximations as a function of volume at three temperatures. The LDA-DMFT calculation at 0.054 eV shows a shallowness, in agreement with the experimental results within the error bars [118]

model. New determination of the p-V isotherms in a large range of temperature (300–800 K) from accurate X-ray diffraction measurements was presented as revealing the existence of the Kondo effect [122]. However, it was remarked that these experimental results could also be explained in the Mott transition model [104]. Recently, measurements of the intensity variation of an X-ray satellite emission have shown a decrease of the correlations between valence and 4f electrons in α-cerium [123]. The authors have concluded that the γ-α cerium transition is not due to intrinsic changes in the 4f localization but to the screening of a valence electron as predicted in the Kondo picture and to a decrease in the number of the f electrons. It should be noted that the same type of X-ray satellite had been observed in the transition metals. The characteristics of this satellite have been explained theoretically from spectroscopic considerations, without the need to invoke a change of the localization of the d electrons [124].

Consequently, while in the γ phase the f electron is known to be localized, it seems difficult to know exactly what is the average occupation number of the f level in α-cerium. Because of the spatially confined nature of the f wave functions, one cannot expect the f-sd hybridization to differ much between both phases [125]. That appears as being in agreement with the Kondo volume collapse model. Among the various experiments interpreted by this model, photoemission studies at low temperature have shown the presence of a structure at the Fermi level, identified as an Abrikosov-Suhl, or Kondo, resonance [111]. However, the presence of a quantum decoherence process could prevent the observation of the Kondo resonance [126]. On the other hand, the 4f electrons have low energies near the Fermi level, in spite of the fact that they remain localized near the cores and splitting of the levels can take place. The energies of the f orbitals and those of the s–d valence band are then similar and vary by only a few percent during the γ-α transition. However, the 4f electron energy needs to change by almost an order of magnitude to justify a strong change in the f electron behaviour. Some self-interfering terms could bring close the energies associated with each of two phases. Thus, it has been suggested that the energy invested in eliminating the local f moment in γ-Ce could be counterbalanced by the energy gained by forming an f band resulting in no change in the f-occupation number. Lastly, it should be mentioned that not all these models generally take into account the temperature. However, the electronic energy levels in solids depend on temperature and pressure, i.e. of electron–phonon interactions. A decrease in pressure is associated with an increase of the entropy, and therefore an increase of the electron–phonon interactions. Inversely, during the γ-α transition, the decrease of the entropy causes the decrease of the electron–phonon interactions, which favours the delocalization of the electrons and the hybridization between 4f and valence electrons. Indeed, hybridization can take place if the 4f band width is greater than the crystal field splitting.

1.3.5.2 Other Rare-Earth Metals

The 4f electrons of the rare-earth metals have recently been treated by using a new calculation technique, named HIA model, considered as suitable for strongly correlated systems [127, 128]. This model is a multi-band generalization of the Hubbard approximation combining an atomic description of the 4f electrons, considered as subjected to on-site quasi-atomic interactions, with a band description of the delocalized spd electrons. Indeed, multiplet structures related to the localized nature of the 4f electrons have been observed spectroscopically. An atomic-type term, which takes into account the multiplet effects within the open 4f shell of a single ion, has been introduced in the Hubbard Hamiltonian. This atomic term describes the two-body part of the f–f interaction and is not included in the calculations concerning the spd electrons. In this model, the open 4f shell is treated in a simplified manner and is subsequently embedded in a bath of itinerant electrons. The appearance of multiplet structures is notoriously known as a many-electron phenomenon observed in the excited levels while theoretical LDA + U treatment resorts to a one-electron picture and leads to various degrees of success for the ground state properties of the rare-earth elements. Consequently, due to the limited spatial extent of the f orbitals, to their strong intra-shell correlation effects and to the appearance of multiplet structures, the HIA treatment of the 4f shell as a part of the core seems to be an approach that describes correctly most of the properties of the rare-earth metals in their ground state. In this case, the spin–orbit interaction of the 4f electrons plays an important role.

The treatment of the open 4f shell using an atomic model had been proposed already a long time ago [129, 130]. Atomic calculations were made for all the rare-earths, without taking into account interaction with the delocalized spd electrons, i.e. by considering the various ions as separate and non-interacting, but by using an ab initio method, in contrast with the previous calculations which simulate multiplet structures with parameters adjusted to reproduce the experimental results. Since the hybridization between the 4f and delocalized electrons, i.e. the direct influence of the valence electrons on the 4f electrons, is negligible, these ab initio calculations are able to predict with success the behaviour of the 4f electrons. According to their predictions, discrete energy levels are associated with the 4f electrons of the rare-earth metals in their STP state and they enable the determination of the number of the 4f electrons present in each rare-earth at the ground state (cf. Chap. 4).

As already mentioned, the magnetic properties of the rare-earth metals are essentially determined by the strongly localized 4f electrons whose number can be deduced from magnetic measurements. Thus, for cerium, a Curie–Weiss-like magnetism is present in the γ phase under low pressure revealing the presence of one localized f electron whereas a Pauli-like paramagnetism is observed in the α phase under high pressure. However, it is necessary to take into account the additional moment arising from the polarization of the valence electrons. Indeed, the valence electrons are spin-polarized by an exchange interaction with the localized 4f electrons. For example, for gadolinium with seven f electrons, a saturation magnetization of 7.6 μ_B per atom is obtained in the metal from LDA + U calculations [131], in good agreement with the experimental value of 7.63 μ_B [132]. This value is higher than that expected for an ^{8}S ion. The remaining magnetization of about 0.6 μ_B is due to the polarization of the s and d valence electrons. Gadolinium is the single ferromagnetic rare-earth at the ambient temperature. That is due to the large stability of the full spin-up $4f^7$ shell, which is close to the core. The magnetization carried by the 4f electrons plays no further direct role in inter-atomic exchange but only by the intermediary of the interactions between 4f-valence electrons. The magnetic interactions lead to the formation of a wide variety of magnetic structures, whose periodicities are often incommensurate with the crystal lattice. For the heavy rare-earth elements, of hexagonal structure, the magnetic structures are known to depend on the ratio of the inter-planar distances c and a (cf. Sect. 1.4), but the theoretical understanding of this fact is not clear. In a general point of view, the magnetic properties are a consequence of the electronic and structural characteristics; they confirm their interpretations but evidently without being responsible for them.

In summary, in a first approximation, the core-like 4f electrons of the rare-earths are responsible for their magnetism and s, d electrons as responsible for the other electronic properties, such as transport properties, Fermi surfaces. This approach is insufficient to describe the totality of the electronic properties of these materials. In a general point of view, it is necessary to consider the 4f electrons as being either localized with a core-like character or partially hybridized with the valence electrons. Each rare-earth ion has an integral number of f electrons depending on the valence and must be described in a 4f core-like framework. Other 4f electrons have energy close to that of the valence electrons but while preserving their ground state orbital characters, they can be present in a fractional number and should be treated in a 4f-band framework. These electrons remain partially localized because most of their wave functions are concentrated around the ion core. An important problem was to determine the number of localized 4f electrons in solids. Spectroscopic methods specially designed have been developed to measure this number showing that it is equivalent to what can be expected from the valence. Thus, an increase of the valence results from decreasing by one unity the number of the localized 4f electrons. However, this delocalized 4f electron is expected to remain strongly correlated with the other 4f electrons and consequently specific treatments must be used to explain its characteristics. The ratio of the Coulomb interaction energy U to the bandwidth energy is considered as an important factor in the treatment of quasi-delocalized 4f electrons because it depends on the 4f–4f interactions and 4f-valence hybridization. In spite of its importance, U is not known in a definitive manner. Its experimental values are about two thirds of their theoretical predictions, indicating that the role of the correlations in the quasi-delocalized 4f electrons is smaller than generally expected. Up to now, the interactions of the electrons with the crystal field and the phonons were regarded as negligible and the electron–electron interactions were considered as governing all the properties of the rare-earth metals and compounds. Indeed, the particular characteristics of the 4f open shell are responsible for the existence of unusual properties of numerous materials, such as the compounds at intermediate valence and the heavy fermions compounds, whose understanding remains a challenging problem in solid-state physics.

1.3.6   The 5f Distributions in Actinides

The electronic properties of the light actinides and their compounds are generally only slightly affected by the ageing of the sample. However, irreproducibility of the data due to self-damage mechanisms can exist for radioactive elements. The majority of self-induced radiation damages disappear upon annealing at standard temperature. This explains why a determination of the electronic structure of actinides can be obtained at ambient temperature. Changes due to self-induced radiation damages are studied at low temperatures. They are smaller in americium as compared to the lighter actinides and this was explained by the lack of 5f bonding. Beyond americium, little is known about metals and compounds because the high radioactivity of the actinides complicates the experimental work and requires cumbersome infrastructures.

The degree of localization of the 5f electrons in the actinides is a crucial parameter for the understanding of the complex physical properties of these elements. Indeed, the electronic structure of the actinides is the dominant factor in the determination of their crystalline and magnetic properties. From the theoretical value of the ratio U/W, the 5f electrons were considered initially as being an intermediate stage between that of the localized 4f electrons of the rare-earths and the itinerant d electrons of the transition metals [133, 134]. The actinide metals are separated into two groups, the light actinides from thorium to plutonium and the heavy actinides starting from americium. In the first group, the 5f electrons were supposed to populate conducting levels while in the second, they occupy non-conducting levels [86]. The valence bands of the light actinides are composed of levels involving the 5f, 6d and 7s electrons. But it was necessary to explain the very low symmetry of crystalline structures in the metals (cf. Sect. 4.3). Indeed, the light actinide metals have very particular structural properties, which do not exist in the elements whose metallic bond is characterized by a classical band model with delocalized s–d valence electrons.

Among the conventional experimental parameters, the atomic radii of the elements in the condensed phase provide interesting data because they vary along each series according to the filling of the sub shells (Fig. 1.5) [135].

For the rare-earths, weak monotonous decrease of the radius with the increasing number of the 4f electrons is observed except for the presence of the two maxima corresponding to the two divalent metals, europium and ytterbium. In contrast, for the transition series, the atomic radius variation is quasi-parabolic; initially, the atomic volume decreases because the d electrons enter the valence band and contribute to increase the cohesion. Beyond the half filling of the d sub shell, the cohesion energy decreases progressively and the atomic volume increases. The variation is regular for the 5d series (not indicated in the Fig. 1.5) where all the elements crystallize in compact crystal structures. For the 3d series, magnetism disrupts this variation and a weak irregularity exists for manganese, which has a complex structure instead of the expected hexagonal close packed structure. All the same, iron is bcc instead of h.c.p and cobalt is h.c.p instead of f.c.c. For the first actinides, strong anomalies are observed. The atomic volume of americium is

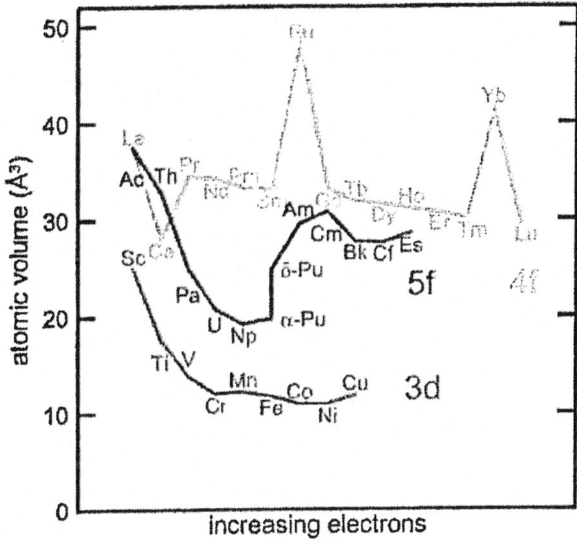

Fig. 1.5 Atomic volumes of the 3d, 4f and 5f elements as a function of the increasing number of electrons. For the 3d series, the variation is parabolic. For the rare-earths, it is linear and decreases slowly, except for the two divalent metals. For the actinides, an unusual variation is observed [135]

almost 50 % larger than that of the preceding element plutonium in its stable α phase and a difference of 25 % exists between the α- and δ-Pu phases. But, for the elements following americium, namely curium, berkelium, californium, the atomic volume decreases slowly and varies as that of the lanthanide series. The actinides heavier than plutonium, often named transplutonium elements, can be considered as forming a second rare-earth series. This view is strongly supported by the fact that all these elements adopt the structure typical of the lanthanides, i.e. the dhcp structure. This situation of the heavy actinides results from the characteristics of the 5f electrons, that are very similar to those of the rare-earth 4f electrons. Atomic volume and compressibility of the heavy actinides are close to values typical for the lanthanides and localization of the 5f electrons is effective starting from americium. In contrast, in the beginning of the actinide series, the atomic radius starts to decrease rapidly with the increasing atomic number. This decrease is similar to that observed for the transition metals. All the electrons contribute to increase the cohesive energy of the solid thus reducing the atomic volume. The 5f spin–orbit interaction is not always large enough to separate the $5f_{5/2}$ and $5f_{7/2}$ levels and the effect of the spin–orbit on the atomic volumes is expected to be weak. Indeed, the $5f_{5/2}$ sub shell fills in first. But this sub shell could be half filled at uranium and the atomic volume could increase beyond. This is not observed. However, the atomic radius begins to increase beyond neptunium and abrupt jumps are observed for δ-plutonium, then for americium. The experimental equilibrium volumes are 20.8 Å^3 for uranium and approximately 20.5 Å^3 for α-plutonium, 25 Å^3 for δ-plutonium and 29.2 Å^3 for americium. These volume variations are characteristic of a change in the 5f electron behaviour.

A parameter, named *critical spacing*, depending on the atomic radius, was introduced to characterize the itinerant versus localized character of the electrons in solids [136, 137]. This parameter was defined as the ratio between the abscissa of the maximum of the relevant atomic wave function and the metallic radius of the element in the considered solid. It was suggested that if the inter-atomic distance between actinide ions in metal, alloy or compound is larger than the critical spacing, the 5f electrons would be localized and the material would be ferromagnetic or antiferromagnetic at low temperature. Otherwise, it would be superconductor. Partial delocalization is necessary to justify the presence of superconductivity but this delocalization must remain sufficiently weak in order to allow for strong electronic interactions. This is the case of thorium, protactinium and uranium, which are superconductors at very low temperature. From this model, the distinction between light and heavy actinides was confirmed. Another parameter is the 5f spin–orbit splitting. It is of the order of 1–2 eV, that is to say larger than for the other valence electrons. As a consequence, one considered generally that only $5f_{5/2}$ electrons are present below E_F for the first actinides up to the $5f^6$ configuration. Indeed, this holds only for cases where the distance of the $5f_{5/2}$ below E_F is smaller than 5f spin–orbit splitting. Generally, the hybridization between the 5f and s–d valence electrons brings the $5f_{5/2}$ far below the Fermi level. In this case the $5f_{7/2}$ level which is 1–2 eV apart moves also below the Fermi level and levels of both spin directions are occupied.

An important characteristic of the light actinides is the absence of magnetic order. The theoretical models such as the density functional theory (DFT) in local density approximation (LDA) or generalized gradient approximations (GGA) used to compute ground state properties predict a long-range magnetic order in the light actinides. This order is not observed experimentally, contrary to that which is seen for most of the rare-earths systems. The tendency of the light actinides to form a magnetic state is also weaker than for the 3d elements like nickel. The presence of long-range order of magnetic moments is associated with a decrease of the wave function overlap between neighbouring atoms. It is consistent with a localized behaviour of the electrons. Indeed, as is well known, bonding and magnetism are incompatible for the same electron and the magnetic moments disappear when the orbitals overlap and bond. The presence or absence of a magnetic order reveals, therefore, the localized or non-localized character of the electrons. The observed absence of localized magnetic moment for the early actinide metals was consequently interpreted as characteristic of the itinerant character of the 5f electrons.

One expects the 5f electrons to have characteristics different from those of the other valence electrons because of the particularities of their wave functions. Due to the shape of the 5f electronic wave functions, fairly similar to that of p electrons, highly directional bonds are present and give rise to low-symmetry crystalline structures and low melting points. Moreover, a large fraction of the 5f wave functions is contained inside the radon core. Their extension decreases with the increasing atomic number more rapidly than that of the radon core and the localization process becomes completed in americium. Consequently, the 5f electrons in the light actinides were supposed to be somewhere between being localized electrons characterized by magnetic moments and being itinerant electrons contributing to the metallic bonds. These latter are considered as being in narrow bands, interacting with the valence sd band. The interaction terms, which contribute to the total energy, depend on the characteristics of the 5f electrons. Consequently, according to whether the 5f electrons are more or less delocalized, the formation energy of 5f bands surpasses or not the polarization energy gained when the 5f levels form a localized multiplet.

Another model was envisaged to treat the light actinides, that of a duality of the character of the 5f electron. Some 5f electrons could partially be localized and the others could form bands and contribute to the bonding. The 5f electrons could oscillate between these two different situations and this dual character could exist in the metal, as well as in metallic compounds [138]. Localized and itinerant 5f electrons could, therefore, be present simultaneously [139]. Many unusual properties of the actinide materials, in particular their complex crystalline structures, could be explained with the help of this model, which explains the absence of localized magnetic moment for these metals. Moreover, unlike the rare-earths, that are trivalent in most solids, the light actinides have several valences. They can be trivalent, tetravalent or even pentavalent. Since, on the average, only two s and one d valence electrons are present, it means that one, or eventually two, f electrons participate in the band valence. However, as already underlined, the movements of the 5f electrons do not have the same extension as that of the other valence electrons because of the characteristics of the 5f wave functions. These delocalized 5f electrons would be present in a relatively narrow band located just below the Fermi level. Another model has been proposed to justify such an energy position of the 5f band. It supposes that this narrow band, initially pinned at the Fermi energy, drops below the Fermi level by the symmetry breaking distortion, which is responsible for the low symmetry of the crystalline structure [140].

Lastly, let us mention that, in order to explain the absence of magnetism in the light actinides as well as the unconventional superconductivity of rare-earth and actinide materials, it has been suggested recently that pairing of itinerant f electrons due to spin–orbit interaction could be formed [141].

The ground configuration of thorium in the free atom is $(Rn)6d^27s^2$ and thorium is tetravalent in the solid. The presence of some delocalized and bonding 5f electrons was expected [142–146]. The 5f electrons are hybridized with the 6d-7s electrons of the valence band. One can suggest that analogy exists between thorium metal and cerium in the high pressure phase, both for electronic structure and crystalline arrangement. Indeed, both thorium and α-cerium are f.c.c. Moreover, they are superconductor at very low temperatures. The transition temperature of thorium, T_C, is 1.4 K and decreases under pressure. For protactinium of ground configuration $5f^26d^17s^2$ in the free atom, delocalized 5f electrons are also expected in the solid and T_C is 1.4 K. Very few works have been done on this element.

Several models have been proposed to describe the uranium, of ground configuration $5f^36d^17s^2$ in the free atom. According to one model, two localized 5f electrons are present and this number remains fixed. The extra 5f electron is found in levels that are mixed with the valence band. In a second model, all the 5f electrons are in narrow bands and strongly coupled with the s–d valence electrons. Another possibility would be the presence of fluctuations between the two configurations $U^{3+}(5f^3)$ and $U^{4+}(5f^2)$. Uranium is superconductor with T_C varying with the crystalline structure and the pressure; it is 0.7 K for high quality single crystal orthorhombic α-uranium at standard pressure and 2.3 K at 10 kbar. The variation of T_C with the pressure is different for thorium and for the other actinides. This is due to the difference in the 5f electron occupation number, which is around three in uranium while it is very small in thorium. Uranium is considered as the first element of the series for which correlations become noticeable. Nevertheless, their effect is expected to be relatively moderate for the pure metal. It becomes noticeable when uranium is present in compounds. The uranium atoms are then pushed apart by the presence of the other elements and localization of the 5f electrons increases. When strong correlations are expected, theoretical tools such as DFT-LDA or DFT-GGA are insufficient for determining the ground state electronic properties. The results deduced from these two computation modes have been compared to those obtained from the quasi-particle self-consistent GW, or QSGW, method [147]. This is the first use of the QSGW method in the treatment of the 5f-orbital electron–electron interactions. Differences between the results obtained from these two types of methods remain rather small for the occupied levels and around the Fermi energy. The number of the 5f electrons is equal to 3.19 in QSGW while it is 3.57 in the LDA calculation. Some differences are expected for the unoccupied levels above

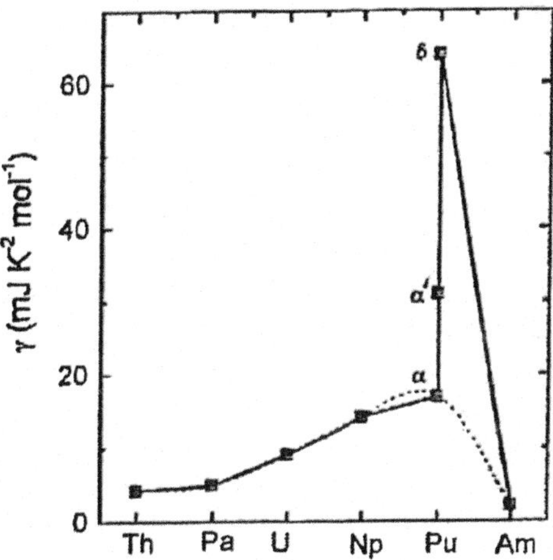

Fig. 1.6 Experimental electronic specific heat γ for the light actinides [135]

the Fermi level. It was concluded that the correlation effects remain moderate in α-uranium because of the low occupation of the 5f levels. They can be observed experimentally from transitions involving excited levels (cf. Chap. 5).

Neptunium is considered as the first element for which 5f electron partial localization starts to appear. Next to α-plutonium, neptunium is among the actinides which have the highest electronic specific heat (Fig. 1.6).

The same applies for the resistivity as a function of the temperature. The decrease of the atomic volume with the atomic number is strongly reduced in the vicinity of neptunium before the spectacular rise of the plutonium δ phase. All the properties of neptunium indicate that the character of these 5f electrons is intermediate between that of the uranium and plutonium 5f electrons. The change of properties is due to the presence of more than three electrons in the $j = 5/2$ level. Indeed, the number of the 5f electrons is about four. No superconductivity was observed. From these characteristics, it was suggested that localization of the 5f electrons appears in neptunium. However, neptunium, as well as plutonium and americium, is paramagnetic.

Plutonium, of configuration $5f^6 7s^2$ in the free atom, is located just between the light and heavy actinides in a range where itinerant versus localized duality is important. This makes it one of the most complex materials known. Its reactivity and radioactivity and also the complexity of its phase diagram make it difficult to obtain reliable experimental data. Among its anomalous properties, both resistivity and Pauli-like magnetic susceptibility are an order of magnitude larger than those of the simple metals. All these particular properties of plutonium result from the presence of six 5f electrons in the free atom. Because the plutonium metal has not more than three non-f electrons [148], approximately five 5f electrons are expected in the solid. The localization of these electrons depends strongly on temperature, dimensionality, atomic arrangement and doping. The energy separation between the $5f_{5/2}$ and $5f_{7/2}$ electrons, due to the 5f spin–orbit interaction, is much larger than the crystal field splitting. The 5f electrons are known as responsible for the low-symmetry monoclinic structure of α-plutonium. The electronic properties of the f.c.c δ phase are most controversial and are discussed in parallel to the electronic properties of the α phase in this paragraph. As for neptunium, partial localization inhibits superconductivity.

The two α and δ phases are energetically very close. However, the δ phase has the largest low-temperature specific heat of any pure element. The substantial volume expansion that occurs between them was explained by an increase of the 5f electron localization, involving a change in the electron–electron interactions. A change from a band-like to an atomic-like behaviour was expected. Numerous theoretical studies employing the usual theoretical methods, LDA, LDA + U, LDA + U combined with DFT, aimed at explaining the volume difference between α- and δ-plutonium. Generally, these calculations predicted a magnetically ordered ground state for δ-plutonium, in contradiction with the observations. In fact, no experimental evidence of magnetic moments is noticed in the δ phase, either ordered or disordered, nor is it noticed in the other phases. Other parameters, like the volume of the δ phase, were also predicted in large error. These results confirmed the inability of the above theoretical methods to treat such a system. Indeed, heavy fermion-type model, i.e. electrons dramatically slowed by the interactions, is necessary to understand the unusual metallic δ-plutonium.

Overlap between two or more configurations was considered possible in the metal. Initially, both Pu^{3+} $5f^5$ configuration and Pu^{4+} $5f^4$ configuration with one 5f electron in the valence band were predicted to be present. The 5f electrons of the Pu^{3+} and Pu^{4+} ions were expected to have spin and orbital contribution to the angular momentum and an associated magnetic moment. Anomalies observed in δ-plutonium were then attributed to ordered magnetism. But other explanations were found later. As an example, anomalies observed in the specific heat were attributed to structural effects. At low temperature, structural effects due to a stabilization of the δ-plutonium phase were identified whereas initially the presence of intra-atomic magnetization had been advanced [149]. Compensation of the spin and orbital moments was also considered in order to explain the absence of magnetism. More recently, an important review of experimental data including magnetic susceptibility, specific heat, electrical resistivity, nuclear magnetic resonance and inelastic neutron scattering for plutonium has been published [135]. It was concluded that no ordered or unordered magnetism involving the 5f electrons exists in plutonium metal in either the α or δ phases above a temperature of about 4 K. The same conclusion was obtained from muon spin rotation spectroscopy [150]. A possibility that plutonium is magnetic only at very low temperature, as observed in alloys, could perhaps be considered. Indeed, the Néel temperature of samarium having the same external configuration, in its normal phase, is only 14.8 K and the samarium 4f electrons are clearly more localized than the plutonium 5f electrons. Therefore, one could expect for plutonium a phase transition close to 0 K. This has not been observed due to experimental difficulties caused by its radioactivity. However, this hypothesis was not followed and other theoretical models were sought to explain the absence of magnetism in plutonium. One possibility was that plutonium has the $5f^6$ configuration with a $J = 0$ level formed out of six localized f electrons but no experimental result has supported this hypothesis.

Several theoretical studies showed that non-magnetic δ-plutonium could be derived from the LDA + U method, reformulated in a spin and orbital rotationally invariant form [151]. This type of formulation was appropriate for metallic phases with Coulomb energy U comparable to the bandwidth W. Non-magnetic δ-plutonium was obtained for values of U from 3 to 5 eV. The number of 5f electrons also depends on U and was ≈5.4 for $U = 4$ eV. Calculated atomic volume was in good agreement with the experimental value. In this model, the non-magnetic character is not due to a cancellation of non-zero spin and orbital parts of the moment but to $S = 0$ and $L = 0$. This approach had a limitation because it did not take into account the dynamical effects, which, independently, had been considered as important. But another study has also predicted the absence of dynamical and static magnetic moments in δ-plutonium in agreement with the experiment [152].

Several complex phase transitions involving changes in the lattice structure and also in the wave functions of the plutonium 5f electrons are associated with relatively small changes in temperature and pressure. It, then, appeared probable that the 5f electrons in plutonium could have several different degrees of localization with nearly degenerate energies. The origin of the volume expansion between the α and δ phases could be explained by a competition between itinerancy and localization of the 5f electrons caused by strong electron–electron interactions, meaning a competition between band-like or atomic-like character of these electrons. The ground state should, then, be a quantum superposition of distinct valences that could wash out the magnetic order by dynamical fluctuation of moments.

A theoretical study using a *mixed-level state* model was developed to follow the evolution of the atomic volume of plutonium and the next elements [153]. In a first calculation, the 5f shell was constrained to have an integral occupation number and to remain un-hybridized with the rest of the valence electrons. The total energy of the ground state of the solid calculated with the localized 5f electrons according to Hund's rule was compared to the energy obtained with all the delocalized 5f electrons. In the ground state, the 5f electrons were found itinerant for uranium and localized for americium. The atomic volume of plutonium with all the 5f electrons localized is larger than the arrangement where they are delocalized. However, the two arrangements have approximately the same energy. Consequently, it was suggested that fluctuations could exist between the two configurations. Another possibility was that the wave function of the ground state contains two components, one from delocalized states and the other from localized states. Using this model of *mixed-level state*, the atomic volume obtained with only four localized electrons is in agreement with the experimental value observed for δ-plutonium (cf. Sect. 1.4). The atomic volume associated with five fully localized electrons is near that observed for americium while the volume associated with itinerant 5f electrons in fcc structure is close to the α-plutonium volume. It was noted that these values could depend on the temperature.

All the experimental results underline the particular character of the 5f electrons in the light actinides, which is intermediate between that in a solid and in an atom. Consequently, the standard approximations cannot be used because they do not allow ground states of systems with energy levels partially populated by localized

5f electrons. A model has been developed specially to make evident the difference in the behaviour of the plutonium 5f electrons between the two α- and δ phases. Incorporation of an atomic 5f shell in a solid-state environment was necessary to treat the real material. Adaptation of the dynamical mean-field theory (DMFT) was made in this aim [154]. Many-body 5f atomic states, hybridized with each other and with the s–d itinerant electrons were introduced as impurities and their Green function and self-energy determined. Simultaneously, a 5f band was introduced. Relativistic the spin–orbit coupling effects were included. The ground state total energy and the phase diagram were determined. When a value equal to 4 eV was used for the Coulomb interaction U among 5f electrons, a double energy minimum was obtained. By assigning one minimum to the α phase, the other to the δ phase, the α-δ phase transition was found to correspond to an increase of volume by 25 %, in agreement with the experiment. The 5f electrons of the α phase are found more bonding than those of the δ phase. However, they must not to be considered as equivalent to other valence electrons. As shown from a series of anomalous transport properties, the α phase is a correlated phase reminiscent of heavy electron systems. For example, the resistivity of α-plutonium around room temperature is temperature independent and anomalously high, its value being higher than the limit obtainable from conventional metals.

According to the same theoretical approach, which treats simultaneously Kondo and magnetic systems and takes into account the atomic multiplet structure and the crystal field splitting, the ground state of metal plutonium was found to be a superposition of multiple valences [155]. This treatment calculates fluctuations in time between atomic configurations with different numbers of 5f electrons, f^4, f^5 and f^6, and exchange of electrons with the surrounding medium. Plutonium was found to be in a mixed valence state with an average 5f occupation of 5.2 electrons. This result shows that a dynamical treatment is needed for plutonium. Indeed, partially delocalized 5f electrons forming narrow bands could be present simultaneously with 5f electrons that keep their atomic character for a very short period. The local moments could disappear over short time scales and their observation would depend on the observational frequency window of the probe used. They could not be observed with the help of static methods such as nuclear magnetic resonance or neutron inelastic scattering. The width of the narrow bands was found to depend on the multiplet splitting rather than on band effects of the type expected in the Hubbard model or seen in the standard model of solids. However, 5f-sd hybridization and f–f hopping are present. Interdependence exists between the degree of itinerancy, the spread of atomic multiplets and the spin–orbit coupling while the absence of magnetism was confirmed theoretically.

In summary, there is no experimental evidence for any ordering of electronic moments in plutonium above 4 K. This is difficult to explain, particularly in the case of the fcc δ phase, which is known to be much less metallic than the α phase. From a theoretical model, named BCN expansion [156, 157], it was shown that taking into account the relativistic effects induces an enhancement of various parameters, among them the spin–orbit interaction parameter. This enhancement could lead to the suppression of magnetism in plutonium, and also in neptunium and americium. However, actually, the preferred interpretation explaining the absence of magnetism is the presence of dynamical fluctuations that suppress the magnetic moments. Another important property of the plutonium is the high sensitivity of its volume to small changes in temperature, pressure or doping. That is explained by considering plutonium as a mixed valence metal. The 5f electrons are considered as fluctuating between different degrees of localization nearly degenerate in energy. Character close to itinerant could also be detected for some 5f electrons present in narrow bands weakly hybridized with the d–s bands close in energy. However, even these quasi-bonding electrons are strongly correlated. The ability of the plutonium ions to change their electronic configurations and adjust their size as their lattice environment changes is the essential characteristic underlying the great diversity of properties of plutonium compounds.

Americium, of $5f^7 7s^2$ configuration in the free atom, is the first heavy actinide. In the metal, the configuration is $5f^6$ with an almost complete $j = 5/2$ arrangement. The 5f electrons are localized, similar to the 4f electrons in the lanthanide series, and contribute no longer to the cohesion. They are present on each lattice site and are chemically inert. Bonding results only from the s–d valence electrons. Physical properties such as atomic volume, crystal structure, cohesive energy, electronic specific heat, resistivity, change considerably from plutonium to americium. Thus an atomic volume expansion of 40 % exists between the two metals. An increase of the symmetry of the crystal occurs in americium, which has the double hexagonal closest packed structure. Reduction of the electronic specific heat and the resistivity is also observed. All these changes are explained by the localization of the 5f electrons. Americium is the actinide analogue of europium. However, it is trivalent in the metal and of configuration $5f^6$. The total angular momentum J of the ground state is zero both in LS and jj coupling and trivalent americium has no localized magnetic moment. Because the well-localized six 5f electrons form a non-magnetic configuration, americium metal is superconducting. The transition temperature T_C is 0.8 K in standard pressure. The americium 5f electrons can be delocalized under a sufficiently large pressure and several phase transitions have been observed between 0 and 100 GPa. A low-symmetry phase appears at 111 kbar, which marks the onset of 5f electron bonding in Am. Consequently, a complex dependence of T_C, which can increase up to 2.2 K, has been observed in this pressure range [158, 159]. On the other hand, if the levels are broadened by inter-atomic interactions then some hint of magnetism should appear and depress T_C. The compounds and alloys where americium has a valence superior to three exhibit a magnetic moment.

The elements beyond americium are chemically inert, have 5f electrons localized on each lattice site, symmetric crystal structures and local magnetic moments. Curium has the configuration $f^7 6d^1 7s^2$ in the free atom and is trivalent. It is analogue to gadolinium and has very similar properties. Its electronic configuration is obtained by adding an electron to the $5f^6$ sub shell of the americium. The $5f^7$ configuration with all the seven spins parallel in the half-filled 5f shell is more stable than the one with six electrons in the $f_{5/2}$ sub shell and one electron in the $f_{7/2}$ sub shell. Curium orders antiferromagnetically with T_N around 60 K. At lower temperatures, it has a large magnetic moment, in contrast with the previous metals that are not magnetic. It is highly resistive like gadolinium and α-plutonium. Its resistivity keeps increasing due to self-damage by α radiation. Starting from berkelium, the metals in the series are divalent and show a behaviour similar to that of europium and ytterbium in standard conditions. The divalence of the heavy actinides results from the high bonding energies of the 5f electrons as was shown experimentally for mendelevium, nobelium, fermium and einsteinium.

1.3.7   Conclusion

Among the theoretical works on the rare-earths, SIC-LSDA has been a prevalent model. It has provided a description of the cohesive properties throughout the series. LDA + U and LDA + DMFT are actually widely employed because concepts like the Hubbard energy U are commonly accepted. The LDA + DMFT model was tested for the cerium $\gamma \Rightarrow \alpha$ transition. DMFT enables the study of the excited states; it is helpful in explaining spectroscopic observations but still rarely utilized. However, these three methods are dependent on LDA and, consequently, have the same formal problems. Thus, the non-local correlations are neglected. That has been partially resolved by adding non-local potentials to treat the localized electron states. However, no improvement exists in the treatment of the electrons considered as itinerant. Another standard model, the GWA, presents the same defects as the LDA for the open f shell. Many-body effects must be taken into account. The combination of many-body and density functional theories could give a better picture. From the recent QSGW approach [160], the filling of 4f states follows rather closely Hund's rule for the atom, in agreement with the atomic model, but the position of unoccupied f levels is predicted systematically too high. Consequently, in spite of the success of these theoretical models, it can be considered that each is marked by some deficiencies.

Concerning the 4f electrons, two different situations can coexist in solids and, consequently, two initially opposite points of view can be unified: the one treats the 4f electrons as core electrons, the other considers the 4f electrons as weakly hybridized with the valence electrons in narrow energy bands. In the latter case, however, the on-site f–f interactions are considered to be strong, pushing the 4f electrons back toward localization. The hybridization of the 4f orbitals with the s–d valence band generally remains very weak, except in some particular cases as α-cerium. One speaks of "strongly correlated 4f electrons". Thus, for the metals in the STP conditions, the 4f electrons are highly localized, the 4f shell retains an integral occupation number verified by spectroscopy. The characteristics of the 4f electrons are related to the small 4f orbital radius. This small radius is associated with a high Coulomb correlation energy U. Let us recall that U is the minimum energy required within the metal to transfer a 4f electron from one atom to the 4f shell of the neighbouring atom. For the rare-earth metals, U is expected to be superior to the width W of the distribution of the 4f levels and the relation $U > W$ prevents the formation of a 4f band. The 4f open shell is known as playing an important role in the magnetic properties while its role in the bonding remains very weak. It is responsible for the very particular properties of the rare-earth compounds that will be discussed in the next chapters. Consequently, suitable experimental techniques are indispensable for clearing up the difficult problem of the localization or not of the 4f electrons and its effects on material properties.

Extension to the case of the actinides is one of the reasons for the intensive development of novel theoretical studies of f electrons. Indeed, the light actinides, particularly plutonium, are characterized by many unusual properties: the presence of crystalline structures of low symmetry, the complete absence of long-range magnetic order, a very high resistivity. Superconductivity is observed in thorium, protactinium, uranium and americium and not in the case of plutonium. Plutonium, considered as the last member of the series where 5f electrons are involved in the metallic bonding in standard conditions, is characterized by an overlap of electronic configurations and by fluctuations of the number of the 5f electrons, which induce spatial and temporal instabilities. These fluctuations between two or more configurations and the screening of the 5f shell moments by the valence electrons were proposed to explain the absence of magnetic moments in plutonium, despite the presence of localized 5f electrons. All these electronic and structural properties as well as their dependence on temperature, pressure and doping, result from the behaviour of the 5f electrons.

In summary, in the light actinides, a distinction exists between itinerant or strongly correlated 5f electrons. The itinerant electrons participate in the bonding whereas the other electrons, present in narrow bands, interact strongly among themselves and determine the crystal symmetry. Localized 5f electrons can also be present in discrete atomic levels. The competition between localization and itinerancy of the 5f electrons and the strong correlation between 5f and valence electrons appear as the major source of the peculiarities of these materials. The same competition exists in α-cerium. The presence of localized f electrons is thus a decisive factor in the knowledge of the physical, chemical as well as structural properties of these materials. This presence can be determined by spectroscopy, by choosing methods that make possible the characterization of the excited states in a time scale compatible with the temporal evolution of the system.

1.4 Crystalline Structures

A crystal lattice is defined by its unit cell (cf Sect. 1.3.1) and the properties of the crystal are identical for any two equivalent points of two unit cells. The geometrical characters of the unit cell, length of the arêtes, angles, number and coordinates of the atoms, are the *lattice parameters*. Crystalline structure and electronic structure are two narrowly interdependent systems and many physical properties such as cohesion, diffusion, elastic and mechanical properties directly depend on the crystal structure.

At the Brillouin zone of the reciprocal lattice corresponds the *cell of Wigner–Seitz* in the real space. This cell is defined by drawing the plane bisectors of the lines connecting an atom to its nearest neighbours; it has the same symmetry as the unit cell of the crystal. Let us note by a_i the unit vector of the real space, or crystal space, along the axis i and by N_i the number of atoms along this axis, with $i = 1, 2, 3$. In the nearly free electron gas approximation, the Bloch wave functions are

$$\psi_k(r) = \psi_k(r + a_i N_i) = u_k(r) \cdot e^{i\mathbf{k}\cdot\mathbf{r}} = \psi_k(r)\, e^{i\mathbf{k}\cdot\mathbf{a_i}N_i}$$

with $e^{i\mathbf{k}\cdot\mathbf{a_1}N_1} = e^{i\mathbf{k}\cdot\mathbf{a_2}N_2} = e^{i\mathbf{k}\cdot\mathbf{a_3}N_3} = 1$

$$\mathbf{k}\cdot\mathbf{a_1} = 2\pi n_1/N_1 \quad \mathbf{k}\cdot\mathbf{a_2} = 2\pi n_2/N_2 \quad \mathbf{k}\cdot\mathbf{a_3} = 2\pi n_3/N_3$$

n_1, n_2, n_3 are integers

The reciprocal lattice, named the *reciprocal space* of the crystal, is defined by a set of unite vectors $\mathbf{b}_j$ perpendicular to planes defined by the axis a_i of the real space. The vectors $\mathbf{b}_j$ satisfy the relations

$$\mathbf{a_i} \cdot \mathbf{b_j} = 2\pi\delta_{ij}$$

δ_{ij} is the Kronecker delta. From the previous relations, the wave vector $\mathbf{k}$ is

$$\mathbf{k} = \mathbf{b_1}n_1/N_1 + \mathbf{b_2}n_2/N_2 + \mathbf{b_3}n_3/N_3$$

The unit cell of the reciprocal space of the crystal, or k-space, is defined by b_1/N_1, b_2/N_2, b_3/N_3. Let $K_m = m_1\mathbf{b_1} + m_2\mathbf{b_2} + m_3\mathbf{b_3}$ be one of the vectors of the reciprocal lattice connecting the origin $k = 0$ to one of its nearest neighbouring atoms. Let us consider the planes perpendicular to the vectors K_m at their middle. The first Brillouin zone is the smallest region of the reciprocal space, located around the origin and limited by such planes. The second Brillouin zone is the region between the first zone and the planes perpendicular to the vectors K_m connecting the origin to the second nearest neighbouring, and so on. Whatever the shape of the basic cell of the reciprocal lattice, the volume of each Brillouin zone is equal to $\mathbf{b_1} \cdot \mathbf{b_2} \cdot \mathbf{b_3} = 8\pi^3/\mathbf{a_1} \cdot \mathbf{a_2} \cdot \mathbf{a_3}$ where $\mathbf{a_1} \cdot \mathbf{a_2} \cdot \mathbf{a_3}$ is the volume of the Wigner–Seitz cell of the crystal. For a cubic lattice of side a, the extension of the first Brillouin zone along one of the reciprocal space axe is thus $b = 2\pi/a$ and it is convenient to study k in an elementary domain of extension $-\pi/a, +\pi/a$. In this domain, there is a series of energy levels, one level for each value of k. The number of electronic levels present in one Brillouin zone is equal to the number of atoms present in the unit cell of the crystal. For a simple cubic crystal, this number is equal to one. The k vectors describe a lattice of unit cell b_1/N_1, b_2/N_2, b_3/N_3. The number of k vectors in a volume dk of the k-space is

$$N_1N_2N_3 \cdot \mathbf{a_1} \cdot \mathbf{a_2} \cdot \mathbf{a_3} \cdot dk/8\pi^3 = V\, dk/8\pi^3$$

where V is the volume of the crystal in the real space and the number of k vectors in one Brillouin zone is $N_1 N_2 N_3$. Conversely, for each atom in the unit cell of the crystal there is associated one electronic level in the first Brillouin zone. This brief recall underlines the interdependence which exists between the valence electron distribution and the crystalline structure.

Each atom in the real space has a number of first neighbouring atoms, characteristic of the crystal structure. The metals crystallize generally in three structures, face-centred cubic (fcc), body-centred cubic (bcc) or hexagonal close-packed (hcp). For an fcc crystal, the unit cell is a cube with one atom in each corner and one in the centre of each face, i.e. a total of 4 atoms per cell. The distance between an atom and its 12 nearest neighbours is $\sqrt{2} \cdot a/2$ where a is the length of the arête of the cube. For a bcc crystal, the unit cell is a cube with one atom in each corner and one in the centre, i.e. a total of two atoms per cell. The distance between an atom and its 8 nearest neighbours is $\sqrt{3} \cdot a/2$. For an hcp crystal, the unit cell is a hexahedron with two arêtes of length a, the third c being defined by $c/a = 2\sqrt{2}/3 = 1.633$. One atom is present in each corner and one atom is on a diagonal, at a position such that each atom has 12 nearest neighbours at the distance a, if c/a is exactly equal to 1.633. This is the ideal hcp structure. If c/a differs from this value, each atom has six neighbours at the distance a and six neighbours at the distance $\sqrt{(a^2/3 + c^2/4)}$. There are two atoms per unit cell.

The fcc and hcp structures are the two common possibilities to arrange hard spheres in the most compact manner. The packing efficiency of the fcc structure is equal to the ratio of the volume of atoms to the volume of the cell, i.e.

$$4.4/3 \cdot \pi r^3 / 16\sqrt{2}r^3 = \pi/(3\sqrt{2}) = 0.7405$$

It is the same for hcp. Two other close-packed structures exist, the double hexagonal close-packed (dhpc) and the α-Sm structure. The number of equidistant nearest neighbours is equal to 12 for the four close-packed structures, only the arrangement of the successive atom layers differs. Thus, in fcc, every third layer is positioned directly above the layer A, giving A,B,C,A,B,C arrangement. In hpc, it is A,B,A,B; in dhpc, A,B,A,C,A,B,A,C and the ratio c/a is 3.266; in α-Sm structure, the period is of nine layers A,B,A,B,C,B,C,A,C. In contrast, for the non-close-packed bcc, the packing efficiency is only equal to 0.68. For the so-called Sm-type crystal structure, the unit cell is rhombohedral but can also be viewed as hexagonal, the atoms being placed in the cell according to a mixture of fcc and hcp positions.

1.4.1   Crystal Structure of Rare-Earths

The rare-earth metals crystallize into close-packed structures, except europium, which has the body-centre structure (Table 1.2).

Table 1.2 Crystalline structures of rare-earths

La	α dhcp, β fcc	
Ce	α fcc, β dhcp, γ fcc	
Pr	α dhcp, β fcc	
Nd	dhcp	
Sm	Rhomb	
Eu	bcc	
Gd	α hcp, β bcc	
Tb	hcp	$c/a = 1.580$
Dy	hcp	$c/a = 1.573$
Ho	hcp	$c/a = 1.570$
Er	hcp	$c/a = 1.569$
Tm	hcp	$c/a = 1.570$
Yb	fcc, hcp	

The light rare-earths crystallize into either fcc or dhcp structures while the heavy rare-earth metals crystallize only into the hcp structure and differ solely in the number of 4f electrons. Structural change is thus present along the series from fcc $\rightarrow$ dhcp $\rightarrow$ Sm-type $\rightarrow$ hcp. The highest theoretically possible c/a ratio is 1.633. This value is obtained for the compact stacking. This ratio decreases from 1.62 and 1.61 for praseodymium and neodymium respectively to 1.57 for thulium. Empirical relation exists between the crystal structure of the rare-earths metals and the volume of the ion core. Thus, for the crystal structure varying from fcc $\rightarrow$ dhcp $\rightarrow$ Sm-type $\rightarrow$ hcp, the increase of the core volume does not follow the increase of the f electron number, on the contrary the core volume decreases along the series.

In a solid, each ion is defined by its radius, or *ionic radius*. This radius decreases with increasing ion charge; in fact, when the ion has lost several electrons, which shielded the charge of the nucleus, the other electrons are much strongly attracted by the nucleus and the ion volume decreases. The *atomic radius*, or *Wigner–Seitz radius*, is defined as the radius of the atomic sphere, i.e. of the sphere having the average volume occupied by an atom, $4\pi r^3/3 = V/N$, where N is the number of the atoms present in the volume V. The experimental atomic radii of all the rare-earth metals including γ and α cerium are compared to the theoretical values in Fig. 1.7.

The model of hard spheres, used to define the atomic radius, does not reproduce the distance between ions to the accuracy with which it can be measured in crystals. In fact, the radii of the metal ions are smaller than the radius of the neutral atom while the negative ions having acquired of electrons, have radii larger than that of the neutral atom. The ionic radii depend on the crystalline structure. Along a column of the periodic table, they increase with the atomic number and with the increasing coordination number. They are larger for ions having a high-spin ground state than for those with a low-spin ground state. The ionic radii vary with the covalence of the binding. If the covalence increases, the length of the bond decreases and the apparent ionic radius increases. A *covalent radius* is defined as the radius of the atom engaged in a pure covalent bond; it is slightly superior to the atomic radius.

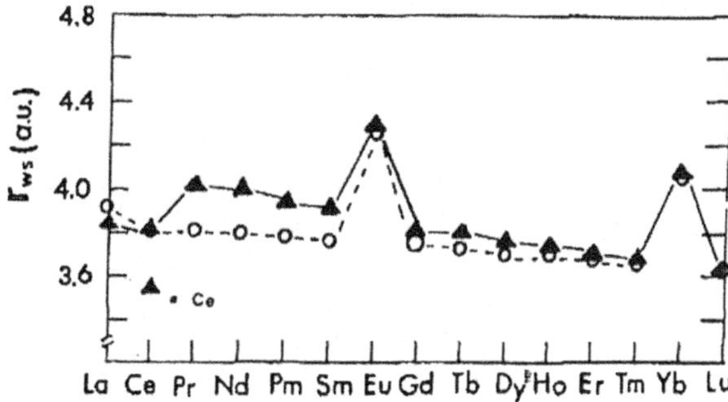

Fig. 1.7 Experimental (circle) and theoretical (triangle) values of the atomic radii for the rare-earth metals [86]

The ionic radii of the rare-earth metals decrease almost linearly along the series with the exception of the bivalent elements, europium and ytterbium, which have an ionic radius about ten percents larger than their neighbours. The same variation is observed for the lattice parameters. The rare-earth ionic radii are relatively small by comparison with the other metals, essentially as compared to metals of groups I and II; they decrease from 1.02 Å for lanthanum to 0.86 Å for ytterbium (Table 1.3).

Various models have been suggested to explain the variation of the core volume along the rare-earth series. This variation was explained by the fact that the screening effect of the 4f sub shell on the more external shell electrons is smaller than the screening effect of the s, p or d internal electrons. Then, when the atomic number increases, the electrons of the sub shells external to the 4f sub shell are more strongly attracted to the nucleus and therefore less easily removed, resulting in higher first ionization energies and a smaller ionic radius [161]. Thus, the ion size is reduced by 15 % from Ce^{3+} to Lu^{3+}. This size decrease is reflected on the unit cell volume of the metals and compounds. The decreasing of all lattice parameters through the series is called the *lanthanide contraction*. This contraction increases

Table 1.3 Ionic radius (angstroms) of rare-earths in their various oxidation degrees

	2+	3+	4+
La		1.172	
Ce		1.15	1.01
Pr		1.13	0.99
Nd	1.43	1.123	
Pm		1.11	
Sm	1.38	1.098	
Eu	1.31	1.087	
Gd		1.075	
Tb		1.063	0.90
Dy	1.21	1.052	
Ho		1.041	
Er		1.03	
Tm	1.17	1.02	
Yb	1.16	1.008	

the localization and the stability of the poorly shielded 4f sub shell across the series. The 4f electrons are located closer to the nucleus. The electron–electron interactions increase, as also the electron density associated with the filled 4f orbitals. However, the light rare-earths, praseodymium and neodymium, preserve characteristics analogous to those of trivalent cerium. Indeed, although localized magnetic moments are associated with the 4f orbitals of these two rare-earths, they are sufficiently spatially extended to be influenced by the surrounding crystal field.

The hcp structure with c/a changing from 1.59 for gadolinium to 1.57 for thulium was associated with the s–d character of the valence band, in analogy with the hcp structure of the trivalent transition elements, scandium and yttrium. The decreasing of the c/a ratio along the rare-earth series was correlated with a change in d band occupancy. It appears that the d-band occupancy is very sensitive to variations of atomic volume and decreases with the ion core size [162]. Indeed, the contraction of the atomic volume increases the Coulomb repulsion between electrons because they are confined to a smaller region of space. For lanthanum, the core size is relatively large and the d band broadens with respect to the bands of the next rare-earths, leading to an increase of the occupied part of the band. The large size of the lanthanum ion core is responsible for the flow of electrons from the almost free electron s band to the d band. The number of d-like electrons was found to be decreasing through the sequence fcc → dhcp → Sm-type → hcp, from about 2.5 in lanthanum to 1.9 in luticium, in agreement with the experimental observations [162]. The presence of the d electrons is considered as governing the crystalline structures of the trivalent rare-earths, which can, therefore, be considered as independent on the 4f electrons. Concerning europium, its bivalence favours the bcc structure. Indeed, since a bivalent ion is larger than a trivalent one, it has a smaller number of neighbours than in the compact hcp and fcc structures. Divalent europium and ytterbium ions have a larger lattice constant than their trivalent counterparts because their additional localized 4f electron brings about an increase of the ionic radius. The crystal arrangement, thus, depends strongly on the valence. Consequently, a conductor, metal or alloy, has the crystal arrangement whose number of valence electrons corresponds to the most stable state, i.e. to the state of thelowest energy.

The structural arrangements change at the surface of solids since the number of first neighbouring atoms is smaller than in the volume. This changes the atomic distances and the density of states and increases the localization. Concerning the surface of the trivalent rare-earth metals, valence change has been suggested for samarium, which becomes divalent or partially divalent [163]. The increase of the number of the 4f electrons increases the electron localization. Changes of the crystalline structure and of the electronic characteristics occur in rare-earth aggregates and increase of the 4f localization is expected. As an example, a discontinuous reduction of the inter-atomic distance was observed when the size of europium and ytterbium aggregates decreased [164, 165]. This reduction is accompanied by a decrease of the localized 4f electron number in ytterbium. Band structure and density of states depend on the atomic arrangement and a close relation exists between these two electronic parameters and the crystalline structure.

Table 1.4 Lattice parameter of gadolinium [131]

		LDA + U	LDA	Experimental
c/a	FM	1.595	1.606	1.547
	AFM	1.610	1.588	

Along the series, due to the incomplete screening of the increased nuclear charge by the additional f electron, the structural parameters of the rare-earths decrease. Their values have been correctly described from LDA + U calculations while LDA or GGA calculations generally lead to too small lattice parameters. As an example, for gadolinium, the LDA + U yields structural parameters for the hcp phase in good agreement with the experimental values (cf. Table 1.4). The calculations reproduce the correct magnetic order, which is ferromagnetic, [131, 166, 167] while from LDA, the anti-ferromagnetism order was found more stable.

The cohesive energy is the energy necessary to dissociate a solid into free atoms at 0 K. This is a macroscopic quantity, expressed in kcal/mol. On the curve giving the potential energy of a solid as a function of the inter-nuclear distance, the cohesive energy is the ordinate of the minimum with respect to the energy of the separated atoms. That is the energy of the system in its equilibrium state. The cohesion is an essential parameter because it determines the conditions of optimum stability in the solid. In the rare-earths, the ions in the solid can be considered as equivalent to free ions if they have the same charge. The overlap between the wave functions of neighbouring ions can be neglected and the interaction between the ions is equivalent to that of spheres of positive charge. The contribution of the 4f electrons to the cohesive energy is thus small. The cohesion is due to the redistribution of the d–s valence electrons in the solid and increases with their number. Consequently, one expects it to be approximately the same in all the rare-earth metals, except in the divalent metals, europium and ytterbium. Cohesive energies of the order of 40, 100 or 145 kcal/mole are expected, respectively, for divalent elements as barium, trivalent as lanthanum or tetravalent ones.[6] However, the cohesive energy of the rare-earths has the particularity to vary irregularly along the series (Table 1.5).

This is due to the existence of two different configurations for the free atom, $4f^n 6s^2$ and $4f^{n-1} 5d^1 6s^2$, while most of the rare-earths are trivalent in the metal. The change of valence from ions of configuration $4f^n 6s^2$ to a trivalent metallic state induces a gain of binding energy of about 60 kcal/mol while the energy required to excite the f electron to the s–d states is smaller. This favours the valence three in the metal, except for europium and ytterbium. Indeed, for these two rare-earths, the excitation energy is larger than the gain in the binding energy because of the high stability of the divalent configurations, $4f^7$ and $4f^{14}$. The cohesive energy of samarium is also relatively small with respect to that of the former elements. Indeed, samarium is trivalent in the bulk metal but it has a divalent or partially

---

[6]The cohesive energy is the difference between the average energy of atoms in the solid and that of the free atoms.

Table 1.5 Cohesive energy

		(Kcal/mol)		(eV/atom)
		(1)	(2)	(3)
	La	103.0		4.47
	Ce	101.0	101.0	4.32
	Pr	85.0	85.2	3.70
	Nd	78.3	76.0	3.40
	Sm	49.4	51.1	2.14
	Eu	42.4	24.0	1.86
	Gd	95.0	95.0	4.14
	Tb	92.9	92.1	4.05
	Dy	69.4	72.9	3.04
	Ho	71.9	72.8	3.14
	Er	75.8	75.9	3.29
	Tm	55.0	55.0	2.42
	Yb	36.4	24.3	1.60

(1) Experimental values; (2) [168]; (3) [169]

divalent surface and the valence two in some compounds. This is because the $4f^6$ configuration becomes more stable than $4f^5$ when the localization increases, i.e. when the number of closest neighbouring atoms decreases, as is the case at the surface or in some compounds. Inversely, samarium metal escapes the divalent crystal state in favour of the trivalent one because the energy required to promote a localized electron from the 4f level to an extended 5d band state is small. Indeed, the energy difference between the trivalent and divalent metallic state is only about 6 kcal/mol [170]. The same is expected for thulium. In a general manner, the stability of the $4f^{n-1}$ sub shell depends on the initial atomic configuration. Thus, the binding energy of the nth 4f electron is small for Ce^{3+} ($4f^1$) and Tb^{3+} ($4f^8$) while it is large for the half filled and filled 4f sub shells of Gd^{3+} ($4f^7$) and Lu^{3+} ($4f^{14}$). The observed cohesive energies of the rare-earth metals (Fig. 1.8) [86] are compared with their calculated values from LMTO method within LDA and LSD functional approximation.

Fig. 1.8 Experimental (*circle*) and theoretical (*triangle*) values of the cohesive energies (in eV) for the rare-earth metals [86]

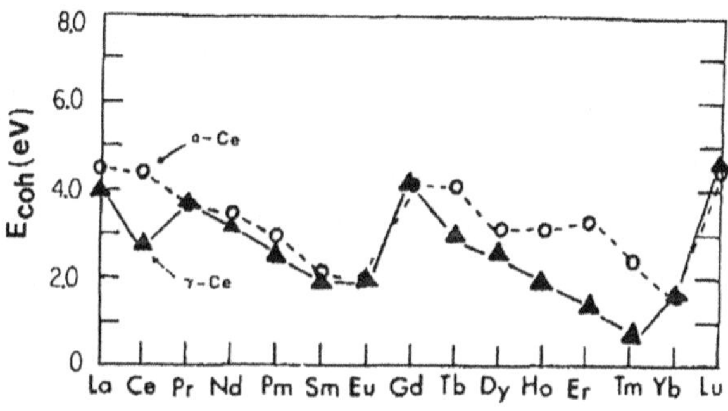

The agreement is better for the light than for the heavy rare-earths, probably due to the theoretical model used. Calculation of cohesive properties taking into account many-body corrections have been made for the light rare-earths [171]. A large part of the 4f electrons are described as localized and separated by the Hubbard energy U from a fraction of 4f electrons hybridized with the valence electrons and located near the Fermi level. Values in better agreement with the experiment have been obtained for the bulk modulus of praseodymium and neodymium.

As already underlined, the crystal potential influences the movement of the valence electrons, changing the valence state density according to whether the structure is crystalline or amorphous. A conductor, metal or alloy, has the structure with the lowest energy, associated with the number of valence electrons in the considered solid. In the presence of "quasi-localized" valence electrons, the electron–lattice interactions can become important, breaking the atomic orbital symmetry in the solid and giving rise to *crystal field splitting*. The crystal field splits the energy levels occupied by these quasi-localized electrons. The splitting increases with the symmetry of the crystal and the localization of the electrons. Crystal field splittings of 3d energy levels have been observed and largely studied in the transition element compounds. In the lanthanides, the 4f electrons are well shielded from the crystal field by filled 5p and 5s sub shells and the energy needed to remove a 4f electron from the $4f^n$ sub shell is relatively independent of the crystalline environment. The splitting of 4f energy levels is mostly due to spin–orbit interaction, which is clearly larger than the crystal field splitting. Then, because the small overlap between the 4f distributions of neighbouring atoms, these distributions are only slightly perturbed by the environment and the f orbitals are generally not quenched by the crystal field.

A generalized pressure–temperature phase diagram was constructed for the trivalent rare-earths and widely discussed. It is shown in Fig. 1.9 [90].

In this figure, the structural phase diagrams of the trivalent rare-earths, except cerium, are assembled in a single pressure–temperature diagram, on the same scale. This presentation clearly shows the close similarity of these elements. It shows also the various possible phase transitions with increasing pressure. It is known that materials with strongly correlated electrons have generally complex phase diagrams. They are very sensitive to small changes in external parameters that bring about many unusual properties. Indeed, the rare-earth and actinide metals exhibit a large variety of structural transitions caused by pressure and temperature changes.

1.4.2 Pressure Effect on the Rare-Earths

The increase of pressure confines partially the valence electrons into a smaller region of space and reduces the Wigner–Seitz radius. The ionic radius is also reduced but it is only slightly sensitive to the pressure. Consequently, the ratio R between the Wigner–Seitz radius and the ionic radius decreases with increasing pressure.

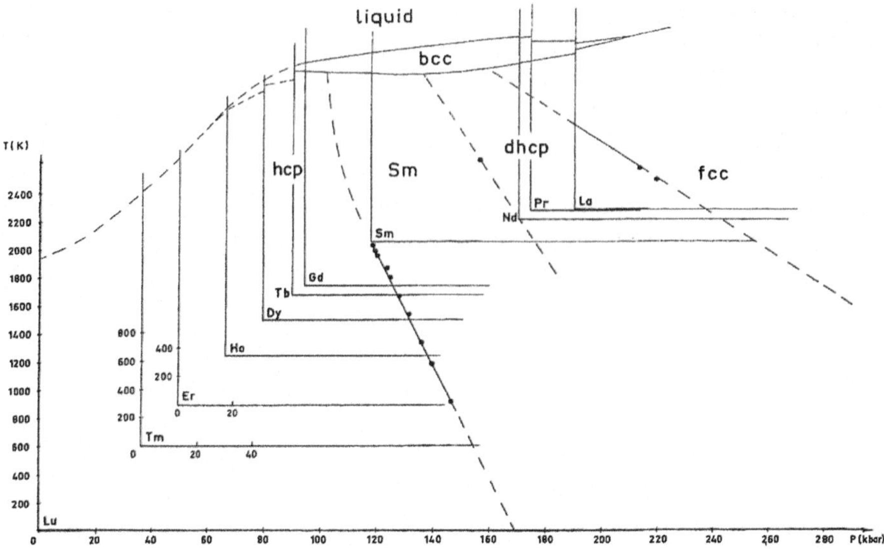

Fig. 1.9 Phase diagram of the rare-earths from [90]

For the rare-earths, as already mentioned, the Wigner–Seitz and ionic radii decrease along the series but, under increased pressure, the ionic radius decreases more rapidly, causing the ratio R to increase along the series. Then, under pressure, a rare-earth reacts in a similar way to the preceding element in the series. This explains why, under high pressure, the heavier rare-earths of hcp structure show the samarium-type structure [172]. It was shown experimentally that for the trivalent rare-earths a bcc phase is observed for temperatures below the melting point. The temperature range of this phase increases with the pressure for the heavy rare-earths while it decreases for the light ones.

Pressure variations can induce changes in the electronic properties. Indeed, increasing of pressure induces a decrease of the crystalline parameters and of the atomic volume, increasing the overlap of electronic wave functions and thereby the electron interactions. As a result, the energy distributions of the valence electrons are broadened. At the same time, the extent of the 4f orbitals increases and the 4f electrons become less localized. The binding energy of the 4f electrons generally decreases with respect to that of the 5d band. Consequently, the number of localized electrons decreases while that of the valence electrons increases. In agreement with [162], one then expects the reverse sequence of that observed when the number of the localized 4f electrons increases, i.e. the sequence hcp $\rightarrow$ Sm-type $\rightarrow$ dhcp $\rightarrow$ fcc. These structural transitions can be accompanied by an s $\rightarrow$ d transition. At very high pressure, complex lower symmetry phases can be formed and are accompanied by volume collapse. The variation in the localized character of the 4f electrons, which accompanies the modification of the crystalline structure, can induce a valence change. Such a change can therefore occur by varying the external parameters, such

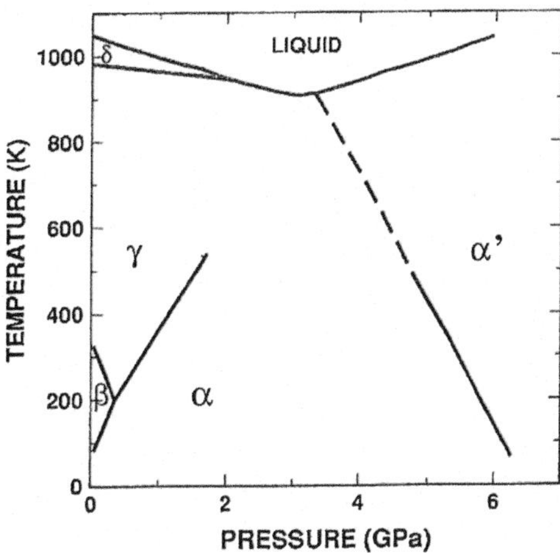

Fig. 1.10 Phase diagram of cerium metal [93]

as the pressure, and this shows how electronic and crystalline properties are strongly related in solids.

Several rare-earth metals undergo phase transitions as a function of pressure characterized by a first-order volume decrease. These structural transitions are accompanied by changes of physical properties [61]. Among the rare-earths, special attention was given to cerium because this metal is unique among elemental solids by its very complex pressure-temperature phase diagram and by its electronic transformation, which accompanies one of these phase transitions. Other compressed rare-earths have been studied, praseodymium [173–175], neodymium [175], investigated by taking into account the effect of the spin–orbit interaction [176], and europium [177, 178] but this list is not exhaustive and practically all the rare-earths have been studied.

The case of the γ-α cerium transition has already been discussed in the Sect. 1.3. Indeed, cerium occupies a special position in the RE series because of the presence of a single 4f electron in its ground state resulting in exceptional properties, as superconductivity in the high-pressure phases. Its behaviour illustrates well the influence of the relations between crystalline arrangement and electron properties. Five distinct solid phases of cerium are known to exist, of which three, α, β and γ, are present in the low pressure and low temperature domain (Fig. 1.10) [93].

At room temperature and atmospheric pressure, cerium is in the γ phase. The first-order iso-structural phase transition fcc γ ⇒ α takes place at a pressure of 7 kbar at room temperature [179], or at about 120 K at atmospheric pressure. In the pressure–temperature diagram, it is possible to go continuously from the large-spacing magnetic γ phase to the more dense non-magnetic α phase. One of the peculiarities of this phase transition is that it ends in a critical point[7] at $T = 600 \pm 50$ K, similar to a liquid–gas phase transition. The existence of a second-order phase transition was

---

[7]A critical point exists only for phases which do not differ in internal symmetry.

proposed to explain the γ-α transition [180]. But that would imply a symmetry difference between the two phases while they are crystallographically identical. The γ phase has a localized magnetic moment of about $2.5\mu_B$ while in the low atomic volume phase α, the magnetic moment is quenched and this phase is characterized by a Pauli-type paramagnetism. The cerium $\gamma \Rightarrow \alpha$ transition is associated with a decrease of the inter-atomic distances, which induces a valence change, i.e. a reduction of the number of the localized 4f electrons. This change is possible because the localized and itinerant state distributions are energetically very close. In the α phase, high density of 4f states is present at the Fermi surface, inducing a large electronic heat capacity coefficient. The low-pressure β phase is dhcp and the $\gamma \rightarrow \beta$ transition takes place at $-16\ ^\circ$C. The atomic volumes of the γ and β phases are found to be similar because the two structures are compact. Like the γ phase, the β phase has a localized magnetic moment; it is anti-ferromagnetically ordered and is described correctly by LSDA-U calculations [131]. The presence of a local moment shows that the f electrons are localized in both these phases. The α' phase, fcc or hcp, stable above 50 kbar, is superconducting with a critical temperature T_c of about 1.7 K at 50 kar. Cerium becomes orthorhombic at high pressure and this transition is also accompanied by a volume collapse. Cerium has the longest liquid range from 795 to 3443 $^\circ$C (2648 $^\circ$C). Its fusion curve is also unique among elemental solids because it exhibits a broad minimum [181].

The lattice parameter of γ-cerium was determined by standard LDA, GGA and LDA + U, GGA + U calculations [131]. Results are compared in Table 1.6 with experimental value a.

LDA and GGA calculations underestimate considerably the volume by 22 % for γ-cerium while LDA calculations with the f electron treated as a core electron and LDA + U are able to describe well both γ- and β-cerium phases. In LDA + U, a magnetic order is imposed. Ferromagnetic order was chosen. Spin–orbit splitting is not taken into account because its effect is negligible. The U energy is 6.1 eV but the lattice parameter is rather insensitive to the precise value of U. A variation of 1 eV induces a variation of 0.04ua in the lattice parameter. The calculated volume for the γ phase is 32.1 Å^3, compared to the experimental value of 34.4 Å^3 and the value calculated by LDA of 23.2 Å^3. Consequently, DFT calculations with either LDA or GGA exchange and correlation functions fail to describe the γ phase. From LDA + U, the equilibrium volume increases because the electrons are localized and no longer participate in the binding. The calculations with the 4f electron in the core underestimate only slightly the lattice parameter.

Table 1.6 Experimental and calculated lattice parameters of γ-cerium

	Exp.	LDA	GGA	LDA + 4f core	LDA + U	GGA + U
$a_{Å}$	5.16[a]	4.49[b]	4.70[b]	5.12[e]	5.20[f]	5.27[d]
		4.52[c]	4.69[c]	5.09[d]	5.04[d]	
		4.52[d]				

[a][182]; [b][183]; [c][184]; [d][131]; [e][103]; [f][185]

Along with the collapse of 15 % of the atomic volume, the ionic radius decreases by approximately 5 %. This is less than the decrease of the radius with the ionization degree, which is of the order of 12 % between Ce^{3+} and Ce^{4+}. The existence of a valence change seems then to depend critically on the orbital degeneracy of the f states and on the hybridization between the 4f orbitals and the s–d valence band. On the other hand, calculation of the phase diagram of the α-γ cerium transition gives the entropy as the driving force in this transition and the finite temperature studies incorporate thermal fluctuations. Since the variation of external parameters, as temperature and pressure, can induce a change in the localization of the 4f electrons, it appears that phonons must be taken into account in the treatment of the localized electrons.

Under high pressure, cerium is in the phase α', analogous to the α-uranium phase, and at still higher pressures it transforms into a monoclinic body-centred structure. These unusual metal structures are present in the phase diagrams of actinides. This suggests a close similarity between the behaviour of the 4f electron of cerium and the 5f electrons of the actinides.

Other rare-earth metals undergo phase changes with atomic volume changes as observed for cerium [186]. The next element, praseodymium, undergoes several phase transitions with increasing pressure: at first, without a volume change, a transition from dhcp to fcc structure at about 40 kbar, then to d-fcc structure at 62 kbar followed by a possible transition to a monoclinic cell at about 100 kbar, finally a transition to an orthorhombic phase, of α-U type, above 200 kbar associated with a volume collapse of about 10 % [172, 187]. Neodymium undergoes under pressure the same phase changes, dhcp to fcc, then to orthorhombic at about 390 kbar with a substantial volume collapse. In both cases, the collapse is accompanied by a crystallographic change, in contrast with cerium. The trivalent heavy rare-earth metals under pressure undergo a structure sequence hcp $\rightarrow$ Sm-type $\rightarrow$ dhcp $\rightarrow$ fcc. This transition is observed in gadolinium. A volume reduction occurs at increasing pressure across the series. This reduction of the volume can cause a delocalization of the 4f electrons as well as a change of the magnetic properties from paramagnetic to non-magnetic. Inversely, an increase of the volume induces a localization of the 4f electrons.

The two divalent rare-earth metals become trivalent at very high pressures. For europium, a phase transition was expected theoretically to occur at about 180 kbar. Between 180 and 350 kbar, europium is an alloy formed of randomized divalent and trivalent ions. From about 350 kbar, the transition is complete and europium is trivalent, like the other rare-earth metals [176]. For ytterbium, the transition occurs at about 140 kbar and is of the same type as for europium [188]. The valence change is accompanied by a decrease of the atomic volume of about 15 %, along with a decrease of the 4f localization. A model was proposed to calculate the volume–pressure equation of state and the electronic structure of such an alloy of the divalent and trivalent ions [189]. This model was applied with success to ytterbium under pressure and good agreement was obtained with the experimental results for both properties.

In summary, the inter-atomic distances vary with pressure and with temperature, and the crystalline arrangements can thus be modified. The energies associated with the localized and itinerant electrons are altered differently. The energy separation between the 4f orbitals and the Fermi level can vary and eventually vanish. The number of the localized 4f electrons can decrease owing to the increase of the interactions between them and the itinerant s–d electrons, thus leading to a valence change.

1.4.3   Crystal Structure of the Actinides

In standard conditions, simple metals, transition metals and lanthanides condense in close-packed high-symmetry structures, such as face-centred cubic, hexagonal close packing or body-centred cubic, while the light actinides have low-symmetry structures, tetragonal, orthorhombic or monoclinic. The actinides and the lanthanides have nf electrons. In the lanthanides, however, the 4f electrons are localized. The binding is determined by the d bands and the crystal structures correspond to those of the transition metals. Similarly in the heavy actinides, from americium, the 5f electrons are localized and close similarity to the lanthanides is observed. These metals have close-packed crystal structures. The same structure is observed in the two first elements of the series, actinium and thorium, which do not have 5f electrons. In the two following elements, the 5f electrons have still little influence and one finds typical metallic crystal structures, few allotropes, and high melting points. As more 5f electrons are added and participate in bonding, the crystal structures become less symmetric, the number of the allotropes increases and the melting points decrease. This continues up to metallic plutonium, which has very specific structural properties associated with unusual mechanical properties, equilibrium volume, thermal expansion, elastic constants, all have values in disagreement with the theoretical predictions. From americium on, simple crystal structures typical of usual metals return, the number of allotropes decreases and the melting points rise, indicating that the 5f electrons become localized or inert. Except for actinium, the actinides have several crystalline phases and the crystal structures of light actinides have little analogy with the lanthanides. The melting point of actinides does not have a clear dependence on the number of 5f electrons. The unusually low melting point of Np and Pu (about 640 °C) is explained by hybridization of 5f and 6d orbitals and the formation of directional bonds in these metals.

From metallic protactinium to plutonium, many low-symmetry crystal structures are present, which have no counterpart among the other metallic elements. This decrease of symmetry is accompanied by an increase of the electron density with respect to the density in the traditional close-packed structures. These features are unique to metals that have an f electron energy band and result from particular characteristics of the 5f valence electron wave functions. Indeed, the symmetry of the f electron bonds is similar to that of the p bonds. It works against high-symmetry structures and brings about higher density obtained in more perturbed structures or liquids. However, if the 5f electron distributions become sufficiently narrow, crystal structures of the same type are expected for all metals—simple, transition or actinide. This change appears abruptly just past plutonium, starting with americium, indicating the transition from the 5f band to localized levels. Transitions to structures of higher symmetry are also observed for the light actinides under pressure.

Strong differences exist between the light actinides and the other metals. The 5f bands are narrow compared to the d bands and especially to the wide sp bands in the metals. However, the light actinides show a weaker tendency to form a magnetic state than do the 3d elements. This is explained by the fact that the exchange integral of the f electrons is only about half of the exchange integral of the d electrons, see in nickel for example. The stability of compact and non compact crystal structures was calculated as a function of the band width for various metals, aluminium, iron, niobium and uranium [140]. It was shown that for the three first metals a high-symmetry structure was stable when the bandwidth used was of the order of the experimental value characteristic of each metal. On the other hand, for uranium a low-symmetry structure is associated with its narrow 5f band. The predominant parameter is not the angular distribution of the valence electron orbitals, but the 5f bandwidth, which is a function of the mobility of the 5f electrons. Indeed, crystal structures of the same type are predicted for all metals having narrow bands. In summary, the parameters that control the crystal structures are essentially the same for all the metals. The important parameter is the spatial distribution of the valence electrons that determines both the crystalline structure and the chemical bonding. The interpretation of the complexity of the structures is therefore one of the major challenges of the actinides. The transformation under pressure of a low-symmetry phase to a high-symmetry phase is considered as due to a broadening of the valence bands. Indeed, if the initially narrow 5f bands become broad enough high-symmetry fcc, bcc and hcp structures can be obtained.

The crystal structural stability is determined by the balance between electrostatic interactions, which induce high symmetry and a Peierls distortion, which induces low symmetry. By means of a crystal structure distortion, the one-particle energy contribution to the total energy, which is the sum of all the occupied valence energy levels, can be reduced. This mechanism is efficient if there are many degenerate energy levels around E_F. Some energy levels are pushed above, whereas others are pushed down and energy is gained. This mechanism is particularly efficient in the case of narrow bands because many energy levels are present in a narrow energy range and it was called upon to explain the structural characteristics of the light actinides.

Thorium crystallizes in the fcc phase, which is stable up to 63 GPa. Above this pressure, it transforms into a body-centred tetragonal structure by a simple distorsion, adopting a low-symmetry crystal structure. Protactinium has a body-centred tetragonal structure like the one seen in thorium under pressure. The decrease of the crystal symmetry is expected to bring about the presence of bonding 5f valence electrons in protactinium and in the high-pressure phase of thorium. Above 77 GPa, protactinium transforms into the orthorhombic α-U structure. A volume collapse of ≈30 % is associated with this phase transition.

Uranium metal has three allotropes. The orthorhombic α-form has 4 atoms per unit cell. It is stable up to 935 K. Between 935 and 1045 K, the β-form is present; it is tetragonal with 30 atoms per unit cell [190]. At 1045 K, uranium transforms into the γ form, which is bcc with two atoms per unit cell. Uranium melts at 1405 K. At temperatures under 43 K the α phase exhibits at least one, and possibly two, charge-density-wave distorsions [191]. The superconducting transition temperature of bcc γ-U is T_C = 1.8 K, i.e. higher than that of the α phase. For uranium under pressure, the predicted crystal structure sequence is α-U (orthorhombic) → body-centred tetragonal → body-centred cubic (bcc). The first transition takes place at 0.8 Mbar. It should be noted that the α-U crystal structure appears often in rare-earth and actinide metals under pressure. It seems be the typical structure for 5f electron bonding. The distance uranium–uranium, equal to 3.50 Å, is close to the limit generally associated with the direct overlap of 5f wave function. Consequently, U metal could be used as a model system in calculations involving strongly bonding f electrons.

Neptunium has three allotropic phases. The orthorhombic α-Np phase is stable up to 52 GPa. Upon further compression, the transition α-Np → β-Np is observed in the case of a 19 % compression and the transition β-Np → bcc γ-Np for a 26 % compression. In a general point of view, the transformation of low-symmetry to high-symmetry structures could be due to a modification of the valence states with pressure.

Plutonium has the most complex phase diagram of all metals (Fig. 1.11) [192].

Fig. 1.11 Phase diagram of plutonium metal [194]

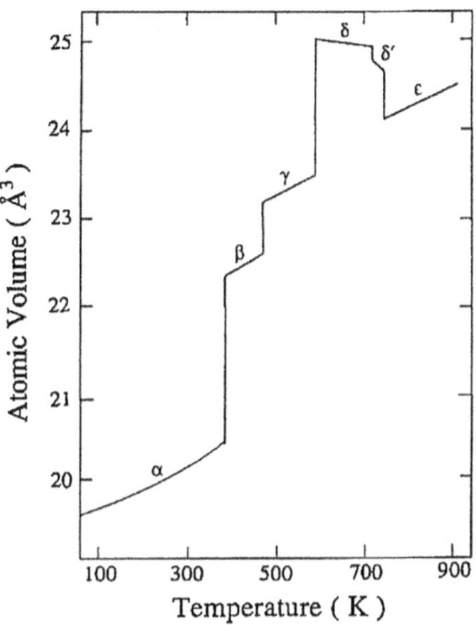

It has six allotropes between absolute zero and its melting point temperature at ambient pressure, some of them of very low symmetry [193]. It is the only metal to have such a large number of allotropic crystal structures. These phases have very different atomic volumes. The physics of the localization-delocalization phenomena was believed to be important for their understanding [103, 154]. The low temperature α phase, stable under 400 K, has a monoclinic structure with 16 atoms per unit cell. This phase, not observed for any other metal, has one of the largest thermal expansion coefficients of all the metals. Other unusual property is the negative coefficient of the resistivity above about 100 K (Fig. 1.12) [194], which reveals the effect of the electron–phonon interactions and structure defects.

The β and γ-Pu phases have unique distorted crystal structures, body-centred monoclinic and face-centred orthorhombic, respectively. The high-temperature fcc δ phase is stable between 592 and 724 K. It has the lowest density in spite of having the only close-packed crystal structure. Its experimental equilibrium volume is approximately 25 Å^3 while it is 20 Å^3 for α-Pu and 29 Å^3 for α-Am. The atomic volume of the δ phase is then 25 % greater than that of the α phase. An agreement with the experimental value was obtained from LDA + U calculation [195]. Such a large thermal expansion is unusual. The δ and δ' phases have negative coefficients of thermal expansion. The δ phase has the largest low-temperature specific heat of any pure element (cf. Fig. 1.6). The δ' and ε phases are body-centred tetragonal and body-centred cubic respectively and are characterized by a small contraction volume. Plutonium in the liquid phase is denser than in the high-temperature solid phases. Therefore it contracts upon melting. One problem is the difficulty to produce large single crystals of plutonium. Generally the experiments use very small samples. This limits the type of possible observations. Specific theoretical treatments are used to study the defects [196].

The monoclinic Pu α phase was known as resulting from the metallic bonds associated with the 5f orbitals. However, it was important to understand why such structural anomalies accompany the presence of 5f orbitals. The low symmetry of the α phase was initially attributed to directional or covalent-like bonding resulting

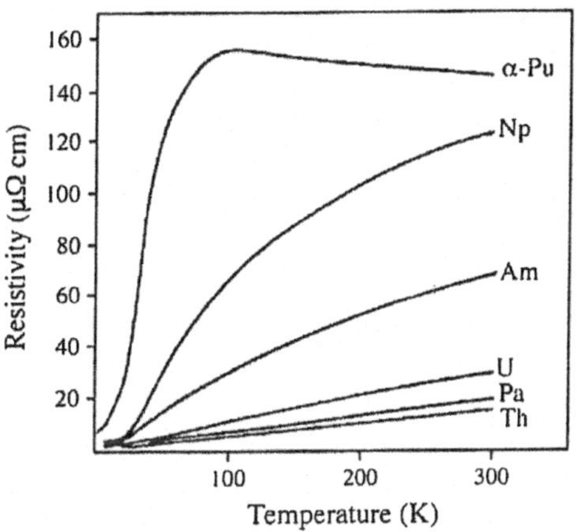

Fig. 1.12 Resistivity as a function of temperature for the α phase of the light actinide metals [194]

from the angular characteristics of 5f wave functions, somewhat analogous to molecular bonding. Other possibility can explain the complexity of the α phase. Indeed, plutonium exhibits several electron configurations of comparable energy in the metal. Consequently, plutonium ions of different sizes coexist. The strain energy required to accommodate different size ions into structures such as bcc and fcc with equivalent atom sites destabilizes these structures. These conditions cause lattice distorsions, which are responsible for the existence of the α phase at the stable ground state. The presence of four or more configurations having nearly equal energy leads also to multiple allotropes.

The Pu fcc δ phase can be stabilized down to temperatures near 0 K by the addition of only a few percent of a trivalent element, Al, Ga, In or Tl [197]. Extremely small deviations from the fcc structure are present. Contraction of the lattice is then observed. It results, for a large part, from the reduction in size of the plutonium host atoms, which accompanies a change of their electronic structure. The number of the localized 5f electrons, expected to be superior to five in the metal δ phase, decreases in the presence of trivalent impurities. In contrast, the presence of gallium expands the lattice of the plutonium α phase. In this case, the localization of the 5f electrons increases. A small region of fcc structure can also occur in pure plutonium.

A pressure of 1 kbar is already sufficient to make the plutonium large volume fcc δ phase disappear. This implies that the total energies of α and δ phases are close. However, these two phases with very similar energies have very different densities, structural properties, thermal expansion coefficients and mechanical properties. Consequently, small pressure or composition changes cause large changes in the physical properties of plutonium. The properties of plutonium in the fcc phase do not fit those of the light actinides. Indeed, from calculations treating the 5f electrons as itinerant, the fcc structure is predicted to be unstable and a lower symmetry structure is expected. Alternatively, for fcc structure, as for hcp or dhcp structure, localized 5f electrons and the presence of a magnetic order are expected [195, 198]. However, the equilibrium atomic volume of δ-Pu is inconsistent with either purely localized or delocalized 5f states. Moreover, the observation of negative thermal expansion is indicative of competing configurations with different associated equilibrium volumes. The electronic properties of δ-Pu involve the presence of strong electron correlations, i.e. a regime of narrow 5f bandwidths, in disagreement with the predictions deduced from the structural properties.

The very different structural and mechanical properties of the α and δ phases reveal their electronic structure to be very different. Stabilization of the two different types of such structures can be obtained. Indeed, the total energy can be lowered, either through bonding energy gained from a structural distortion or through correlation energy gained by localization of the electrons. Many theoretical studies have been made using the different approximations, LDA + U, LDA + U combined with DMFT, GGA. It was predicted that in the α phase the 5f electrons are delocalized, of band-like and responsible for the low-symmetry monoclinic structure. As already underlined, the LDA + DMFT was applied with success to calculate the volume increase in the δ phase of plutonium [154]. Strong electronic correlations and Coulomb interaction value of $U \approx 4$ eV were used to describe the δ phase; magnetic moments were predicted. However, the magnetic susceptibility is independent of the temperature and no evidence of magnetic moments is observed in the various phases.

While the light actinide metals have low-symmetry crystal structures, which are found only for the light lanthanides under pressure, i.e. for elements with itinerant 4f electrons, the heavy actinides have simple structures, double hexagonal packed or cubic close-packed, analogous to those of the rare-earths, i.e. to those of metals with localized f electrons. Thus, americium, curium and berkelium, have a dhcp phase and an axial ratio of about 1.62, both found for the light lanthanides. Evidence has also been found for a high temperature bcc phase in americium and its phase diagram is close of that of praseodymium.

The variation curve of the cohesive energies of the actinides in the beginning of the series follows that of the rare-earths. However, beyond the mid-sub shell the two curves are parallel but the cohesion energy of the actinides is lower (Fig. 1.13) [168].

1.4.4   Pressure Effect on the Actinides

The actinide metals undergo phase transitions as a function of pressure characterized by a first-order volume decrease. Rich phase diagrams already mentioned in part in the previous paragraph, accompany the volume collapse. Unlike the γ-α transition in cerium, which is iso-structural, the transitions with change of volume in actinides are accompanied by changes in the structure. Thus, in plutonium, as in cerium, the higher temperature phases have atomic volumes 15–20 % higher than in the zero temperature phase. However, distorted structures are present in plutonium and not in cerium, which has only one f electron. It was interesting to compress strongly the transplutonium elements, thereby forcing the individual 5f electrons wave functions into strong contact with each other, in order to observe eventual structural changes. Recently high pressure experiments have been performed for Am [199] and Cm with this aim. Contraction of the many-body wave functions accompanies decrease of lattice spacing and dramatic crystal structure changes were observed. These results and other high pressure data have been widely discussed [200].

As for the rare-earths, volume changes under pressure induce differences in the nature of bonding and can cause changes in the electronic structure. Decreasing of the inter-atomic distances induces an increase of the f-orbital overlap. Delocalization of the 5f electrons and their hybridization with the valence electrons can then occur. Thus, in uranium, the α-U structure is stable up to 100 GPa. Consequently, it was deduced that itinerant 5f electrons were present. Indeed, localized and itinerant electrons are simultaneously present.

For Am, as the pressure increases, the 5f electrons are pushed from a localized to an itinerant state. For pressure going from standard up to 100 GPa, americium metal undergoes three phase transitions between four crystal structures: AmI (dhcp), AmII (fcc), AmIII and AmIV (orthorhombic). AmIII has the same crystal structure as

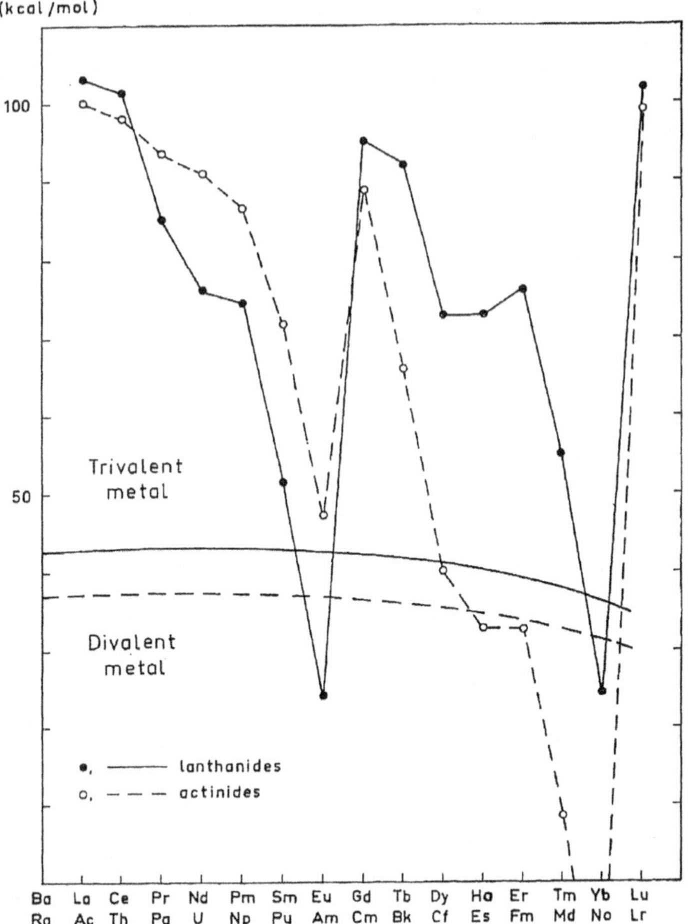

Fig. 1.13 Comparison of the cohesive energy for the rare-earths (*full curve*) and for the actinides (*dashed curve*) [168]

γ-Pu and AmIV is very close to α-Np. The low symmetry phases are present above 110 kbar. These complex structures, similar to those of lighter actinides, mark the onset of 5f electron bonding in Am. The α-U orthorhombic structure was observed in americium at 152 kbar and standard temperature. This structure is typical for the presence of delocalized 5f electrons. No dramatic volume change accompanies the delocalization, conforming to what has already been discussed for the γ-α cerium transition. The atomic volume variation of the metal, which accompanies these structural transitions, is shown in Fig. 1.14 [199] as a function of pressure.

It is indicative of variations in the physical properties of americium metal under high pressure. It suggests a change of the electronic structure and the presence of 5f bonding. These observations shed a light on the relation between volume collapse and 5f electron delocalization under pressure.

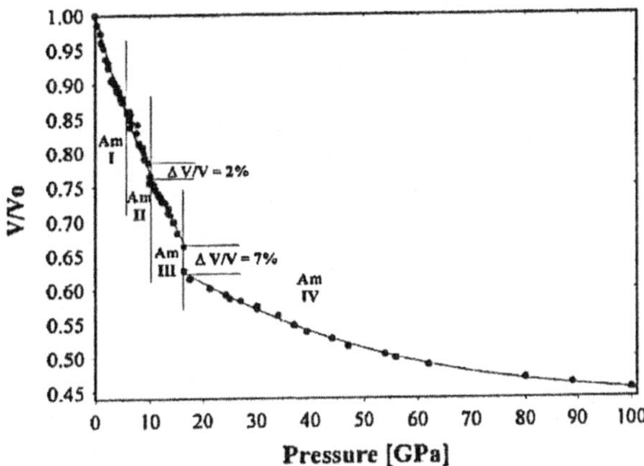

Fig. 1.14 Relative volume of americium metal as a function of pressure up to 100 GPa [199]

The effect of the pressure is analogous for curium metal. When the pressure is varied from normal to 90 GPa, phase transitions take place between the crystal structures CmI (dhcp), CmII (fcc), CmIII (monoclinic), CmIV and CmV (orthorhombic). CmIV has the same structure as AmIII and CmV as AmIV. In contrast, the structure of CmIII is not observed in any other actinide. This structure together with α-Pu and β-Pu are the only monoclinic phases observed for the metals. The standard dhcp phase of berkelium (BkI) is also transformed to the fcc phase under pressure at 8 GPa, then probably to α-U crystal structure at 22 GPa. As for the previous elements, several phase transformations with small volume reductions take place, rather than a single large one. These volume reductions accompany a change of the 5f electrons from localized to delocalized.

Finally, in these materials, the applied pressure may affect properties such as superconductivity, heavy fermion behaviour, mixed valence, which accompany the delocalization. As an example, americium is superconductor under pressure.

1.4.5   Conclusion

The progressive decrease of the crystalline parameters through the lanthanide series reflects the contraction of the 4f wave functions, located closer to the nucleus, the decrease of the ionic radii and also the increase of the electron–electron interactions. The presence of localized f electrons stabilizes high-symmetry phases associated with large atomic volumes as well in the rare-earths as in the actinides. Consequently, the number of localized f electrons can stay the same or be reduced by lowering the equilibrium volume, for example by pressure application. On the other hand, the presence of 5f electrons in narrow bands, i.e. electrons highly sensitive to small perturbations, enhances polymorphism and causes instability of the atomic arrangement in the solid. Structural anomalies are present in the actinide series as, for example, the very large increase of atomic volume between plutonium and americium. The great structural complexity of the allotropes of the lighter actinide metals up to plutonium and the presence of low symmetry phases is recognized to be due to the participation of the 5f electrons in the metallic bonding. High density of 5f states is present in narrow bands at or very near the Fermi energy. In the absence of these special conditions high-symmetry structures appear. Similar to the case of the 4f electron of cerium, the localized character of the 5f electrons decreases under pressure but the 5f electrons remain strongly correlated. Conversely, alloying disturbs the distribution inside the 5f bands, thus reducing the bonding energy and leading to at least partial localization of the 5f electrons. The role of defects is important since defects are generated in radioactive materials by radiation-induced self-damage.

In summary, electronic structure, crystalline arrangement and structural properties are closely related. Thus, the narrow bands characteristic of 5f bonds bring about lower symmetry structures and can explain the structural properties of plutonium and neptunium. However, understanding of the structure/property relationship remains a challenge for the light actinides, which have still numerous unexplained properties, as, for example, their magnetic characteristics.

References

1. G. Racah, Private communication (1963)
2. R.D. Cowan, *The Theory of Atomic Structure and Spectra* (University of California Press, 1981)
3. J. Friedel, in *The Physics of Metals*, ed. by J.M. Ziman (Cambridge University Press, NewYork, 1969)
4. C. Herring, Phys. Rev. **57**, 1169 (1940)
5. J.C. Slater, Phys. Rev. **51**, 846 (1937)
6. J.C. Slater, Phys. Rev. **92**, 603 (1953)
7. M.M. Saffren, J.C. Slater, Phys. Rev. **92**, 1126 (1953)
8. J. Korringa, Physica **13**, 392 (1947)
9. W. Kohn, N. Rostoker, Phys. Rev. **94**, 1111 (1954)
10. O.K. Andersen, Solid State Comm. **13**, 133 (1973)
11. L. Kleinman, J.C. Phillips, Phys. Rev. **116**, 880 (1959)
12. V. Heine, I.V. Abarenkov, Philos. Mag. **9**, 451 (1964)
13. M.H. Cohen, V. Heine, Phys. Rev. **122**, 1821 (1961)
14. C.Y. Fong, M.L. Cohen, Phys. Rev. Lett. **24**, 306 (1970)
15. J.C. Slater, G.F. Koster, Phys. Rev. **94**, 1498 (1954)
16. D. Spanjaard, J. dePhysique **39**, 397 (1978)
17. O.K. Andersen, Phys. Rev. B **12**, 3060 (1975)
18. D. Bohm, D. Pines, Phys. Rev. **92**, 609 (1953)
19. D. Pines, in *Solid State Physics*, ed. by F. Seitz, D. Turnbull, vol 1 (Academic Press, New York, 1955), p 368
20. M. Gell-Mann, K.A. Brueckner, Phys. Rev. **106**, 364 (1957)
21. E.P. Wigner, Trans. Faraday Soc. **34**, 678 (1938)
22. P. Nozières, D. Pines, Phys. Rev. **111**, 442 (1958)
23. D. Pines, *The Many-Body Problem* (Benjamin, New York, 1962)
24. L.J. Sham, Proc. Roy. Soc. (London) **A283**, 33 (1965)

25. S. Lundqvist, C.W. Ufford, Phys. Rev. **A1**, 139 (1965)
26. F. Bassani, J. Robinson, B. Goodman, J.R. Schrieffer, Phys. Rev. **127**, 1969 (1962)
27. L. Landau, Soviet Phys. JETP **8**(70) (1959)
28. P. Nozières, J.M. Luttinger, Phys. Rev. **127**, 1423 (1962)
29. J. Hubbard, Proc. Roy. Soc. A **276**, 238 (1963)
30. W. Metzner, D. Vollhardt, Phys. Rev. Lett. **62**, 324 (1989)
31. J.H. VanVleck, Rev. Mod. Phys. **25**, 220 (1953)
32. P. Hohenberg, W. Kohn, Phys. Rev. **136**, B864 (1964)
33. W. Kohn, L.J. Sham, Phys. Rev. **140**, A1133 (1965)
34. U. Barth, L. Hedin, J. Phys. **C5**, 1629 (1972)
35. R.G. Parr, W.T. Yang, Annu. Rev. Phys. Chem. **46**, 701 (1995)
36. W. Kohn, A.D. Becke, R.G. Parr, J. Phys. Chem. **100**, 12974 (1996)
37. O. Gunnarson, B.I. Lundqvist, Phys. Rev. B **13**, 4274 (1976)
38. J.P. Perdew, A. Zunger, Phys. Rev. B **23**, 5048 (1981)
39. S.F. Sousa, P.A. Fernandes, M.J. Ramos, J. Phys. Chem. A **111**, 10439 (2007)
40. J.P. Perdew, Phys. Rev. B **33**, 8822 (1986)
41. A.D. Becke, J. Chem. Phys. **84**, 4524 (1986)
42. S. Kurth, J.P. Perdew, P. Blaha, Int. J. Quantum Chem. **75**, 889 (1999)
43. O.K. Andersen, T. Saha-Dasgupta, R.W. Tank, C. Arcangeli, O. Jepsen, G. Krier, Lecture Notes in Physics, ed. by H. Dreysse (Springer, Berlin, 1999)
44. L. Hedin, B.I. Lundquist, J. Phys. **C4**, 2064 (1971)
45. A. Svane, O. Gunnarsson, Phys. Rev. Lett. **65**, 1148 (1990)
46. V.I. Anisimov, J. Zaanen, O.K. Andersen, Phys. Rev. B **44**, 943 (1991)
47. C. Duan, R.F. Sabiryanov, J. Liu, W.N. Mei, P.A. Dowben, J.R. Hardy, Phys. Rev. Lett. **94**, 237201 (2005)
48. A. Georges, G. Kotliar, W. Krauth, M.J. Rozenberg, Rev. Mod. Phys. **68**(1), 13 (1996)
49. P. Coleman, Phys. Rev. B **29**, 3035 (1984)
50. Z. Zou, P.W. Anderson, Phys. Rev. B **37**, 627 (1988)
51. P.W. Anderson, Phys. Rev. B **78**, 174505 (2008)
52. M. Biagini, Phys. Rev. Lett. **77**, 4066 (1996)
53. A. Georges, G. Kotliar, Phys. Rev. B **45**, 6479 (1992)
54. V.I. Anisimov, P. Kuiper, J. Nordgren, Phys. Rev. B **50**, 8257 (1994)
55. V.I. Anisimov, A.I. Poteryaev, M.A. Korotin, A.O. Anokhin, G. Kotliar, J. Phys.: Condens. Matter **9**, 7359 (1997)
56. A.I. Lichtenstein, M.I. Katsnelson, Phys. Rev. B **57**, 6884 (1998)
57. P.W. Anderson, Phys. Rev. **124**, 41 (1961)
58. F. Aryasetiawan, M. Imada, A. Georges, G. Kotliar, S. Biermann, A.I. Lichtenstein, Phys. Rev. B **70**, 195104 (2004)
59. K. Held, I. Nekrasov, N. Blümer, V. Anisimov, D. Vollhardt, Int. J. Mod. Phys. B **15**, 2611 (2001)
60. G. Kotliar, S.Y. Savrasov, K. Haule, V.S. Oudovenko, O. Parcollet, C.A. Marianetti, Rev. Mod. Phys. **78**, 865 (2006)
61. K. Held, Adv. Phys. **56**, 829 (2007)
62. S.Y. Savrasov, G. Kotliar, Phys. Rev. Lett. **90**, 056401 (2003)
63. L. Hedin, Phys. Rev. A **139**, 796 (1965)
64. P. Sun, G. Kotliar, Phys. Rev. B **66**, 085120 (2002)
65. S. Biermann, F. Aryasetiawan, A. Georges, Phys. Rev. Lett. **90**, 086402 (2003)
66. M.S. Hybertsen, S.G. Louie, Phys. Rev. B **34**, 5390 (1986)
67. A.N. Chantis, M. vanSchilfgaarde, T. Kotani, Phys. Rev. B **76**, 165126 (2007)
68. M. vanSchilfgaarde, T. Kotani, S.V. Faleev, Phys. Rev. Lett. **96**, 226402 (2006)
69. T. Kotani, M. vanSchilfgaarde, S.V. Faleev, Phys. Rev. B **76**, 165106 (2007)
70. J. Friedel, Can. J. Phys. **34**, 1190 (1956)
71. A. Blandin, J. Friedel, J. Phys. Rad. **20**, 160 (1959)
72. J. Friedel, Phil. Mag. **43**, 153 (1952)

73. J.O. Dimmock, A.J. Freeman, Phys. Rev. Lett. **13**, 750 (1964)
74. S.C. Keeton, T.L. Loucks, Phys. Rev. **168**, 672 (1968)
75. T.L. Loucks, *Augmented Plane Wave Method* (W.A. Benjamin Inc, New-York, 1967)
76. D.D. Koelling, B.N. Harmon, J. Phys. C **10**, 3107 (1977)
77. B. Coqblin, *The Electronic Structure of Rare Earth Metals and Alloys* (Academic Press, New York, 1977)
78. B. Johansson, Phys. Rev. B **20**, 1315 (1979)
79. B.N. Harmon, A.J. Freeman, Phys. Rev. B **10**, 1979 (1974)
80. D. Glôtzel, J. Phys. F **8**, L163 (1978)
81. A. Svane, Phys. Rev. B **53**, 4275 (1996)
82. A. Delin, L. Fast, B. Johansson, J.M. Wills, O. Eriksson, Phys. Rev. Lett. **79**, 4637 (1997)
83. B. Johansson, P. Munck, J. Less-Common Met. **100**, 49 (1984)
84. W.M. Temmerman, A. Svane, Z. Szotek, H. Winter, *in Electronic Density Functional Theory: Recent Progress and New Directions*, in ed. by J.F. Dobson, G. Vignale, M.P. Dax (Plenum Press, New-York, 1998)
85. B.N. Harmon, J.dePhys. **40**, C5-65 (1979)
86. B.I. Min, H.J.F. Jansen, T. Oguchi, A.J. Freeman, J. Magn. Magn. Mat. **61**, 139 (1986)
87. K.A. Gschneidner, J. Less Common. Metals **25**, 405 (1971)
88. P. Strange, A. Svane, W.M. Temmerman, Z. Szotek, H. Winter, Nature **399**, 756 (1999)
89. W.M. Temmerman, L. Petit, A. Svane, Z. Szotek, M. Lüders, P. Strange, J.B. Staunton, I.D. Hughes, B.LGyorffy, *Handbook on the Physics and Chemistry of Rare Earths*, vol. 39, ed. by K.A. Gschneidner, J-C.G. Bunzli, V.K. Pecharsky (2009)
90. B. Johansson, A. Rosengren, Phys. Rev. B **11**, 2836 (1975)
91. G.N. Chestnut, Y.K. Vohra, Phys. Rev. Lett. **82**, 1712 (1999)
92. W. Bi, Y. Meng, R.S. Kumar, A.L. Cornelius, W.W. Tipton, R.G. Hennig, Y. Zhang, C. Chen, J.S. Schilling, Phys. Rev. B **83**, 104106 (2011)
93. D.C. Koskenmaki, K.A. Gschneidner, in *Handbook of the Physics and Chemistry of Rare Earths*, ed. by K.A. Gschneidner, L. Eyring, North-Holland, Amsterdam, p. 337 (1978)
94. B. Coqblin, A. Blandin, Adv. Phys. **17**, 281 (1968)
95. C.F. Ratto, B. Coqblin, E. Galleani D'Agliano, Adv. Phys. **18**, 489 (1969)
96. R. Ramirez, L.M. Falicov, Phys. Rev. B **3**, 2425 (1970)
97. D.R. Gustafson, A.R. Mackintosh, J. Phys. Chem. Solids **25**, 389 (1964)
98. D. Glötzel, J. Phys. **F8**, L163 (1978)
99. U. Kornstädt, R. Lässer, B. Lengeler, Phys. Rev. B **21**, 1898 (1980)
100. R. Podloucky, D. Glötzel, Phys. Rev. B **27**, 3390 (1983)
101. D.R. Gustafson, J.D. McNutt, L.O. Roellig, Phys. Rev. **183**, 435 (1969)
102. B. Johansson, Philos. Mag. **30**, 469 (1974)
103. B. Johansson, I.A. Abrikosov, M. Alden, A.V. Ruban, H.L. Skriver, Phys. Rev. Lett. **74**, 2335 (1995)
104. B. Johansson, A.V. Ruban, I.A. Abrikosov, Phys. Rev. Lett. **102**, 189601 (2009)
105. M. Lavanga, C. Lacroix, M. Cyrot, Phys. Lett. **90A**, 210 (1982)
106. M. Lavanga, C. Lacroix, M. Cyrot, J. Phys. **F13**, 1007 (1983)
107. J.W. Allen, R.M. Martin, Phys. Rev. Lett. **49**, 1106 (1982)
108. J.W. Allen, L.Z. Liu, Phys. Rev. B **46**, 5047 (1992)
109. J. Loegsgaard, A. Svane, Phys. Rev. B **59**, 3450 (1999)
110. F. Patthey, J.-M. Imer, W.D. Schneider, H. Beck, Y. Baer, B. Delley, Phys. Rev. B **42**, 8864 (1990)
111. M. Garnier, K. Breuer, D. Purdie, M. Hengsberger, Y. Baer, B. Delley, Phys. Rev. Lett. **78**, 4127 (1997)
112. L.Z. Liu, J.W. Allen, O. Gunnarsson, N.E. Christensen, O.K. Andersen, Phys. Rev. B **45**, 8934 (1992)
113. J.W. van der Eb, A.B. Kuz'menko, D. van der Marel, Phys. Rev. Lett. **86**, 3407 (2001)
114. A.C. Hewson, *The Kondo Problem to heavy Fermions* (Cambridge University Press, Cambridge, 1993)

115. J.J. Joyce, A.J. Arko, J. Lawrence, P.C. Canfield, Z. Fisk, R.J. Bartlett, J.D. Thompson, Phys. Rev. Lett. **68**, 236 (1992)
116. J.W. Allen, O. Gunnarsson, Phys. Rev. Lett. **70**, 1180 (1993)
117. M.B. Zölfl, I.A. Nekrasov, T. Pruschke, V.I. Anisimov, J. Keller, Phys. Rev. Lett. **87**, 276403 (2001)
118. K. Held, A.K. McMahan, R.T. Scalettar, Phys. Rev. Lett. **87**, 276404 (2001)
119. A.K. McMahan, K. Held, R.T. Scalettar, Phys. Rev. B **67**, 075108 (2003)
120. K. Haule, V.S. Oudovenko, S.Y. Savrasov, G. Kotliar, Phys. Rev. Lett. **94**, 036401 (2005)
121. B. Amadon, S. Biermann, A. Georges, F. Aryasetiawan, Phys. Rev. Lett. **96**, 066402 (2006)
122. M.J. Lipp, Phys. Rev. Lett. **101**, 165703 (2008)
123. M.J. Lipp, A.P. Sorini, J. Bradley, Phys. Rev. Lett. **109**, 195705 (2012)
124. P. Jonnard, G. Giorgi, C. Bonnelle, Phys. Rev. A **65**, 032507 (2002)
125. R. Podloucky, D. Glötzel, Phys. Rev. B **27**, 3390 (1983)
126. A. Kaminski, YuV Nazarov, L.I. Glazman, Phys. Rev. Lett. **83**, 384 (1999)
127. S. Lebègue, A. Svane, M.I. Katsnelson, A.I. Lichtenstein, O. Eriksson, Phys. Rev. B **74**, 045114 (2006)
128. S. Lebègue, A. Svane, M.I. Katsnelson, A.I. Lichtenstein, O. Eriksson, J. Phys.: Condens. Matter **18**, 6329 (2006)
129. C. Bonnelle, Struct. Bind. **31**, 23 (1976)
130. C. Bonnelle, *Advances in Xray Spectroscopy* (Pergamon Press, 1982) p. 104
131. B. Amadon, F. Jollet, M. Torrent, Phys. Rev. B **77**, 155104 (2008)
132. Ph Kurz, G. Bihlmayer, S. Blügel, J. Phys.: Condens. Matter **14**, 6353 (2002)
133. J.F. Herbst, R.E. Watson, I. Lindgren, Phys. Rev. B **14**, 3265 (1976)
134. J.F. Herbst, Phys. Scr. **21**, 553 (1980)
135. J.C. Lashley, A. Lawson, R.J. McQueeney, G.H. Lander, Phys. Rev. B **72**, 054416 (2005)
136. H.H. Hill, in *Plutonium 1970 and Other Actinides*, ed. by W.N. Miner (The Metallurgical Society of the AIME, NewYork, 1970), p. 2
137. J.L. Smith, Physica **102B**, 22 (1980)
138. B. Johansson, M.S.S. Brooks, in *Handbook on the Physics and Chemistry of Rare Earths*, vol. 17, ed. by K.A. Gschneidner, Jr., L. Eyring, G.H. Lander, G.R. Choppin (Elsevier, Amsterdam, 1993), p. 1
139. P. Thalmeier, G. Zwicknagl, *Handbook on the Physics and Chemistry of Rare Earths*, vol. 34 (Elsevier, Amsterdam, 2004), p. 135
140. P. Söderlind, O. Eriksson, B. Johansson, J.M. Wills, A.M. Boring, Nature **374**, 524 (1995)
141. G. Chapline, M. Fluss, S. McCall, J. Alloys Compd **444–445**, 142 (2007)
142. D.D. Koelling, A.J. Freeman, Solid State Comm. **9**, 1369 (1971)
143. A.J. Freeman, D.D. Koelling, J. Phys. (Paris), Colloq. **33**, C3–57 (1972)
144. A.J. Freeman, Phys. Rev. Lett. **30**, 1061 (1973)
145. D.D. Koelling, A.J. Freeman, Phys. Rev. B **7**, 4454 (1973)
146. D.D. Koelling, A.J. Freeman, Phys. Rev. B **12**, 5622 (1975)
147. A.N. Chantis, R.C. Albers, M.D. Jones, M. vanSchilfgaarde, T. Kotani, Phys. Rev. B **78**, 081101R (2008)
148. L. Brewer, J. Opt. Soc. Am. **61**, 1101 (1971)
149. L. Nordström, D.J. Singh, Phys. Rev. Lett. **76**, 4420 (1996)
150. R.H. Heffner et al., Phys. Rev. B **73**, 094453 (2006)
151. A.B. Shick, V. Drchal, L. Havela, Europhys. Lett. **69**, 588 (2005)
152. L.V. Pourovskii, M.I. Katsnelson, A.I. Lichtenstein, L. Havela, T. Gouder, F. Wastin, A.B. Shick, V. Drchal, G.H. Lander, Europhys. Lett. **74**, 479 (2006)
153. O. Eriksson, J.D. Becker, A.V. Balatsky, J.M. Wills, J. Alloys Compd **287**, 1 (1999)
154. S.Y. Savrasov, G. Kotliar, E. Abrahams, Nature **410**, 793 (2001)
155. J.H. Shim, K. Haule, G. Kotliar, Nature **446**, 513 (2007)
156. F. Cricchio, F. Bultmark, L. Nordström, Phys. Rev. B **78**, 100404 (2008)
157. F. Bultmark, F. Cricchio, O. Granäs, L. Nordström, Phys. Rev. B **80**, 035121 (2009)
158. P. Link, D. Braithwaite, J. Wittig, U. Benedict, R.G. Haire, J. Alloys Comp. **213**, 148 (1994)

159. J.-C. Griveau, J. Rebizant, G.H. Lander, G. Kotliar, Phys. Rev. Lett. **94**, 097002 (2005)
160. L. Petit, A. Svane, Z. Szotek, W.M. Temmerman, Phys. Rev. B **72**, 205118 (2005)
161. K.N.R. Taylor, K.C. Darby, *Physics of Rare Earth Solids 60–62* (Chapman and Hall, London, 1972)
162. J.C. Duthie, D.G. Pettifor, Phys. Rev. Lett. **38**, 564 (1977)
163. G.K. Wertheim, G. Crecelius, Phys. Rev. Lett. **40**, 813 (1978)
164. C. Bonnelle, F. Vergand, J. Phys. Chem. Solids **36**, 575 (1975)
165. F. Vergand, Philos. Mag. **31**, 537 (1975)
166. A.B. Shick, A.I. Liechtenstein, W.E. Pickett, Phys. Rev. B **60**, 10763 (1999)
167. Ph Kurz, G. Bihlmayer, S. Blügel, J. Phys.: Condens. Matter **14**, 6353 (2002)
168. B. Johansson, A. Rosengren, Phys. Rev. B **11**, 1367 (1975)
169. L.J. Nugent, J.L. Burnett, L.R. Morss, J. Chem. Thermodyn. **5**, 665 (1973)
170. B. Johansson, Phys. Rev. B **19**, 6615 (1979)
171. U. Lundin, I. Sandalov, O. Eriksson, B. Johansson, Solid State Comm. **115**, 7 (2000)
172. L.-G. Liu, W.A. Basset, M.S. Liu, Science **180**, 298 (1973)
173. A. Svane, J. Trygg, B. Johansson, O. Eriksson, Phys. Rev. B **56**, 7143 (1997)
174. P. Söderlind, Phys. Rev. B **65**, 115105 (2002)
175. A.K. McMahan, Phys. Rev. B **72**, 115125 (2005)
176. A.K. McMahan, R.T. Scalettar, M. Jarrell, Phys. Rev. B **80**, 235105 (2009)
177. B. Johansson, A. Rosengren, Phys. Rev. **14**, 361 (1976)
178. W. Bi, Y. Meng, R.S. Kumar, A.L. Cornelius, W.W. Tipton, R.G. Hennig, Y. Zang, C. Chen, J.S. Schilling, Phys. Rev. B **83**, 104106 (2011)
179. A.W. Lawson, T.Y. Tang, Phys. Rev. **76**, 301 (1949)
180. G.M. Eliashberg, H. Capellmann, JETP Lett. **67**, 125 (1998)
181. A. Jayaraman, Phys. Rev. **137**, A179 (1965)
182. I.-K. Jeong, T.W. Darling, M.J. Graf, Th Proffen, R.H. Heffner, Y. Lee, T. Vogt, J.D. Jorgensen, Phys. Rev. Lett. **92**, 105702 (2004)
183. P. Söderlind, O. Eriksson, B. Johansson, J.M. Wills, Phys. Rev. B **50**, 7291 (1994)
184. P. Ravindran, I. Nordström, R. Ahuja, J.M. Wills, B. Johansson, O. Eriksson, Phys. Rev. B **57**, 2091 (1998)
185. A.B. Shick, W.E. Pickett, A.I. Liechtenstein, J. Electron Spectrosc. Relat. Phenom. **114–116**, 753 (2001)
186. G.D. Mahan, *Many-particle Physics* (1993)
187. G.S. Smith, J. Akella, J. Appl. Phys. **53**, 9212 (1982)
188. G.N. Chesnut, Y.K. Vohra, Phys. Rev. Lett. **82**, 1712 (1999)
189. M. Colarieti-Tosti, M.I. Katsnelson, M. Mattesini, S.I. Simak, R. Ahuja, B. Johansson, C. Dallera, O. Erksson, Phys. Rev. Lett. **93**, 096403 (2004)
190. A.C. Lawson, C.E. Olsen, J.W. Richardson, M.H. Mueller, G.H. Lander, Acta Cryst. B **44**, 89 (1988)
191. S. Raymond, J. Bouchet, G.H. Lander, M. LeTacon, G. Garbarino, M. Hoesch, J.P. Rueff, M. Krisch, J.C. Lashley, R.K. Schulze, R.C. Albers, Phys. Rev. Lett. **107**, 136401 (2011)
192. P. Söderling, Europhys. Lett. **55**, 525 (2001)
193. M.S.S. Brook, H.L. Skriver, B. Johansson, in *Handbook on the Physics and Chemistry of the Actinides*, ed. by A.J. Freeman, G.H. Lander (North-Holland, Amsterdam, 1984)
194. K.T. Moore, G. Laan, Rev. Mod. Phys. **81**, 235 (2009)
195. J. Bouchet, B. Siberchicot, F. Joliet, A. Pasturel, J. Phys.: Condens. Matter **12**, 1723 (2000)
196. J.E. Klepeis, J. Mater. Res. **21**, 2979 (2006)
197. S.S. Hecker, Metall. Mater. Trans. **35A**, 2207 (2004)
198. S.Y. Savrasov, G. Kotliar, Phys. Rev. Lett. **84**, 3670 (2000)
199. A. Lindbaum, S. Heathman, K. Liftin, Y. Méresse, R.G. Haire, T.L. Bihan, H. Libotte, Phys. Rev. B **63**, 214101 (2001)
200. B. Johansson, Am. Phys. Soc. APS March Meeting, 13–17 Mar 2006

Chapter 2
Electron Distributions and Physicochemical Properties

Abstract Electron distributions in the main compounds of rare-earths and actinides are described in relation with the valence of these elements. Their changes as a function of the physical conditions, temperature, pressure and some potential applications of these materials are discussed. Cerium compounds are particularly considered.

Keywords Chemical binding · Valence fluctuation · Rare-earth compounds · Actinide compounds

2.1 Rare-Earth Compounds

Densities of valence states are strongly modified in the compounds with respect to those in the elements and all the electronic levels are shifted. The shifts are a consequence of the chemical binding. Valence electrons of the metal and the ligand are present in bond orbitals, forming the valence band. The bonds are often described as a "charge transfer" from metal to ligand orbitals. This terminology can suggest the transfer of one or eventually more electrons and the existence of ionic bonding. However, the compounds having an ionic bonding are rare and concern mostly alkaline metal compounds. The chemical bondings result from the overlap of the relevant wave functions and their strength depends on the inter-atomic spacing of elements in the solid. Thus, in the oxides, the reduction of the number of valence electrons in the atomic sphere around each metal ion is small for the elements having s and p valence electrons like magnesium and this reduction concerns essentially the s valence electrons [1]. These results can be generalized to the transition elements, which have a fractional number of s, d electrons contributing to the cohesion and to the chemical bonding. All the same, in lanthanides and actinides, the s, d electrons contribute to bonds, which are due to *electron redistribution* rather than a transfer. This feature of the charge redistribution has been discussed by Slater [2]. This redistribution increases with the covalent character of the bond.

© Springer Science+Business Media Dordrecht 2015
C. Bonnelle and N. Spector, *Rare-Earths and Actinides in High Energy Spectroscopy*, Progress in Theoretical Chemistry and Physics 28, DOI 10.1007/978-90-481-2879-2_2

Despite frequent coincidence of the 4f energy levels with the broad s, p, d bands, the interaction of the 4f electrons with the valence electrons of the ligand is negligible and the 4f electrons remain localized as in the metal. In some cases, one of the localized 4f electrons of the metal delocalizes and becomes similar to a valence electron. This is the case of tetravalent cerium compounds.

For an element in a solid state, the valence is defined as the number of electrons per atom contributing to the valence band formation. As previously mentioned, the rare-earth metals are trivalent and have the configuration $4f^n(5d6s)^3$ under SPT conditions, except for europium and ytterbium, which are divalent and have the configuration $4f^n(5d6s)^2$. In a chemical compound, the valence of an element is equal to the number of its bonds with the neighbouring atoms. It was initially suggested that the rare-earth valence was equal to the number of their 6s–5d electrons in the compound and the 4f electrons were not taken into account in the bonding. In the case of oxidation, for example, all the rare-earth elements oxidize readily [3] and, including the two divalent europium and ytterbium, form sesquioxides RE_2O_3 [4]. The rare-earths are trivalent in these compounds. Each rare-earth atom donates three 5d–6s electrons to the bonds with oxygen and has the configuration RE^{3+}; its *oxidation state* is equal to three. The oxidation state characterizes the number of electrons, which participate in the bonding, but not the ionic or covalent character of the bonding. As is well known, the bond is partially covalent in the oxides, it is often defined as having an *intermediate* character.

Lanthanum, gadolinium and lutetium, having a configuration $4f^n5d^16s^2$ in the free atom with $n = 0, 7$ and 14, are considered as forming very stable trivalent compounds. Thus, gadolinium is designated as the archetypal trivalent rare-earth owing to the stability of the half-full 4f sub shell, which remains the same in the free atom and the metal. Other rare-earths, which precede, or follow, these elements, can be, respectively, divalent or tetravalent. Thus samarium, europium, thulium and ytterbium have the +2 oxidation state in numerous compounds [5]; whereas, cerium, praseodymium and terbium can have the +4 oxidation state. In numerous rare-earth compounds, the valence varies with external parameters like pressure, temperature and alloying and this variation is accompanied by changes in the physical properties such as resistivity, optical reflectivity and scattering phenomena.

In the metals and compounds of trivalent rare-earths, whose free atom configuration is $4f^n6s^2$, the integral number of localized 4f electrons is $(n - 1)$ and the three valence electrons are of the s–d type. The electronic configuration changes are induced by the need to obtain an energetically stable solid. Indeed, the cohesion energy increases with the number of the valence electrons. It varies also with their character and is higher for the 5d than for the 4f electrons. In the light rare-earths, the electron configurations are less stable than in the heavy rare-earths and the number of the valence electrons can change. Thus, the number of the 5d electrons is higher in light rare-earth compounds and that increases the cohesion. Concurrently, the number of the 6s electrons decreases to maintain the trivalence. The valence state of the metallic elements is determined theoretically from the difference of the total energies calculated for the divalent and trivalent configurations [6].

If the rare-earths have the same valence in the metal and in its compounds, so is the number of their 4f electrons. These 4f electrons are localized and of atomic-like character and their interactions with the valence electrons are negligible. However, an increase of the valence or a variation of the temperature–pressure conditions can bring about a change in the character of the 4f electrons owing to the interactions between the strongly correlated 4f electrons and the delocalized valence electrons in the solid. Such a change occurs in the light rare-earths because their 4f electrons are less tightly bound and in terbium because the single 4f electron outside the half-filled shell is also weakly bound. In a valence change from a trivalent to a tetravalent rare-earth, the number of the localized 4f electrons decreases while the number of the valence electrons increases. Because the number of the localized electrons is integral, it is reduced by one unit while the number of 4f electrons that participate in the chemical bonding is unspecified. When a localized electron makes a transition to a delocalized state, this delocalization increases the valence as well as the *cohesive energy*. Thus increase of the cohesive energy following the arrival of a 4f electron into a bonding orbital is supplied by the delocalization process.

Compounds with a non-integral number of 4f electrons were considered as formed from a rare-earth with two different valences and were designated as *mixed, or intermediate, valence compounds* [7, 8]. Among the 4f-electron materials with atypical properties, compounds and alloys in which the rare-earth is a mixing of divalent and trivalent ions have been particularly studied [9–11]. In the notion of mixed valence, the number of 4f electrons in each ion is integral and a non-integral number of 4f electrons results from the mixing of ions having n 4f electrons with those having $(n - 1)$ 4f electrons. The presence of ions with only $(n - 1)$ 4f electrons is explained by the transfer of one 4f electron into the 5d6s band. The ions fluctuate between the two pure configurations of very close energies. Such coexistence of two different configurations was considered possible only if the energy difference between their two lowest levels is smaller than energy of the order of Δ, which is the width of the Friedel virtual bound state. Δ is around 10^{-2} eV for rare-earth metals [12]. Systems fluctuating between two configurations of very close energies are also named *valence fluctuation compounds*. Initially limited to a few monochalcogenides, this representation now concerns many compounds of cerium, samarium, europium, thulium and ytterbium, that exhibit non-integral valence at ambient conditions.

Various models have been developed to treat this type of compounds. It was initially suggested that if the 4f levels were sufficiently close to the valence band, transfer of a non-integral number of 4f electrons/ion could take place to this band [13]. In this case, 4f electrons would be present in the chemical bonds and a covalent ligand nl-4f mixing would exist. A strongly correlated band model has then to be used to treat such a system. More generally, some of the 4f electrons arrive into the 5d6s band and fluctuations between the $4f^n$ and $4f^{n-1}$ configurations render the distribution homogeneous. For this transition to take place, the $4f^n$ localized levels are expected to be very close to the Fermi level. The $4f^{n-1}$ levels are localized and only a non-integral number of the 4f electrons, equal to the number of rare-earth ions having the $4f^{n-1}$ configuration, are delocalized. The time of valence fluctuations is inversely proportional to the energy width of the configurations. Spectroscopic observations of the compound on a time scale shorter than the fluctuation time can analyze each configuration. Generally, a mixing of configurations is observed.

The existence of intermediate valence phases depends on the lattice constant, whose values can vary between those expected for pure divalent and trivalent compounds. It also depends on the cohesive energy and the chemical bonding, which vary according to the rare-earth valence, hence the contribution of the 4f electrons to the valence. Particular properties of the mixed valence phases are the high electronic specific heat and, in some cases, the absence of magnetic ordering. The delocalization of a non-integral number of 4f electrons per ion accompanies a decrease of atomic volume and compressibility. When atomic volume and compressibility change in a continuous way, the electronic transition takes place gradually. When they change discontinuously, an $f \rightarrow d$ discrete transition is expected. Collapse in the atomic volume is thus characteristic of a change in the electronic structure of the rare-earth ions (cf. Sect. 1.4.1).

The ground state of systems exhibiting valence fluctuations was difficult to treat because the interaction energies between electrons in open shells are not known with precision. A theoretical method was developed to obtain the energy differences between the two configurations with an accuracy better than a few tenths of electronvolt [6]. Initially, it was applied to divalent–trivalent systems and tested for the rare-earth metals. In this method, the energy difference between the divalent and trivalent compounds is calculated from energy differences between a solid and an atom having the same valence because many interaction energies cancel between the atom and the solid. Indeed, the interaction energies within the 4f shell are essentially the same in the solid and the atom having the same valence and cancel. The 4f–5d inter-shell interactions can be considered as being the same for the divalent and trivalent configurations in the solid above the magnetic ordering temperature. The coupling between 4f electron and the electrons in the valence band is weak in the solids and can be neglected. No coupling is present in the divalent atom because only the 4f shell is open. In contrast, the 4f–5d inter-shell interaction is to be taken into account in the trivalent atom. Its energy is to be added to terms calculated without interaction. Thus, the energy difference between the divalent and trivalent atoms is only the 4f–5d excitation energy. The energies of the two valence states in the solid have been calculated using the LMTO method with either the SIC-LDA or GGA density functional. This original calculation method was tested by comparison between the theoretical and experimental energy differences. The valence state is correctly predicted for all the trivalent metals and difference of only some percents is obtained between the GGA calculations and the experimental data. This method, suitably tested for the metals, has been used to predict the rare-earth valence in various series of compounds.

In the heavy rare-earths, the 4f electrons are chemically inert and set up localized magnetic moments. The total magnetic moments have both orbital and spin components and spin–orbit interactions are strong for the rare-earth elements and compounds, in comparison with the Pauli paramagnetism of the itinerant electrons.

As one proceeds along the series adding electrons to the localized 4f ones, the 4f electrons become more tightly bound in the ions, reducing the inter-ion overlap of the 4f wave functions, whose spatial extension remains inside the 5s and 5p core sub shells. Then, despite the fact that the ionization energies of the 4f sub shell are comparable to those of the 5d and 6s valence electrons, the 4f electrons retain an atomic character in the solid and do not participate in the chemical bonding. Therefore, as already underlined for the valence band in the metals, those are the 5d and 6s valence electrons, hybridized with the ligand orbitals, that form the valence band in a series of trivalent compounds. They are responsible for the chemical bonding and their densities of states remain practically unchanged along the series. Thus, the trend toward bivalence of the rare-earths, associated with divalent ligands, is due to the localized nature of the 4f electrons and their well-known lanthanide contraction.

In compounds, the charge distribution due to anions surrounding the metal ion induces a static electric field, named ligand-field, or *crystal field*, which produces a splitting of the energy levels occupied by localized electrons. The angular part of the crystal field parameters reflects the influence of the topological arrangement of the anions. The crystal field splitting tends to decrease with increasing size of the coordination polyhedron. The rare-earth 5d electron interacts strongly with the surrounding anions, making the 4f–5d excitation energy depend on the ligands. Owing to the crystal field, the energy difference between d orbitals varies; it increases with oxidation number. Inversely, the energies of the 4f orbitals are barely sensitive to the crystal field and the 4f orbital moments are not quenched by it. In regard to 4f energy levels in rare-earth ions, their crystal field splitting is generally not taken into account because it is much smaller than the splitting due to the spin–orbit interaction, which is strong for many of the rare-earth compounds. Indeed, partly filled 4f orbitals behave like localized magnetic moments. However, in order to estimate the electron distributions with sufficient accuracy, it appears necessary, in some cases, to take into account the interaction of the 4f orbitals with the crystal field.

Owing to the localization of the 4f electrons, the overlap between 4f orbitals centered on adjacent atoms is very small or negligible. Thus, the dipole–dipole and exchange interactions between these 4f orbitals constitute a negligible portion of interactions between rare-earth ions. An exchange mechanism due to interactions between localized electrons via the other valence electrons, called indirect exchange, was developed by Ruderman and Kittel for transition metal alloys. Kasuya has shown the validity of this indirect exchange model as a basis for describing the interactions in gadolinium metal and more generally in the rare-earths. Consequently, the dominant interaction in the rare-earths is a coupling between the localized 4f electrons and the itinerant valence electrons, named *RKKY* (Ruderman–Kittel–Kasuya–Yoshida) *interaction*. This interaction increases with a decrease of the direct interactions between 4f electrons, i.e. when U decreases. Initially developed for metallic materials, it varies with a change in the bonds and the presence of ligands introduces more complicated interaction mechanisms such as *super-exchange interactions* [14, 15]. Thus, in rare-earth compounds, both super-exchange interaction and indirect RKKY-type interaction coexist and their relative contribution varies with the ligand radius, the super-exchange increasing with it. In contrast, the RKKY interactions increase considerably with the density of the charge carriers. They increase when the lattice constant decreases and the metallic character of the compounds increases. The RKKY interactions are responsible for the existence of *half-metallic ferromagnets* in intermetallic compounds [16]. In this particular class of materials, an energy gap exists between the valence and conduction bands for electrons of one spin polarization, while this gap is absent for electrons of the other spin polarization. A remarkable consequence is that the electrons at the Fermi level are 100 % spin-polarized in these materials. Half-metallic materials are rare because they require a large exchange splitting and a small band gap.

Theoretical models based on one-electron theories have been used to describe the cohesive properties, such as lattice constant, crystal structure, elastic constant enthalpy. Valence configurations as well as electronic configurations of the ground state can be deduced by minimizing the total energy. Strong on-site Coulomb repulsions exist among the 4f electrons and the independent particle approximation is no longer valid. Essentially, three models have been largely used. In first, the valence spd electrons are treated in the LDA using the LMTO method in the atomic sphere approximation and the 4f electrons are considered as core electrons. Another model, the SIC-LSDA, that has the advantage of being parameter free, introduces a self-correction term to take into account the 4f electron localization. An alternative model is LDA + U. It corresponds to a Hartree–Fock treatment of the 4f electron configuration with a screened effective U parameter. The essential feature of this model is that the energy functional is orbital-dependent rather than density-dependent. SIC-LSDA has been used to predict the valence of numerous rare-earth compounds [17]. In contrast, densities of states and spectral analysis of electron transitions involving the localized 4f electrons must be described by theoretical models taking into account the many-body effects due to localization of 4f electrons. An approximation has been used in which the energy bands are treated with LDA and additional atomic multiplets are included to describe the localized $4f^n$ configurations. The use of multiplets to describe the 4f levels underlines their atomic character. Agreement was found between this description and photoemission spectra. Further theoretical developments were needed to explain the entire spectroscopic observations. They are discussed in Chap. 3.

When the 4f electrons are treated as core electrons and the valence electrons by LDA, both electron systems are considered as independent. The 4f electrons do not interact directly with the valence electrons because they are concentrated on the core but they contribute to the ion field with which the valence electrons interact. Consequently, interactions exist between them. A model has been developed in which the 4f shell retains its atomic-like character but is coupled to the surrounding solid. This approach is based on the LDA + DMFT method in the LMTO basis set [18]. Taking into account intra-atomic Coulomb interactions, spin–orbit coupling and crystal field effects, a *many-body self-energy* is computed within the Hubbard approximation and inserted into the function describing the solid. That is equivalent to treating the quasi-localized 4f electrons as quantum impurities. Physical properties for only a few compounds have been calculated up to now with this model. The agreement with the observations is improved compared to the LDA + U approach, where interactions in the 4f shell are described by a self-consistent, orbital- and spin-dependent, one-electron potential. In a one-electron description, either Coulomb interactions or lattice symmetry are minimized while both requirements are satisfied in a many-body treatment of the electron correlations. A usual treatment of the interaction effects in the 4f shell improves the description of the quasi-free valence electrons and thus of the electronic properties.

As already mentioned, a particularity of the rare-earths is that they form mixed valence, or *intermediate valence*, compounds, in which the rare-earth is considered as having two distinct configurations, $4f^n$ and $4f^{n-1}$. In these compounds, an integral number of 4f electrons are present on each ion while the number of ions containing each configuration appears to be a fractional number and the two electron configurations have nearly the same energy. The valence of a divalent metal rare-earth present in a mixed valence compound is equal to $2 + n_f$ where n_f is the participation number of the trivalent configuration. In a mixed valence compound of a trivalent rare-earth, the valence is $3 + n_f$ where n_f is the participation number of the tetravalent configuration.

Initially, the presence of mixed valence was associated with the relation $RE^{2+} \rightarrow RE^{3+} + e^{-}5d$ [9]. All the same, Yuan and others have considered the cerium valence as fluctuating between 3 and 4, with the cerium configuration varying between $4f^1$ and $4f^0$ + (5d6s) electrons. This model predicts a large energy shift of the 4f levels from well below to above the Fermi level and a shift in the opposite direction for the 5d bands. These changes have not been confirmed. The presence of an integral number of 4f electrons reveals a strong 4f localization and only very weak interactions between the 4f and the valence electrons.

In a general point of view, a valence change does not correspond to an excitation or ionization but to a partial *delocalization* of a bound electron by transfer into the valence band. This delocalized electron thus interacts with the other valence electrons. This description is in agreement with the characteristics exhibited at low temperature by some intermetallic compounds where two different regimes exist, one where the 4f electrons are all localized and another where one 4f electron per atom is delocalized and interacts with the valence electrons. Cerium clearly illustrates the double character of the 4f electrons, which can be localized and treated as core electrons or strongly correlated with the valence electrons. Trivalent Ce^{3+} ions of $4f^1$ configuration associated with three 5d6s valence electrons are present in the metal and in a large number of compounds. However, in numerous other solids, tetravalent Ce^{4+} ions are associated with four valence electrons having a partial 4f character. The 4f electron is then partially mixed with the 5d6s valence electrons of cerium. The 4f electrons are localized or not as a function of their level positions with respect to the 5d–6s valence band. Nevertheless, the cerium 4f electron always remains relatively near the ion, due to the characteristics of its wave function. Its mobility is not equivalent to that of the other valence electrons. The dual character of the 4f electrons is an important factor, that influences the valence, and thereby the physicochemical properties of the cerium compounds.

Many rare-earth compounds are mixed valence compounds and show numerous anomalous physical properties [10, 19]. In these compounds, two local electron configurations are present with nearly the same energy. Thus, a valence change can occur between the divalent and trivalent states for samarium, europium, thulium and ytterbium. As an example of such compounds, let us mention here SmS, SmB_6 and TmSe. A change from insulating to metallic conduction occurs at a low temperature when $n_f \approx 0.8$. In the divalent phase, the ground state is an insulator. Through a first-order transition, its valence changes to a mixed valence phase, which is still insulating. Under pressure, a valence change from divalent to trivalent metallic ions can be induced in the monochalcogenides, SmS and TmSe. This change takes place gradually and stable non-integral valences are observed for a large range of pressures. Discontinuous jumps from one mixed state to another have also been reported. Metallic conduction appears for $n_f \approx 0.8$ at a volume that is rather larger than the volume calculated for a pure trivalent configuration.

The hexaborure SmB_6 is mixed valence type in STP conditions, of the same type as that of SmS [20, 21]. The electronic properties of samarium with valence $v < 2.8$, are characteristic of a divalent rare-earth insulator compound. For $v \geq 2.8$, the physical properties become characteristic of a trivalent metallic compound. The pressure-induced insulator–conductor transition appears around 4–6 GPa while the changes in the magnetic properties appear at a higher pressure. That is because each samarium ion is surrounded by a cage of boron ions and the Sm–Sm exchange interactions remain weak. Differences between SmS and SmB_6 are due to the difference of their crystalline structures, thereby also of their band structures. In SmB_6 [22], the resistivity increases strongly with decreasing temperature up to 4 K but below it, a residual low temperature conductivity still exists. Recent interest has been raised by the observation of metallic surface states [23] (cf. Sect. 4.3).

Change in the atomic environment, in the bonds, or in the cohesive properties can cause a valence change. Under pressure, transformation from a divalent ion semiconductor to trivalent ion metal is generally possible. It depends on the energy gap between the 4f levels and the conduction band and on the rate at which this gap decreases with the pressure. The ionic radius of the trivalent ion is much smaller than that of the corresponding divalent ion and such a volume decrease can induce the delocalization of a 4f electron into the valence band. The analysis of the pressure–volume relationship gives a direct measure of the 4f delocalization, i.e. of the valence change. The theoretical treatment is that of the transfer of a localized electron into the valence band where important relaxation effects must be suitably treated. Conversely, the 4f electrons are pushed toward localization when the distances between rare-earth ions increase, i.e. when the on-site f–f interactions are the strongest. These electronic changes induce large changes in the electrical, optical and magnetic properties. Because of the varied and complex structures of many of the rare-earth compounds, there is no complete theory concerning the $4f^n$ electrons that gives theoretical good predictions of the valence.

2.1.1 Oxides

Oxides are the most wide-spread rare-earth compounds. They form a contamination layer at the surface of the metals resulting from the reaction with the atmosphere. They have important applications in catalysis and in electronics, where they are associated with silicon and semiconductors. All the rare-earths form a sesquioxide, [4] the most stable oxygen compound except for cerium, praseodymium and terbium (Table 2.1). In these compounds, each rare-earth donates three electrons to the bonds with oxygen and the 4f electrons stay strongly localized at the rare-earth site. Three different structures exist, hexagonal for the light rare-earths, cubic or monoclinic distortion of the cubic for middle rare-earths and cubic of bixbyite-type for heavy rare-earths. In the first case, each rare-earth ion is surrounded by seven oxygen ions, of which four are closer. In the cubic or distorted structures, only six oxygen first neighbours are present. Following the well-known lanthanide contraction, the lattice parameters decrease along the series. The crystal lattices of the oxides have low symmetry, for which ab initio electronic structure calculations are not adequate.

Stable oxides with an oxidation number superior to three exist in the light rare-earths and terbium (Table 2.2) [24]. Thus, for cerium, praseodymium and terbium, the stable oxides are, respectively, CeO_2, Pr_6O_{11} and Tb_4O_7. Metastable NdO_2 is also predicted to exist. The number of the rare-earth valence electrons is larger in these compounds than in the metal and, consequently, the number of the localized 4f electrons smaller.

Cerium oxide has a well-known particular property: a continuously ongoing transformation exists between the two oxides, the oxygen-rich CeO_2, or ceria and the oxygen-poor Ce_2O_3, depending on the external oxygen concentration [25]. Indeed, cerium that oxidized completely to (IV) oxide in the presence of

Table 2.1 All the rare-earths form a sesquioxide, the most stable oxygen compound except for cerium, praseodymium and terbium

Metal	Oxides
Ground configuration	
La $(6s5d)^3$	La_2O_3
Ce $4f^1$ $(6s5d)^3$	Ce_2O_3, CeO_2
Pr $4f^2$ $(6s5d)^3$	Pr_2O_3, Pr_6O_{11}, PrO_2
Nd $4f^3$ $(6s5d)^3$	Nd_2O_3, NdO_2
Sm $4f^5$ $(6s5d)^3$	SmO, Sm_2O_3
Eu $4f^7$ $(6s5d)^2$	EuO, Eu_2O_3
Gd $4f^7$ $(6s5d)^3$	Gd_2O_3
Tb $4f^8$ $(6s5d)^3$	Tb_2O_3, Tb_4O_7, Tb_6O_{11}, TbO_2
Dy $4f^9$ $(6s5d)^3$	Dy_2O_3
Ho $4f^{10}$ $(6s5d)^3$	Ho_2O_3
Er $4f^{11}$ $(6s5d)^3$	Er_2O_3
Tm $4f^{12}$ $(6s5d)^3$	TmO, Tm_2O_3
Yb $4f^{14}$ $(6s5d)^2$	Yb_2O_3

Table 2.2 a Sesquioxide data. **b** Dioxide data

a

Compound	$E_{IV} - E_{III}$ (eV)	V_{hexag} (Å^3)		V_{cubic} (Å^3)	
	Hexagonal	Theory	Expt.	Theory	Expt.
Ce_2O_3	0.38	76.4	79.4	88.18	87.0
Pr_2O_3	0.78	75.6	77.5	86.77	86.7
Nd_2O_3	0.94	74.0	76.0	85.28	85.0
Pm_2O_3	0.97	72.9	74.5	83.89	83.0
Sm_2O_3	1.09	72.0		82.50	81.7
Eu_2O_3	1.13	70.5		82.14	80.2
Gd_2O_3	1.29	68.8		80.70	79.0
Tb_2O_3	1.12	67.6		79.49	77.2
Dy_2O_3	1.21	66.3		78.62	75.9
Ho_2O_3	1.36	65.2		77.30	74.7

b

Compound	$E_{IV} - E_{III}$ (eV)	V_{theo} (Å^3)	V_{exp} (Å^3)
CeO_2	−2.40	39.61	39.6
PrO_2	−1.44	39.22	39.4
NdO_2	−0.65	39.37	–
PmO_2	−0.18	39.15	–
SmO_2	0.49	42.19	–
EuO_2	2.31	41.87	–
GdO_2	1.22	41.08	–
TbO_2	−0.27	36.50	35.6
DyO_2	0.05	39.65	–
HoO_2	0.46	39.06	–

Sesquioxide data $E_{IV} - E_{III}$ is the energy difference between tetravalent and trivalent ground configurations in electron volts. $V_{hexagonal}$ is the calculated volume in (Å)^3 for this structure and the corresponding experimental volume in (Å)^3. V_{cubic} is the calculated volume in (Å)^3 for the cubic structure and the corresponding experimental volume in (Å)^3 (from [24])
Dioxide data $E_{IV} - E_{III}$ is the energy difference between tetravalent and trivalent ground configurations in electron volts. V_{theo} is the volume calculated for the cubic fluorite structure and V_{exp} the corresponding experimental volume. A negative energy difference indicates that the tetravalent configuration is the more stable one (from [24])

atmosphere, can release oxygen by forming various reduced oxides up to the (III) oxide. Inversely, the stoichiometric oxide, Ce_2O_3, can take up oxygen and return to the (IV) oxide. This reversibility is used for the storage and the transport of oxygen and makes ceria interesting as a catalytic converter.

Concerning the structural properties, one expects that such a reversible addition or removal of oxygen atoms should involve a minimal rearrangement of the cerium ions during the transition. This suggests the presence of a common unit cell for both oxides. Ce_2O_3 has a hexagonal lattice with a c/a ratio equal to 1.55. Stoichiometric CeO_2 has the cubic fluorite structure characteristic of compounds such as CaF_2.

However, it was possible to construct a unit cell of the type Ce_2O_3 from eight unit cells of CeO_2 by increasing their volume by 3 % and removing 25 % oxygen ions along four (111) diagonals [26]. The reversibility of the reduction or oxidation processes is thus established and the reduction–oxidation transition can be described as an almost isostructural transition with a 10 % volume contraction, by analogy with the γ-α transition in pure metal with a volume reduction of about 16 %.

A proper description of the Ce_2O_3 4f electron characteristic is difficult with standard band methods. A good estimate of the magnetic and electronic ground state properties of Ce_2O_3 was obtained using the full-potential LMTO method in the framework of the DFT and by treating simply the 4f electron as localized on the cerium ion [27]. Cerium is therefore close to the trivalent state in this oxide. Ce_2O_3 is antiferromagnetic insulator with a magnetic moment of $2.17\mu_B$ by unit cell [28].[1] As two cerium ions are present in the unit cell, one expects a magnetic moment of $1\mu_B$ from each 4f electron, the difference resulting from the polarization due to the valence electrons. This value of the magnetic moment confirms the localized character of the 4f electron, which, therefore, does not contribute to the bond in the Ce_2O_3 oxide.

Various opposite points of view have been formulated to treat the 4f orbital of cerium in CeO_2. In a first model, the dioxide at the ground state was described as a mixed valence compound with a 4f occupation of about 0.5. A specific mechanism was introduced to explain the coexistence of this model with the fact that CeO_2 is insulator [29–31]. Another model considered cerium as having four valence electrons and the 4f orbital as unoccupied [32]. Other pictures were introduced depending on the occupation of the 4f orbital. From an electronic structure calculation by the LAPW method, the 4f and 5d orbitals in tetravalent cerium were described as hybridized with the oxygen 2p band [33]. The magnetic and electronic properties of CeO_2 were found correctly described when the 4f electron was treated as a valence electron. By opposition to the localized character of the 4f electron in cerium metal at room temperature, the 4f electron becomes delocalized in cerium tetravalent compounds. This characteristic had been observed from spectroscopy experiments, specially designed to determine the localization of the valence electrons (cf. Chap. 4). Some cerium orbitals of 5d and 4f character are mixed with the 2p band of oxygen and form the valence states (Fig. 2.1) [27]. This shows that cerium 4f levels are partially occupied. A very narrow empty band of a 4f character is present just above the Fermi level.

The CeO_2–Ce_2O_3 transition involves a localization–delocalization process of the cerium 4f electron. The reduction of CeO_2 was described as due to the formation of an oxygen vacancy and the localization of 4f electrons on the neighbouring cerium ions. In CeO_2, every oxygen atom is surrounded by four cerium ions. A vacancy of

---

[1]Antiferromagnetism is due to coupling between spin moments carried by near ions in the solid and it is present below the Néel temperature, i.e. at low temperatures. This atomic description is well adapted to rare-earth compounds because of the presence of incomplete 4f sub shell localized on each rare-earth ion.

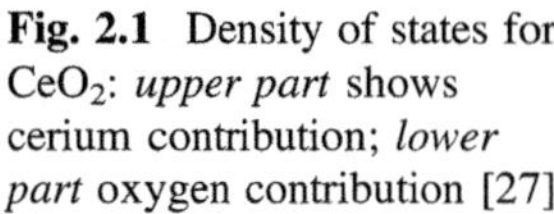

Fig. 2.1 Density of states for CeO_2: *upper part* shows cerium contribution; *lower part* oxygen contribution [27]

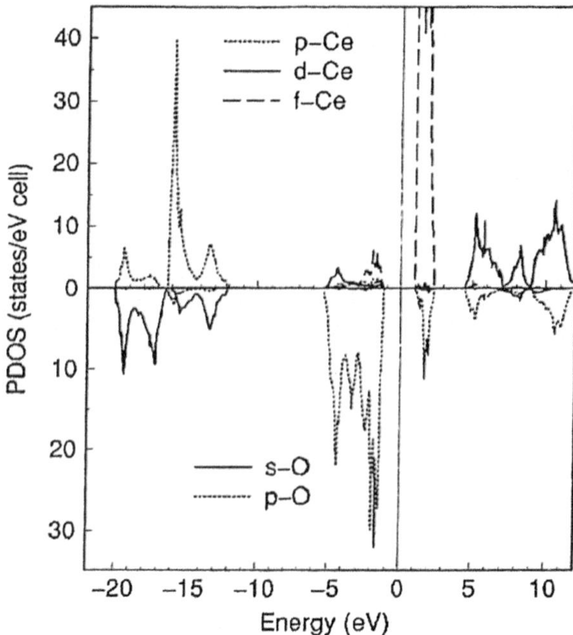

oxygen is created when an oxygen atom quits the lattice and leaves other electrons in the vicinity of the vacancy in the lattice. These electrons fill the empty levels of lowest energy, which are the 4f empty ones. Indeed, in CeO_2, narrow empty 4f levels are present in the band gap between the valence and conduction band [34]. Then, electrons are localized on cerium ions in the immediate surrounding of the vacancy and, conversely, a vacancy may be created. The energy for the formation of a vacancy was calculated and it was shown that a minimum energy situation corresponds to two ions Ce^{3+} surrounding an oxygen vacancy [26]. The oxygen vacancy formation depends on the localization–delocalization of the 4f electron in cerium. The localization of the 4f electrons in Ce_2O_3 leads to a volume increase. By analogy with a treatment used for the γ and α phases of the metal, calculations of the electronic, structural and magnetic properties were made for Ce_2O_3 and CeO_2 by considering the 4f electron either as a localized core electron or as being in the valence band [27]. The method used was the full-potential LMTO method in the framework of the LDA–GGA for exchange and correlation. An improved agreement with the experimental data was obtained using the localized model for Ce_2O_3 and the band model for CeO_2. The 4f electron behaves, then, differently according to the different valences of cerium in the compound.

Praseodymium and terbium are also naturally present in several oxidation forms. Praseodymium occurs naturally as Pr_6O_{11}, exhibiting a slightly oxygen-deficient fluorite structure. The stoichiometric fluorite structure PrO_2 exists under oxygen pressure. Terbium behaves as praseodymium: it exhibits an oxygen-deficient fluorite structure Tb_4O_7 and oxidizes, giving the dioxide TbO_2 under oxygen pressure. Praseodymium and terbium were considered as tetravalent [32, 35] or of mixed valence in these compounds [36, 37]. The REO_2 dioxides have a cubic fluorite structure; each rare-earth has eight first neighbours at a distance a little shorter than the one between first neighbours in the sesquioxides. Consequently, all the rare-earth ions have the same surrounding and their ionic volume is smaller in the dioxides than in the sesquioxides. From this decrease of the ionic volume, one deduces that the number of localized electrons associated with each ion decreases too. It is usual to consider that in all tetravalent rare-earth compounds, the ion RE^{4+} lost one 4f electron with respect to the trivalent ion RE^{3+}. Then, in Pr_6O_{11} and Tb_4O_7, one quasi-delocalized 4f electron is present in the valence band, hybridized with the s–d electrons, while the rest of the 4f electrons of Pr and Tb stay strongly localized at the rare-earth site and their number is integral [24, 38]. In contrast, the number of delocalized electrons is non-integral and these electrons participate in the bonding. The valence band acquires some f character, creating a type of 4f electron that can participate in electron bonding. Theoretically, 4f levels move down into the oxygen 2p band. To conclude, for elements as praseodymium or terbium, which have several 4f electrons, coexistence of localized and partially delocalized f electrons occurs in the dioxide, and more generally, in the tetravalent configuration. This was confirmed by high energy spectroscopy experiments [32, 39].

The two divalent rare-earth metals, europium and ytterbium, maintain their divalent character in some compounds, in particular in the mono-oxides. In YbO, the two Yb valence electrons are transferred into the anion 2p band, which is separated from the conduction band by several electronvolts. The localized 4f orbitals fall in the gap between the valence and conduction bands. YbO is stable under STP conditions. The same is valid for EuO mono-oxide. However, owing to its half-filled 4f shell, EuO is ferromagnetic with a practically complete electron spin polarization. That makes EuO interesting for spin electronic devices [40]. Under pressure, the energy difference between divalent and trivalent ground states of Yb and Eu decreases. Overlap between the 4f and 5d energies induces the formation of a ground state designated as having a mixed valence, obtained from the $4f^n$ and $4f^{n-1}5d$ configurations. High pressures are necessary to induce the 4f–5d energy overlap and transform the divalent rare-earth oxide into a trivalent compound. The transformation takes place gradually and the change is observed through a wide range of pressures. For YbO, the pressure–volume variation shows an anomalous behaviour from about 100 kbar. The volume reaches a lower limit value near 350 kbar for an ytterbium mean valence of the order of 2.6 [41].

Samarium and thulium have also stable divalent oxides. Both metals are trivalent in the bulk because the energy required to displace a localized electron from the 4f level to an extended 5d band state is small. The energy difference between the trivalent and divalent metallic states is only about 6 kcal/mol for samarium. This makes the two valences nearly degenerate and facilitates a valence change for this metal [42]. Thus, at the surface, due to the decrease of the coordination number, samarium has a divalent or partially divalent character. The bonding energy is smaller than in the bulk and the 4f electrons are more strongly bound. Consequently, samarium and thulium frequently appear as divalent in compounds and SmO and TbO are stable under the SPT conditions. Some other monoxides have been synthesized under high pressure, CeO, PrO and NdO [43].

All the oxides are insulators. This results also from the calculated density of states. They are not ionic compounds and, therefore, the charge density and the valence cannot be directly connected. The densities of states and band structures are similar for all the sesquioxides.

As an example, the densities of states calculated for Nd_2O_3 is presented Fig. 2.2 [24]. The valence band originating from the O-2p and rare-earth 5d–6s electrons is completely filled. An unoccupied 4f level is situated in the band gap above the Fermi level between the valence and conduction band. The latter corresponds essentially to the unoccupied rare-earth d–s levels. The position of the occupied and unoccupied $4f^n$ and $4f^{n+1}$ levels with respect to the band edges is important because they can contribute to the evaluation of the band gap width. If the 4f electrons are more strongly bound than the low-energy valence electrons, the band gap takes place between the valence and conduction bands. However, if unoccupied 4f levels are present in the forbidden band, the band gap is between the valence band and these 4f levels. Finally, if occupied 4f levels are situated above the valence band, the band gap is between these 4f levels and the lower unoccupied levels. According to the position of the 4f levels, the band gap takes place either between the valence and conduction bands or between the valence band and empty 4f levels or between occupied 4f levels and the conduction band. The band gap widths vary with the 4f level positions with respect to the valence and conduction band and, consequently, they vary with the atomic number. For the sesquioxides, the largest band gaps, about 5.5 eV, are associated with the rare-earths having more stable electronic configuration, lanthanum, gadolinium and lutecium. The lowest band gaps, between 2.3 and 4 eV, belong to rare-earths immediately following the two previous ones, namely cerium, praseodymium and terbium. For the rare-earths immediately preceding the more stable, samarium, europium, thullium and ytterbium, the band gaps are slightly bigger than 4 eV while for dysprosium, holmium and erbium, they are of the order of 5 eV. For the rare-earths with a small or medium gap, the 4f levels

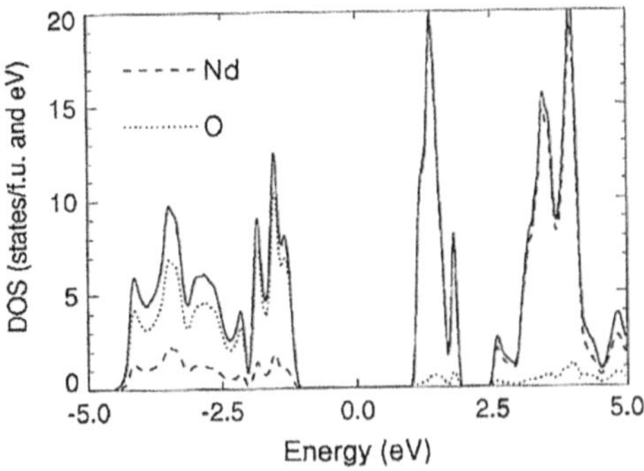

Fig. 2.2 Density of states for Nd_2O_3 in the hexagonal structure: Neodymium is in the trivalent f^3 configuration. The zero of the energies corresponds to the midgap position [24]

are located in the band gap while for a large gap the 4f levels are in the same energy region as the bands. The variation of the band gap value along the series was calculated by considering only the valence and conduction bands and not the 4f levels. Such a calculation is relatively simple, whereas it is complicated to calculate simultaneously the energies of the 4f levels because it necessitates taking into account all the screening and relaxation effects and the use of quasi-atomic model. It should be noticed that the band gap in a thin film is different from the one in the bulk.

Among the applications in microelectronics, there is the possibility to use the dielectric properties of the rare-earth oxides instead of those of SiO_2 in the metal-oxide semiconductor (MOS) electronic devices. In order to increase the performance of these devices, it is necessary to decrease the dielectric thickness of the insulating layer, d/ε, where d is the mechanical thickness and ε the dielectric constant. A high dielectric constant is then a significant condition to be a good candidate for microelectronics, in particular for memory applications [44]. Several rare-earth oxides have a high dielectric constant. Thus, ε is about ten times larger for some sesquioxides, in particular for Sm_2O_3, than for SiO_2. This constant depends on the infrared vibrations, i.e. the crystalline structure [45]. It is higher for structures with the main absorption band at lower energies. Consequently, the light RE oxides of hexagonal structure have higher dielectric constants than the heavy RE oxides crystallized in cubic and monoclinic structures. However, the dielectric constant depends strongly on the conditions of preparation of the sample. Another property, important for an oxide to be usable in MOS electronic devices, is that there should be no interfacial layer between the RE oxide and the semiconductor. As an example, such a well defined interface exists for Pr_2O_3 on Si(111) [46] and Gd_2O_3 on GaAs (100) [47] but generally an interfacial layer cannot be completely avoided.

2.1.2   Chalcogenides

These are, in principle, compounds formed with elements X of the sixth column of the periodic table, X = O, S, Se, Te. However, the term chalcogenide is commonly reserved for the compounds with X = sulphur, selenium and tellurium. Among these, the most studied are the monochalcogenides, of formula RE-X. The electronegativity of the ligand decreases with increasing size, i.e. from oxygen to tellurium. Consequently the ligand p bands move towards lower binding energies. These compounds are semiconductors and have interesting optical, electronic and magnetic properties. In addition to the two divalent rare-earths, europium and ytterbium, two other rare-earths, samarium and thulium, appear in a divalent state in the monochalcogenides. But samarium and europium undergo isostructural transition to intermediate valence state with the pressure. All the rare-earths form mono-sulphides, RE-S and have then a divalent behaviour. The other compounds are expected to be trivalent.

Europium and ytterbium monochalcogenides are stable under STP conditions because divalent rare-earth ions lead to exact charge compensation with divalent anions. Two valence electrons of the rare-earth, having the s–d character, hybridize with the S 3p electrons and form the valence band. These compounds crystallize in the cubic rock-salt, or NaCl-type, structure. The 4f levels lie in the energy gap between the valence band and the conduction band. The conduction band consists of the rare-earth 5d levels, which are split by the cubic crystal field into lower energy t_{2g} and higher energy e_g levels. From optical absorption, the energy separation between the 4f levels and the conduction band increases in going from the oxide to the telluride.

A NaCl-CsCl-type crystalline transition observed under pressure for the three compounds, EuS, EuSe and EuTe, did not involve a change of the valence, contrary to what is observed for EuO [48]. Valence change is due to the delocalization of one 4f electron per ion into the conduction band and it induces a semiconductor to metal transition. A striking change in the reflectivity has been observed for EuO but not for EuS, EuSe and EuTe. Thus a semiconductor-metal transition occurs only in EuO in the pressure ranges studied. However, from LSDA total energy calculations, europium and samarium chalcogenides showed isostructural transitions into an intermediate valence state with the pressure [49].

YbS, YbSe and YbTe undergo an electronic transition under pressure. Thus, for YbS, near 400 kbar, the mean value of the valence was estimated to be 2.5 [50]. These results are deduced from the experimental pressure-volume data and from optical absorption. The compression curves (Fig. 2.3) [41, 50, 51], show that electronic collapse due to a gradual change in the valence state of ytterbium from two towards three is a continuous function of the pressure in all the ytterbium monochalcogenides. Striking changes in the optical reflectivity are observed above the 200 kbar. YbS, YbSe and YbTe usually of black colour become, respectively, golden yellow, copper-like and purple. These observations confirm the presence of a semiconductor to metal transition at high pressure.

Increase of the valence is easier for divalent samarium and thulium than for europium and ytterbium because the ionization potentials of Sm^{2+} and Tm^{2+} are smaller than those of Eu^{2+} and Yb^{2+} respectively. Thus, all the samarium monochalcogenides can undergo a valence change from divalent semiconductor to trivalent metallic ions. This change is induced by a reduction of the lattice constant under pressure at room temperature [52, 53] or by chemical alloying under SPT conditions [9, 10]. Generally, no change of the crystalline structure accompanies the contraction of the lattice. The valence change observed in the samarium monochalcogenides is associated with a semiconductor to metal transition. This transition is manifested experimentally by the rapid decrease of resistivity with the increase of the pressure. Metallic conductivity is reached under pressure. It is accompanied by a metal-like reflectivity, observed for photons of energy lower than 2 eV [54, 55].

A specially well-known case is the sharp decrease in volume observed in samarium sulphide SmS at 0.65 GPa, due to a first-order phase change. This abrupt change is isostructural and reversible when the pressure decreases below 0.1– 0.15 GPa. The crystalline arrangement remains NaCl-type up to very high pressures. At 1 GPa, the volume decrease is about 15 %. Compressibility also decreases with decreasing lattice constant and shows an abrupt reduction at 0.65 GPa. This

Fig. 2.3 Comparison of the
volume-pressure relations:
a for fcc Yb metal:
experiment (*full line*),
isothermal equation of states
for divalent (*dotted line*) and
trivalent (*dashed line*) [51];
b for YbO: experiment with
or without medium pressure
(*point*), calculated (*solid line*)
[41]; **c** for YbS: experiment
(*closed circles and triangles*),
pressure-volume relations
(*dashed lines*). Near 400 kbar,
the mean value of the valence
was estimated to be 2.5 [50]

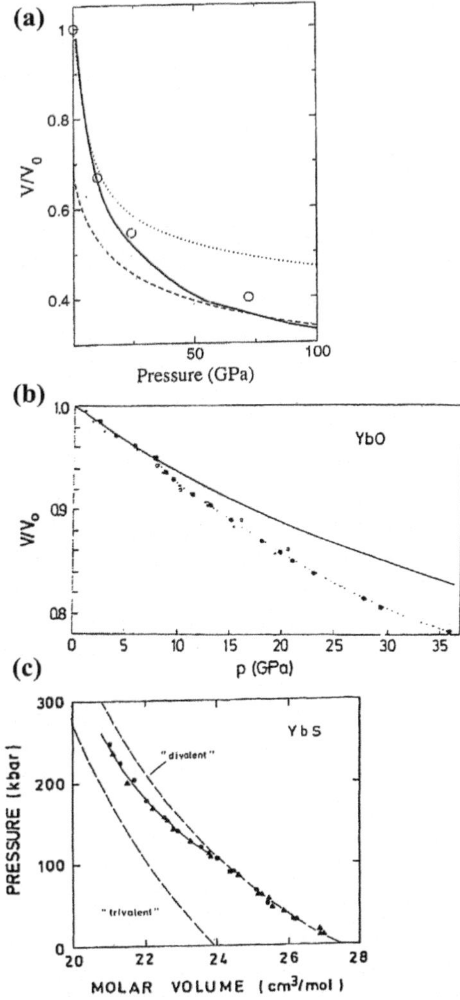

large volume decrease under pressure triggers a valence change of samarium ions
from Sm^{2+} to Sm^{3+}. Under SPT conditions, SmS is a mixed valence-type compound. Abrupt decrease in resistivity by a factor of 5 is observed at 0.65 GPa and
transition to the metallic state takes place above 2 GPa. From spectroscopy
experiments divalent f^6 ions are present in the ground state and mixed f^5–f^6 ions in
the high-pressure metallic phase [56, 57]. The pressure-induced semiconductor to
metal transition has been observed from measurement of the optical reflectivity
[58]. SmS was found to be opaque to photon radiation of energy higher than
0.37 eV under pressure. Semiconductor to metal transition has also been observed
after polishing of the surface of a SmS semiconductor crystal. This treatment is
considered as the first phase of a pressure effect. The valence transition is evident
from the change of the surface. Black semiconductor at STP conditions, SmS

exhibits after polishing or under pressure a golden yellow metallic coloration [59]. The valence transition takes place at relatively low pressure in this compound because the levels of Sm^{2+} associated with the $4f^6$ ground configuration are located only 0.2 eV below the conduction band [60] whereas this distance is about 1.6 eV for EuS. Valence transitions can also be obtained by alloying with trivalent ions such as Y, La, Ce and Gd.

For SmTe and SmSe, the resistivity decreases by about seven orders of magnitude between 0 and 6 GPa and the $4f^n$–$4f^{n-1}$ transition takes place in a continuous manner over a broad pressure range at room temperature. Strong reduction of the volume is observed. For SmTe, as for SmSe, a crystallographic transition from NaCl-type to a tetragonal structure, equivalent to a distorted CsCl structure was also observed around 11 GPa and for PrTe around 9 GPa. Volume discontinuities accompany these transitions and both phases are present over a large pressure range [61]. There is complete delocalization of one 4f electron and conversion of Sm^{2+} to Sm^{3+}. This behaviour is contrary to that seen for europium because the $4f^6$ sub shell of Sm^{2+} is weakly bound whereas the half-filled $4f^7$ shell of Eu^{2+} is very stable. SmSe and SmTe exhibit under pressure the same changes of coloration as YbSe and YbTe, i.e. a copper-like metallic coloration for SmSe and a deep purple one for SmTe. The energy gap between the 4f levels and the conduction band decreases in going from SmTe to SmSe then to SmS. This variation of the energy gap is in the same direction as the variation observed in the Eu chalcogenides. In contrast, the valence transition of samarium is abrupt in SmS while it is gradual in SmSe and SmTe. This is related to the decrease of the lattice constant with pressure, which is much more pronounced in SmS than in SmSe and SmTe, leading to a discontinuous adjustment and to first-order structural transition. There exists, therefore, a critical inter-atomic distance between the Sm ions which governs the process.

Theoretical studies of the samarium monochalcogenides have been made by using models described in Chap. 1. Conventional band structure calculations cannot be used because they describe the 4f levels as narrow bands. Only models treating the 4f electrons as localized can be considered. Total energy calculations have been made with the help of the SIC-LSD approximation associated with the LMTO method in ASA. The spin–orbit interaction is included in the Hamiltonian [62]. In this scheme the s–d electrons are described as delocalized and either 5 or 6 localized 4f electrons are considered on each samarium ion, corresponding to a trivalent or divalent samarium ion, respectively. From total energies per unit cell volume calculated for the $4f^6$ and $4f^5$ configurations of SmS, it is seen that a configuration of mixed character appears with increasing pressure (Fig. 2.4) [62]. At high pressure, the $4f^5$ configuration is stable. As shown from calculated densities of states, this f^6 to f^5 valence transition is made possible because the $4f^6$ energy levels are energetically very near to the conduction band in the SPT conditions.

In summary, in the samarium chalcogenides, reduction of the lattice constant induces a gradual diminishing of the energy gap until it vanishes, an energy decrease of the 4f level energies and a delocalization of one 4f electron into the

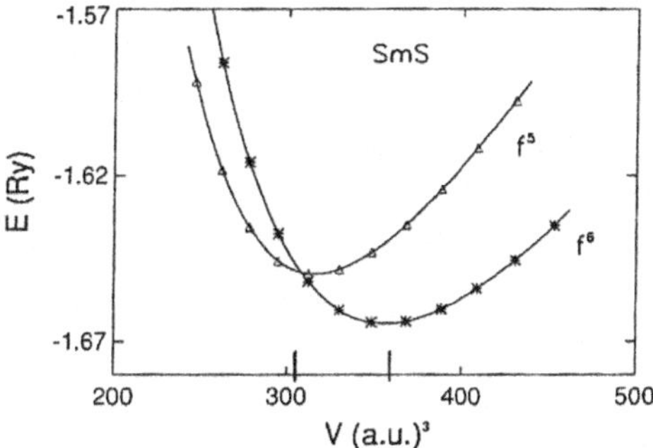

Fig. 2.4 SmS: SIC-LSDA total energy calculated as function of unit cell volume; the position of the minimum of the f^6 *curve* gives an equilibrium lattice constant in STP conditions, 11.25 a.u., in agreement with the experimental value [62]

conduction band, leading to a transition from semiconductor to metal. The probability of the ensuing valence change depends on the matrix elements connecting the localized and extended states. Except for SmS, these transitions are continuous. Stable non-integral valence values are observed over a large pressure range [63]. For SmS, the valence change is abrupt and the reverse metal to semiconductor transition thermally induced is also observed, for example in thin films [64]. In this case, the number of the 4f localized electrons increases. A significant charge transfer is imposed on the system, its energetic cost being balanced by the gain in the localization energy. When SmS becomes metallic, its magnetic properties change from paramagnetic to magnetic. This transformation also is reversible. The above variations are due to the Sm–Sm exchange interactions.

The properties of thulium monochalcogenides vary with the ligand. TmS appears to be a metallic compound at ambient pressure where thulium is trivalent [65]. Its resistivity shows a Kondo-like behaviour [66]. TmSe is a mixed valence compound, with a valence between 2.5 and 2.7. TmTe is a divalent semiconductor [67]. The experimental values of the lattice constants confirm that the valence is 3 for TmS, 2.75 for TmSe and 2 for TmTe, showing a trend towards a greater 4f localization with the increase of the ligand size. Agreement between the lattice constant value calculated in the SIC-LSD approximation and its experimental value was obtained by describing TmS and TmSe with the help of a localized $4f^{12}$ configuration while TmTe is characterized by a localized f^{13} configuration (Fig. 2.5) [68]. The level distribution calculated for the localized $4f^{12}$ configuration consists, in addition to the twelve localized 4f electrons, of a narrow band partly occupied having a 4f character, located just at the Fermi level. This description fits well TmSe, which is known as a mixed valence-type compound. But, this one-electron theoretical model does not give a correct value for the total energy of the f^{13} configuration. Agreement with the valence values has also been obtained from spectroscopic experimental results (cf. Chap. 4). The valence variation reveals an increase of the localization with the increase of the lattice constant, i.e. with the increase of the ligand size. In TmTe under pressure at room temperature, semiconductor to metal transition takes

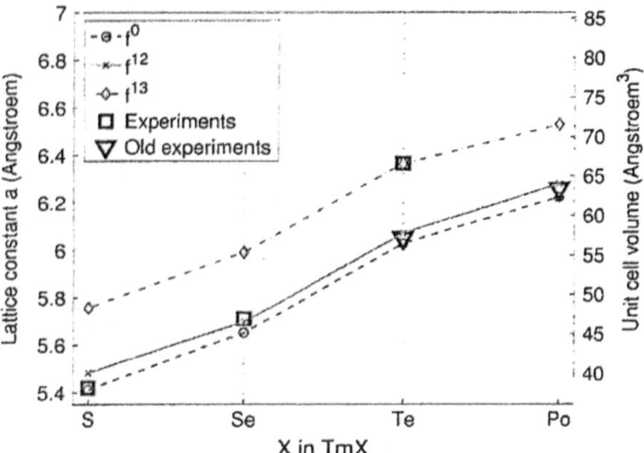

Fig. 2.5 Calculated and experimental lattice constants for Tm chalcogenides [68]

place [69]. This transition is ascribed to the gap decrease and the transfer of 4f electrons into the valence band as this band moves to lower energies and crosses the 4f levels. Finally the gap vanishes above 6GPa and one 4f electron per Tm ion is delocalized into the 5d–6s band. This change of Tm^{2+} to Tm^{3+} results from the decrease of the atomic volume under pressure. In addition, a crystallographic transition from the NaCl-type structure to a tetragonal phase was observed at 8GPa [70].

One of the very first LDA + DMFT calculations was for the mixed valence 4f material TmSe [71]. Energy difference between divalent and trivalent configurations was calculated for the metals [6]. The same method was used to calculate the energy differences between the di- and trivalent samarium and thulium chalcogenides as a function of the pressure [60]. An analogous curve was also obtained for the rare-earth sulphides, showing its universal character (Fig. 2.6) [72].

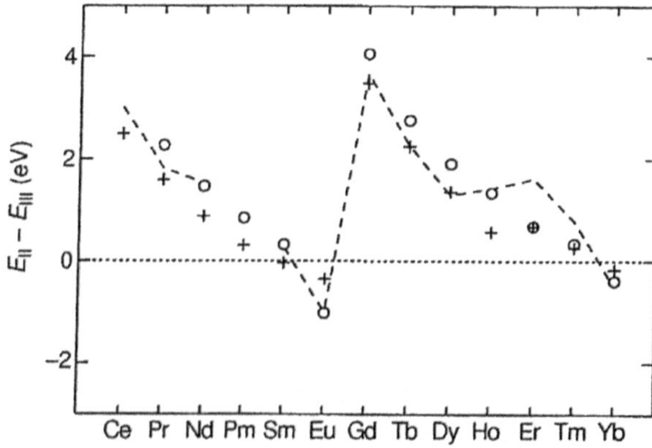

Fig. 2.6 Energy difference in eV between the divalent and trivalent states: experimental values for the rare-earth metals (*dashed line*); calculated values for the metals (*open circles*) and for the sulphides (*crosses*) [72]

From this curve, the divalent to trivalent energy difference is large and positive in the beginning of the series and the trivalent state is predominant. The energy difference drops and becomes negative for Eu and for SmS and EuS. A large jump to positive value occurs at Gd and GdS, which are trivalent. Then the difference gradually falls and becomes negative again for Yb metal and sulphide. Thus a trend to a divalent configuration exists for the heavy chalcogenides under SPT conditions. However, the trivalent ground configuration prevails in a large number of compounds. Calculations of energies as a function of lattice parameters for relevant valence configurations were generalized for numerous materials by using the SIC-LSDA. The calculated values of the lattice parameters in the ground state configuration are within about 1.5 % of the experimental values.

The sulphides constitute a special category of the chalcogenides. Thus, the mono-sulphides RE-S exist for all the rare-earths. They are the monochalcogenides for which the bond between the rare-earth and ligand is strong. Then, the bond strength decreases with the increasing volume of the X element, i.e. down column VI of the periodic table. The valence band is composed of the S 3p electrons hybridized with two valence electrons of the rare-earth. The divalent phase predominates. The rare-earths are divalent and the number of the localized 4f electrons increases. The calculated lattice parameters are compared to their experimental values in Fig. 2.7 [17]. The agreement is good and the calculations reproduce well the sudden increase of the lattice parameter at SmS, EuS and YbS, which accompanies the increasing of the 4f electron localization. It was shown that the

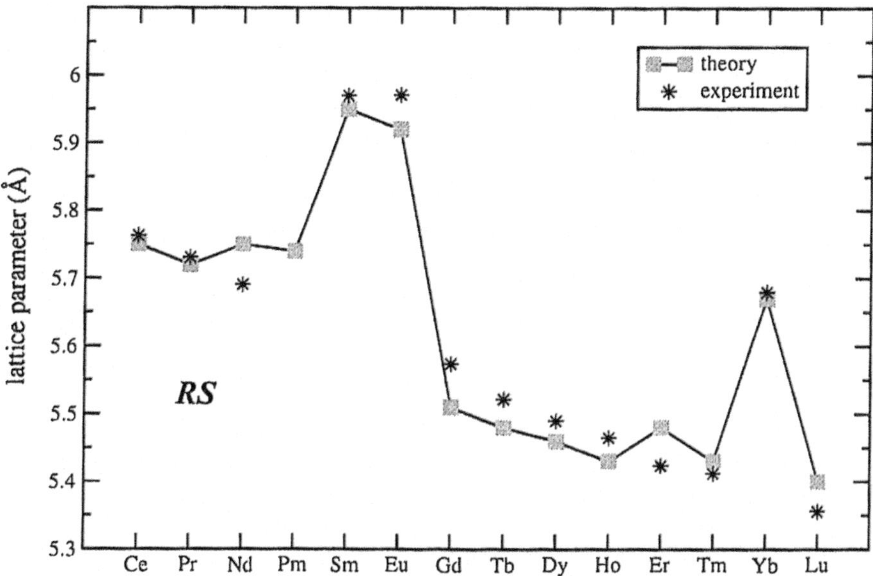

Fig. 2.7 Lattice parameters of the rare-earth sulphides: comparison between values calculated in the ground state configuration and experimental data [17]

energy difference between the valence band and the conduction band is constant throughout the series but that the position of the $4f^n$ energy levels with respect to the conduction band changes along the series. From LaS to NdS and GdS to ErS, the transfer of a 4f electron into the conduction band leads to metallic behaviour, while for EuS and YbS, the $4f^n$ levels are well below the conduction band, resulting in semiconductor-like behaviour. For SmS, the $4f^n$ levels are situated very close to the conduction band and the material can vary between metallic and semiconducting phase under critical perturbation. Thus, under pressure, the bottom of the conduction band overlaps the 4f levels, causing each Sm^{2+} ion to yield one electron to the conduction band and become Sm^{3+}. The material undergoes a discontinuous isostructural transition to a metallic phase. This transition has also been observed in thin films, the pressure effect being replaced by a polishing effect. Reverse metal to semiconductor transition can be induced by relaxing the perturbation. This has been realized by heating, i.e. by thermal lattice expansion of the metallic phase, for films whose thickness varies between 50 and 800 nm [64]. This thermal expansion decreases the crystal field splitting until the bottom of the 5d conduction band rises above the 4f levels.

The rare-earth sulphides, unlike the other chalcogenides, are refractory solids. They are thermally stable with high melting points and are mechanically strong with moderate expansion coefficients. Their electronic properties and infrared transmission characteristics made them good candidates for numerous applications. Changes in the structural and electronic properties induce changes in the electrical conductivity. When the material undergoes a high-pressure structural change, it displays a higher conduction state. This change of the conductivity can be of several orders of magnitude. Thus, the high-pressure phases of SmS exhibit electrical resistivities of about 10^{-4} Ω cm, i.e. smaller by two orders of magnitude than in the SPT conditions. The sulphides can be used as switching circuit devices. If they undergo an irreversible structural change, they can be used as memory switching devices. Owing to their high thermal stability and the existence of compound series of the form $RE_{3-x}S_4$ with optimal thermo-electrical parameters, the rare-earth sulphides can be used as efficient high-temperature materials for thermoelectric energy converters [73]. Among these sulphides, GdS_{1+x} has the highest thermo-electrical efficiency [74].

The optical properties of sulphides originate from the weak RE-S bonds, giving rise to characteristic low vibrational frequencies and resulting in an infrared absorption cut-off. Consequently, the cubic sulphides are transparent in the long wavelength spectral region, in contrast to the oxides. The excellent far-infrared transmission of the rare-earth sulphides together with their glass-forming ability make these materials very good candidate for far-infrared glasses. Mixed with other glass-forming materials, such as gallium, germanium or arsenic sulphides, rare-earth sulphides form systems that are more thermally stable than the other chalcogenides glasses, with transition temperatures higher than 500 °C. Their optimal thermal and mechanical properties, high melting points, high hardness, make them useful in harsh environmental conditions, in particular aerodynamic heating, thermal shock and rain erosion. Associated with their high transmission in the far-infrared region (3–5 μm), these glass sulphides can play a major role in equipments deployed on aircraft and guided weapons, such as optical windows for heat-seeking missiles, surveillance equipment alerting devices, sensing low temperature objects. Doped with Nd^{3+}, the rare-earth sulphide glasses show fluorescent properties and are good candidates for laser applications. The crystallization of amorphous glasses requires high temperatures and leads to sulphide ceramics that are mechanically and thermally resistant materials with good optical transmission. Chemical processing of the rare-earth sulphides is complex and their extreme affinity for oxygen makes their production difficult. However, because of their useful properties both in the amorphous and crystalline state, rare-earth sulphides emerge as promising optical systems.

2.1.3   Pnictides

These are the compounds formed with the elements of the fifth column of the periodic table, X = N, P, As, Sb, Bi. The X atoms have three unoccupied p-orbitals and can accommodate three electrons from the rare-earth, resulting in a filled hybridized valence band. The rare-earths are in a trivalent state in these compounds, with the exception of cerium, which is tetravalent in CeN. The difference of rare-earth valence between pnictides and chalcogenides is due to the difference in the occupation of the ligand p-orbitals. Pnictides, as chalcogenides, crystallize in the simple rock-salt, or NaCl-type, structure, except EuAs, EuSb and EuBi. The lattice constants of the monopnictides increase from N to Bi and decrease from La to Yb. This variation is explained simply by the increase of anion sizes and the decrease of cation sizes with the increase of atomic number. Pressure can induce structural phase transition of the monopnictides from NaCl structure to a tetragonal structure, which may be considered as a distorted CsCl structure. Change in the lattice parameter can be accompanied by a valence change. Consequently electronic structures and physical properties of these compounds are sensitive to external pressure, strain, impurities. Pnictides, as chalcogenides, have been very widely studied because they offer a wide range of applications and they crystallize in a simple cubic structure, facilitating the theoretical studies [75].

The rare-earth pnictides are semiconductors or semimetallic. The energy gap is between the valence band and the rare-earth unoccupied 5d levels. Overlap between these distributions can give rise to a small density of states. The light rare-earth pnictides tend to have a larger energy gap than the heavy pnictogen compounds because the metallic behaviour increases with decreasing ligand electronegativity. The variation of the energy gap along the rare-earth series is not well defined. However, the energy gap appears to decrease with increasing rare-earth contraction [17]. High values of the dielectric constant are associated with these small values of the gap. These compounds have a large variety of electronic and magnetic properties. The magnetic interactions gradually increase from lanthanum to gadolinium, then decrease from gadolinium to lutecium. Strong exchange interactions exist between localized 4f electrons and valence electrons and can lead to possible changes of the electronic properties in the presence of an external magnetic field. Most monopnictides are antiferromagnetic in the ground state with some exceptions among the phosphides and especially the nitrides, which are ferromagnetic. A combination of ferromagnetic and antiferromagnetic configuration is also present. Some monopnictides are expected to be half-metallic ferromagnets.

Initially, the 4f electrons were treated as part of the core and the other valence electrons in a band model. Gadolinium monopnictides have been particularly investigated because gadolinium is situated in the middle of the rare-earth group and the Gd 4f orbitals are exactly half-filled. The ground state of Gd^{3+} is $^8S_{7/2}$, where the orbital angular momentum is zero. Under such symmetric conditions the contribution of the spin–orbit interaction and that of the f–p and f–d electrostatic interactions to the Hamiltonian are small, greatly simplifying the problems associated with the electronic structure. The energy distribution of the valence electrons has been calculated by using an APW method with Slater $X\alpha$ exchange potential [76]. GdN was found to be a semiconductor and the other Gd monopnictides semimetallic. However, it was remarked that by taking into account the exchange interaction, the characteristics of GdN could be different. A model that treats the 4f electrons as core electrons but uses the LMTO method within the LSDA for the valence electrons was applied to cerium, praseodymium and neodymium antimonides [77], then to the gadolinium and erbium pnictides [78, 79]. Concerning these compounds, it was expected that a narrow set of occupied 4f levels would be present several eV below the Fermi level with a narrow set of unoccupied 4f levels several eV above it. Interactions between the 4f multiplets and the bands located in the neighbourhood of the Fermi level were expected to be weak and were neglected. In this model, large overlap exists between valence levels of the rare-earth and pnictogen and the valence band is formed from three pnictogen np orbitals bonding with a rare-earth sd-like band. The conduction band is formed from five bands mainly rare-earth 5d-like, antibonding with pnictogen np orbitals. Gradual increase of the metallic character from N to Bi appears in all the compounds. From the calculated bands, the nitrides have a small gap while the phosphides and arsenides are *semimetals*, i.e. have "holes" in a region of the Fermi surface and electrons in another. This situation is due to fact that the lowest rare-earth 5d band dips below one of the pnictogen np orbital. These calculations concern the non-magnetic case and the perfect solid.

Treatment considering the 4f electrons as core-like electrons was found to be valid for numerous purposes. It provides lattice constants in good agreement with experimental values confirming that the 4f electrons are not significantly involved in the bonding in these compounds (Table 2.3) [79]. It accounts well for the major features of the electronic structure and magnetic properties of this family of materials. In particular, it leads to a semimetallic band structure of the pnictides, accompanied by a possible transition to semiconductors for the nitrides, in agreement with the experiment.

From calculations treating the 4f electrons as localized and not as valence electrons, the praseodymium monopnictides are predicted from LDA + U

Table 2.3 Theoretical and experimental equilibrium lattice constants in (Å) [79]

Compound	a (Å), theor.	a (Å), exp.
GdN	4.977	4.999
GdP	5.704	5.729
GdAs	5.843	5.854
ErN	4.789	4.839
ErP	5.557	5.595
ErAs	5.700	5.732

Table 2.4 Experimental lattice constants in (Å) compared to values calculated in the NaCl-type phase for pentavalent (f^0) and trivalent (f^2) praseodymium (from [81])

Compound	$a(f^0)$ (a.u.)	$a(f^2)$ (a.u.)	a (Expt.) (a.u.)
PrP	10.87	11.13	11.15 [17]
PrAs	11.06	11.43	11.34 [18]
PrSb	11.79	11.99	12.00 [19]
PrBi	11.99	12.14	12.21 [28]
PrS	10.66	10.86	10.83 [28]
PrSe	11.10	11.26	11.23 [28]
PrTe	11.75	11.92	11.93 [28]

calculation to be semimetals with the occupied 4f levels situated about 4 eV below the Fermi level [80], in agreement with spectroscopic results. The effects of pressure on the stability of the NaCl and CsCl phases of praseodymium pnictides and chalcogenides were investigated by using SIC-LSD method and by treating the 4f electrons as localized or itinerant [81]. Under SPT conditions, all the compounds have the NaCl structure. The lattice constants calculated by considering two localized 4f electrons in praseodymium are in good agreement with experimental values (Table 2.4). PrP reveals a crystallographic transition from NaCl-type to distorted CsCl around 26 GPa with a volume collapse of 12.1 % [82]. Similar structural transitions were observed in PrAs, PrSb and PrBi at pressures of 27, 13 and 14 GPa, respectively. These experimental observations are in agreement with the structural transitions expected at 16, 12, 8 and 8 GPa, respectively, for PrP, PrAs, PrSb and PrBi. The pressures predicted to induce the change of the lattice constants are lower than the experimental ones but the experimental and calculated volume reductions are similar. The praseodymium ion remains in the trivalent configuration through the transition and the calculated lattice constants are in agreement with the experimental data.

Calculation for NdSb, and also PrSb, treating also the ions as trivalent and the 4f orbitals as localized, gave good agreement with experiment both for the specific heat and for the photoemission spectrum [77]. Europium and ytterbium, generally divalent, undergo a valence change and are trivalent in the monopnictides. Samarium ion is $4f^5$ in the monopnictides, which are antiferromagnetic with very low Néel temperatures (Table 2.5) [62] while the $4f^6$ ion or mixed valence is present in the chalcogenides, which are non magnetic. The interval $4f^5$–$4f^6$ decreases strongly for the heavier pnictide ligands. From total energies calculated by the

Table 2.5 Experimental lattice constants in (a.u.) compared to values calculated in the NaCl structure with ground state configuration of f^5 for the pnictides and SmO and of f^6 for the other chalcogenides (from [62])

Compound	Lattice constant (a.u.)	
	Calc.	Expt.
SmN	9.46	9.52
SmP	10.99	10.88
SmAs	11.18	11.16
SmSb	11.90	11.84
SmBi	12.14	12.01
SmO	9.41	9.34
SmS	11.25	11.25
SmSe	11.70	11.66
SmTe	12.43	12.46
SmPo	12.64	12.71

SIC-LDA method for the trivalent rare-earth ions, the equilibrium lattice constants have been deduced for all the monopicnides except the cerium compounds and they agree within one percent with experimental values. A comparison between the experimental and calculated lattice parameters is presented in Fig. 2.8 [17] for all the rare-earth monochalcogenides and monopnictides.

The experimental values are plotted along the x-axis and the calculated values from SIC-LSDA along the y-axis. All the data are approximately on the $x = y$ line and that is a measure of the agreement between theory and experiment in the determination of crystalline data. The difference between calculated and experimental values is ≤ 1.5 % except for CeN and DyTe. Concerning CeN, this anomaly can be explained by the unusual behaviour of the 4f electrons, described below.

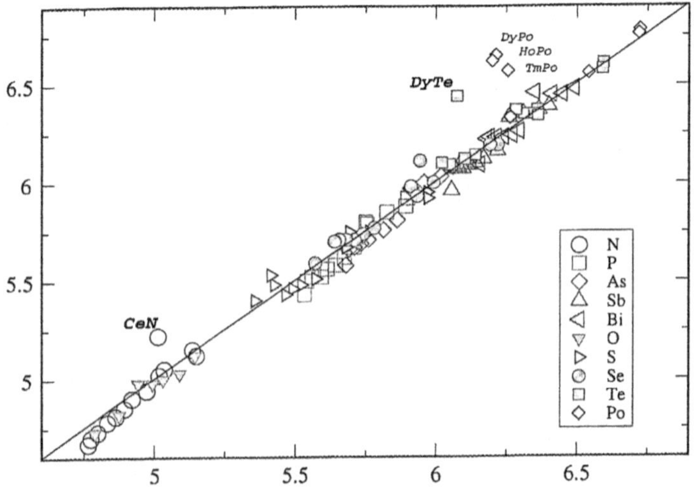

Fig. 2.8 Experimental and calculated lattice parameters (in Å) of the rare-earth monochalcogenides and monopnictides [17]

Therefore, crystallographic data can be predicted with a sufficient precision from the SIC-LSDA model provided the 4f electrons be considered as localized.

Intensive theoretical studies have been made for ErAs because of its technological interest. Calculations based on the LDA and treating the 4f electrons as atomic core electrons correctly obtained a semimetal and the theoretical 4f multiplet structure agreed with photoemission experiments [78, 79]. However, this treatment did not give a sufficiently accurate description of the electronic structure of ErAs near the Fermi level. A new theoretical model has been developed recently. It uses a many-body model to treat the interaction of the 4f electrons with the crystal field and with an external magnetic field [83]. The Slater integrals have been deduced from optical measurements on erbium ions embedded in a rock-salt LaF_3 host. Agreement has been obtained between prediction and experiment for the densities of states near the Fermi level, in particular the repartition of the holes and the electrons. Nevertheless, it must be underlined that possible hybridization between the valence band and 4f orbitals located near the Fermi level is to be expected from the calculations, contrary to the idea that the 4f electrons are thought to be localized.

The peculiar behaviour already found in cerium metal and in some compounds is manifested also in CeN. The observed lattice constant of CeN is 5.007 Å at 4.2 K, 5.019 Å at room temperature, while the calculated value for pure trivalent CeN is 5.24 Å. From an interpolation between the lattice constants of LaN, PrN and NdN, the valence of CeN at room temperature was expected to be 3.46. Three valence electrons of the cerium are present in the valence band mixed with the N 2p electrons. The additional electron of cerium is a localized 4f electron in Ce^{3+}. In Ce^{4+}, this is a delocalized electron [7]. It was, then, suggested that CeN was a mixed valence compound [8] and the additional electron fluctuated between the d and f symmetries. In this model, the 4f levels are considered as located at the vicinity of E_F. However, the energy difference between the f^0 and f^1 configurations is much larger than the hybridization energy. Moreover, the lattice constant calculated by assuming itinerant 4f electrons was 4.91 Å, i.e. relatively close to the experimental value [84]. This result suggested that CeN was not a mixed valence compound. It was, then, considered as a narrow band compound [85, 86]. The cerium 4f levels were treated in a band model. Calculation of the lattice constant was used as a test. From relativistic calculations using LMTO method in the GGA, a value of 5.04 Å was obtained [87]. It is very close to the experimental value. The lattice constant calculated by treating the 4f electron in a band model is thus in better agreement with the anomalously small experimental value than the value obtained by considering localized 4f orbitals. Comparison between experimental and calculated optical reflectivity spectra showed that agreement is good if the 4f electron is treated as itinerant (Fig. 2.9) [87]. These results show that the on-site Coulomb repulsion is strongly screened in this compound resulting in unusual situations of the 4f electrons in CeN under SPT conditions.

In CeN at $T > 1200$ K, cerium becomes trivalent. It is also trivalent in CeP [88], CeAs and CeSb [89] under SPT conditions. These compounds are semimetallic. The 4f electron is localized and the 4f levels are located a few eV below E_F. From

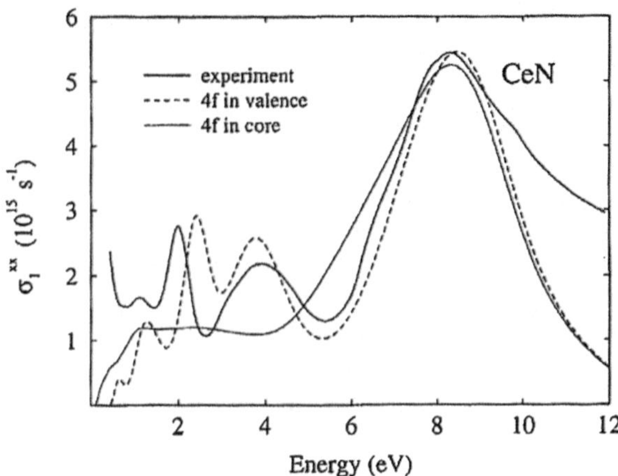

Fig. 2.9 Comparison between calculated real part of the optical conductivity and experimental data for CeN. The calculations are made assuming the 4f electrons as being core electrons or as forming a band [87]

LSDA + U calculation, in CeSb, the Ce 4f levels were found 2 eV below the Fermi level and energetically mixed with the Sb p bands. Anomalous volume contraction was observed in CeP at a pressure of 100 kbar [88]. Valence modification was seen in CeAs and CeSb at low temperatures [89]. More recently, extensive theoretical studies on high-pressure behaviour of cerium pnictides and their electronic structure have been carried out by using the SIC-LSD approximation [90]. Structural phase transition from NaCl to CsCl was shown to occur in parallel with a localized to delocalized transition of the 4f electrons. Consequently, a proper description of the 4f electrons is difficult in the cerium compounds because the on-site electron repulsion, or Hubbard U, is high compared to the bandwidth and these systems are strongly correlated [91, 92]. Nevertheless, the calculated lattice constants are in agreement with the experimental data.

In summary, from calculation methods such as LSDA + U, the positions of the occupied and unoccupied 4f levels have been found several eV below and above the Fermi energy, respectively, in agreement with experiments. The origin of the semimetallic properties in the rare-earth monopnictides was explained by the close proximity of the unoccupied rare-earth 5d states to a point of high symmetry of the Brillouin zone. On the other hand, the metallicity of the rare-earth pnictides was found to increase gradually from N to Bi. Thus, it was showed experimentally from high-quality stoichiometric samples that GdN was ferromagnetic semiconductor and GdP, GdAs and GdSb antiferromagnetic semimetals [93, 94]. However, metal-semiconductor transition observed in some rare-earth pnictide compounds or alloys might be due to impurity doping, temperature or pressure effects, thus explaining the eventual disagreement between predictions and experimental results. Moreover, the uncertainty remained on the electronic structure of GdN because of the spread of the obtained results.

Particular attention must then be given to the rare-earth nitrides, which have been studied specially because of their interesting magnetic properties. Most of them are ferromagnetic with Curie temperature extremely low, of a few degrees

Kelvin. Indeed, the electron exchange splittings of these compounds are significantly larger than those of arsenides and phosphides. This can be explained by a difference in the character of bonds. The latter have a weaker metallic character in the nitrides and the overlap between the orbitals is smaller, predicting these compounds generally to be semiconductors. GdN has been the subject of an intense controversy. Its strong ferromagnetism was explained theoretically [95]. But from resistivity measurements, it was determined to be either semimetallic [96] or insulator [97]. Conductivity was also observed in the presence of nitrogen vacancies. The same equivocal results were obtained theoretically; however, these predictions concerned the non-magnetic case. The smallness of the gap suggested that it would disappear in the magnetic phase. This could be expected for GdN because the exactly half-filled 4f shell makes this material particularly sensitive to magnetic exchange interactions. Spin-polarized ab initio calculations using the LMTO method within LSDA and treating the 4f electrons as localized core-like electrons with various spin orientations were performed [79]. From this model, GdN was found to be semiconductor in the paramagnetic phase and in the minority spin channel of the ferromagnetic phase whereas it was metallic in majority spin channel of the ferromagnetic phase. It appeared thus as being a half-metal. In contrast, with the same theoretical treatment, ErN was found to be a semiconductor in both phases. However, in another theoretical description, GdN was predicted to be a narrow indirect band gap insulator [98], in agreement with the experimental results. The electronic structure of this nitride still remains inscrutable.

Systematic calculations of the thirteen rare-earth nitrides from CeN to YbN were then performed by using the SIC-LSDA methodology [99–101]. From these calculations the rare-earths are trivalent in the mononitride ground state, except for cerium which is tetravalent. These rare-earth nitrides display a wide range of electronic properties, despite their equal structure and similar lattice constants. In the ground state calculations, TbN, DyN and HoN were found to be narrow gap insulators and CeN, ErN, TmN and YbN metallic in both spin channels, while PrN, NdN, SmN, EuN and GdN were found to be half-metallic ferromagnets. From more recent calculations by LSDA + U method [75, 102, 103], GdN has also been predicted to be half-metallic with a gap about 0.5–0.6 eV at the ground state. The occupied 4f levels are expected about 6 eV below the Fermi level and the unoccupied 4f levels about 5 eV above, in agreement with the experimental results. The change of the magnetic order from ferromagnetic in GdN to antiferromagnetic in the other pnictides is a result of the increased ionic radius of the ligand. As the exchange parameters depend strongly on the lattice constant, large changes of the electronic and magnetic properties is expected upon applying stress or by doping impurities. Thus, changes in the conductivity are associated with volume increase and GdN is expected to undergo a phase transition from half-metallic to a semiconductor with lattice expansion. Reduction of the exchange parameters with increasing volume also suggested a decrease of the Curie temperature of GdN with an increase in lattice constant. According to the quasiparticle self-consistent GW method, GdN has also been predicted to be very near a critical point of a metal–insulator transition [104].

Other calculations of electronic structure for all the rare-earth nitrides have been made from the LSDA + U approach in the full-potential LMTO model [105]. Magnetic moments were employed as a validity test. They are given in Table 2.6 [105]. The 4f spin magnetic moments are almost an integral number depending on the number of 4f electrons. The 4f orbital magnetic moments are comparable to the spin moments. Orbital and spin moments are opposite in the first half of the series and parallel in the second half, thus obeying Hund's third rule. The magnetic moments of the d and valence electrons compensate approximately. A quasi-zero total magnetic moment is obtained for SmN. This explains why SmN is not ferromagnetic; it was suggested that it is anti-ferromagnetic at the ground state. Concerning the electronic structure, CeN and EuN are metallic. The other light members of the series, with the exception of NdN, are found to be semimetallic in the majority spin channel only; they are thus half-metals. The heavier members of the series from gadolinium, but also NdN, show semiconducting behaviour. It was suggested that the gap had a tendency to increase through the series. But these suggestions depend on the choice of U_f. The value of U_f used for each rare-earth is equal to the radial Slater-Coulomb integral F^0, obtained from Hartree–Fock calculations. Fitting of the calculated values was made only for GdN using accurate photoemission data obtained for this compound. As the values of U_f increase going from lanthanum to ytterbium, the 4f orbitals are pushed away from the Fermi level, both below and above. The energy interval between occupied and empty 4f levels increases along the series. Except for the cerium compounds, the 4f electrons are

Table 2.6 Experimental and calculated lattice constants a compared

	a (Å)		$m_f^{\sigma}(\mu_B)$	$m_f^{L}(\mu_B)$	$m_d(\mu_B)$	$m_N(\mu_B)$
	Expt.	Theory				
LaN	5.30	5.38	0	0	0	0
CeN	4.87	4.90	0	0	0	0
PrN	5.17	5.29	−1.96	−4.87	0.049	−0.077
NdN	5.15	5.24	2.96	−5.88	0.062	−0.084
PmN[a]		5.19	3.96	−5.88	0.089	−0.140
SmN	5.04	5.10	4.91	−4.85	0.105	−0.136
EuN	5.00	5.14	5.96	−1.61	0.087	−0.129
GdN	4.98	5.08	6.93	0	0.081	−0.083
TbN	4.92	5.05	5.93	2.96	0.066	−0.083
DyN	4.90	5.03	4.95	4.96	0.064	−0.065
HoN	4.87	4.98	3.95	5.94	0.040	−0.026
ErN	4.83	5.00	2.97	5.93	0.032	−0.027
TniN	4.80	4.90	1.98	4.88	0.020	−0.013
YbN	4.78	4.79	0.99	1.54	0.006	0.046
LuN	4.76	4.87	0	0	0	0

4f spin and orbital magnetic moments, 5d spin magnetic moments and N 2p magnetic moments, all in μ_B (from [105])

localized and not hybridized with the rare-earth 5d–6s and pnictogen p bands, contrary to suggestions made in some of the previously cited references.

Experimentally, stoichiometric GdN was found to be semiconductor with an optical gap of about 1.5–2 eV [106] and ferromagnetic with a low Curie temperature. In the presence of nitrogen vacancies, the importance of the charge carrier dynamics for both the transport and magnetism properties of GdN was underlined and the formation of magnetic polarons centred on the vacancies was suggested [107]. Because GdN is on the border between semiconductor and semimetal states, its electronic and transport properties are very sensitive to non-stoichiometry, impurities, defects, external pressure, temperature, magnetic field, etc. High quality of the samples is necessary to obtain semimetals and its absence could explain the difficulty in obtaining the experimental characteristics of GdN. All the same, variance between the various theoretical data could be explained by an ambiguity in the characteristics of this compound, making difficult their exact determination. However, these results also show that theoretical models can be convenient to obtain some properties, as the spatial arrangements, but are incapable of describing the real electronic distributions.

From a technological point of view, the rare-earth monopnictides are materials of great interest due to their remarkable structural, electronic and magnetic properties. Unusual combinations of a semimetallic or semiconducting character associated with the magnetism induced by the 4f localized moments present numerous advantages. Semimetallic ferromagnetic materials at ground state have large potential in electronics. The most promising materials would be the half-metal ferromagnets, which are possible candidate for spin-dependent transport devices [108]. Another interesting materials would be half-metallic antiferromagnets [109] because they would be completely spin-polarized at the Fermi surface but with zero magnetization, that could exist only in mixed compounds. The presence of very small energy gaps in these materials induces large values of the dielectric constant. They are of the order of 22–26 for the gadolinium and erbium pnictides (Table 2.7).

The nitrides have particularly attracted the attention because some of them were predicted to be half-metallic ferromagnets lying on the boundary between metals and insulators. However, experimentally, the electronic structure of the nitrides is not always well determined, partially because it is difficult to obtain pure single crystals.

Table 2.7 Dielectric constants ε of gadolinium and erbium pnictides. Bands gaps E_g (eV) calculated from LSDA, LDA and with self-energy corrections included (from [79])

Compound	ε	$E_g^{P,LSDA}$	$E_g^{P,LDA}$	$E_g^{P,est}$
GdN	26.5	−0.49	−0.18	0.10
GdP	23.9	−1.50	−1.04	−0.77
GdAs	22.8	−1.66	−1.19	−0.91
ErN	22.9	−0.26	−0.15	0.18
ErP	22.9	−1.61	−1.53	−1.24
ErAs	22.1	−1.65	−1.63	−1.41

Among the interesting pnictides for potential spin electronic applications, GdN, GdAs, ErAs, YbN, should be mentioned. They have a semiconductor-like character, the same as conventional semiconductors like AlAs and GaAs, associated with exceptional electronic and magnetic properties. Thus, they can be used as a dopant without losing their magnetic behaviour. They can be grown on III–V semiconductors, making these materials interesting for the development of electronic devices [110]. The first experiments concerned the single crystal ErAs films deposited on GaAs [111]. Epitaxial growth of ErAs on GaAs and AlAs has been developed. ErAs is a semimetal with electron and hole concentrations of about 3×10^{20} cm^{-3} and a large exchange splitting induced by the open 4f shell. GaAs/ErAs/GaAs hetero-structures have been obtained from ultra thin layers of ErAs. The electrons can tunnel through hole states of the semimetal ErAs quantum well and these structures form resonant tunnelling diodes whose response depends on the magnetic field orientation [112]. Hetero-structures consisting of layers of ErAs islands separated by GaAs were also found to be ultra-fast photoconductors [113]. Carrier trapping occurs on a sub-picosecond time scale. As the structure period controls the carrier trapping time, it is possible to obtain a desired carrier trapping time and thus a photoconductor with a response time compatible with time-resolved differential reflectance measurement.

Spintronics, or spin transport electronics, has opened up a new class of applications, where the information is not carried by the charge of the electron but by its spin. New devices combining standard microelectronics with spin-dependent effects are being developed. Initially, this technique was limited by the lack of polarized current sources for spin injection into semiconductors. The use of half-metallic ferromagnets, i.e. fully spin-polarized ferromagnets, can circumvent the problem. But such materials are difficult to obtain. However, polarization near 100 % can be created by using the phenomenon of *spin filtering* [114]. For a magnetic semiconductor, ferromagnetic below the Curie temperature due to the spin splitting of the band, the height of the tunnel barrier is spin-dependent. As the tunnel current depends exponentially on the barrier height, one of the spin channels has a larger tunnelling probability than the other, resulting in a nearly 100 % spin-polarized current [115]. Non-magnetic metal/ferromagnetic semiconductor/ferromagnetic metal devices have been obtained. The ferromagnetic semiconductor layer acts as spin filter. From this type of hetero-structures, resulting from the combination of a spin filter tunnel barrier and a ferromagnetic electrode, large magneto-resistance effects are created. Numerous hybrid devices utilizing the typical properties of rare-earth monopnictides are to be expected in future applications.

2.1.4   Intermetallics

During the last fifteen years, numerous studies of intermetallic compounds with cerium [116] and ytterbium [117] have been made and have revealed an unusual and interesting behaviour of these materials and a variety of their crystal structures.

Among these compounds, the best known are the binary R_mT_n and ternary $R_mT_nX_p$ ones with R = rare-earth, T = transition element and X = fourth or fifth group elements. Most of these materials are metallic. However, at low temperatures, they show remarkable deviations from the Landau-Fermi liquid theory of metals.

In rare-earth intermetallic compounds, the localized 4f orbitals generally form a magnetically ordered subsystems interacting with the valence electron distribution. The dominant interaction is the 4f-valence electron interaction RKKY. This long range magnetic interaction can lead to complex magnetic structures. Indeed, while the RKKY interaction is responsible for magnetic ordering, this ordering is nearly always incommensurate with the crystal lattice. Another interaction to be considered is the Kondo interaction (cf. Chap. 1). Let us recall briefly that the Kondo effect refers to the changes undergone, at very low temperatures, by the valence electrons of a metallic compound when they are in strong interaction with magnetic impurity ions. The Kondo effect is generally treated as the exchange interaction between a local spin and the valence orbitals present in every cell of the crystal; this is the *Kondo singlet interaction* [118]. In the case of the rare-earths, the Kondo exchange interaction results from the coupling between the spin associated with the localized 4f sub shell and the spin density associated with the 6s5d valence electrons. That is equivalent to treating the 4f and 6s5d electrons as strongly interacting. In the presence of Kondo interaction, i.e. at strong coupling, the valence electrons are considered as *heavy fermion*. Within the framework of Landau-Fermi liquid theory, the heavy fermion systems are described as formed by quasi-particles with an itinerant character. They have the charge and quantum numbers of the non-interacting electrons but their effective mass m* exceeds the free electron mass by a factor that can attain a thousand at low temperatures. As a consequence of the high effective mass of these particles, the Fermi energy is reduced to the order of the magnetic energy and the Fermi temperature becomes of the order of the Néel temperature.

The relative importance of the RKKY and Kondo interactions varies with parameters such as pressure, chemical composition or magnetic field. When the Kondo interaction prevails, the system can be understood in terms of a Fermi liquid of heavy quasi-particles, with narrow band-like states located in the region near the Fermi level. The ions carry magnetic moments but are not ordered magnetically. Let us recall that in the Landau's Fermi-liquid theory, the excitations are analogous to those of a non-interacting Fermi gas and the system is in the paramagnetic phase. The specific heat C, which is the sum of electron and phonon contributions, varies with the temperature according the relation $C = \gamma T + aT^3$, where γT is the term due to electrons and aT^3 that due to phonons. The Sommerfeld coefficient γ is independent of the temperature T and proportional to the effective electron mass m*. Since the effective mass becomes bigger than the real mass of the electrons at low temperatures, the first term becomes dominant. Then the measurement of the specific heat gives an estimate of m*. The Pauli susceptibility χ is also independent of T. The electrical resistivity contribution, due to particle–particle collisions, is $\Delta\rho = AT^2$ with A independent of T. The three terms, specific heat, Pauli susceptibility, electrical resistivity, are connected by the relation $\gamma \propto \chi \propto \sqrt{A}$. Moreover, from the Wiedemann–Franz law, applicable to all systems characterized by particles of spin 1/2 and of charge e, the thermal conductivity/electrical conductivity ratio, or Lorentz ratio, is a universal constant, $\pi^2/3\,(k_B/e)^2$ (equal to 2.44×10^{-8} WΩ K^{-2}) in the $T \rightarrow 0$ limit. Violation of the Wiedemann-Franz law implies a massive breakdown of the Fermi liquid model.

Numerous rare-earth intermetallic compounds are considered as being heavy fermion materials, i.e. materials whose valence electrons are slowed down dramatically by scattering with the magnetic ions below the Kondo temperature, i.e. at temperatures much inferior to T_{Debye} and T_{Fermi}. An important question is to what extent these compounds can be described as Fermi-liquids. Indeed, at very low temperature, breakdown of the Fermi-liquid model can occur and non-Fermi-liquid behaviour, or *heavy fermion behaviour*, is found in a large variety of materials including metallic, superconducting, insulating and magnetic ones. This heavy fermion behaviour is manifested by a large deviation of the thermodynamic and transport properties from those predicted for a Fermi-liquid. The transport properties are very indicative in this type of studies because they can be measured under extreme conditions, very high pressure, high magnetic field and very low temperature. Observations were made initially for the cerium alloy $CeCu_{6-x}Au_x$ at temperature $T \rightarrow 0$ K [119]. In the calculations for this alloy, the electronic term of the specific heat can be up to 1000 times bigger than the value expected from the free electron theory. The Sommerfeld coefficient γ acquires an unusual temperature dependence with $C/T = \gamma(T) = -\ln(T/T_0)$. The mass of the electrons present at the Fermi surface tends to become infinite and their energies to vanish with the lowering of temperature. The susceptibility χ depends on the inverse of the temperature and the temperature dependent part of the electrical resistivity $\Delta\rho$ varies as T^b with b close to unity. The anomalous logarithmic temperature dependence of the specific heat and of the thermo-power, observed at temperature close to zero, is explained by a gradual loss of the magnetic moments. The associated drop of entropy is compensated by an unusual increase of the entropy carried by "hot" electrons at the Fermi surface of the heavy fermion compound [120]. In the alloy $CeCu_{6-x}Au_x$, at least two among the fundamental properties characteristic of the normal metals, specific heat and electrical resistance, undergo dramatic changes with the temperature, showing a very different characteristics of this metallic compound with respect to the normal metals. This is found in materials typically containing cerium and ytterbium and also uranium and neptunium.

At very low temperature, cerium and ytterbium intermetallic compounds exhibit properties unaccounted for by the present physical models. The change of fundamental properties suggests that the characteristics of the electrons are strongly modified in these compounds. Physical properties, like conductivity, magnetism, arise from the collective motion of electrons and ions in the solid. In order to eliminate discrepancies between theory and experiment, a quantum model had been developed to explain the intricate complexities of electron and ion motions at very low temperature [121]. This model led to suggest the existence of a new kind of phase transition, called *quantum phase transition*. It takes place between an ordered and a disordered phase around zero temperature. There are continuous second-order transitions driven not by the thermal motions, i.e. the phonons, but by quantum fluctuations associated with Heisenberg's uncertainty principle. Indeed, electrons and ions cannot be at rest at the absolute zero temperature because their position and their velocity would, then, be simultaneously determined. Each quantum phase transition is dependent on an *order parameter*. This parameter characterizes the symmetry breaking of the ordered phase and it vanishes in the disordered phase. Near-zero-point fluctuations are present and susceptible to modify the system. These fluctuations depend on the non-thermal parameters. When such quantum fluctuations are sufficiently strong, quantum phase transition is expected to occur in the vicinity of a *quantum critical point* located between the two stable phases [122]. The *quantum critical phase* at this point, distinct from the two previous phases, is characteristic of a strongly correlated electron system, i.e. heavy fermion system. A quantum critical point is thus present whenever a material undergoes a second-order quantum phase transition, i.e. phase transition where the system changes in a continuous fashion at zero temperature. New interest in this field has come because of accurate studies of the quantum critical points and the unusual effects observed in their vicinity. The quantum phase transitions have been considered as responsible for particular properties such as the zero resistivity in the superconducting metals and the zero viscosity in the superfluids.

According to this model, the abnormal characteristics of the rare-earth intermetallic compounds, observed at very low temperature, were interpreted as due to a continuous phase transition in the limit of absolute zero, taking place between a long range magnetically ordered phase, often antiferromagnetic, and a paramagnetic phase [122, 123]. A few examples of ferromagnetic cerium compounds or alloys are also known to exist, for example $CeSi_{1.81}$ under pressure [124]. Here, the order parameter characterizes the magnetic phase and is zero for the paramagnetic phase. The ordered magnetic phase is described via magnetic RKKY interactions. The disordered phase is a paramagnetic Fermi liquid. The quantum critical point separates the magnetically ordered phase, antiferromagnetic, from the paramagnetic Fermi-liquid regime (Fig. 2.10).

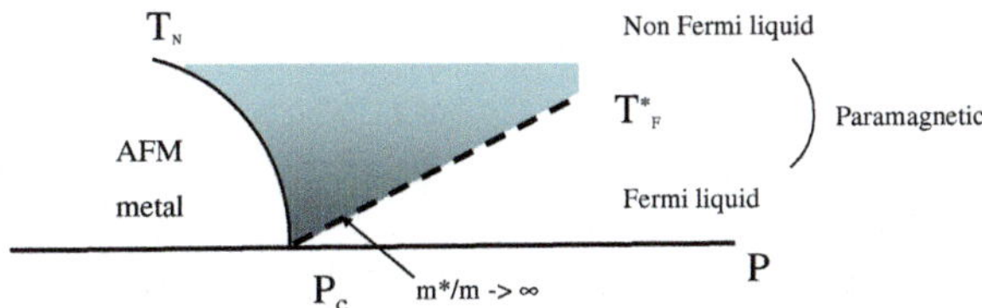

Fig. 2.10 A schematic representation of quantum critical point in heavy fermion metals: the pressure drives the system from the antiferromagnetic into the paramagnetic phase. But the reversed situation is possible. P_c is the critical point. If the Fermi temperature goes to zero at the quantum critical point, a new class of excitation processes is predicted above the Fermi temperature

These two regimes occur at very low temperatures, below the Néel and Fermi temperatures, respectively, and both are expected to disappear at the quantum critical point. A second-order quantum phase transition takes place in the vicinity of most quantum critical points. A new phase with non-Fermi-liquid characteristics is then present. In this quantum critical phase, also named *Kondo phase*, the 4f and valence electrons are considered as forming composite quasi-particles of very large masses. A phase diagram was calculated in a mean-field model as a function of the J_K/W ratio where J_K is the Kondo coupling between the localized and valence electrons and W the band width of the valence electrons (Fig. 2.11) [125]. On the other hand, phase diagrams have been obtained experimentally as a function of non-thermal parameters, like pressure, chemical composition or magnetic field. Indeed, it is possible to obtain a quantum critical state by reducing the transition temperature to zero using one of these non-thermal parameters.

As already underlines, the existence of a Kondo phase implies the presence of a critical point at $T \approx 0$ and is obtained whenever $T \rightarrow 0$ regardless of how this is achieved [119]. Secondly, a quantum critical transition depends on the presence of quantum critical fluctuations of the order parameter, i.e. in this case, fluctuations of the antiferromagnetic moment. Consequently, the quantum transition, i.e. the emergence of the Kondo phase, is associated with magnetic fluctuations, i.e. with a large number of low-energy magnetic excitations. A description of this type of phase transition was given based on the physics of electron fluids. In this non-local model, an itinerant antiferromagnet is developed and described in terms of a spontaneous spatial modulation of the spins associated with heavy quasi-particles of the paramagnetic phase, i.e. in terms of a *spin density wave* [126, 127]. The conditions for an ordered antiferromagnet to transit to such a heavy fermion system are given by the coordinates of the quantum critical points in the phase diagram. The

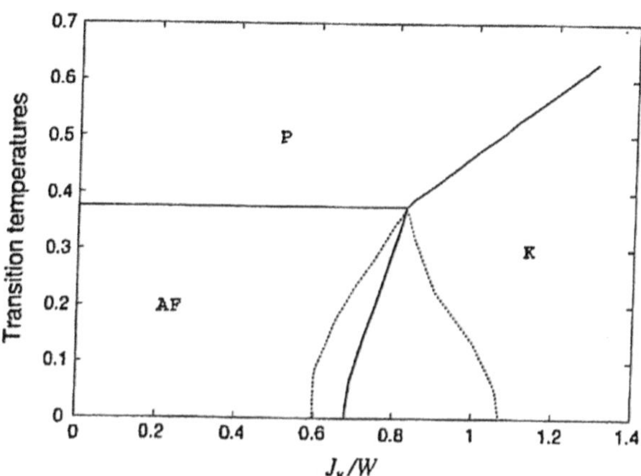

Fig. 2.11 Phase diagram showing the antiferromagnetic (AF), Kondo (K) and paramagnetic (P) regions. The first-order transition line between the AF and K phases is within a region limited by the two dashed lines, where metastable Kondo and magnetic solutions can coexist [125]

energy scale of these processes is much lower than the energy associated with the Kondo temperature.

In the spin density wave theory of the quantum phase transitions [128–131], the non-Fermi liquid properties observed in the vicinity of the magnetic–non-magnetic transition is therefore controlled by the fluctuations of the antiferromagnetic moment. The intensity of these fluctuations diverges at the quantum critical point and the Kondo temperature goes to zero. The quantum critical point between the localized antiferromagnetic and Kondo phases was observed in some cases along with a small *domain* in which the system is superconductor. The phase transition, induced by a parameter as pressure or composition, may, in principle, occur at $T = 0$. It is driven by quantum fluctuations instead of thermal fluctuations, present in classical phase transitions. Spin fluctuations in various Ce, Yb and U intermetallic compounds have been observed in many experiments, among them inelastic neutron scattering.

However, the standard spin density wave theory did not explain the entire experimental data and other models have been proposed. In the standard model, the Kondo phase partially overlaps the antiferromagnetic ordered phase. When the Kondo breakdown happens just at the onset of the antiferromagnetic phase, separation between the two phases occurs abruptly at the quantum critical point. The Kondo state exists only in the paramagnetic phase and, consequently, the presence of a new class of quantum critical point is expected. The order parameter is characteristic of the antiferromagnetic phase. The fluctuations of this antiferromagnetic order parameter can be considered as slow, in contrast with the dynamical nature present in the Kondo model. Various mechanisms, named Kondo breakdown models, have then been introduced in order to account for the presence of such critical quantum phase transitions. One of them, the spin-liquid formation among local moments, was proposed but not substantiated. Local magnetic fluctuations were then considered as explaining the entire experimental data [132–136].

Independently, the observation of a superconducting phase, clearly separated from the antiferromagnetic phase, in a cerium compound under high pressure [137], has lead to a search for an alternative model using another type of non-magnetic quantum fluctuations (Fig. 2.12).

Fig. 2.12 A schematic representation of temperature versus lattice density phase diagram for cerium intermetallics of the $CeCu_2Si_2$-type

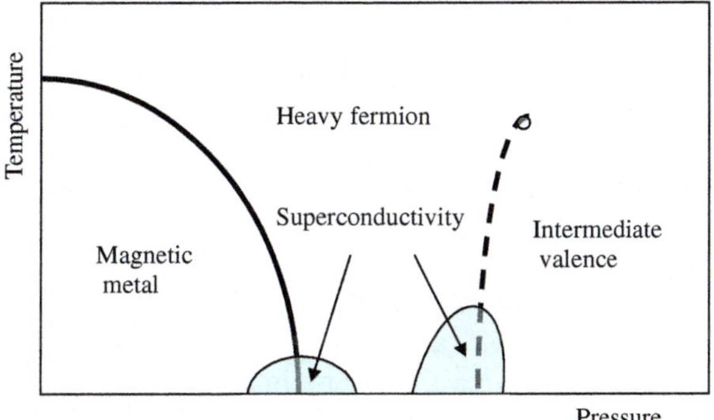

While in the Kondo model the heavy particles are composite objects arising from the 4f-valence electron strong interaction, in the presence of a Kondo breakdown these heavy particles disintegrate and this can be accompanied by charge density fluctuations [138]. Quantum transition to a superconducting state can then be driven by continuous localization–delocalization of electrons. Critical local fluctuations of valence have consequently been proposed to explain the presence of superconductivity at high pressure [139]. The interdependence of magnetic and electronic fluctuations will be discussed in the next paragraph.

Let us mention that a sudden variation of the magnetization in the absence of externally applied magnetic field is designated as *metamagnetic transition* [140]. Metamagnetism has been associated with various processes, first-order phase transitions, continuous phase transitions at a critical classical or quantum point, itinerant magnetism as well as antiferromagnetism. Here, it is generally associated with the change in the 4f electron character from localized to heavy fermion or itinerant, which accompanies a variation of temperature or pressure.

2.1.4.1 Cerium Intermetallic Compounds

In the cerium compounds, at low temperature, effective mass m* about two orders of magnitude heavier than the free electron mass was found experimentally. The first studied compound was $CeAl_3$ [141]. Later on $CeCu_2Si_2$ [142] and $CeCu_6$ [119] retained the attention, along with various uranium compounds. The Ce–Ni and Ce–Pt systems were also widely studied because the low symmetry of these noncubic compounds made possible the observation of anisotropic effects associated to the presence of the cerium 4f orbitals [143].

At standard and high temperatures, the cerium 4f electron is localized on each atomic site and does not contribute to the conductivity. At very low temperatures, the 4f electron can lose its localized character. This change induces an uncommon behaviour of heavy fermion type since a small variation of the number of 4f holes can have huge effects on the properties. More generally, compounds of the form CeT_2X_2 with T = Ni, Cu, Rh, Pd, Au, X = Si or Ge, were found to be of the heavy fermion type with a large variation of magnetic orderings and superconductivity. Among these compounds, $CeCu_2Si_2$ appeared as the prototype of heavy fermion superconductor. It was the first unconventional superconductor to be discovered [144, 145]. The superconducting state was induced by decreasing of the temperature down to 0.5 K or by the presence of an external magnetic field at a very low temperature. The transition temperature, T_c, is the temperature at which the electrical resistance of a superconducting material drops to zero. It should be remarked that the energy associated with a temperature of 1 K is less than 1 meV. With the increase of the external magnetic field, the superconducting state disappears and an antiferromagnetic state takes place (Fig. 2.13) [146]. From the value of the specific heat (1.1 J/mol K^2) at low temperature, $CeCu_{2.02}Si_2$ forms a heavy fermion system whose effective mass is approximately $100m_0$. Similar results have been obtained for stoichiometric samples with a high degree of lattice perfection [147]. In contrast

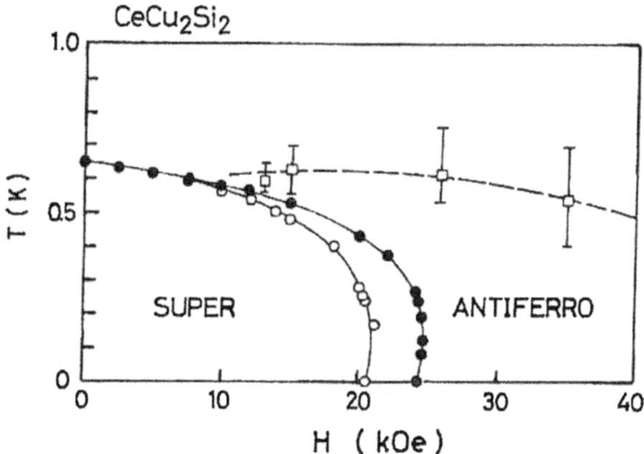

Fig. 2.13 Temperature versus magnetic field phase diagram for CeCu$_{2.02}$Si$_2$ [146]

with the conventional Bardeen–Cooper–Schrieffer (BCS) theory according to which the superconductivity is due to electron-phonon interactions, the superconductivity of this metallic system is due to electron–electron interactions [144].

Pressure is the parameter used for moving the antiferromagnetic ordered lattice Kondo compounds throughout their magnetic instability. Indeed, external pressure is a suitable parameter for investigating the physics of strongly correlated electron systems. As for pure cerium metal, the long distance magnetic interactions are modified by the application of pressure and one expects the localization of the cerium 4f electron to decrease with increasing pressure. The 4f energy level goes down and the compounds follow the sequence: magnetically ordered trivalent, non-magnetic trivalent or Kondo system, mixed valence system and eventually tetravalent. In the Kondo system, the 4f electron remains strongly correlated at very low temperature and has heavy fermion character, thus offering the possibility to induce a zero-temperature magnetic–non-magnetic quantum transition. The presence of such a transition is substantiated by the increase of the Sommerfeld coefficient when the temperature tends to zero. In the temperature-pressure phase diagrams, a narrow superconducting domain can be observed near the quantum critical point where magnetic order is suppressed and Kondo interaction is present.

In high-purity samples of CeNi$_2$Ge$_2$ with well-ordered atoms, superconductivity was observed at standard pressure and $T \approx 0.9$ K and this made this compound specially attractive. The observations were well described in the spin density wave model [148]. In contrast, at standard pressure, CePd$_2$Si$_2$ is antiferromagnetic below a Néel temperature of about 10 K [149, 150]. With increasing pressure, the Néel temperature decreases. The antiferromagnetic order is suppressed progressively and the system turns into a paramagnetic metal. The quantum critical point has not explicitly been observed. However, a small domain in which the system is a superconductor exists for a critical high pressure around 2.8 GPa and a very low temperature (Fig. 2.14) [136, 151].

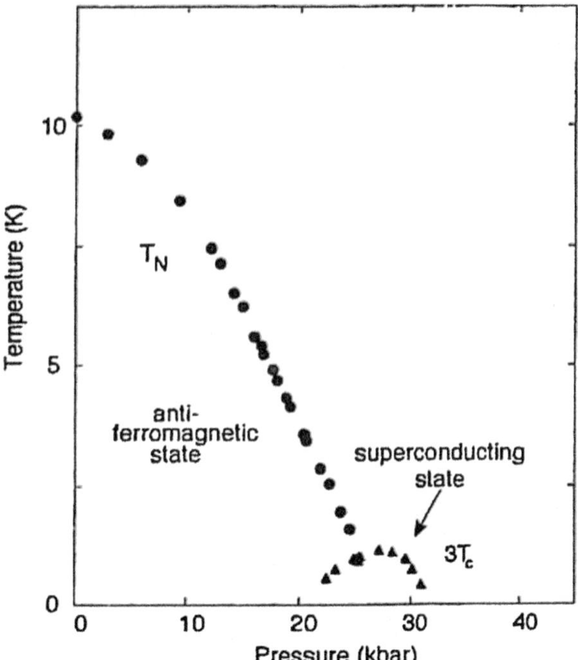

Fig. 2.14 Temperature versus pressure phase diagram of high-purity single crystal $CePd_2Si_2$ [151]

Contrary to previous examples, for some cerium compounds, either a relatively extended supraconducting domain or two clearly separate superconducting domains were observed, one near the magnetic phase and centred around the quantum critical point, the other widely separated from this point and observable at higher pressure [137]. Such two domains were observed for $CeCu_2Si_2$. The first domain was interpreted using the three-dimensional spin density wave model [152]. This model was supported by measurements made at 0.06 K [153]. The second superconducting domain appeared at high-pressure and was located far from the threshold of the magnetic phase. Complementary study was therefore necessary to explain it.

From measurements of specific heat and resistivity at low temperature and high pressure, the characteristic temperature of the appearance of superconductivity, named superconducting critical temperature, T_c, was determined for $CeCu_2Si_2$ [139]. This critical temperature varies with the pressure. At standard pressure, T_c is around 0.7 K. When pressure is applied, T_c increases suddenly at about 3 GPa, reaches a maximum value of 2 K at about 4.5 GPa then goes down to zero. This variation corresponds to the presence of the second domain of supraconductivity at high pressure. Moreover, resistivity measurements have shown a drastic decrease of the ratio m*/m, i.e. of the effective mass. This drop is characteristic of a decreasing of the localized character of the 4f electron and is associated with a change of the cerium valence. Considered as trivalent with the $4f^1$ configuration in the antiferromagnetic phase, then characterized as heavy fermion in the Kondo phase, cerium becomes a mixed valence element at very high pressures. Both the heavy fermion and mixed valence states are nearly degenerate in the middle pressure region. At very high pressure, the cerium valence is considered as fluctuating between three and four. The 4f electron is then either localized on the ion of the $4f^1$ configuration or in orbitals near the Fermi level. The increasing of T_c can be related to the dual character of the 4f electron. This dual character induces the presence of critical valence fluctuations responsible of the quantum transition, leading to the super-conductor domain observed at high pressure [139]. Consequently, two different processes are considered as responsible for the quantum phase transitions in $CeCu_2Si_2$: magnetic fluctuations, present in the low-pressure domain near the antiferromagnetic quantum critical point and charge fluctuations, responsible for the well separated high-pressure domain. However, not all the properties of $CeCu_2Si_2$ are fully understood and studies are going on. Very high-purity samples must be used because the results depend strongly on the stoichiometry.

For similar high-purity compounds such as $CeAu_2Si_2$ [154] or $CeCu_2Ge_2$ (Fig. 2.15) [138, 155], the pressure dependence of the various properties is almost quantitatively identical to that of $CeCu_2Si_2$ when pressure shifts are taken into account. The comparison is less straightforward with compounds in which Cu is replaced by a transition metal. In contrast, important variations of T_c take place according to the perfection of the sample. Indeed, large variations in the electronic properties result from extremely small differences in composition. This shows the important role of the chemical and structural conditions. Thus, the presence of germanium in $CeCu_2Si_2$ creates disorder and consequently decreases the super-conductivity. However, for $CeCu_2Si_{2-x}Ge_x$, partial substitution of silicon by germanium is equivalent to a slight negative pressure effect and the same change is observed but at higher pressure. Thus, as for $CeCu_2Si_2$, two separate supercon-ductivity domains are present (Fig. 2.16) [139]. The first is near the magnetic phase

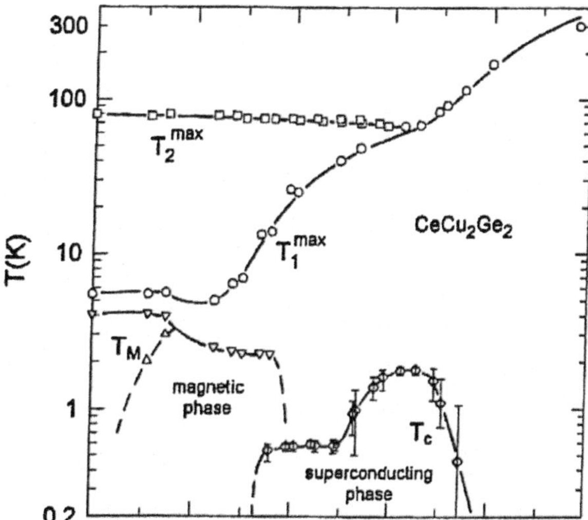

Fig. 2.15 Temperature versus pressure phase diagram for $CeCu_2Ge_2$ [138]

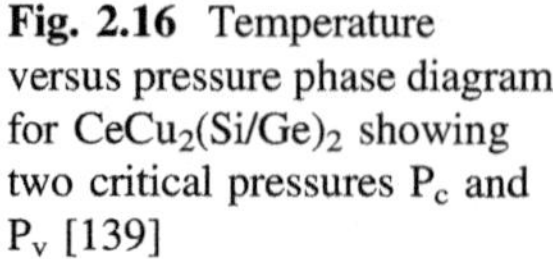

Fig. 2.16 Temperature versus pressure phase diagram for CeCu$_2$(Si/Ge)$_2$ showing two critical pressures P$_c$ and P$_v$ [139]

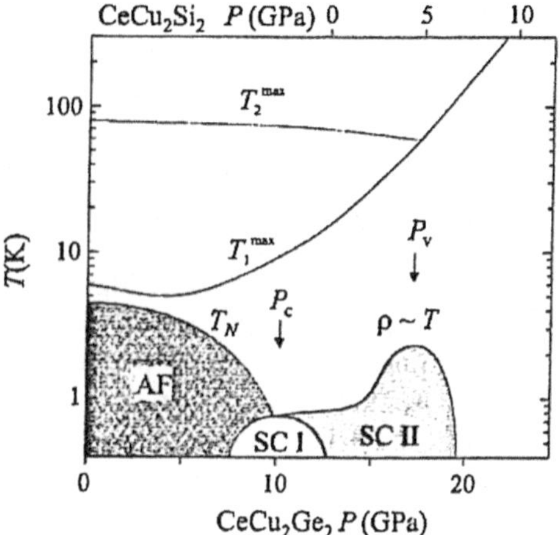

and is interpreted by spin fluctuations, the second is explained by an electronic configuration change of the same type as that described above, taking place at higher pressure.

Among the materials described by spin dynamics, it is possible to cite the heavy fermion system Ce$_{1-x}$La$_x$Ru$_2$Si$_2$. Pure CeRu$_2$Si$_2$ is paramagnetic in the ground state. Lanthanum doping can be considered as equivalent to applying a negative pressure, thus as increasing the localization of the 4f cerium electron. The initially non-magnetic heavy fermion compound becomes antiferromagnetic. Investigation around the quantum critical point was performed using inelastic neutron scattering [156]. This was the first analysis made on a single crystal where pressure is the control parameter of the magnetic–non-magnetic phase diagram. The compound Ce$_{0.925}$La$_{0.075}$Ru$_2$Si$_2$ is characterized by a strong increase of the Sommerfeld coefficient, indicating the presence of a quantum critical regime. Indeed, the x-T phase diagram presents a paramagnetic regime for $x < x_c = 0.075$, accompanied by a Fermi liquid at low temperatures, and an RKKY antiferromagnetic phase for $x > x_c$. The quantum phase transition at x_c and $T \rightarrow 0$ K is characterized by an enhancement of the antiferromagnetic fluctuations [157]. All parts of the phase diagram were analyzed. From the observation of an enhancement at the critical point (x_c, $T \rightarrow 0$) it was shown that the antiferromagnetic fluctuations are responsible for the quantum transition between the paramagnetic and antiferromagnetic phases at x_c. It was, moreover, suggested that this could be a first-order quantum transition. The validity of the conventional spin fluctuation model appears to be confirmed by this compound.

Study of the quantum critical point by single-crystalline neutron scattering has been extended to the Kondo system CeRu$_2$Si$_2$ doped with rhodium, CeRu$_{2-x}$Rh$_x$Si$_2$ [158]. The lattice mismatch between CeRu$_2$Si$_2$ and CeRh$_2$Si$_2$ induces an important change of the magnetic field-temperature phase diagram at low temperatures. The system changes from an antiferromagnetic phase at field zero to a polarized paramagnetic phase at very high magnetic fields. Transitions among antiferromagnetic, paramagnetic and polarized paramagnetic phases depend on temperature, magnetic field and pressure. It was shown, as in the previous study, that the quantum critical behaviour of $CeRu_{2-x}Rh_xSi_2$ is controlled by the spin density wave model [159].

Some compounds such as $CeCu_{6-x}Au_x$, seem to have more complex properties than the previously described ones. For $x = 0$, Fermi-liquid phase is observed. $CeCu_6$ is not magnetically ordered [160] and shows a strong Pauli paramagnetism. No localized 4f electrons are present. The large radius of the gold atom causes a lattice dilatation and the exchange interaction between 4f and s–d valence electrons varies with the composition. Au doping induces a non-Fermi-liquid phase, responsible for the anomalous logarithmic temperature dependence of specific heat observed in $CeCu_{6-x}Au_x$ [119]. The 4f localization increases and this leads to a stabilization of localized magnetic moments, which interact via the RKKY inter-action. For a critical concentration $x \approx 0.1$, a long range antiferromagnetism, incommensurate with the crystalline arrangement, is induced. This can be under-stood as due to the partial 4f re-localization, having as consequence the change of the balance between dominant Kondo and RKKY interactions. The Fermi-liquid phase reappears in a sufficiently large magnetic field. At the critical concentration and zero temperature, a quantum phase transition from the long range magnetic order to a non-magnetic phase takes place due to the competition between RKKY and Kondo effects. Anomalies of the thermodynamic and transport properties are present at low temperatures and *unusual magnetic fluctuations* have been probed by inelastic neutron scattering [161, 162]. This complex spatial dependence of the magnetic fluctuations was already present in pure $CeCu_6$ and is therefore not due to disorder. The observed numerous low-energy magnetic excitations do not show the variations expected for spin fluctuations. The fluctuations are quasi-2D. This low dimensionality can be related to a strong anisotropy of the RKKY interaction in the presence of doping and underlines the importance of the atom arrangement around the rare-earth for the electronic properties. It has been suggested that the disorder in phase transitions may mask a first-order transition between magnetic and non-magnetic ground states thus mimicking a quantum critical point. Direct iden-tification of critical fluctuations is therefore essential for a good understanding of the phenomena.

Generally, in the intermetallic compounds of the type CeX_3, the rare-earth mag-netic moments are rather weakly coupled to each other or to the X element moments. Thus, compounds with X = Pd, Sn, In, have been generally found to be mixed valence compounds [163]. 4f levels would distribute in a narrow band for each of the three compounds. Upon increase of the thermoelectric power at high pressure, the 4f distributions have been found to widen with pressure as for metallic cerium (cf. Chap. 1) and the effect grows along the sequence $CeSn_3 \rightarrow CeIn_3 \rightarrow CePd_3$. $CePd_3$ shows Pauli paramagnetism. In some compounds, the 4f level lies very close to the valence band. No long range magnetic ordering is present in $CeAl_3$ but heavy fer-mions are present. The above suggests that cerium has an analogous behaviour to that of α-cerium in these compounds.

In contrast, $CeAl_2$ is an antiferromagnet with the magnetic moments localized on the cerium ions. The RKKY interaction is responsible for magnetic ordering, which is only slightly perturbed by the Kondo effect. A localized 4f electron is expected to be present on each cerium ion, which is trivalent in this compound. Among the other compounds of the type CeX_2, $CeIr_2$ and $CeRu_2$ are supraconductors at very low temperature, 0.21 and 6 K, respectively [164]. Except for this particularity, they are analogous to α-cerium metal. No localized 4f electron is present and these materials are both non magnetic. The valence electron distribution has a 4f character mixed with a 5d one. These materials are mixed valence compounds in which the relative proportion of trivalent cerium ions is much less than unity. Below 3.5 K, $CeCu_2$ has a Kondo lattice with antiferromagnetic order and a Kondo temperature of about 6 K. CeNiSn and CeRbSb are Kondo semiconductors at 2–4 K with gaps of 8–10 and 20–27 meV, respectively, which are comparable to the Kondo temperatures. They move to a Kondo-metallic state upon increase of the temperature [165].

Intermetallic compounds of the type $CeTIn_5$ with T = Co, Rh, Ir, often designated as Ce-115, are actually actively studied because they exhibit all the typical properties. They are antiferromagnetic at standard pressure with Néel temperature of a few Kelvins. The Néel temperature decreases with increasing pressure and the antiferromagnetic ordering disappears progressively. The antiferromagnetic state vanishes when the Néel temperature equals the superconducting transition temperature and the material becomes superconductor. Antiferromagnetism and superconductivity coexist in a narrow pressure range and fluctuations of the magnetic order parameter, i.e. spin fluctuations, are expected to prompt the quantum transition. As an example, the heavy fermion compound $CeRhIn_5$ shifts under pressure from the antiferromagnetic phase to the superconducting phase at the temperature $T_c = 2.1$ K [166]. The two phases coexist in the pressure range between 1.6 and 1.9 GPa [167]. Above 2GPa the antiferromagnetic order vanishes suddenly leading to a pure superconducting ground state (Fig. 2.17) [168]. When a magnetic field is applied to $CeRhIn_5$ in the superconducting phase, a critical line is induced between the purely superconducting phase and the phase in which superconductivity and antiferromagnetism coexist. This line terminates at a point where the high magnetic field completely suppresses the superconductivity [169]. The effects of doping and of pressure were compared and it was shown that superconductivity is suppressed at

Fig. 2.17 Combined temperature, pressure and field phase diagram of $CeRhIn_5$ [168]

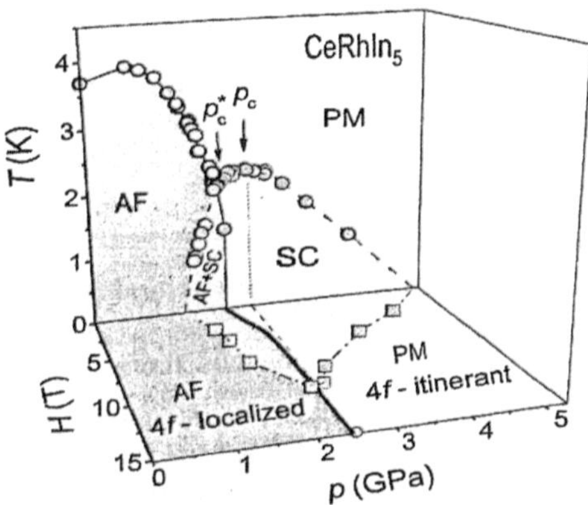

standard pressure by the substitution of In by Sn [168]. The crystalline structure of CeRhIn$_5$ was described as a quasi-2D structure formed by a stack of alternating layers of CeIn$_3$ and RhIn$_2$ The same structure was found in CeCoIn$_5$, which is a superconductor at standard pressure and at T_c = 2.3 K, i.e. at a relatively high temperature for a heavy fermion material [170]. In contrast, T_c is only 0.2 K for cubic CeIn$_3$. This temperature difference was attributed to the layered crystal structure of CeCoIn$_5$. Indeed, the presence of planes of magnetic Ce^{3+} ions renders CeCoIn$_5$, and more generally CeTIn$_5$, a quasi two-dimensional character, analogous to the layered structure of the superconductor cuprates. Consequently, it was suggested that their superconducting state could have a d-wave symmetry [171, 172]. A superconducting order parameter of d$_{x2-y2}$ symmetry was proposed as the most probable [173]. Suppression of the superconductivity has been observed in single crystals of Ce$_{1-x}$R$_x$CoIn$_5$ with R = Ca^{2+}, Y^{3+}, Eu^{2+}, Yb^{2+}, La^{3+}, Gd^{3+}, Er^{3+}, Lu^{3+}, Th^{4+}. Substitution of cerium with another lanthanide or alkaline-earth element gives the same results, independently of the valence and the magnetic nature of the substituent [174]. Further calculations are necessary to explain these observations. The properties of CeIrIn$_5$ at very low temperature have also been investigated. T_c was found equal to 0.4 K [175] and magnetotransport measurements under pressure were explained by spin fluctuations [176]. Owing to the quasi-2D character of the crystalline structure, it was suggested that other members of this compound family with higher T_c could, perhaps, be found as long as dimensionality effects were considered.

More recently, first-order valence transitions, analogous to the cerium metal $\gamma \rightarrow \alpha$ transition, have been considered in Ce-115 compounds and their possible influence on the observed critical phenomena was investigated. Let us recall that the Ce $\gamma \rightarrow \alpha$ transition terminates at the *critical end point* characterized in the temperature-pressure plane by $T_V \approx$ 600 K, P $\approx$ 2GPa where T_V is the temperature of the valence transition. But, in this case, no quantum criticality is present because at zero pressure T_V is relatively high, $\approx$120 K. The critical end point of a valence transition can be controlled by a magnetic field. In the presence of such a field, the critical end points form in the temperature-pressure-magnetic field space a continuous first-order transition line that can reach zero temperature. This critical line then emerges in the temperature–magnetic field phase diagram, where the effects due to magnetism and to valence coexist. It has been suggested theoretically that valence fluctuations could initiate the superconductivity when the first-order transition line is sufficiently close to the quantum critical line associated with the magnetic–non-magnetic transition [177]. Resistivity measurements under pressure and magnetic field showed that m*/m increased considerably in CeIrIn$_5$. This suggests the presence of local correlation effects, distinct from long range magnetic fluctuations. Under pressure, as already seen for CeCu$_2$Si$_2$ and CeCu$_2$Ge$_2$, the superconducting transition temperature increases in the region where the cerium valence is changing, i.e. in the region where the antiferromagnetic spin fluctuations are suppressed. Proximity exists between the critical points associated with the field-induced first-order valence transition and the quantum magnetic–non-magnetic transition. Consequently, the cerium valence transition can explain the temperature–magnetic field phase diagram and the non-Fermi liquid phase observed in $CeIrIn_5$, and also the origin of the superconductivity.

Similar observations were made for $CeRhIn_5$, which at zero pressure is a heavy-fermion antiferromagnet with the Néel temperature ≈ 3.5 K. At zero magnetic field no evidence of first-order transition is found and a superconducting phase is detected at a pressure >2 GPa. Under magnetic field, antiferromagnetism is present up to ≈ 2.4 GPa where superconductivity and antiferromagnetism are mixed. From resistivity measurements, the temperature-pressure-magnetic field phase diagram was explained by the presence of valence fluctuations [177]. For this compound, the effective mass enhancement is due to local correlation and not to antiferromagnetic fluctuations. First-order valence transitions induced by the magnetic field appear necessary for understanding the entire experimental results. Actually, a difference seems to exist between $CeCoIn_3$ and the two previous compounds, thus their interpretation could be different.

Similar phase diagrams accompanied by remarkable physical properties have been observed also in the series $CeMSi_3$ with M = Rh or Ir [178]. Suppression of the antiferromagnetic ordering and appearance of the superconductivity have equally been observed under pressure. However, this is not a general phenomena and a variety of cerium intermetallic compounds do not exhibit a second-order phase transition, among them $CeRhIn_3$, $CeIn_3$.

2.1.4.2 Ytterbium Intermetallics

Ytterbium intermetallic compounds were studied at low temperature in order to establish a parallel between the 4f electron of trivalent cerium and the hole of trivalent $4f^{13}$ ytterbium. In the two cases, the pressure effect induces an increase of the valence. Thus, under pressure, ytterbium changes from divalent to trivalent according the transition $Yb^{2+}4f^{14} \rightarrow Yb^{3+}4f^{13}$. As well for ytterbium as for cerium, one 4f electron, initially localized in the vicinity of the ion core, is partially delocalized under pressure. However, for ytterbium, the ground configuration is non magnetic with an ionic radius larger than in the configuration under pressure while the inverse holds for cerium. Indeed, in ytterbium at the ground state, the 4f sub shell is full with all the fourteen 4f electrons localized on each Yb^{2+} ion while under pressure, the configuration is magnetic with only thirteen 4f electrons localized on each Yb^{3+} ion and the other non-localized 4f electron strongly correlated with the valence electrons. This shows that strong interconnection exists between valence and magnetic properties. The localization of the non-filled 4f shell and its quasi-atomic character induces an increase of the RKKY interaction. Consequently, unlike the cerium case, the magnetic order can be stabilized by means of volume compression. The competition between the Kondo effect and magnetic ordering is expressed by the ratio T_{RKKY}/T_K, which shows a very broad maximum for ytterbium compounds while it is monotonously decreasing for cerium compounds. Relatively few ytterbium compounds have been studied up to now because it is difficult to obtain their pure single crystals. Moreover, the experiments are more difficult for ytterbium because the localized 4f electrons are closer to the nucleus when the sub shell is full, and therefore higher pressures have to be applied to obtain the valence change. On the other hand, an ytterbium valence change has also been predicted under magnetic field at low temperatures and a first-order valence transition was observed. The critical end point is controlled by the magnetic field. Heavy fermion states are present in a wider range of valence values in ytterbium than in cerium compounds.

The first observation of non Fermi liquid in an ytterbium compound was made at standard pressure in the absence of magnetic field for $YbRh_2Si_2$ [179]. Initially, $YbRh_2Si_2$ was considered as a 4f shell mixed valence compound like the others silicides or germanides. At 300 K and standard pressure, the valence of ytterbium in this compound is $\approx 2.9 \pm 0.05$. A fully localized Yb^{3+} state is reached at high pressures of about 8.5 GPa. In contrast, when the temperature decreases at standard pressure, the valence becomes of the order of 2.8 near $T = 0$ K. Slow valence fluctuations may be expected due to the number of the trivalent ions being different from unity. Dramatic changes of the Fermi surface as a function of temperature and magnetic field were observed using Hall effect measurements [180, 181]. These changes can be associated with the transition from the configuration where all 4f electrons are localized to the configuration with one itinerant-like 4f electron included in the Fermi volume. At low temperature, the electrical resistivity and the Sommerfeld coefficient show temperature dependence characteristic of a compound of heavy fermion type. Below the Néel temperature of 70 mK, $YbRh_2Si_2$ is anti-ferromagnetic with a magnetic moment of only $0.02\mu_B$ (Fig. 2.18) [182]. When a magnetic field is applied at standard pressure, this compound can pass through an experimentally accessible quantum critical point [181]. A weak critical field is sufficient to suppress the weak magnetic ordering and above this field Fermi-liquid (FL) is recovered. That seems to exclude the presence of spin density waves and, inversely, to favorize a local description in which the magnetic order is due to localized 4f moments. From the electrical resistivity and the Sommerfeld coefficient

Fig. 2.18 Temperature versus field phase diagram of $YbRh_2Si_2$ [182]

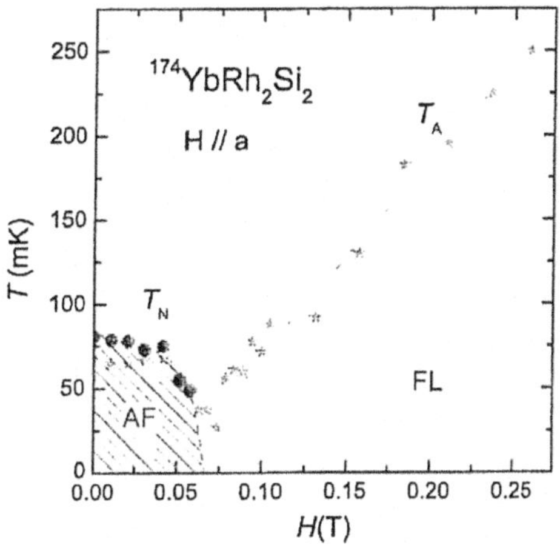

measured at sufficiently low temperatures, the Wiedemann-Franz law was found valid for values of the magnetic field far from the critical field, i.e. near-zero, while it is not satisfied for a magnetic field of about $0.3T$ [183]. This reveals the presence of a non-Fermi liquid phase in the temperature—magnetic field phase diagram, coexisting with the antiferromagnetic and Fermi-liquid phases. Electronic quantum critical fluctuations, more often designated as valence fluctuations, could be responsible for all these observations.

Comparison was made between $YbRh_2Si_2$ and $CeRh_2Si_2$ [182]. The two compounds have the same crystalline structure. Pressure effect increases the valence in both but in ytterbium it stabilizes the magnetic order instead of destabilizing it in cerium. Indeed, the magnetic Yb^{3+} $4f^{13}$ state is stable with an ionic radius smaller than in the non-magnetic $4f^{14}$ state. The radial extension of the 4f orbital is much larger for cerium (0.37 Å) than for ytterbium (0.25 Å). In ytterbium long range magnetism is still conserved in the presence of a small amount of $4f^{14}$ ion while in cerium long range magnetism disappears even for a slight departure from Ce^{3+} configuration. The local atomic character of ytterbium is also reinforced by the size of the spin orbit interaction which splits the $j = 1 - s$ and $j = 1 + s$ individual 4f level (5/2 and 7/2) and is almost five times stronger for ytterbium than for cerium. Consequently, some of the usual rare-earth properties are found in $YbRh_2Si_2$. The degree of hybridization of the delocalized 4f electron with the valence electrons is smaller for ytterbium than for cerium and the pressure effect is less marked. Thus, even at 20 K, the number of the delocalized 4f electrons differs notably from unity and the valence fluctuations are slower in ytterbium. The Kondo temperatures of the two compounds have rather similar values while the ordering temperatures are quite different, 36 K for $CeRh_2Si_2$ and 70 mK for $YbRh_2Si_2$. This confirms how close $YbRh_2Si_2$ is to a quantum critical point at pressure equal to zero.

$YbCu_2Si_2$ is a heavy fermion system without magnetic order under standard pressure; it acquires a long range magnetic order for pressures higher than 8 GPa, whose magnetic moments values tend toward that of the free ion Yb^{3+}. The temperature dependence of the magnetic susceptibility of a single crystal was measured for a magnetic field parallel to the (100) direction and a maximum was observed around 50 K [184]. This maximum reveals the presence of a metamagnetic transition at this temperature. The observation of this transition reveals a change of the 4f orbital character from heavy fermion to an almost localized 4f electron. At pressure about 8 GPa and temperature equal to 1.2–1.3 K, high-quality single crystals of $YbCu_2Si_2$ could order ferromagnetically by first-order transition [185]. The transition temperature increases with the pressure, the applied magnetic field and also the quality of the crystal. Ytterbium valence under pressure was measured by RIXS (cf. Chap. 4) at room temperature and at 15 K (Fig. 2.19) [186]. The trivalent state appears only at very high pressure. Consequently the study of this fluctuating valence compound is difficult.

In several ytterbium compounds, ferromagnetic correlation dominates. This is the case, for example, in $YbInCu_4$ where magnetic order is induced at rather low pressure, about 3GPa [187]. Among the $YbXCu_4$ compounds [177], an isostructural first-order valence transition, analogous to that of cerium metal, is observed for

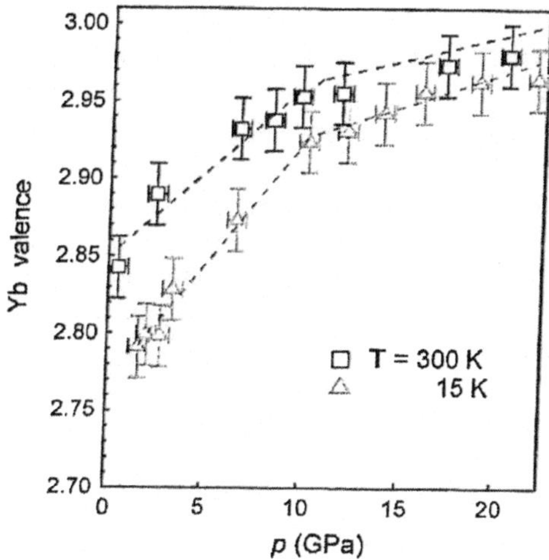

Fig. 2.19 Variation of Yb valence in YbCu$_2$Si$_2$ as function of the pressure at 300 and 15 K [186]

X = In. The transition temperature is 42 K. The ytterbium valence changes from 2.97 at high temperatures to 2.84 at low temperatures, with a volume expansion of about 0.5 %, i.e. much smaller than in cerium metal. The electron density is smaller in the high-temperature phase because of the higher entropy. On the other hand, the volume is expected to be higher when the 4f electron number increases. First-order valence transition is present with volume expansion when the temperature decreases (Fig. 2.20) [188]. These results have been confirmed by X-ray microscope experiments. YbInCu$_4$ is a clear example where a strong interplay exists between valence and magnetic fluctuations. In compounds such as YbAgCu$_4$ and YbCdCu$_4$, neither first-order valence transition nor magnetic transition has been observed. These two compounds have the paramagnetic metal ground state. Their magnetization curves are different in spite of the fact that both have nearly the same Kondo temperature, $\approx$200 K. A valence change of ytterbium occurs in the absence of magnetic field at $T = 40$ K in YbAgCu$_4$ but not in YbCdCu$_4$. Each compound has thus a specific character and its theoretical predictions appear difficult.

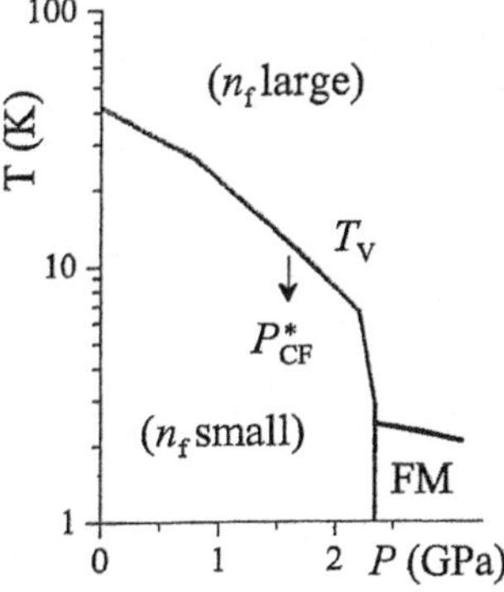

Fig. 2.20 Schematic temperature versus pressure phase diagram of YbInCu$_4$ [188]

Among the other studied ytterbium intermetallics, we can cite compounds of the type YbX_2Ge_2 with $X = Cu$, Ni and YbT_2Zn_{20} with $T = Co$, Rh, Ir and also $YbAl_3$ and $YbNi_3Al_9$ but actually their properties have not yet been completely established. Fundamental physicochemical data, such as valence and electronic distribution, depend on the atomic surroundings in the crystal and, consequently, these surroundings are responsible for the new physical properties observed in these artificial compounds. As already underlined, the 4f electrons are more localized in ytterbium than in cerium. Furthermore, it seems that strong ferromagnetism is favoured in ytterbium compounds, rather than antiferromagnetism, in contrast with the cerium compounds. This is related to the almost filled 4f sub shell that imposes the electron spin orientation, in contrast with the almost empty 4f shell in cerium. The presence of ferromagnetism is responsible for the order of the transition at low temperature and first-order transitions should be less favourable for magnetically induced superconductivity. However, it has been suggested that superconductivity could be present at extremely low temperature, in $YbCu_2Si_2$ for example. It should be noted that the ordering temperature and many other important properties depend strongly on the perfection of the crystal.

2.1.4.3 Comparison Between Cerium and Ytterbium Compounds

Over the past two decades, the cerium and ytterbium intermetallics have attracted much attention but these artificial materials are very difficult to obtain in pure crystalline phase. Consequently, only in the last few years, experimental studies have been performed on high-quality single crystals, resulting in a considerable recent progress of this research field. Known as non-Fermi liquids at low temperatures, these heavy fermion metallic compounds have physical properties that do not follow the laws applicable to metals. Violation of the Wiedemann-Franz law was observed and this implied that, in the low temperature range, the low-energy excitations could no longer be described as characteristic of particles with electron charge, obeying Fermi statistics. It appeared that low-temperature physics of many of these systems was controlled by second-order quantum phase transitions taking place between an ordered phase and a non-Fermi liquid.

These phase transitions at temperatures near absolute zero depend, among others, on the lattice density or the presence of a magnetic field. They result from the existence of critical fluctuations. Two models based on two different types of fluctuations have been developed. The more widely used one is the spin density wave model, based on long distance magnetic fluctuations. This model discusses the transitions between the antiferrromagnetic phase and the Kondo or supraconducting phases in several cerium and ytterbium compounds, among them $CePd_2Si_2$. The model is valid only in the vicinity of the magnetic phase. Consequently, the magnetic excitations were initially considered as playing a fundamental role in the unconventional supraconductivity of the heavy fermion alloys, contrary to the incompatibility existing between magnetism and conventional BCS supraconductivity. Another model, based on valence fluctuations, has been introduced to explain the presence of superconducting phase in a more extended domain than the previous one, i.e. in a domain where the spin fluctuations are absent. It appears that these two apparently different models, based on critical magnetic or valence fluctuations, are necessary in order to interpret the experimental data of cerium compounds such as $CeCu_2Si_2$. In contrast, it has been suggested that in the absence of magnetic field, critical valence fluctuations are widely predominant in the heavy fermion metals having strong local electron correlations, such as, for example, $YbRh_2Si_2$ and $CeIrIn_5$ [189]. From a general point of view, if T_V is comparable to any characteristic temperature, T_N, T_{Curie} or T_C, or if the valence transition line intercepts the antiferromagnetic-paramagnetic boundary, the valence fluctuations must be considered in addition to the spin fluctuations.

The antiferromagnetic-Kondo phase transition, present in the vicinity of the critical quantum point in numerous intermetallics, is accompanied by a change of the 4f electron behaviour. Let us recall that the 4f electron of cerium metal is localized with properties equivalent to that of a core electron in the antiferromagnetic phase, and delocalized but keeping a strongly correlated character in the α-cerium Kondo phase. In the antiferromagnetic phase of an intermetallic compound, the 4f electron is localized on the Ce^{3+} ion with a quasi-atomic character. In the Kondo phase, on the contrary, the 4f electron is not localized but couples very strongly with the valence electrons to produce heavy fermions. A 4f narrow band is present. At the antiferromagnetic-Kondo phase transition, near the critical quantum point, electron localization–delocalization accompanies then the spin fluctuations. Therefore, magnetic and electronic parameters are strongly related and the two above models are not independent because both involve the character of the 4f electrons.

Consequently, the experimental results can be explained by a change in the characteristics of the 4f electrons, which are included in the total Fermi volume, i.e. have always their energy levels present below the Fermi level [177]. Experimentally, the most studied transitions were the second-order quantum phase transitions, taking place in the vicinity of the absolute zero, because this quantum criticality can generate unconventional superconductivity. Other studies concerned the first-order valence transitions, observed under high pressure. These transitions are associated with a strong reduction of the effective mass of the heavy fermions [139] and, therefore, a strong decrease of the electron correlations. The cerium 4f electron becomes similar to a valence electron and the cerium valence increases. Superconductor-paramagnetic transitions have been observed at high pressure in various cerium and ytterbium compounds and a relation has been established between these phase transitions and the presence of valence fluctuations.

In cerium and ytterbium, the same peculiarity related to their electronic configuration exists. In cerium, Ce^{3+} is more stable than Ce^{4+}. The number of 4f electrons remains always close to unity even in the absence of magnetic ordering and with the formation of heavy fermions. In ytterbium, the number of holes in the 4f sub shell is also close to unity. The valence change is larger and the ytterbium configuration is close to $Yb^{3+} 4f^{13}$. Variation of the pressure and the atomic surroundings induces changes in the character of the 4f electrons of these two elements. For example, in cerium, a volume reduction tends to squeeze the 4f electron out of the 4f sub shell. The character of the cerium 4f electron changes from almost localized in the antiferromagnetic phase to the heavy fermion state in the Kondo phase and to quasi-itinerant electron in the paramagnetic phase. Then, charge fluctuations have a role equivalent to that of the magnetic fluctuations present near the critical point because in the two cases there is a change of the 4f electron character. The fundamental problem is, therefore, the particularities of the 4f electrons, which can be localized and responsible of magnetic ordering or have Kondo or Fermi-liquid character according the considered system. In a magnetic field, first-order valence transition of the previously described type can also influence the electronic properties. The presence of the superconducting phase seems to be independent of the type of transition by which it is obtained: magnetic quantum critical transition near a quantum critical point or first-order valence transition. However, the superconductivity temperature is higher when the phase is obtained from valence fluctuations. The temperature parameter is an important factor and the research of a higher transition temperature is at the base of the actual activity in the domain.

In summary, cerium and ytterbium intermetallic compounds are subject to extensive fundamental research because they present particular physical effects, such as spin fluctuations, heavy fermions, coexistence of magnetism and superconductivity. This variety of properties is associated with the common feature of these two rare-earths whose number of localized 4f electrons is in accordance with the surrounding of the rare-earth in the material. Actually, superconductivity is observed for an increasing number of cerium and ytterbium compounds and efforts are being made in order to search compounds with higher critical temperatures. This research range is a very active field both experimentally and theoretically. There is a vast variety of these modern rare-earth materials. However, there is no complete theory that can treat the complexity of their properties. Further experimental results at very low temperatures using judicially selected spectroscopies are actually necessary.

2.1.4.4 Other rare-earth Compounds

Initially limited to cerium and ytterbium compounds, the study of the intermetallic compounds was extended to all the rare-earths. In the ternary compounds, the lattice is sufficiently complex for that two types of microscopic interaction such as magnetism and superconductivity could coexist. These compounds form large families of isostructural or nearly isostructural materials [190]. One of the most extensively investigated is the series of RA_4 or $RA_2A'_2$ compounds that crystallize in the $BaAl_4$-type or $ThCr_2Si_2$-type structures. Strong interest in this family of compounds was related to the intriguing properties of the cerium compounds such as $CeCu_2Si_2$. The layered arrangement of atoms in the lattice was considered as a critical factor contributing to the existence of remarkable properties. Studies of the RT_2X_2 intermetallics where R = rare-earth, T = transition element, X = Si or Ge, involved mainly their magnetic properties [191]. One can cite $PrNi_2Si_2$, non magnetic at the ground state, which becomes ferromagnetic under an external magnetic field [192]. In the RMn_2Ge_2 compounds, the magnetic moments associated with the transition ions of the same atomic layer have a parallel arrangement at low temperature. The magnetic moments of the different layers are either subject to an antiferromagnetic coupling or adopt a parallel arrangement. Phase transitions between these two different magnetic states take place at a critical temperature and are first order. This is the case, for example, for $SmMn_2Ge_2$, which varies several times from ferromagnetic to antiferromagnetic as a function of the temperature [193, 194]. Calculations for this compound show that the interlayer coupling constant depends strongly on the unit cell volume, in particular on the interlayer distance, and varies linearly with the thermal expansion. Application of this model to other layered R-Mn intermetallics has been planned. Magnetic phase transitions can also take place under external pressure. These magnetovolume effects could be responsible for resistance anomalies and eventually the existence of giant magneto-resistance [195, 196]. Europium intermetallics are of a special interest because europium is divalent in the ground state $4f^7$ half-filled 4f shell, and can transit to trivalent europium with the $4f^6$ configuration, giving rise to mixed valence behaviour. The compounds EuT_2X_2 are generally antiferromagnetic. In germanides, europium is trivalent. In contrast, in the silicides, it shows mixed valence. The fluctuation time between the two valence states is short, of the order of 10^{-13} s.

A class of materials RT_4X_{12} with R = rare-earth, T = Fe, Ru, Os, X = P, As, Sb, named filled skutterudite compounds, have been widely studied because they reveal the properties characteristic of heavy fermion compounds and have the same crystal structure. Among them, $PrFe_4P_{12}$, $PrRu_4P_{12}$ and $PrOs_4P_{12}$ are especially interesting because they exhibit unusual features, not observed in the other praseodymium compounds. Thus, $PrFe_4P_{12}$ undergoes at 6.5 K a transition attributed to "the coupling between spin and quadrupole degrees of freedom" [197]. This quadrupole ordering phase is suppressed by a magnetic field. $PrFe_4P_{12}$ transits then to a heavy fermion state as indicated by specific heat and resistivity measurements under magnetic field at low temperature. In the ordered quadrupolar, or *antiferro-quadrupolar*, phase, the 4f electrons are localized. In the heavy fermion phase, the effective mass of the electrons is of the order of $80m_0$, indicating the presence of strongly correlated electrons. In the paramagnetic phase at the standard temperature, hybridization of 4f electrons with valence electrons has been considered. The presence of a non-magnetic order parameter with a multipolar component has been confirmed more recently [198]. In the $PrOs_4Sb_{12}$ compound, unconventional superconductivity is present below 1.8 K. An antiferroquadrupolar order was considered to be induced under magnetic field [199] and reveals the importance of the crystal field. From measurements of the Sommerfeld coefficient and the specific heat, heavy quasi-particles participate in the superconductivity. Compared to other heavy fermion supraconductors, this material is characterized by rather well localized 4f electrons.

Among the compound series having remarkable physical properties, one can cite the series R_mTIn_{3m+2} with T = Co, Rh or Ir and $m = 1, 2$. These materials grow in a tetragonal variant of the Cu_3Au-type structure and can be described as layers of cubic cells RIn_3 stacked sequentially along the c axis with layers of TIn_2. Unconventional superconductivity was discovered in several of these compounds and was believed to be dependent on magnetic fluctuations and also on relationship with dimensionality and crystal structures. The interplay between crystalline electric field effects and exchange magnetic interaction was investigated. Linear dependence was found between T_C and the ratio c/a of the tetragonal lattice parameters at standard pressure: an increase of the quasi two-dimensional character of the crystal would favorize superconductivity. Among the $RRhIn_5$, $TbRhIn_5$ is antiferromagnetically ordered at the highest temperature $T_N = 45.5$ K while T_N is about 32 K for $TbIn_3$ [200]. It was shown that T_N depends on the crystalline electric field along the series. When the magnetic ordered moments are not aligned along the axis c, T_N is smaller than the value for the cubic TIn_3 parent compound. This is the case for $CeRhIn_5$. In contrast, when the moments point along the c axis, T_N is bigger than the value for TIn_3. This is verified for $NdRhIn_5$. Lastly, when the crystal effects are small, the magnetic properties remain nearly the same as those of TIn_3.

Another family of compounds of the R_5X_4-type or $R_5X_2X'_2$-type was developed [190, 201]. These compounds are built from weakly interacting slabs, each slab being formed by several strongly interacting monolayers of atoms. The study of gadolinium compounds was particularly developed following a giant magnetocaloric effect observed in $Gd_5Si_2Ge_2$ [202]. This alloy shows striking properties. The application of magnetic field induces a very large magnetic entropy change, concentrated over a narrow temperature interval in the vicinity of 276 K. This change is accompanied by a first-order ferromagnetic-ferromagnetic transition, which is reversible. Simultaneous contributions of the structural and magnetic entropy change can bring out a temperature change of the alloy. The two binary compounds Gd_5Si_4 and Gd_5Ge_4 crystallize in the same Sm_5Ge_4-type orthorhombic structure. However, there exist significant differences between the atomic arrangement of these two compounds. These differences were deduced from large displacements of atoms and the breaking of some Si_2 pairs during the transition from Gd_5Si_4 to Gd_5Ge_4 [203]. Difference exists also between the magnetism of the silicide and germanide of gadolinium: Gd_5Si_4 is ordered ferromagnetically at 335 K, i.e. at a temperature higher than that for Gd metal while Gd_5Ge_4 is ordered antiferromagnetically at much lower temperature [204]. Consequently, the solubility of silicon in solid Gd_5Ge_4 depends strongly on the temperature. The orthorhombic structure of the binary compounds undergoes a monoclinic distortion in the ternary compounds of $Gd_5(Si_xGe_{1-x})_4$-type. All these changes are due to significant differences in the inter-atomic interactions. Indeed, the hybridization between Ge 4p orbitals and spin-polarized Gd 5d orbitals leads to a magnetization of Ge and a ferromagnetic coupling between 4f Gd moments belonging to adjacent Gd slabs [205]. This coupling is very weakened by the break of the Ge-Ge or Si–Si bonds and the ferromagnetic order is then destroyed. The large differences in the bonding characters of Si and Ge in the Gd_5Si_4-Gd_5Ge_4 pseudobinary system are responsible for the variation of properties depending on the Si/Ge ratio.

Up to now, the studies of the R_5X_4 or $R_5X_2X'_2$ compounds concerned essentially gadolinium alloys [206]. They have been extended to $Gd_5(Si_xGe_{4-x})_{1-y}T_{2y}$ with T = Fe, Co, Ni, Cu, C, Al, Ga. Other R_5X_4 compounds with R = rare-earth and X = Si, Ge, Sn, Ga, In, Sb, are widely studied at present. Atomic and crystal structures are analyzed because they play a major role in the magnetic properties and much work remains in order to obtain a general description of these compounds. However the studies are mainly oriented towards compounds presenting a large magnetocaloric effect. Besides this family of compounds, this effect is very strong for other rare-earth intermetallics such as $La_{0.8}Ce_{0.2}Fe_{11.4}Si_{1.6}$, $La(Fe_{1-x}Si_x)_{13}$, $PrNi_5$, $TbCo_2Al$ [207]. Applications had initially concerned the low-temperature refrigeration. The extension to room temperature is actually an important challenge because of its potential impact on energy and environmental problems.

Among the intermetallics of interest for the study of the unconventional superconductivity, one can cite the compounds of the ROMPn-type with R = La-Nd, Sm or Gd, O = oxygen, M = transition metal and Pn = element of the group V (P, As, ...). These compounds have a layered structure and are superconductors with transition temperatures T_c of 3–5 K. However, a transition temperature of the order of 26 K was obtained for LaOFeAs by dopage of F ions at the oxygen sites [208]. This material is composed of alternating LaO and FeAs layers, positively and negatively charged. The doping of the La-O layer with F ions increases the charge transfer and, consequently, increases T_c. Application of an external pressure was suggested to increase T_c because the pressure enhances charge transfer between the insulating and conducting layers. Indeed, a maximum of 43 K was obtained for F-doped LaOFeAs under 4 GPa [209]. The effect is expected to increase with the electronic polarisability of the ions. By replacing lanthanum by samarium, transition temperatures as high as 55 K were observed for $SmO_{1-x}F_xFeAs$ or $SmO_{1-x}FeAs$ at standard pressure [210, 211]. These temperatures are higher than the maximum predicted from Bardeen–Cooper–Schrieffer theory hence the interest of this type of materials for the development of unconventional superconductors.

Several remarkable physical properties of the rare-earth intermetallic compounds have lead to interesting technical applications. Thus, the fabrication of permanent magnets is possible from numerous hard rare-earth-transition metal intermetallics. Compounds with Curie temperatures between 260 and 650 K, of the type $RT_{12-x}T'_x$ with T = 3d transition metal, T' = Ti, V, Cr, Mn, Mo, W, Al, Si, have been produced [212]. $SmCo_5$ and $SmCo_7$ have also a good thermal stability because their Curie temperatures are superior to 1000 K. However, compounds of the type Nd–Fe-B have preferentially been developed because they are cheaper. Nanocomposites formed from two magnetic phases, one hard $Nd_2Fe_{14}B$ and another soft Fe_3B, have been produced [213]. Certain transition metals, such as cobalt, were added to these

composites to improve the coercivity.[2] More recently, nanocrystalline alloys Sm
$(Fe,Ga)_9C$ have been studied because of their high coercivity owing to their crystalline arrangement [214].

At low temperatures, the rare-earths are magnetostrictive materials, i.e. undergo
a relatively large deformation in a magnetic field. In the RFe_2 Laves phase[3] compounds, the effect exists at high temperature and becomes useful for numerous
applications. These compounds are used to obtain captors that change electrical
energy to mechanical energy or vice versa.

Among the other potential applications of the rare-earth intermetallic compounds
one can mentioned hydrogen storage, for example in $LaNi_5$ [215], a property of the
spin glass type, for example in $Dy_xY_{1-x}Ru_2Si_2$ [216]. The presence of disorder in
alloys such as $CeNi_{1-x}Cu_x$ [217] or $CePd_{1-x}Rh_x$ [218] can create cluster spin glass
states at low temperatures [219]. Effects related to crystalline electric field can also
be present. Magnetic frustration mechanism driven by this field can create, in a
plane, magnetic fluctuations that can induce quasi-2D unconventional superconductivity. This list is not exhaustive. Numerous other alloys, in which interplay
between heavy fermion, magnetism and superconductivity can exist, are being
studied in order to understand how magnetic fluctuations stimulate the superconductivity. The aim is to discover new superconductor materials, to obtain a better
understanding of their physics and to find new applications.

2.2 Actinide Compounds

The properties of the actinide compounds depend on the spatial extent of the 5f
orbitals. By normalizing this extent taking into account the metal lattice spacing,
one obtains for the 5f orbitals of the lighter actinides values intermediate between
those of the 4f orbitals in the rare-earths and of the 3d orbitals in the transition
elements. In the light actinides, the 5f orbitals are then less localized than the 4f
orbitals and they overlap slightly, like in uranium metal. This overlap is reduced
when the separation between the metal ions is increased by the introduction of
various anions. However, some 5f electrons can participate in the bondings. In
contrast, in the heavy actinides, where the 5f orbitals contract significantly, the 5f
electrons are localized as shown by the rapid increase of the atomic volume in
americium. Moreover, for a given actinide in various compounds, a small change in
the chemical surrounding can cause a change of the 5f electrons from localized to
itinerant. This has a large effect on the 5f electron contribution to the bonding.

---

[2]The coercivity of a material is the value of the magnetic field required to reduce to zero the
magnetization.

[3]Laves phases are structures characteristic of AB_2 type intermetallic compounds. The unit cell is
cubic or hexagonal. Tetrahedrally close packet structure is obtained if the ratio of the spheres
characteristic of the A and B atoms is equal to $\sqrt{(3/2)}$.

Interchange between localization and delocalization makes then the electronic structure of these compounds difficult to anticipate.

Correlations effects of the 5f electrons play an important role in determining the properties of most of the compounds of the actinides. These properties cannot be predicted by a conventional band-structure approach. Strong electron-electron correlations enhance the effective electron mass and favorize the presence of Mott insulator in the 5f electron systems. Electronic structure is then strongly dependent on the on-site Coulomb repulsion U of the 5f electrons. Although an order of magnitude larger than for the rare-earths, the crystal field interactions still remain weak. They are clearly smaller than the 5f spin orbit interaction and are often neglected. Valence fluctuations are largely present; they are strongly dependent on the pressure because of the overlapping of the 5f wave functions. Consequently the hybridization increases when the volume decreases. Quasi-band model was considered as convenient for the calculation of several compounds of the lighter anions and actinides, among them the nitrure of uranium [220].

The IV valence is the most stable in the actinides up to californium. The stability of the III valence increases with Z. The VI valence is observed for the four first elements of the series. In contrast, most lanthanide compounds are trivalent and the IV valence occurs only in the beginning of the series, for cerium and praseodymium, and then for terbium. Model systems that have received a large attention are the simple cubic uranium compounds, such as the dioxide, monochalcogenides and mononitrides.

2.2.1 Oxides

The actinide dioxides are representatives of the tetravalent actinide compounds. They exist for all the elements thorium through californium and crystallize in fcc structures. Beyond that, the sesquioxides appear as the stable oxides. As already underlined (cf. Chap. 2, p. 87), it is important not to confuse oxidation state and ionicity. Indeed, the bonds are known to have a partially covalent character in the oxides. The number of 5f electrons on each actinide ion is not an integer and the presence of a fraction number of 5f electrons is not to be confused with the notion of intermediate valence. For the light actinides, most of the 5f electrons are expected to concentrate in narrow bands with narrow overlap, located just below the Fermi level E_F while a wide band of O 2p electrons having the same energy as the valence electrons of the actinide is located several eV under E_F. These dioxides are considered as Mott insulators. However, it has been suggested from recent spectroscopic measurements that the lowest energy transition is of the 5f–6d type and is associated with a band gap of about 4.4 eV [221]. Photoexcitation then involves charge carrier hopping between localized levels. When Z increases, the binding energy of the 5f electrons increases, the 5f-O 2p band distance decreases and a small overlap can be present, except if the radial contraction of the 5f wave functions becomes the dominating factor. When passing from one element to the next the increase of the number of 5f electrons is accompanied by the contraction of the 5f sub shell. This contraction increases the stability of the 5f sub shell (cf. Chap. 1). Similarly, a monotonic decrease of the experimental lattice constant with increasing Z was observed from thorium to californium with a small deviation at curium, which corresponds to the half-filled shell [222]. As discussed in Chap. 1, this is due to a change in the characteristics of 5f wave functions around the middle of the series.

Thorium is tetravalent in all its compounds and no 5f electron is present in the ground state. In the stable oxide ThO_2, the O 2p valence band is large and extends between about -9 to -4 eV below the Fermi level, in agreement with the theoretical predictions.

Uranium has several oxides. The most stable is U_3O_8. It is obtained from UO_2 in contact with oxygen according to the reaction $3UO_2 + O_2 \rightarrow U_3O_8$ at 970 K. The dioxide UO_2 has the fluorite structure (CaF_2) like the dioxide of cerium and those of neptunium and plutonium. The experimental lattice constant is 5.47 Å. Each uranium ion is at the centre of a cube and coordinated to eight oxygen atoms located at the corners of the cube. The crystal field splits the f orbitals into three sub-orbits. Under normal conditions only the more stable would be occupied with two electrons, leading to an unsplit triplet ground term according to Hund's rules. However, this simple description is altered by the spin–orbit interaction and the triply degenerate ground state is split into three adjacent levels. Indeed, the $5f_{7/2}$–$5f_{5/2}$ spin–orbit splitting is expected to be of the order of 1–1.5 eV for UO_2 while the magnitude of the cubic crystal field splitting is 0.5–0.1 eV for $5f_{7/2}$ and $5f_{5/2}$, respectively. However, the inclusion of the spin–orbit interaction has only a small effect on the properties of this compound. UO_2 is a simple paramagnet above the Néel temperature $T_N = 30.8$ K and shows a first-order antiferromagnetic transition at T_N with a moment of $1.74\mu_B$. Pressure-induced weak ferromagnetism is also present [223]. UO_2 is an insulator with an optical gap of about 2 eV. Conventional LSDA and LSDA + U calculations had incorrectly predicted a non-magnetic metallic ground state. By including in LSDA + U a term describing the Hubbard on-site repulsion between 5f electrons, improvement was obtained, in particular for the equilibrium lattice constant [224]. More recently, an energy gap in reasonable agreement with the experimental value has been predicted from a hybrid DFT-type calculation [225].

From this theoretical approach, the 5f distribution has an energy width of about 1 eV and located very close to E_F. Experimentally this distribution was observed at about 1 eV below E_F. In the oxides, the bonds are often described as due to "charge transfers" from the metal to the ligand. Indeed, in the metal, 6d, 7s electrons are mixed with the ligand 2p valence electrons and charge density is present on and also between uranium and oxygen ions. The valence band associated with these electron distributions is located between -3 and -8 eV below E_F. A slight U 5f character is present in this band. The number of U 5f electrons engaged in the bonding is of the order of 0.2 while two 5f electrons are localized on each uranium site. There are more than two 5f electrons associated with each uranium ion and the effective charge of the uranium ions is less than 4 owing to the partially covalent character of the bond. A charge close to 3.5 was observed. The levels at the onset of the conduction band are predicted to be unoccupied 5f levels and 5f–5f non radiative jump is expected from about 2.5 eV above E_F in a range of several eV. In the other uranium oxides, U_3O_8 and UO_3, the uranium oxidation state progressively increases while the number of the localized 5f electrons decreases (cf. Chap. 5).

PuO_2 and Pu_2O_3 were studied theoretically by using the same approach used for UO_2 [226, 227]. This approach does not take into account the spin–orbit interaction. The two compounds were found antiferromagnetic at the ground state with magnetic coupling relatively weak. They were predicted to be insulators with energy gaps a slightly superior to 2 eV. The lattice constant of PuO_2 was calculated to be 5.39 Å, in agreement with the experimental value of 5.40 Å. For the two oxides, energy distributions characteristic of 5f electrons are located in the -1 to -3 eV energy range while distributions characteristic of the valence electrons are situated approximately in the 4–8 eV range. More recently, very detailed study of PuO_2 and particularly Pu_2O_3 was made using LDA/GGA + U formalism [228]. PuO_2 has the fluorite structure in standard conditions and the $PbCl_2$ structure beyond 39 GPa. It is insulator with a conductivity band gap of only 1.8 eV. Pu_2O_3 has several phases, two non-stoichiometric cubic α and α'-phases and a stoichiometric hexagonal β-phase. β-Pu_2O_3 is insulating and antiferromagnetic below 4 K. The atomic volumes of the two oxides increase with the Coulomb interaction U, which appears as a parameter. Indeed, increase of the correlations, due to the localization of the 5f electrons, has as consequence the decrease of the cohesion of the crystal and the increase of the lattice parameter. In PuO_2, the Pu 5f and O 2p distributions are very close in energy. In contrast, the Pu 5f levels are energetically separated from the valence band in Pu_2O_3. As expected, the Pu 5f electrons are more localized in the trivalent oxide than in the tetravalent one. In Pu_2O_3, only three plutonium valence electrons are necessary for the plutonium-oxygen bonds and, in principle, the 5f electrons do not participate. That is different for the tetravalent oxide because in this case one 5f electron contributes to the chemical bonds with the oxygens. The differences between these two oxides are somewhat analogous to the differences seen between Ce_2O_3 and CeO_2.

In going from UO_2 to PuO_2, one expects a reduction of the 5f orbital radius and a stabilization of the 5f orbitals. Stabilization is, indeed, obtained in PuO_2 and consequently the 5f-O 2p mixing is expected to be greater in PuO_2 than in UO_2. Thus, the Pu 5f peak is observed at -2.5 eV, while the calculated peak is at -0.5 eV. Theoretically, the 5f orbitals in PuO_2 show an intermediate character between being hybridized or not. Experimentally, they manifest partially hybridized electronic structure. PuO_2 is in the crossover between the Mott insulator family and the dioxides characterized by a normal band gap. The same behaviour is seen for AmO_2. The two compounds, PuO_2 and AmO_2, are characterized by a weak energy mixing of the 5f and oxygen 2p orbitals.

The stoichiometric oxide CmO_2 has a magnetic moment equal to $3.36\mu_B$. The magnetic moments of the ions Cm^{4+} $5f^6$ and Cm^{3+} $5f^7$ are, respectively, 0 and $7.94\mu_B$. The value of the observed magnetic moment in the oxide confirms the partially covalent character of the bond. Covalent picture has been predicted theoretically [229]. It must be noted that if the spin–orbit interaction were dominant, the expected ground configuration of curium in CmO_2 would be mainly $5f^6$ because this configuration corresponds to the complete filling of the lower $f_{5/2}$ subband. Such a ground configuration is predicted to have a zero magnetic moment, contrary to the large moment found experimentally. In fact, curium has a number of 5f electrons between 6 and 7, but only a small part of them contributes to the bonds with oxygen, as in the preceding elements.

The actinide oxides are of great interest. The most important is PuO_2. It is a component of nuclear reactor fuels and an important compound for the very long-term storage of plutonium. Elemental plutonium rapidly oxidizes to PuO_2 when exposed to air. The products of the chemical reactivity of plutonium metal, oxides and hydrides, are very complex. Non-stoichiometric oxide, PuO_{2+x} with $x \leq 0.27$, was believed to take part in the fast corrosion of the metal [230]. More recently, the existence of such a compound has been questioned and the stability of any higher binary plutonium oxide has been excluded [231]. The presence of one extra oxygen would probably distort the lattice too much. It would seem that plutonium oxidation chemistry still needs further research.

With its various technical applications possible, UO_2 is sometimes compared to Si and GaAs. Its intrinsic conductivity at standard temperature is about the same as that of single crystal silicon. Its dielectric constant is about 22, i.e. twice as high as that of Si (11.2) and GaAs (14.1). This is an advantage in the construction of integrated circuits. Stoichiometry dramatically influences its electrical properties. This is a ceramic material resistant at high temperatures. It has a very small thermal conductivity and its applications for photovoltaic device can be considered.

2.2.2 *Monochalcogenides*

The actinide monochalcogenides have the NaCl-type cubic structure. In spite of their highly symmetrical crystal structure, the uranium monochalcogenides have very large anisotropy. Thus, the anisotropy constant of the cubic uranium compound, US, near 0 K, was found to be an order of magnitude greater than that of $TbFe_2$, which was the largest known anisotropy constant of a cubic material [232].

The uranium monochalcogenides US, USe and UTe, exhibit a ferromagnetic ground state with the uranium magnetic orbital moment more than twice larger than the magnetic spin moment and a very large magnetic anisotropy. Their Curie temperatures are 177, 160 and 104 K, respectively. They are higher than those of other ferromagnetic uranium compounds. The monochalcogenides are metallic. Their density of valence states is characterized by a broad feature located between -2 eV and E_F and centred around -1 eV and a rather small peak at the Fermi level, attributed to the 5f electrons. These data, deduced from spectroscopic experiments (cf. Chap. 5), were in disagreement with the densities of states calculated from LDA + U method but were reproduced approximately from DMFT associated with LDA + U calculations using U equal 2 eV [233]. The calculated chalcogen p band is located between −6 and −3 eV in USe and UTe. The 5f distribution is split into a narrow occupied part crossing the Fermi level and a broader unoccupied band.

The plutonium monochalcogenides PuSe and PuTe are paramagnetic semiconductors with narrow energy gaps of the order of 10 meV and temperature independent magnetic susceptibility. For temperatures higher than the energy gap, the gap is irrelevant and these compounds can be considered as highly correlated metals. From spectroscopic experiments, energy distribution of 5f electrons was observed in a range between −1.5 eV and the Fermi level, the large part of the 5f electron spectral weight being concentrated in a narrow peak near E_F. Similar features were observed in δ-Pu and PuN and this 5f distribution appears to be independent of the chemical environment. As has already been often seen, the usual LDA-GGA-based and LDA + U calculations fail to provide the correct description of the electronic properties of the actinide compounds. However, the magnetic properties are correctly described by LDA + U calculations. In contrast, DMFT-LDA + U calculations with U equal 3 eV give a picture rather similar to the experimental 5f electron distribution [233]. The chalcogen p band is calculated to be in the range between −6 and −2.5 eV in PuSe and PuTe.

In summary, on the basis of the DMFT-LDA + U calculations including the spin–orbit interaction, the ground state magnetic properties of the ferromagnetic USe and UTe and non-magnetic PuSe and PuTe compounds have been correctly reproduced and an acceptable description of the electronic distributions has been obtained. In contrast, the electronic structure is not described successfully in the LDA + U approach except for the magnetic moments. These computational results underline the decisive role played by dynamic correlations in improving the agreement between theoretical 5f densities of states and experimental results.

2.2.3 Mononitrides and Monocarbides

Both nitrides and carbides have values of thermophysical constants higher than the corresponding ones in the oxides. They have higher melting point, higher heavy atom density, higher thermal conductivity [234]. This makes then possible alternatives to the oxide based fuels. Studies on their physics and chemistry are beginning to be developed.

Mononitrides and monocarbides, like a great number of actinide compounds, crystallize in the NaCl structure. Experimentally, the 5f orbital overlap was observed to decrease with increasing anion size. Consequently, the 5f electron localization increases. Thus, the 5f electron character and the details of the electronic structure are strongly interrelated. Generally, the 5f–5f overlap controls the characteristics of the compound. However, f–d and f–p hybridizations can also be present and influence those characteristics.

In UN and UC, owing to the small radii of the nitrogen and carbon atoms and the large extent of the uranium 5f orbitals, the uranium 5f electrons were considered initially as itinerant. The lattice parameters calculated by using a band model for the two compounds were found in agreement with the experimental values. However, from SIC-LSD calculations, the electronic structure is dominated by a narrow uranium 5f density peak located at the Fermi level [235]. The 2p band is centred at about -4 eV in UN but only at -2.7 eV in UC. Overlap exists between the U 5f orbitals and the N, or C, 2p orbitals and it is bigger for UC. Finally, the ground configurations are predicted to be f^0 in UC and f^1 in UN. The electrons 5f are then considered as delocalized only in UC and it appears clearly that the influence of electron–electron correlations increases from UC to UN. However, it is not sure whether a localized picture is suitable to describe the UN properties.

In PuN, both $5f^3$ and $5f^4$ configurations contribute almost equally to the ground state electronic structure whereas in PuC, the f^3 configuration slightly dominates. This is due to the larger electronegativity of nitrogen. The f–p hybridization is then less pronounced. The p states are further separated from the 5f states in PuN than in PuC. In PuC, the $5f^3$ and $5f^4$ configurations have been found energetically close. The fact that $5f^3$ corresponds to the energy minimum and the localization of an additional electron is slightly less favourable has lead to suggest that three 5f electrons were localized and one 5f electron was itinerant in this compound.

From UN to CmN, the number of localized 5f electrons present in the ground state gradually increases. The same is true from NpC to CmC. With Z increasing, the 5f orbitals contract and their overlap with neighbouring sites decreases. The energy necessary for the localization becomes higher than the band formation energy. The valence decreases with the increase of the number of localized 5f electrons. The effect is stronger for the nitride than for the carbide compounds. Moreover, different valence configurations, closely separated in energy, are expected in these compounds and their presence characterizes the coexistence of localized and delocalized 5f electrons. Contribution from the more localized electrons gradually increases as one moves along the series. Abrupt increase of the lattice parameter was observed between PuN and AmN and reproduced by the calculation (cf. Fig. 5 in [235]). This variation was associated with the localization of the 5f electrons in Am. The ground state configuration in the nitride becomes the trivalent configuration Am^{3+} $5f^6$. In the carbides, the presence of fully localized 5f electrons is expected from CmC with Cm^{3+} f^7. The decrease of the number of itinerant electrons with increasing actinide atomic number was confirmed by the observed decrease of the thermal conductivity from UN to PuN [236].

2.2.4 Intermetallics

Spectacular properties, similar to those existing in rare-earth intermetallic compounds, were observed in the actinide intermetallics. Let us recall that at elevated temperatures, in the rare-earth intermetallics with a high 3d metal, the 4f–3d exchange interactions are strong and have led to the discovery of excellent materials used as permanent magnets and magnetostrictors. This is not the case for uranium intermetallics containing high 3d metals because 5f–3d hybridization reduces the magnetic moment of the 3d ions. As an example, UFeAl is paramagnetic down to low temperatures. However, both uranium and 3d metal sublattices are known to be magnetic in some potentially interesting U-3d intermetallics.

As has already been shown in the case of the rare-earths, the actinide intermetallics at low temperature show various ground states revealing interplay between magnetism and heavy fermion states. These are non-Fermi liquid states characterized by a very large Sommerfeld coefficient, magnetically ordered phases, unconventional superconductor states and eventually some systems with both magnetic order and unconventional superconductivity. $Np_{1-x}Mo_6Se_8$, of critical temperature 5.6 K, was the first superconductor actinide compound, to be discovered [237]. Later on, heavy fermion phenomena and superconductivity were observed at low temperatures as a function of pressure or applied magnetic field, in a large number of intermetallic compounds of actinides, U, Np and Pu.

Heavy fermion phenomena and superconductivity were observed at low temperatures as a function of pressure or applied magnetic field, in a large number of intermetallic compounds of actinides, U, Np and Pu. The above characteristics result from the presence of the 5f electrons. Among the theoretical models used to describe the 5f electrons in these compounds, we cite a calculation which takes into account a weak 5f delocalization by considering simultaneously a finite 5f band-width within the Anderson Lattice Hamiltonian and localized f-spins S = 1 without f-band width to describe the U^{4+} $5f^2$ [238]. However, this model does not explain all the very particular properties of these compounds. Initially, heavy fermion phenomena were considered to occur only with very narrow 5f bands that did not order magnetically at all or else showed very small ordered moments. However, magnetic ordering was found to coexist at low temperature with heavy fermion state in numerous compounds. Whereas in the cerium intermetallics the ordering temperatures are typically of the order of 5–10 K, in some uranium compounds, such as UTe, $UCuSb_2$, a ferromagnetic order is present at Curie temperatures T_{Curie} as high as 102 K or 113 K. The same is observed in neptunium compounds, $NpNiSi_2$ and Np_2PdGa_3 with T_{Curie} equal, respectively, to 51.5 and 62.5 K and in plutonium compounds. Among the more studied compounds, it is important to mention UBe_{13}, UPt_3, URu_2Si_2, UFe_2Si_2, which were among the first ones whose unconventional superconductivity was observed and studied and also UGe_2, URhGe, UCoGe, which are the first discovered ferromagnetic superconductors.

Unconventional superconductivity of actinide intermetallic compounds was discovered for the first time in UBe_{13} [239]. The U–U distance is large, about 5.13 Å, in this compound. The uranium ions are tetravalent with two localized 5f electrons. Measurements of the electronic specific heat, the magnetic susceptibility and the electrical resistivity as a function of the temperature for a single crystal were made. Those three parameters increase with decreasing temperature in zero magnetic field and their values at about 1 K are characteristic of a heavy fermion system. At lower temperature, the electrical resistivity decreases rapidly and this variation reveals a superconducting transition at 0.86 K in agreement with the strong variations of specific heat and magnetic susceptibility in this same temperature range. This is heavy fermion superconductivity, in which no magnetic ordering or magnetic correlation was found. At lower temperature transition in the superconductor phase of $U_{1-x}Th_xBe_{13}$ was observed [240, 241]. At about 0.25 K another transition was observed in UBe_{13}, initially attributed to the presence of an antiferromagnetic phase. Studies are being pursued to understand the mechanisms leading to these observations [242].

Superconductivity was also seen in UPt_3. Resistivity measurements showed that the resistance drops to zero at about 0.54 K [243]. The transition width is of about 0.030 K and is decreased by annealing the sample, namely reducing existing defects. However, the variation of the resistivity with the temperature was clearly different for UPt_3 and UBe_{13} (Fig. 2.21) [243].

Measurements of specific heat and magnetic susceptibility give the same transition temperature. The variation of the specific heat with the temperature in a zero magnetic field was interpreted as following the law predicted for a spin fluctuation system by Doniach and Engelsberg [244] and confirmed the presence of unconventional superconductivity in UPt_3. Improvements in the quality of the samples increased the resolution and two transitions were observed at temperatures between 0.3 and 0.4 K [245]. This splitting of the superconducting transition was compared with the second structure seen in UBe_{13} [240, 242] and attributed to an ordered phase.

In contrast with UPt_3, which is described as being of the heavy fermion type, UPd_3 has the localized $5f^2$ configuration [246, 247]. This material has a double hexagonal close-packed structure with two different sites for the uranium ions. Two transitions take place at low temperature; the one at 6.7 K involves the quadrupolar ordering of the uranium ions, the second at 4.5 K involves magnetic ordering with a very small magnetic moment. The quadrupolar moments are large in f compounds and a quadrupolar ordering has already been found at higher temperatures than those of magnetism for lanthanide compounds. This is not observed in systems where the interactions involving the f-valence electron states are stronger than the quadrupolar moments.

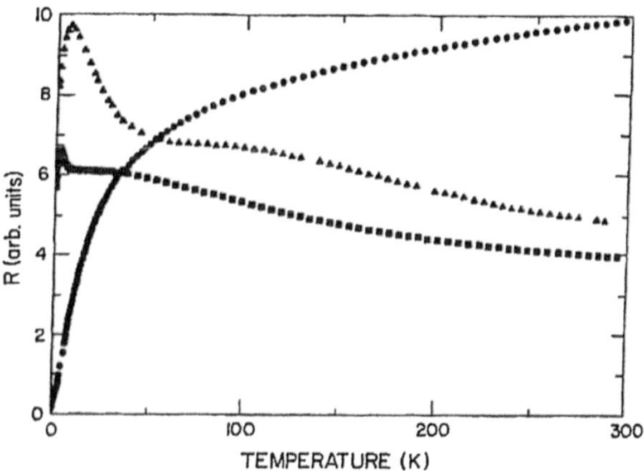

Fig. 2.21 Resistance versus temperature for $CeCu_2Si_2$ (*triangle*), UBe_{13} (*squares*) and UPt_3 (*dots*) [243]

Among the UT_2Si_2 compounds with T = transition metal, URu_2Si_2 was studied intensively [248–252]. In this heavy fermion compound, ordered phase and unconventional superconductivity at very low temperatures were obtained from thermal, electric and magnetic measurements. In fact, rapid decrease of the electrical resistivity was observed at standard pressure below 50 K but a sharp peak was observed at $T_0 = 17.5$ K. This peak was associated with a freezing of both the charge and the spin scattering and with a considerable decrease of the charge carrier number. Initially, an extremely weak antiferromagnetic moment was detected at a temperature near 0 K and associated with the presence of an ordered phase. But this moment was much too weak to account for the large anomaly observed in the vicinity of T_0. Another abrupt drop of resistivity was also observed below $T_c = 1.7$ K and associated with the onset of unconventional superconductivity, coexisting with the ordered state present in this range. These two temperatures changed differently with pressure, T_0 increased linearly while T_c decreased. From the measured Sommerfeld coefficient, the electron effective mass was found equal to about $25m_0$, where m_0 is the free electron mass. The nature of the order governing the phase observed below 17.5 K is actually a controversial subject and, for this reason, it was named a *hidden order*. Attributed at first to the development of a spin or charge density wave, this phase was associated with other order parameters, multipolar ordering, orbital antiferromagnetism, helicity. Actually, unexpected order parameters are still researched and this compound is the subject of a large variety of experiments. An anisotropic inelastic term of resistivity was observed in the hidden-order phase [253]. This suggests that an anisotropic scattering mechanism is present in this phase. Temperature versus pressure phase diagram is presented in Fig. 2.22 [252].

The unconventional superconductivity, present up to 1.2 K at P = 0 disappears at about 0.5 GPa for $T \rightarrow 0$ K. Simultaneously, the pressure induces a first-order phase transition from the hidden-order (HO) phase to a large moment antiferromagnetic (LMAF) phase. The border between the hidden-order phase and this antiferromagnetic phase follows a transition line up to a tricritical point located at

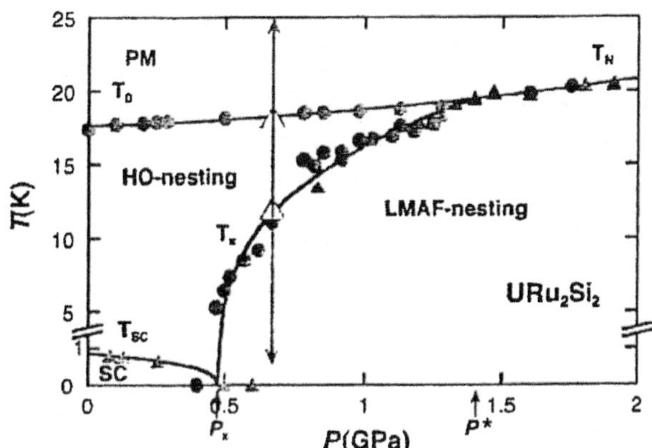

Fig. 2.22 Temperature versus pressure phase diagram for URu_2Si_2 from resistivity (*circles*) and ac calorimetry (*triangles*) measurements [252]

$T = 19.3$ K and P $= 1.36$ GPa corresponding to the transition to the normal paramagnetic (PM) phase. Above 1.36 GPa, a single transition occurs on cooling from the paramagnetic state to the large moment antiferromagnetic phase. The presence of a tricritical point suggests that the various phases have different symmetries and, consequently, no superconductivity could be present in the normal antiferromagnetic phase. Numerous experimental studies are pursued on high-quality single crystals in order to explain the unusual characteristics of the superconductor state as well as the hidden-order phase in which this state is embedded [188, 251].

The unusual phase diagram of URu_2Si_2 was attributed to the localization and delocalization of the uranium 5f electrons. In a metallic environment model, fluctuations occur between the $U^{3+}(5f^3)$ configuration and the $U^{4+}(5f^2)$ configuration with an extra 5f electron hybridized with the other valence electrons. In this model, the number of localized 5f electrons varies and one expects the presence of valence fluctuations. The phase diagram is influenced by a competition between these two different configurations. The decrease of the carrier number at the transition to the hidden-order phase suggests that this transition could be due to a partial localization of 5f electrons. In another possibility no valence fluctuations exist but only atoms situated in a particular space direction would have an increasing number of localized 5f electrons, thus explaining the anisotropy in the observed properties and the eventual presence of a multipolar order. In contrast, the pressure would favour the $5f^2$ configuration. Analogy had been researched with intermetallics of a rare-earth that has the same external electronic configuration as uranium and a parallel had been established with $PrFe_4P_{12}$ in which a hidden-order phase had been observed [254]. Initially identified with a antiferroquadripolar phase [197], it is expected to be a non-magnetic order phase with a multipolar component [251].

The coexistence of superconductivity and a ferromagnetic phase was observed for the first time in UGe_2 under pressure. This compound crystallizes in the orthorhombic structure. The uranium ions form chains with the distance between nearest neighbours $d_{U-U} = 3.85$ Å. UGe_2 is an itinerant ferromagnetic at standard pressure with a Curie temperature T_{Curie} equal to 52 K and a magnetic moment of $1.5\mu_B$. It is superconductor under pressure at about 1.2 GPa with a critical temperature T_C of 0.7 K [255]. Since T_C is lower than T_{Curie}, one deduces that the superconductor phase is present in the ferromagnetic phase. Two ferromagnetic phases have been identified, one at low pressure with the moment of $1.5\mu_B$, another at higher pressure with a moment of $1\mu_B$ (Fig. 2.23) [256].

It was shown that the pressure at the superconductivity maximum corresponds to the pressure for which the magnetic moment drops from $1.5\mu_B$ to $1\mu_B$. When the pressure increases, T_C decreases and ferromagnetism and superconductivity disappear simultaneously at the critical pressure of about 1.5 GPa through a first-order transition. This first-order transition was attributed to critical magnetic fluctuations [257].

Similar observations were made in compounds of the family UTGe with T = transition metal. Similar to UGe_2, the crystal structure of these compounds is orthorhombic with uranium chains ordered along the same axis. Their magnetic ordered temperatures vary as a function of the distance d_{U-U} between nearest

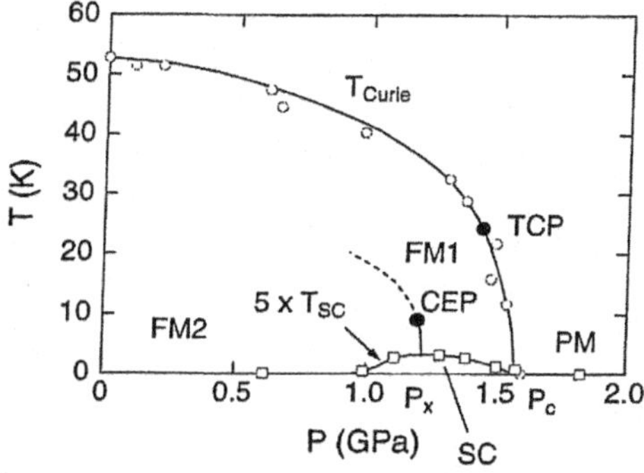

Fig. 2.23 Temperature versus pressure phase diagram for UGe$_2$. Two ferromagnetic phases, FM$_1$ and FM$_2$, have been identified, one at low pressure with the moment of 1.5μ_B, another at higher pressure with a moment of 1μ_B [256]

neighbour uranium ions. For d_{U-U} equal to 3.5 Å, the compounds are ferromagnetic; below this value, they are paramagnetic and above it antiferromagnetic. In URhGe, d_{U-U} is equal to 3.50 Å. This value is very close to the distance for which there is a direct overlap of the 5f wave functions associated with two neighbouring atoms. Ferromagnetism and superconductivity are observed at standard pressure. URhGe has a magnetic moment of 0.4μ_B and T_{Curie} is equal to 9.5 K. From temperature vs magnetic field phase diagram at standard pressure, superconductivity with T_C of about 0.25 K is observed at low field. As the field increases one observes a re-appearance of the superconductivity directly associated with an increase of the effective mass (Fig. 2.24) [188, 258, 259].

Fig. 2.24 Superconductivity directly associated with an increase of the effective mass for URhGe [188]

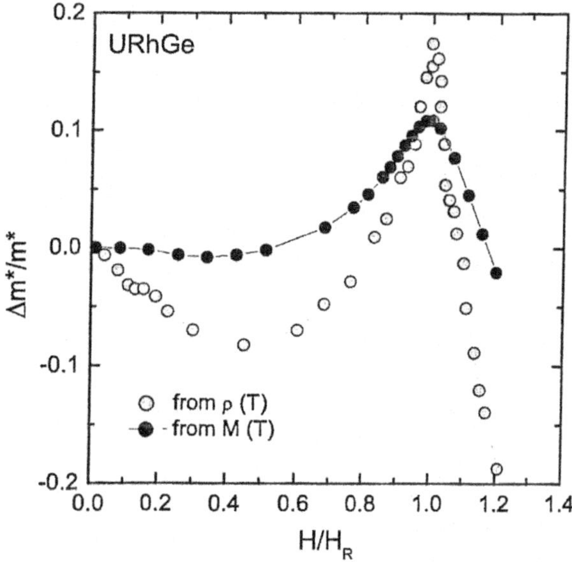

For the first time a link between the effective mass enhancement and the appearance of supraconductivity is directly observed. As in UGe$_2$, supraconductivity should be due to critical magnetic fluctuations.

The compound UCoGe is a weak itinerant ferromagnet with T_{Curie} equal to 2.8 K and an unusually small magnetic moment of about 0.05μ_B. From electronic specific heat measurements, the electron interactions are found relatively weak and it was shown that at standard pressure supraconductivity coexists with metallic ferromagnetism below the superconducting transition temperature T_C = 0.7 K. T_C increases with the pressure. Because T_{Curie} and T_C are close, interplay between ferromagnetism and superconductivity is strong. In contrast with the two precedent compounds, superconductivity persists in the paramagnetic regime above a critical pressure of 1.40 GPa. Pressure-temperature phase diagram was determined for high-quality single crystals (Fig. 2.25) [260].

In fact, the temperatures T_{Curie} and T_C depend on the quality of the sample and decrease when the quality decreases. Ferromagnetism vanishes above 1.3 GPa. Near this ferromagnetic critical point, superconductivity is enhanced and this diagram is different from that of other superconducting ferromagnets. Consequently, this compound presents an unusual behaviour and is considered as particularly interesting for studying the relation between unconventional superconductivity and magnetic interactions. But it is essential that high-quality single crystals be available. This has not yet been achieved for UCoGe nor for URhGe. It is important to underline that up to now the known ferromagnetic superconductors are all uranium compounds.

Many other uranium intermetallics have been studied. Thus, in UPd$_2$Al$_3$ and UNi$_2$Al$_3$, antiferromagnetism and superconductivity coexist at low temperature [261–263] like in Chevrel phase compounds (of the type RMo$_6$Se$_8$ with R = rare-earth). These systems are antiferromagnetic and have two separate classes of localized and delocalized electrons responsible either for magnetic properties or superconductivity. For UPd$_2$Al$_3$, it was deduced from the quasi-atomic magnetic moment measured below $T_N(0.85\mu_B)$ and from the large Sommerfeld coefficient, that two 5f electrons are localized in the U^{4+} ion, while the other 5f electrons are

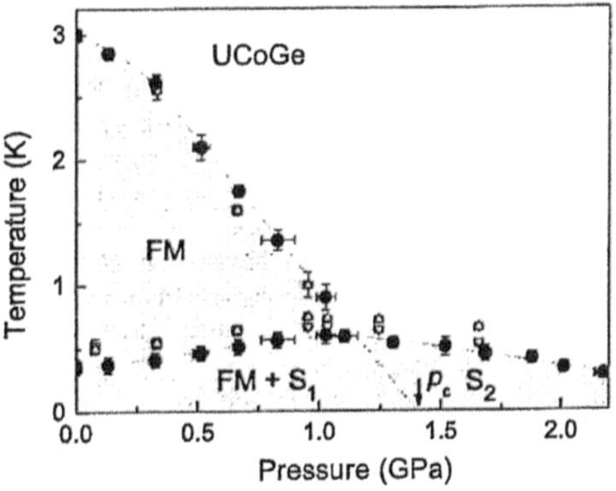

Fig. 2.25 Temperature versus pressure phase diagram of UCoGe [260]. FM: ferromagnetism; S$_1$, S$_2$: superconductivity; critical pressure P$_c$ = 1.40 ± 0.05 GPa

hybridized with the valence electrons. In the alloy system $Y_{1-x}U_xPd_3$, no Fermi liquid phase was found from measurements of specific heat, magnetic susceptibility and electric resistance [264]. Later on, this phase was observed in a number of other f-electron alloy systems, including $Sc_{1-x}U_xPd_3$, $UCu_{3.5}Pd_{1.5}$, $Th_{0.1}U_{0.9}Be_{13}$, $Th_{1-x}U_xRu_2Si_2$, $Th_{1-x}U_xPd_2Al_3$. Let us mention also that uranium and neptunium compounds of the $U_2(Np_2)T_2X$ type have been obtained with nearly all transition metals of the Fe, Co and Ni column. X represents Sn or In. From these materials, interaction of the 5f electrons with the d electrons of the transition metal was studied and found decreasing with the gradual filling of the d band.

The discovery of superconductivity in $PuCoGa_5$ [265] has lead to make this family of 115 compounds the best studied among the plutonium intermetallics. The transition temperature of $PuCoGa_5$, T_C, is 18.5 K, i.e. about an order of magnitude greater than that of the heavy fermion superconductors of cerium and uranium. $PuCoGa_5$ crystallizes in the tetragonal structure and is formed of alternating layers of $PuGa_3$ and $CoGa_2$ stacked along the c axis. This structure is similar to that of the 115 analogous rare-earths compounds that include several unconventional super-conductors. Electrical resistivity, magnetic susceptibility and specific heat of $PuCoGa_5$ were measured as function of the temperature and the magnetic field. Zero resistivity transition was observed around 18.2 K and a sharp diamagnetic transition was observed slightly above 18 K. A local magnetic moment close to that expected for Pu^{3+} was found at higher temperature. Its Sommerfeld coefficient value of 95 mJ mol^{-1} K^{-2} [266], is about one order of magnitude lower than that of the isostructural $CeCoIn_5$ and its spin fluctuation temperature T_{cf}, which is inversely proportional to Sommerfeld coefficient, is therefore one order of magnitude higher. Analogous variation is predicted for T_C in case the model of magnetically mediated superconductivity proves applicable. This variation was proposed in order to explain the high value of T_C. The presence of antiferromagnetic spin fluctuations just above T_C as well as that of a d-wave pairing below T_C were deduced from nuclear magnetic resonance measurements [267]. Analogy appears, then, with the properties of the $CeMIn_5$ heavy fermion superconductors. However, for the latter, superconductivity is observed near the antiferromagnetic quantum critical point, making the use of the same model for $PuCoGa_5$ difficult. Increase of T_C with increase of the c/a ratio of the tetragonal lattice parameters was observed in the $PuTGa_5$ and $CeTIn_5$ (T = Co, Rh, Ir) superconductors and the existence of a common mechanism of superconductivity related to the structure was, then, sug-gested [266]. Independently, a value of the Coulomb interaction U equal to 3 eV was found necessary in order to provide a good description of the phonon spectrum [268]. On the other hand, since the spin fluctuation temperature is inversely pro-portional to the effective mass, it was suggested that the increase of the transition temperature implies that a hybridization of the f electrons with the valence electrons in the plutonium compounds is more complete than in the cerium ones. Charge fluctuations associated with a change in the 5f configuration as well as density fluctuations associated with different ionic radii of the $5f^n$ and $5f^{n-1}$ configurations have, then, been proposed to account for the unconventional superconductivity, and could also be responsible for spin fluctuations. For $PuRhGa_5$, the critical transition temperature T_C is 8.7 K [269]. The two $PuRhGa_5$ and $PuCoGa_5$ compounds display very similar properties. The quasi two-dimensional structure of tetragonal compounds is advantageous for superconductivity [270]. Indeed, this structure was shown to be preferable to the formation of pairs by density fluctuations than a three-dimensional structure [271].

More recently, properties of $PuCoIn_5$ have been investigated [272]. Its unit cell is about 30 % larger than that of $PuCoGa_5$. This volume expansion is associated with a change in the electronic properties. The unit cell volume of $PuCoIn_5$ is nearly identical to that of $SmCoIn_5$, i.e. to the volume expected for Sm ions with localized 4f electrons. In contrast, the unit cell volume of $PuCoGa_5$ is smaller than that of $SmCoGa_5$, suggesting a mixed valence ground state of the 5f electrons. $PuCoIn_5$ is a superconductor with $T_C = 2.5$ K. Its Sommerfeld coefficient is 200 mJ mol^{-1} K^{-2}. $PuCoIn_5$ can then be classified as a heavy fermion compound with 5f electrons more localized than those of $PuCoGa_5$ for which a possibility of a mixed valence state exists. Consequently, it was suggested that the high superconducting transition temperature of $PuCoGa_5$ could be due to valence fluctuations while the superconductivity of $PuCoIn_5$ would be associated with antiferromagnetic spin fluctuations. Schematic phase diagram, based on such a scenario, was proposed (Fig. 2.26) [272].

This model presents similarities with that describing $CeCu_2Si_2$ under pressure, according to which a second domain of superconductivity is observed at higher T_C,

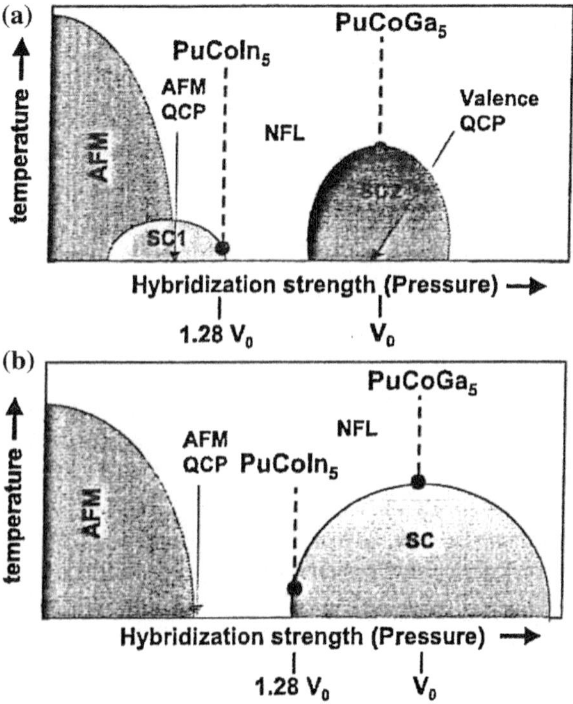

Fig. 2.26 Schematic Temperature versus pressure phase diagram of $PuCoX_5$ (X = In, Ga) [272]. AFM: antiferromagnetic; QCP: quantum critical point; SC: superconductor; NFL: non Fermi liquid; **a** SC1 corresponds to $PuCoIn_5$ and SC_2 to $PuCoGa_5$; **b** $PuCoIn_5$ and $PuCoGa_5$ have the same superconducting dome

along with valence instability, and is attributed to critical valence fluctuations. However, these interpretations have not been substantiated and other alternative models have been proposed. Further experiments would be necessary in order to reach a conclusion. Actually, the superconductivity in $PuCoGa_5$ is considered as directly dependent on the plutonium anomalous electronic properties. This material is considered as an intermediate addition, in terms of T_c, to the two other classes of "magnetic well known" superconductors: the heavy-fermion materials, which have T_cs of about 1 K and the copper oxides, which have T_cs of about 100 K.

In contrast with $PuCoGa_5$, temperature independent paramagnetism was seen in the isostructural $UCoGa_5$ compound. No local moment was detected and low Sommerfeld coefficient of about 10 mJ mol^{-1} K^{-2} was measured. A value of Coulomb interaction U close to zero provided a good theoretical description of the phonon spectrum. This compound is not a superconductor within the observed temperature limits. From these results, the localization of the 5f electrons is expected to be weaker in this compound than in $PuCoGa_5$.

In spite of numerous studies, the interpretation of the phase diagrams of 5f electrons systems and of their superconductivity remains an open problem. Application of pressure, doping or presence of a magnetic field can lead to quantum phase transitions between a magnetically ordered state and a paramagnetic state. These transitions are attributed to critical quantum fluctuations, often considered as having a magnetic origin. Nevertheless, no clear evidence has been given. Quantum critical phenomena, which are not explained by the conventional quantum critical model with spin fluctuations, are explained by a model with local critical valence fluctuations. This is extensively discussed [189]. In the case of heavy fermion materials, the effective electron mass can be sufficiently large for the electronic energies to be of the same order of magnitude or superior to the magnetic energies. The localization or delocalization of the 5f electrons is, then, an important parameter because it governs the electronic energies and can give a predominant role to the valence fluctuations. An intensive research into magnetic supercon-ductors is conducted actually in order to understand the bases for stimulating the superconductivity, since that is crucial in the search for new superconducting materials.

References

1. P. Jonnard, F. Vergand, C. Bonnelle, E. Orgaz, M. Gupta, Phys. Rev. B **57**, 12111 (1998)
2. J.C. Slater, Symmetry and Energy bands in crystals, vol. 2 (McGraw-Hill, New York, 1965)
3. E. Holland-Moritz, Z. Phys. B:Condens. Matter **89**, 285 (1992)
4. L. Eyring, Handbook on the physics and chemistry of rare earths, vol. 3, ed. by K.A. Gschneider, L. Eyring (North-Holland, Amsterdam, 1979), p. 337
5. G. Scarel, A. Svane, M. Fanciulli, Topics Appl. Phys. **106**, 1 (2007)
6. A. Delin, L. Fast, B. Johansson, J.M. Wills, O. Eriksson, Phys. Rev. Lett. **79**, 4637 (1997)
7. J. Danan, C. deNovion, R. Lallement, Solid State Commun **7**, 1103 (1969)
8. Y. Baer, C. Zürcher, Phys. Rev. Lett. **39**, 956 (1977)

9. C.M. Varma, Rev. Mod. Phys. **48**, 219 (1976)
10. J.M. Lawrence, P.S. Riseborough, R.D. Parks, Rep. Prog. Phys. **44**, 1 (1981)
11. P. Wachter, *Handbook on the physics and chemistry of rare earths*, vol. 19, Chap. 132, eds. by K.A. Gschneidner, L. Eyring, G.H. Lander, G.R. Choppin (Elsevier Science, Amsterdam, 1994)
12. M.B. Maple, D. Wohlleben, Phys. Rev. Lett. **27**, 511 (1971)
13. J.R. Iglesias Sicardi, A.K. Bhattacharjee, R. Jullien, B. Coqblin, Solid State Commun. **16**, 499 (1975)
14. P.W. Anderson, Phys. Rev. **79**, 350 (1950)
15. P.W. Anderson, Phys. Rev. **115**, 2 (1959)
16. R.A. de Groot, F.M. Mueller, P.G. van Engen, K.H.J. Buschow, Phys. Rev. Lett. **50**, 2024 (1983)
17. L. Petit, R. Tyer, Z. Szotek, W.M. Temmerman, A. Svane, New J. Phys. **12**, 113041 (2010)
18. L.V. Pourovskii, B. Amadon, S. Biermann, A. Georges, Phys. Rev. B **76**, 235101 (2007)
19. J.D. Thompson, J.M. Lawrence, *Handbook of the physics and chemistry of rare earths*, vol. 19 (1994), p. 383
20. T. Kasuya, Europhys. Lett. **26**, 283 (1994)
21. J. Derr, G. Knebel, G. Lapertot, B. Salce, M.-A. Méasson, J. Flouquet, J. Phys.: Condens. Matter **18**, 2089 (2006)
22. A. Menth, E. Buehler, T.H. Geballe, Phys. Rev. Lett. **22**, 295 (1969)
23. S. Wolgast, C. Kurdak, K. Sun, J.W. Allen, D.-J. Kim, Z. Fisk, Phys. Rev. B **88**, R180405 (2012)
24. L. Petit, A. Svane, Z. Szotek, W.M. Temmerman, Phys. Rev. B **72**, 205118 (2005)
25. P. Patsalas, S. Logothetidis, L. Sygellou, S. Kennou, Phys. Rev. B **68**, 035104 (2003)
26. N.V. Skorodumova, S.I. Simak, B.I. Lundqvist, I.A. Abrikosov, B. Johansson, Phys. Rev. Lett. **89**, 166601 (2002)
27. N.V. Skorodumova, R. Ahuja, S.I. Simak, I.A. Abrikosov, B. Johansson, B.I. Lundqvist, Phys. Rev. B **64**, 115108 (2001)
28. H. Pinto, M.N. Mintz, M. Melamud, H. Shaked, Phys. Lett. **88**, 81 (1982)
29. A. Fujimori, Phys. Rev. B **27**, 3992 (1983)
30. A. Fujimori, Phys. Rev. B **28**, 2281 (1983)
31. F.L. Normand, J.E. Fallah, L. Hiliaire, P. Légaré, A. Kotani, J.C. Parlebas, Solid State Commun. **71**, 885 (1989)
32. E. Wuilloud, B. Delley, W.-D. Schneider, Y. Baer, Phys. Rev. Lett. **53**, 202 (1984)
33. D.D. Koelling, A.M. Boring, J.H. Wood, Solid State Commun. **47**, 227 (1983)
34. R. Gillen, S.J. Clark, J. Robertson, Phys. Rev. B **87**, 125116 (2013)
35. T. Hanyu, H. Ishii, M. Yanagihara, T. Kamada, T. Miyahara, H. Kato, K. Naito, S. Suzuki, T. Ishii, Solid State Commun. **56**, 381 (1985)
36. A. Bianconi, A. Kotani, K. Okada, R. Giorgi, A. Gargano, A. Marcelli, T. Miyahara, Phys. Rev. B **38**, 3433 (1988)
37. H. Ogasawara, A. Kotani, K. Okada, B.T. Thole, Phys. Rev. B **43**, 854 (1991)
38. L. Petit, A. Svane, Z. Szotek, W.M. Temmerman, Top. Appl. Phys. **106**, 331 (2007)
39. R.C. Karnatak, J.-M. Esteva, H. Dexper, M. Gasgnier, P.E. Caro, L. Albert, Phys. Rev. B **36**, 1745 (1987)
40. J. Lettieri, V. Vaithyanathan, S.K. Eah, J. Stephens, V. Sih, D.D. Awschalom, J. Levy, D.G. Schlom, Appl. Phys. Lett. **83**, 975 (2003)
41. A. Werner, H.D. Hocheimer, A. Jayaraman, J.M. Leger, Solid State Commun. **38**, 325 (1981)
42. B. Johansson, Phys. Rev. B **19**, 6615 (1979)
43. J.M. Leger, N. Yacoubi, J. Loriers, J. Solid State Chem. **36**, 261 (1981)
44. M. Alessandri, A. Del Vitto, R. Piagge, A. Sebastiani, C. Scozzari, C. Wierner, L. Lamagna, M. Perego, G. Ghidini, M. Fanciulli, Microelectron. Eng. **87**, 290 (2010)
45. E. Bonera, G. Scarel, M. Fanciulli, P. Delugas, V. Fiorentini, Phys. Rev. Lett. **94**, 27602 (2005)
46. J.P. Liu, P. Zaumseil, E. Bugiel, H.J. Osten, Appl. Phys. Lett. **79**, 671 (2001)

47. M. Hong, J. Kwo, A.R. Kortan, J.P. Mannaerts, A.M. Sergent, Science **283**, 1897 (1999)
48. A. Jayaraman, A.K. Singh, A. Chatterjee, S. Usha Devi, Phys. Rev. B **9**, 2513 (1974)
49. A. Svane, G. Santi, Z. Szotek, W.M. Temmerman, P. Strange, M. Horne, G. Vaitheeswaran, V. Kanchana, L. Petit, H. Winter, Phys. Status Solidi B **241**, 3185 (2004)
50. K. Syassen, H. Winzen, H.G. Zimmer, H. Tups, J.M. Leger, Phys. Rev. B **32**, 8246 (1985)
51. M. Colarieti-Tosti, M.I. Katsnelson, M. Mattesini, S.I. Simak, R. Ahuja, B. Johansson, C. Dallera, O. Eriksson, Phys. Rev. Lett. **93**, 96403 (2004)
52. A. Jayaraman, V. Narayanamuri, E. Bucher, R.G. Maines, Phys. Rev. Lett. **25**, 368 (1970)
53. A. Jayaraman, V. Narayanamuri, E. Bucher, R.G. Maines, Phys. Rev. Lett. **25**, 1430 (1970)
54. B. Batlogg, E. Kaldis, A. Schlegel, B. Wachter, Phys. Rev. B **14**, 653 (1976)
55. K. Syassen, J. Phys. (Paris) Colloq. **45**, C8-123 (1984)
56. M. Campagna, E. Bucher, G.K. Wertheim, L.D. Longinotti, Phys. Rev. Lett. **33**, 165 (1974)
57. S.-J. Oh, J.W. Allen, Phys. Rev. B **29**, 589 (1984)
58. E. Kaldis, P. Wachter, Solid State Commun. **11**, 907 (1972)
59. V.P. Zhuze, A.V. Golubkov, E.V. Goncharova, T.I. Komarova, V.M. Sergeeva, Sov. Phys. Solid State **6**, 213 (1964)
60. A. Jayaraman, in *Handbook on the physics and chemistry of rare earths*, Chap. 9, vol. I, eds. by K.A. Gschneidner, Jr, I. Eyring (North-Holland, Amsterdam, 1979)
61. A. Chatterjee, A.K. Singh, A. Jayaraman, Phys. Rev. B **6**, 2285 (1972)
62. A. Svane, V. Kanchana, G. Vaitheeswaran, G. Santi, W.M. Temmerman, Z. Szotek, P. Strange, L. Petit, Phys. Rev. B **71**, 045119 (2005)
63. A. Rosengren, B. Johansson, Phys. Rev. B **13**, 1468 (1976)
64. E. Rogers, P.F. Smet, P. Dorenbos, D. Poelman, E. VanderKolk, J. Phys, Condens. Matter **22**, 015005 (2010)
65. E. Bucher, K. Andres, F.J. Disalvo, J.P. Maita, A.C. Gossard, A.S. Cooper, G.W. Hull, Phys. Rev. B **11**, 500 (1975)
66. A. Berger, E. Bucher, P. Haen, F. Holzberg, F. Lapierre, T. Penney, R. Tournier, in *Valence instabilities and related narrow band phenomena*, ed. by R.D. Parks (Plenum, New York, 1977), p. 491
67. R. Suryanarayanan, G. Guntherodt, J.L. Freeouf, F. Hultzberg, Phys. Rev. B **12**, 4215 (1975)
68. S. Lebègue, G. Santi, A. Svane, O. Bengone, M.I. Katsnelson, A.I. Lichtenstein, O. Eriksson, Phys. Rev. B **72**, 245102 (2005)
69. T. Matsumura, T. Kosaka, J. Tang, T. Matsumoto, H. Takahashi, N. Mori, T. Susuki, Phys. Rev. Lett. **78**, 1138 (1997)
70. P. Link, I.N. Goncharenko, J.M. Mignot, T. Matsumura, T. Suzuki, Phys. Rev. Lett. **80**, 173 (1998)
71. A.I. Lichtenstein, M.I. Katsnelson, Phys. Rev. B **57**, 6884 (1998)
72. P. Strange, A. Svane, W.M. Temmerman, Z. Szotek, H. Winter, Nature **399**, 756 (1999)
73. G.G. Gadzhiev, ShM Ismailov, KhKh Abdullaev, M.M. Khamidov, Z.M. Omarov, High Temp. **38**, 875 (2000)
74. G.G. Gadzhiev, ShM Ismailov, M.M. Khamidov, KhKh Abdullaev, Z.M. Omarov, High Temp. **39**, 407 (2001)
75. C.-G. Duan, R.F. Sabirianov, W.N. Mei, P.A. Dowben, S.S. Jaswal, J. Phys.: Condens. Matter **19**, 315220 (2007)
76. A. Hasegawa, A. Yanase, J. Phys. Soc. Jpn **42**, 492 (1977)
77. M.R. Norman, D.D. Koelling, A.J. Freeman, Phys. Rev. B **32**, 7748 (1985)
78. A.G. Petukhov, W.R.L. Lambrecht, B. Segall, Phys. Rev. B **50**, 7800 (1994)
79. A.G. Petukhov, W.R.L. Lambrecht, B. Segall, Phys. Rev. B **53**, 4324 (1996)
80. D.B. Ghosh, M. De, S.K. De, Phys. Rev. B **67**, 035118 (2003)
81. G. Vaitheeswaran, L. Petit, A. Svane, V. Kanchana, M. Rajaagopalan, J. Phys. Condens. Matter **16**, 4429 (2004)
82. T. Adachi, I. Shirotani, J. Hayashi, O. Shimomura, Phys. Lett. A **250**, 389 (1998)
83. L.V. Pourovskii, K.T. Delaney, C.G. Van de Walle, N.A. Spaldin, A. Georges, Phys. Rev. Lett. **102**, 096401 (2009)

84. M.S.S. Brook, J. Magn. Mater, **47–48**, 260 (1985)
85. F. Patthey, Europhys. Lett. **2**, 883 (1986)
86. F. Patthey, Phys. Rev. B **42**, 8864 (1990)
87. A. Delin, P.M. Oppeneer, M.S.S. Brooks, T. Kraft, J.M. Wills, B. Johansson, O. Erksson, Phys. Rev. B **55**, R10173 (1997)
88. A. Jayaraman, W. Lowe, L.D. Longinotti, E. Bucher, Phys. Rev. Lett. **36**, 366 (1976)
89. G. Busch, O. Vogt, Phys. Lett. **20**, 152 (1968)
90. A. Svane, W. Temmerman, Z. Szotek, Phys. Rev. B **59**, 7888 (1999)
91. J. Laegsgaard, A. Svane, Phys. Rev. B **58**, 12817 (1998)
92. N. Kioussis, D. Swearinger, B.R. Cooper, J.M. Wills, J. Appl. Phys. **69**, 5475 (1991)
93. P. Wachter, E. Kaldis, R. Hauger, Phys. Rev. Lett. **40**, 1404 (1978)
94. D.X. Li, Y. Haga, H. Shida, T. Suzuki, T. Koide, G. Kido, Phys. Rev. B **53**, 8473 (1996)
95. T. Kasuya, D.X. Li, J. Magn. Mater. **167**, L1 (1997)
96. E. Kaldis, C. Zurcher, Helv. Phys. Acta **47**, 421 (1974)
97. J.Q. Xiao, C.L. Chien, Phys. Rev. Lett. **76**, 1727 (1996)
98. W.R.L. Lambrecht, Phys. Rev. B **62**, 13538 (2000)
99. C.M. Aerts, P. Strange, M. Home, W.M. Temmerman, Z. Szotek, A. Svane, Phys. Rev. B **69**, 045115 (2004)
100. M. Home, P. Strange, W.M. Temmerman, Z. Szotek, A. Svane, H. Winter, J. Phys.: Condens. Matter **16**, 5061 (2004)
101. Z. Szotek, W.M. Temmerman, A. Svane, L. Petit, P. Strange, G.M. Stocks, D. Kodderitzsch, W. Hergert, H. Winter, J. Phys.: Condens. Matter **16**, S5587 (2004)
102. C.-G. Duan, R.F. Sabirianov, J. Liu, W.N. Mei, P.A. Dowben, J.R. Hardy, Phys. Rev. Lett. **94**, 237201 (2005)
103. C.-G. Duan, R.F. Sabiryanov, W.N. Mei, P.A. Dowben, S.S. Jaswal, E.Y. Tsymbal, Appl. Phys. Lett. **88**, 182505 (2006)
104. A.N. Chantis, M. van Schilfgaarde, T. Kotani, Phys. Rev. B **76**, 165126 (2007)
105. P. Larson, W.R.L. Lambrecht, A. Chantis, M. van Schilfgaarde, Phys. Rev. B **75**, 045114 (2007)
106. S. Granville, B.J. Ruck, F. Budde, A. Koo, D.J. Pringle, F. Kuchler, A.R.H. Preston, D.H. Housden, N. Lund, A. Bittar, G.V.M. Williams, H.J. Trodahl, Phys. Rev. B **73**, 235335 (2006)
107. F. Natali, B.J. Ruck, H.J. Trodahl, Do Le Binh, S. Vezian, B. Damilano, Y. Cordier, F. Semond, C. Meyer, Phys. Rev. B **87**, 035202 (2013)
108. S.A. Wolf, D.D. Awschalom, R.A. Buhrman, J.M. Daughton, S. van Molnar, M.L. Roukes, A.Y. Chtchelkanova, D.M. Treger, Science **294**, 1488 (2001)
109. H. van Leuken, R.A. de Groot, Phys. Rev. Lett. **74**, 1171 (1995)
110. S.J. Allen, D. Brehmer, C.J. Palmstrom, *Rare earth doped semiconductors*, eds. by G.S. Pomrenke, P.B. Klein, D.W. Langer. Material Research Society symposia proceedings, vol. 301 (1993), p. 307
111. C.J. Pamstrom, N. Tabatabaie, S.J. Allen, Appl. Phys. Lett. **53**, 2608 (1988)
112. D.E. Brehmer, K. Zhang, ChJ Schwarz, S.P. Chau, S.J. Allen, J.P. Ibbetson, J.P. Zhang, C.J. Palmstrom, B. Wilkens, Appl. Phys. Lett. **67**, 1268 (1995)
113. C. Kadow, A.W. Jackson, A.C. Gossard, S. Matsuura, G.A. Blake, Appl. Phys. Lett. **76**, 3510 (2000)
114. L. Esaki, P. Siles, S. van Malnar, Phys. Rev. Lett. **19**, 852 (1967)
115. P. LeClair, J.K. Ha, H.J.M. Swagten, J.T. Kohlhepp, C.H. van de Vin, W.J.M. de Jonge, Appl. Phys. Lett. **80**, 625 (2002)
116. L. Degiorgi, Rev. Mod. Phys. **71**, 687 (1999)
117. K.A. Gschneidner, L.R. Eyring, G.H. Lander (eds.), *Handbook on the physics and chemistry of rare earths*, vol. 32 (2002)
118. S. Doniach, Phys. B **91**, 231 (1977)
119. H.V. Löhneysen, T. Pietrus, G. Portisch, H.G. Schlager, A. Schröder, M. Sieck, T. Trappmann, Phys. Rev. Lett. **72**, 3262 (1994)

120. I. Paul, G. Kotliar, Phys. Rev. B **64**, 184414 (2001)
121. J. Hertz, Phys. Rev. B **14**, 1165 (1976)
122. G.R. Stewart, Rev. Mod. Phys. **73**, 797 (2001)
123. H.V. Löhneysen, A. Rosch, M. Vojta, P. Wölfle, Rev. Mod. Phys. **79**, 1015 (2007)
124. H. Yashima, H. Mori, T. Satoh, Solid State Commun. **43**, 193 (1982)
125. B. Coqblin, M.A. Gusmao, J.R. Iglesias, A.R. Ruppenthal, C. Lacroix, J. Magn. Magn. Mat. **226–230**, 115 (2001)
126. G.R. Stewart, Rev. Mod. Phys. **56**, 755 (1984)
127. A.C. Hewson, *The Kondo problem to heavy fermions* (Cambridge University Press, Cambridge, 1993)
128. T. Moriya, *Spin fluctuations in itinerant electron magnetism* (Springer, Berlin, 1985)
129. A. Millis, Phys. Rev. B **48**, 7183 (1993)
130. T. Moriya, K. Ueda, Adv. Phys. **49**, 555 (2000)
131. T. Moriya, K. Ueda, Rep. Prog. Phys. **66**, 1299 (2003)
132. A. Schroder, G. Aeppli, R. Goldea, M. Adams, O. Stockert, H.v. Löhneysen, E. Bucher, R. Ramazashvill, P. Coleman, Nature **407**, 351 (2000)
133. Q. Si, S. Rabello, K. Ingersent, J.L. Smith, Nature **413**, 804 (2001)
134. P. Coleman, C. Pépin, Q. Si, R. Ramazashvili, J. Phys.: Condens. Matter **13**, R723 (2001)
135. P. Coleman, A.J. Schofield, Nature **433**, 226 (2005)
136. Q. Si, F. Steglich, Science **329**, 1161 (2010)
137. H.Q. Yuan, F.M. Grosche, M. Deppe, C. Geibel, G. Sparn, F. Steglich, Science **302**, 2104 (2003)
138. D. Jaccard, H. Wilhelm, K. Alami-Yadri, E. Vargoz, Phys. B **259–261**, 1 (1999)
139. A.T. Holmes, D. Jaccard, K. Miyake, Phys. Rev. B **69**, 024508 (2004)
140. E. Stryjewski, N. Giordano, Adv. Phys. **26**, 487 (1977)
141. K. Andres, J.E. Graebner, H.R. Ott, Phys. Rev. Lett **53**, 1779 (1975)
142. W. Franz, A. Griessel, F. Steglich, D. Wohlleben, Z. Phys. B **31**, 7 (1978)
143. D. Gignoux, J.C. Gomez-Sal, J. Appl. Phys. **57**, 3125 (1985)
144. F. Steglich, J. Aarts, C.D. Bredl, W. Lieke, D. Meschede, W. Franz, H. Schäfer, Phys. Rev. Lett. **43**, 1892 (1979)
145. W. Assmus, M. Herrmann, U. Rauchschwalbe, S. Riegel, W. Lieke, H. Spille, S. Hom, G. Weber, F. Steglich, Phys. Rev. Lett. **52**, 469 (1984)
146. H. Nakamura, Y. Kitaoka, H. Yamada, K. Asayama, J. Magn. Magn. Mater. **76–77**, 517 (1988)
147. P. Gegenward, C. Langhammer, C. Geibel, R. Helfrich, M. Lang, G. Spam, F. Steglich, R. Hom, L. Donnevert, A. Link, W. Assmus, Phys. Rev. Lett. **81**, 1501 (1998)
148. P. Gegenward, F. Kromer, M. Lang, G. Sparn, C. Geibel, F. Steglich, Phys. Rev. Lett. **82**, 1293 (1999)
149. R.A. Steeman, E. Frikkee, B. Helmholdtr, J. Menovsky, J. Vandenberg, G.J. Nieuwenhuys, J.A. Mydosh, Solid State Commun. **66**, 103 (1988)
150. S.R. Julian, C. Pfleiderer, F.M. Grosche, N.D. Mathur, G.J. McMullan, A.J. Diver, I.R. Walker, G.G. Lonzarich, J. Phys.:Condens. Matter **8**, 9675 (1996)
151. N.D. Mathur, F.M. Grosche, S.R. Julian, I.R. Walker, D.M. Freye, R.K.W. Haselwimmer, G.G. Lonzarich, Nature **394**, 39 (1998)
152. F. Steglich, J. Arndt, O. Stockert, S. Friedemann, M. Brando, C. Klingner, C. Krellner, C. Geibel, S. Wirth, S. Kirchner, Q. Si, J. Phys.: Condens. Matter **24**, 294201 (2012)
153. O. Stockert, J. Arndt, E. Faulhaber, C. Geibel, H.S. Jeevan, S. Kirchner, M. Loewenhaupt, K. Schmali, W. Schmidt, Q. Si, F. Steglich, Nat. Phys. **7**, 119 (2011)
154. P. Link, D. Jaccard, Phys. B **230–232**, 31 (1997)
155. K. Miyake, O. Narikiyo, Y. Onishi, Phys. B **259–261**, 676 (1999)
156. S. Raymond, L.P. Regnault, J. Flouquet, A. Wildes, P. Lejay, J. Phys.: Condens. Matter **13**, 8303 (2001)
157. W. Knafo, S. Raymond, P. Lejay, J. Flouquet, Nat. Phys. **5**, 753 (2009)

158. H. Kadowaki, Y. Tabata, M. Sato, N. Aso, S. Raymond, S. Kawarazaki, Phys. Rev. Lett. **96**, 016401 (2006)
159. D. Aoki, C. Paulsen, H. Kotegawa, F. Hardy, C. Meingast, P. Haen, M. Boukahil, W. Knafo, E. Ressouche, S. Raymond, J. Flouquet, J. Phys. Soc. Jpn. **81**, 034711 (2012)
160. Y. Onuki, et al., J. Magn. Magn. Mater. **54–57**, 389 (1986)
161. H.V. Löhneysen, J. Magn. Magn. Mater. **200**, 532 (1999)
162. H.V. Löhneysen, H. Bartolf, S. Drotziger, C. Pfleiderer, O. Stockert, D. Souptel, W. Löser, G. Behr, J. Alloys Compd. **408**, 9 (2006)
163. N.V. ChandraShekar, M. Rajagopalan, J.F. Meng, D.A. Polvani, J.V. Badding, J. Alloys Compd. **388**, 215 (2004)
164. M. Wilhelm, B. Hillenbrand, J. Phys. Chem. Solids **31**, 559 (1970)
165. T. Ekino, T. Takabatake, H. Tanaka, H. Fujii, Phys. Rev. Lett. **75**, 4262 (1995)
166. H. Hegger, C. Petrovic, E.G. Moshopoulou, M.F. Hundley, J.L. Sarrao, Z. Fisk, J.D. Thompson, Phys. Rev. Lett. **84**, 4986 (2000)
167. T. Mito, S. Kawasaki, Y. Kawasaki, G.-Q. Zheng, Y. Kitaoka, D. Aoki, Y. Haga, Y. Onuki, Phys. Rev. Lett. **90**, 077004 (2003)
168. G. Knebel, J. Buhot, D. Aoki, G. Lapertot, S. Raymond, E. Ressouche, J. Flouquet, J. Phys. Soc. Jpn. **80**, SA001 (2011)
169. T. Park, F. Ronning, H.Q. Yuan, M.B. Salamon, R. Movshovich, J.L. Sarrao, J.D. Thompson, Nature **440**, 65 (2006)
170. C. Petrovic, P.G. Pagliuso, M.F. Hundley, R. Movshovich, J.L. Sarrao, J.D. Thompson, Z. Fisk, P. Monthoux, J. Phys.: Condens. Matter **13**, L337 (2001)
171. C. Stock, C. Broholm, J. Hudis, H.J. Kang, C. Petrovic, Phys. Rev. Lett. **100**, 087001 (2008)
172. B.B. Zhou, S. Misra, E.H. da Silva Neto, P. Aynajian, R.E. Baumbach, J.D. Thompson, E.D. Bauer, A. Yazdani, Nat. Phys. **9**, 474 (2013)
173. F. Ronning, J.-X. Zhu, T. Das, M.J. Graf, R.C. Albers, H.B. Rhee, W.E. Pickett, J. Phys.: Condens. Matter **24**, 294602 (2012)
174. C. Capan, G. Seyfarth, D. Hurt, A.D. Bianchi, Z. Fisk, J. Phys.:Conf. Ser. **273**, 012027 (2011)
175. C. Petrovic, R. Movshovich, M. Jaime, P.G. Pagliuso, M.F. Hundley, J.L. Sarrao, Z. Fisk, J.D. Thompson, Europhys. Lett. **53**, 354 (2001)
176. Y. Nakajima et al., Phys. Rev. B **77**, 214504 (2008)
177. S. Watanabe, A. Tsuruta, K. Miyake, J. Flouquet, J. Phys. Soc. Jpn. **78**, 104706 (2009)
178. R. Settai, K. Katayama, D. Aoki, I. Sheikin, G. Knebel, J. Flouquet, Y. Onuki, J. Phys. Soc. Jpn. **80**, SA069 (2011)
179. O. Trovarelli, C. Geibel, S. Mederle, C. Langhammer, F.M. Grosche, P. Gegenwart, M. Lang, G. Spam, F. Steglich, Phys. Rev. Lett. **85**, 626 (2000)
180. P. Gegenwart et al., New J. Phys. **8**, 171 (2006)
181. P. Gegenwart, Q. Si, F. Steglich, Nat. Phys. **4**, 186 (2008)
182. G. Knebel, R. Boursier, E. Hassinger, G. Lapertot, P.G. Niklowitz, A. Pourret, B. Salce, J.P. Sanchez, I. Sheikin, P. Bonville, H. Harima, J. Flouquet, J. Phys. Soc. Jpn. **75**, 114709 (2006)
183. H. Pfau, S. Hartmann, U. Stockert, P. Sun, S. Lausberg, M. Brando, S. Friedemann, C. Krellner, C. Geibel, S. Wirth, S. Kirchner, E. Abrahams, Q. Si, F. Steglich, Nature **484**, 493 (2012)
184. K. Sugiyama et al., Phys. B **403**, 769 (2008)
185. A. Fernandez-Panella, D. Braithwaite, B. Salce, G. Lapertot, J. Flouquet, Phys. Rev. B **84**,134416 (2011)
186. D. Braithwaite, A. Fernandez-Panella, E. Colombier, B. Salce, G. Knebel, G. Lapertot, V. Baldent, J.-P. Rueff, L. Paolasini, R. Verben, J. Flouquet, Proceedings of international conference on superconductivity and magnetism, ICSM (2012)
187. T. Mito, M. Nakamura, M. Otani, T. Koyama, S. Wada, M. Ishizuka, M.K. Forthaus, R. Lengsdorf, M.M. Abd-Elmeguid, J.L. Sarrao, Phys. Rev. B **75**, 134401 (2007)

188. J. Flouquet, D. Aoki, F. Bourdarot, F. Hardy, E. Hassinger, G. Knebel, T.D. Matsuda, C. Meingast, C. Paulsen, V. Taufour, J. Phys. Conf. Ser. **273**, 012001 (2011)

189. S. Watanabe, K. Miyake, J. Phys.: Condens. Matter **24**, 294208 (2012)

190. V.K. Pecharsky, K.A. Gschneidner Jr, Pure Appl. Chem. **79**, 1383 (2007)

191. A. Szytula, in *Ferromagnetic materials*, vol. 5, ed. by K.H.J. Buschow (North-Holland, Amsterdam, 1991)

192. J.A. Bianco, B. Fak, E. Ressouche, B. Grenier, M. Rotter, D. Schmitt, J.A. Rodriguez-Velamazan, J. Campo, P. Lejay, Phys. Rev. B **82**, 054414 (2010)

193. E.M. Gyorgy, B. Batlogg, J.P. Remeika, R.B. vanDower, R.M. Fleming, H.E. Bair, G.P. Espinosa, A.S. Cooper, R.G. Maihes, J. Appl. Phys. **61**, 4237 (1987)

194. M. Duraj, R. Duraj, A. Skytula, Z. Tomkowicz, J. Magn. Magn. Mater. **73**, 240 (1988)

195. J.H.V.J. Brabers, A.J. Nolten, F. Kaysel, S.H.J. Lenczowski, K.H.J. Buschow, F.R. deBoer, Phys. Rev. B **50**, 16410 (1994)

196. H. Fujii, T. Okamoto, Solid State Commun. **53**, 715 (1985)

197. J.-G. Park, D.T. Adroja, K.A. McEwen, M. Kohgi, K. Iwasa, Y.S. Kwon, Phys. B **359–361**, 868 (2005)

198. E. Hassinger, J. Derr, J. Levallois, D. Aoki, K. Behnia, F. Bourdarot, G. Knebel, C. Proust, J. Flouquet, J. Phys. Soc. Jpn. **77**(supplement A), 172 (2008)

199. M. Kohgi, K. Iwasa, K. Kuwahara, Phys. B **385–386**, 23 (2006)

200. R. Lora-Serrano, C. Giles, E. Granado, D.J. Garcia, E. Miranda, O. Aguero, L. Mendoça Ferreira, J.G.S. Duque, P.G. Pagliuso, Phys. Rev. B **74**, 214404 (2006)

201. G.S. Smith, A.G. Tharp, Q. Johnston, Acta Crystallogr. **22**, 940 (1967)

202. V.K. Pecharsky, K.A. Gschneidner Jr, Phys. Rev. Lett. **78**, 4494 (1997)

203. V.K. Pecharsky, K.A. Gschneidner Jr, Alloys. Compd. **260**, 98 (1997)

204. F. Holtzberg, R.J. Gambino, T.R. McGuire, J. Phys. Chem. Solids **28**, 2283 (1967)

205. D. Haskel, Y.B. Lee, B.N. Harmon, Z. Islam, J.C. Lang, G. Srajer, Y. Mudryk, K.A. Gschneidner Jr., V.K. Pecharsky, Phys. Rev. Lett. **98**, 247205 (2007)

206. J.H. Belo, A.M. Pereira, C. Magen, L. Morellon, M.R. Ibarra, P.A. Algarabel, J.P. Araujo, J. Appl. Phys. **113**, 133909 (2013) and references cited

207. R. Tetean, E. Burzo, I.G. Deac, J. Opt. Adv. Mat. **8**, 501 (2006)

208. Y. Kamihara, T. Watanabe, M. Hirano, H. Hosono, J. Am. Chem. Soc. **130**, 3296 (2008)

209. H. Takahashi, K. Igawa, K. Arii, Y. Kamihara, M. Hirano, H. Hosono, Nature **453**, 376 (2008)

210. X.H. Chen, T. Wu, G. Wu, R.H. Liu, H. Chen, D.F. Fang, Nature **453**, 761 (2008)

211. Z.-A. Ren et al., Europhys. Lett. **83**, 17002 (2008)

212. W. Suski, in *Handbook on the physics and chemistry of rare earths*, vol. 22, eds. by K.A. Gschneidner Jr, L. Eyring (Elsevier, Amsterdam, 1996)

213. E. Burzo, Rep. Progr. Phys. **61**, 1099 (1998)

214. L. Bessais, E. Dorolti, C. Djega-Mariadassou, Appl. Phys. Lett. **87**, 192503 (2005)

215. L. Shlapbach, Top. Appl. Phys. **67** (1992)

216. Y. Tabata, K. Matsuda, S. Kanada, T. Waki, H. Nakamura, K. Sato, K. Kindo, J. Phys. Conf. Ser. **200**, 022063 (2010)

217. N. Marcano, J.C.G. Sal, J.I. Espeso, L.F. Barquin, Phys. Rev. B **76**, 224419 (2007)

218. T. Westerkamp, M. Deppe, R. Kuchler, M. Brando, C. Geibel, P. Gegenwart, A.P. Pikul, F. Steglich, Phys. Rev. Lett. **102**, 206404 (2009)

219. F.M. Zimmer, S.G. Magalhaes, B. Coqblin, J. Phys. Conf. Ser. **273**, 012069 (2011)

220. G.H. Lander et al., Phys. Rev. B **43**, 13672 (1991)

221. P. Roussel, P. Morrall, S.J. Tull, J. Nucl. Mat. **385**, 53 (2009)

222. L.R. Morss, J. Fuger, N.M. Edelstein, *The chemistry of the actinide and transactinide elements*, 3rd edn (Springer, Dordrecht, 2006)

223. H. Sakai, H. Kato, Y. Tokunaga, S. Kambe, R.Z. Walstedt, A. Nakamura, N. Tateiwa, T.C. Kobayashi, J. Phys. Cond. Matt. **15**, S2035 (2003)

224. S.L. Dudarev, D. Nguyen Manh, A.P. Sutton, Phil. Mag. **B75**, 613 (1997)

225. K.N. Kudin, G.E. Scuseria, R.L. Martin, Phys. Rev. Lett. **89**, 266402 (2002)

226. I. Prodan, G. Scuseria, J. Sordo, K. Kudin, R. Martin, J. Chem. Phys. **123**, 014703 (2005)
227. M.T. Butterfield et al., Surf. Sci. **600**, 1637 (2006)
228. G. Jomard, B. Amadon, F. Bottin, M. Torrent, Phys. Rev. B **78**, 075125 (2008)
229. I.D. Prodan, G.E. Scuseria, R.L. Martin, Phys. Rev. B **76**, 033101 (2007)
230. J.M. Haschke, T.H. Allen, L.A. Morales, Science **287**, 285 (2000)
231. T. Gouder, A. Seibert, L. Havela, J. Rebizant, Surf. Sci. **601**, L77 (2007)
232. G.H. Lander, M.S.S. Brooks, B. Lebech, P.J. Brown, O. Vogt, K. Mattenberger, Appl. Phys. Lett. **57**, 989 (1990)
233. L.V. Pourovskii, M.I. Katsnelson, A.I. Lichtenstein, Phys. Rev. B **72**, 115106 (2005)
234. D. Srivastava, S.P. Garg, G.L. Goswami, J. Nucl. Mater. **161**, 44 (1989)
235. L. Petit, A. Svane, Z. Szotek, W.M. Temmerman, G.M. Stocks, Phys. Rev. B **80**, 045124 (2009)
236. Y. Arai, K. Nakajima, Y. Suzuki, J. Alloys Compd. **271–273**, 602 (1998)
237. D. Damien, C.H. de Novion, J. Nucl. Mater. **100**, 167 (1981)
238. C. Thomas, A.S.R. Simies, J.R. Iglesias, C. Lacroix, N.B. Perkins, B. Coqblin, J. Phys.: Conf. Ser. **273**, 012028 (2011)
239. H.R. Ott, H. Rudigier, Z. Fisk, J.L. Smith, Phys. Rev. Lett. **50**, 1595 (1983)
240. H.R. Ott, H. Rudigier, Z. Fisk, J.L. Smith, Phys. Rev. B **31**, 1651 (1985)
241. B. Batlogg, D. Bishop, B. Golding, C.M. Varma, Z. Fisk, J.L. Smith, H.R. Ott, Phys. Rev. Lett. **55**, 1319 (1985)
242. Y. Shimizu, Y. Ikeda, T. Wakabayashi, K. Tenya, Y. Haga, H. Hidaka, T. Yanagisawa, H. Amitsuka, J. Phys.:Conf. Ser. **273**, 012084 (2011)
243. G.R. Stewart, Z. Fisk, J.O. Willis, J.L. Smith, Phys. Rev. Lett. **52**, 679 (1984)
244. S. Doniach, S. Engelsberg, Phys. Rev. Lett. **17**, 750 (1966)
245. R.A. Fisher, S. Kim, B.F. Woodfield, N.E. Phillips, L. Taillefer, K. Hasselback, J. Flouquet, A.L. Giorgi, J.L. Smith, Phys. Rev. Lett. **62**, 1411 (1989)
246. U. Steigenberger et al., J. Magn. Magn. Mater. **108**, 163 (1992)
247. K.A. McEwen et al., Phys. B **186**, 670 (1993)
248. T.T.M. Palstra, A.A. Menovsky, J. van den Berg, A.J. Dirkmaat, P.H. Kes, G.J. Nieuwenhuys, J.A. Mydosh, Phys. Rev. Lett. **55**, 2727 (1985)
249. M.B. Maple, J.W. Chen, Y. Dalichaouch, T. Kohara, C. Rossel, M.S. Torikachvili, M.W. McElfresh, J.D. Thompson, Phys. Rev. Lett. **56**, 185 (1986)
250. C. Broholm, J.K. Kjems, W.J.L. Buyers, P. Matthews, T.T.M. Palstra, A.A. Menovsky, J.A. Mydosh, Phys. Rev. Lett. **58**, 1467 (1987)
251. E. Hassinger, G. Knebel, K. Izawa, P. Lejay, B. Salce, J. Flouquet, Phys. Rev. B **77**,115117 (2008) and references cited
252. A. Villaume, F. Bourdarot, E. Hassinger, S. Raymond, V. Taufour, D. Aoki, J. Flouquet, Phys. Rev. B **78**, 012504 (2008)
253. Z.W. Zhu, E. Hassinger, Z.A. Xu, D. Aoki, J. Flouquet, K. Behnia, Phys. Rev. B **80**, 172501 (2009)
254. M.A.L. De La Torre, P.Visani, Y. Dalichaouch, B.W. Lee, M.B. Mapple, Phys. B **179**, 208 (1992)
255. S.S. Saxena, P. Agarval, K. Ahilan, F.M. Grosche, R.K.W. Haselwimmer, M.J. Steiner, E. Pugh, L.B. Walker, S.R. Julian, P. Monthoux, G.G. Lonzarich, A. Huxley, L. Shelkin, D. Braithwaite, J. Flouquet, Nature **106**, 587 (2000)
256. D. Aoki, J. Flouquet, J. Phys. Soc. Jpn. **81**, 011003 (2012) and references cited
257. K.G. Sandeman, G.G. Lonzarich, A.J. Schofield, Phys. Rev. Lett. **90**, 167005 (2003)
258. F. Levy, I. Sheikin, B. Grenier, A.D. Huxley, Science **309**, 1343 (2005)
259. A. Miyake, D. Aoki, J. Flouquet, J. Phys. Soc. Jpn. **77**, 094709 (2008)
260. E. Slooten, T. Naka, A. Gasparini, Y.K. Huang, A. de Visser, Phys. Rev. Lett. **103**, 097003 (2009) and references cited
261. C. Geibel et al., Z. Phys. B **81**, 1 (1991)
262. A. Krimmel et al., Z. Phys. B **86**, 161 (1992)

263. A. Hiess, N. Bernhoeft, N. Metoki, G.H. Lander, B. Roessli, N.K. Sato, N. Aso, Y. Haga, Y. Koike, T. Komatsubara, Y. Onuki, J. Phys. Cond. Matter **18**, R437 (2006)
264. C.L. Seaman et al., Phys. Rev. Lett. **67**, 2882 (1991)
265. J.L. Sarrao, L.A. Morales, J.D. Thompson, B.L. Scott, G.R. Stewart, F. Wastin, J. Rebizant, P. Boulet, E. Colineau, G.H. Lander, Nature **420**, 297 (2002)
266. E.D. Bauer, J.D. Thompson, J.L. Sarrao, L.A. Morales, F. Wastin, J. Rebizant, J.C. Griveau, P. Javorsky, P. Boulet, E. Colineau, G.H. Landre, G.R. Stewart, Phys. Rev. Lett. **93**, 147005 (2004)
267. N.J. Curro, T. Caldwell, E.D. Bauer, L.A. Morales, M.J. Graf, Y. Bang, A.V. Balatsky, J.D. Thompson, J.L. Sarrao, Nature **434**, 622 (2005)
268. S. Raymond, P. Piekarz, J.P. Sanchez, J. Serrano, M. Krisch, B. Janousova, J. Rebizant, N. Metoki, K. Kaneko, P.T. Jochym, A.M. Oles, K. Parlinski, Phys. Rev. Lett. **96**, 237003 (2006)
269. F. Wastin, P. Boulet, J. Rebizant, E. Colineau, G.H. Lander, J. Phys.: Condens. Matter **15**, S2279 (2003)
270. P. Monthoux, G.G. Lonzarich, Phys. Rev. B **59**, 14598 (1999)
271. P. Monthoux, G.G. Lonzarich, Phys. Rev. B **69**, 064517 (2004)
272. E.D. Bauer et al., J. Phys.: Condens. Matter **24**, 052206 (2012)
273. F. Hulliger, *Handbook on the physics and chemistry of rare earths*, vol. 4, eds. by K.A. Gshneidner, L. Eyring (North-Holland, Amsterdam, 1979), p. 153

Chapter 3
High Energy Spectroscopy and Resonance Effects

Abstract The various radiative and non-radiative processes in the rare-earths irradiated by photons and electrons are discussed. Notions as resonant lines, transitions in the presence of a spectator electron, Auger transition, scattering and fluorescence are explained. The accent is put on the experimental techniques, which can establish the presence of localized nf electrons. Interpretation of the experimental results using a theoretical model is presented. The spectra of lanthanum are given as an example.

Keywords X-ray emission · X-ray absorption · Photoemission · Auger emission · Inelastic X-ray scattering · Electron energy loss

3.1 Basic Principles

Spectroscopic methods in the X and X-UV ranges are powerful tools for studying the electronic structure of matter, in particular for the determination of the energy levels of the core electrons and for the analysis of the distributions of the valence electrons and of the empty conduction levels.

The energy levels of each electronic sub shell belonging to atoms or ions present in any material can be determined from photon or electron spectroscopy. *Diagrams of energy levels* associated with each element have initially been obtained by determining the energy of a chosen level from its absorption threshold and by deducing the energies of the other levels relatively to this level from the energies of the emission lines [1, 2]. The energy levels have also been determined by photoemission [3]. The diagrams of energy levels are plotted on a negative energy scale, whose zero corresponds to the limit between the negative potential energies and the positive kinetic energies. This limit is labelled the *vacuum level*. The absolute value of an energy level is equal to its *ionization energy*, i.e. the energy necessary to remove one electron from this level into the vacuum and leave it with

© Springer Science+Business Media Dordrecht 2015

C. Bonnelle and N. Spector, *Rare-Earths and Actinides in High Energy Spectroscopy*, Progress in Theoretical Chemistry and Physics 28, DOI 10.1007/978-90-481-2879-2_3

zero kinetic energy. In solids, the energy of the levels is measured with respect to the Fermi level energy E_F, taken as zero. The energies of the core levels are important data to interpret the transitions between energy levels.

The absorption and emission transitions in the X and X-UV regions are treated as *one vacancy transitions*. In absorption, a hole is created in a core sub shell. In emission, an initial hole transits from a level to another less deep level. The hole can be considered either as a localized positive charge or as a positive charge free to move in the valence band. It can also form an electron–hole pair, or *exciton*, to which are associated excitonic levels. In this case, the electron and the hole are bound together by Coulomb interaction. Core holes are created under irradiation by any ionizing particles, electrons, ions or photons, leading to different spectroscopies according to the characteristics of the incident and detected particles.

Extensive research has been done in order to investigate the perturbations accompanying the creation of a core hole. This creation induces a change of the potential in the core region, where such a change is relatively large. It is accompanied by a modification of the electron density. Initially, the transitions were treated in the approximation of the Koopmans theorem, which states that the change in the Hartree–Fock total energy due to the removal of an electron from an unrelaxed orbital is simply related to its Hartree–Fock eigenvalue. In this approximation, also designated *independent electron model*, only the electron that makes the transition is considered. The perturbations due to electron–hole interactions are not taken into account. This introduces systematic errors in the calculated energies and densities of states. Indeed, the ejection of an electron from an inner sub shell induces a change in the potential seen by the $(Z - 1)$ electrons remaining in the atom and by those in the surrounding atoms. An excess of positive nuclear charge pulls down the filled and unfilled orbitals making their energies higher in the ion than in the neutral system. All the electrons respond to the potential change and the entire system then relaxes inducing a charge redistribution [4]. For free atoms, the difference between the energy of a level calculated in the independent electron model and its observed value is called the *intra-atomic relaxation energy*. In a solid, the screening by the valence electrons is more effective than in the free ion. It is more effective in a good metal than in an insulator or semiconductor. An additional term called the *inter-atomic relaxation energy* has to be added; it represents the decrease in energies of the levels in the solid with respect to those of the free ion; it increases with the itinerant character of the valence electrons and varies with the chemical binding. The energy of the electronic transitions is lowered in the solid with respect to their value in the free ion and the difference is occasionally called *red shift*.

The modification of all the electron energy levels by the presence of the extra Coulomb and exchange interactions associated with the core hole are considered as *correlations*. In the case of good metals, these correlations perturb the valence electron distributions in two narrow energy ranges, at the bottom of the band and in the vicinity of the Fermi energy in emission and in absorption. Calculations using the N-body diagrammatic theory and taking into account the electron–electron interactions and the interactions due to the excited states that accompany the creation of the core hole have led to results in agreement with the experimental emission and absorption spectra observed for simple metals such as magnesium and aluminium, i.e. metals with s and p valence electrons [5–8]. Simultaneously, the emission process from the valence band creates a mobile hole, i.e. a positive charge, in the valence electron distribution. The electron distribution, labelled electron plasma, is perturbed and this perturbation can create collective oscillations of the plasma [9]. Satellite transitions due to simultaneous emission of a *plasmon* are then observed. All these N-body effects are weak. In the case of semiconductors or insulators with s and p valence electrons, the correlations have little effect on the spectra. However, a weak decrease of the band gap width can be observed.

The response of the valence electrons to the creation of a core hole depends on their orbital momentum. The f electrons are expected to be more perturbed than the s, p or d valence electrons. In the presence of a core hole, the wave functions become more localized and the hybridization among valence electrons is reduced. For big enough perturbations, charge redistribution between the f and d–s orbitals leading to an increase of the number of the localized f electrons is possible but remains little probable.

Charge redistributions accompanying the creation of a hole in an inner sub shell can also create additional excitations or ionizations, which necessitate an energy supplement and are the origin of satellite processes. Among them, the most probable are shake-up and shake-off. In a shake-up process, the primary ionization is accompanied by the excitation of a second electron and the atom is in a final doubly ionized–excited configuration. In a shake-off process, the atom is in a final doubly ionized configuration because the primary ionization is accompanied by the creation of an additional ionization. Transitions from such configurations are possible and are situated towards the higher energies of the normal transitions. In semiconductors and insulators, the screening of the valence electrons is less efficient than in a conductor and can be described by an excitonic model, i.e. by considering the simultaneous creation of an electron–hole pair of low energy.

A core hole is unstable, its lifetime τ is short and its energy is not well determined. If no interaction exists between the hole and the surrounding system, the hole is rearranged exponentially with time and the energy distribution characteristic of this decay is a Lorentz curve. The full width at half-maximum (FWHM) of the Lorentz curve, Γ, is equal to the width of the energy distribution associated with the core hole. It is inversely proportional to its lifetime τ and is related to τ by the uncertainty principle

$$\Gamma \cdot \tau \sim \hbar \ (h \text{ is the Planck constant})$$

τ is defined from all the decay processes of the core hole: the inverse of the lifetime τ is equal to the sum of the probabilities per second of all its decay processes. A quantum state with a core hole of width Γ equal to 1 eV has a lifetime equal to 6.6×10^{-16} s.

The lifetime of the ground state is infinite in the absence of a perturbation, making its energy width zero. In contrast, the creation or annihilation of a core hole is very rapid and the abruptness of the perturbation can prevent the system from reaching the ground level of the perturbed configuration. An electronic transition between the ground state and a quantum state A with a core hole is a Lorentz curve, whose width Γ_A is characteristic of the hole. A transition between two non-interacting quantum states, A and B, is the convolution of the energy distributions associated with the two quantum states. This is a Lorentz curve of half-height width equal to the sum of widths of each state, $\Gamma = \Gamma_A + \Gamma_B$. The widths of the energy distributions of the excited and ionized configurations are connected to their lifetime. When the energy of the levels increases, their lifetime decreases and the widths of the transitions increase. This can limit the spectral resolution and this is the reason that the energy of the studied transitions generally does not exceed several keV.

Phonon broadening is smaller than the widths of the core levels. Indeed, this broadening can reach 0.1 eV [10, 11], while the width of the core levels varies between a few tenth eV and several eV. Consequently, the transitions of the X-ray region are not influenced by crystal fields and by phonon–electron interactions because the lifetime of the core holes is shorter than the phonon relaxation time. The recoil following electron emission is not taken up by the surrounding lattice before the filling of the hole occurs and the process is adiabatic. Nonadiabatic processes can exist in the UV region where the lifetime of the perturbed states is longer.

A hole in a nl core sub shell is localized on an atomic site. It is designated by $(nl)^{m-1}$. In contrast, a hole created in the valence band of a metal or of a ligand is not localized on a site. The associated quantum state corresponds to an average configuration in the solid and is designated by $(V)^{-1}$. If the wave functions describing the core hole and the associated excited electron are approximately in the same spatial region, the excited electron still partially screens the core hole. The relaxation effects remain weak and the changes in the average distribution of the charges are small. In contrast, if the electron is excited towards the continuum, its wave function is spread out. The relaxation to the core hole is strong and modifies the transition energy. In solids, core holes are generally associated with the creation of ionized configurations. However, excited and ionized configurations can be created simultaneously if an open sub shell with localized electrons is present in the solid. This was observed only when an nf open sub shell is present.

In high energy spectroscopies, since the lifetime of an inner hole is much shorter than the valence fluctuation time, the observed spectrum is an instantaneous description of the electronic configuration. These spectroscopies can be separated into two classes, one that deals with the occupied level distributions, another which gives information about the distributions of the unoccupied levels. They analyze the electronic distributions of the target material subject to an electromagnetic radiation or a corpuscular bombardment by studying the observed transitions, which depend on the characteristics of the incident irradiation, energy, spectral width, excitation and ionization cross sections.

The most important spectroscopic methods to investigate the occupied electronic distributions are X-ray and soft X-ray emission (XES), X-ray and XUV photoemission (XPS and UPS) and resonant X-ray scattering. It is from the observation of the valence electron distribution by X-ray emission that the band model was verified for the first time [12].

XES is the only emission that can be stimulated either by electrons or by radiation [13]. The quantum states created under electron bombardment and under monochromatic radiation of the same initial energy can be different. These quantum states, named *intermediate states*, can be characterized by comparing the results obtained from the two different stimulation modes. From XES, emission characteristics of each element and of each level of different symmetry can be analyzed independently. The localized or itinerant character of the analyzed electron distributions can be determined. These various characteristics make this method particularly rich in information. The intermediate states can also decay with emission of a secondary electron: this is Auger spectroscopy, which is a complementary method of XES and can supply additional information [14, 15].

In photoemission, the incident X-ray and VUV radiation ionizes the core or valence electrons and the energy of these photoelectrons is measured with respect to the Fermi level E_F in the solids [16]. XPS and UPS have been widely used to study the distribution of the valence electrons in lanthanides and actinides and, specially, the energy of the nf electrons with respect to that of the other valence electrons in metal and compounds. Various types of experimental methods based on the choice of a constant parameter, as the observation angle, the collected kinetic energy, were developed.

With the availability of high brightness third generation synchrotron sources, X-ray emissions induced by photons were analyzed in the vicinity of the nd thresholds of the rare-earths. According to the energy of the incident photons with respect to the threshold, the observed secondary emission can be attributed either to a Raman scattering process or a fluorescence emission. As for EXES, a non-radiative Auger process, named Auger Raman scattering, is associated with the radiative Raman process. Inelastic X-ray scattering was also employed to study the magnetic properties. Circularly polarized light was then used to observe X-ray magnetic circular dichroism [17].

Among the spectroscopic methods available to probe the unoccupied level distributions, the most used are X-ray, or X-UV, absorption spectroscopy (XAS) [18] and electron energy loss spectroscopy (EELS) [19, 20]. Reflectivity measurements in the specular conditions were also developed [21]. In these methods, the spectra are related to a final configuration with a core hole. The transitions are generally discussed in the adiabatic approximation, i.e. by considering that the atom with a core hole adjusts its energy to the effective atomic potential in an instantaneous, self-consistent way. Comparison between the results obtained by using as incident particles photons or electrons can provide interesting results. Measurements in the optical range can also contribute to characterize the unoccupied level distributions. As an example, reflectivity measurements have been used to show the localized character of the electrons in the final configuration of the nd absorption transitions. From XAS, two different types of information can be obtained. The "X-ray absorption near edge structures", named XANES, gives information on the density of unoccupied levels and, therefore, on the electronic structure [22]. "Extended X-ray absorption fine structures", named EXAFS, gives information on levels lying about twenty electronvolts beyond the absorption threshold that depend on the atomic arrangement [23, 24]. The EXAFS structures are analyzed in the aim to obtain the distribution of the radial distances between the considered ion and its nearer neighbours and information connected with diffraction analysis. Only XANES is considered here.

Other methods exist to obtain the densities of unoccupied levels, among them bremsstrahlung isochromat spectroscopy (BIS) and characteristic isochromat spectroscopy (CIS) [25]. In these methods, the unoccupied levels are populated with the help of an incident electron beam and their distribution is analyzed by observing radiative transitions among them. BIS has the advantage that the system is not perturbed by the creation of a hole. CIS is considered as process inverse of photoemission and is often named inverse photoemission spectroscopy (IPS). The two processes are equally sensitive to the surface. Spin-polarized electron source is used for spin-polarized inverse photoemission.

By using XES and Auger emission, information on the dynamics of the excited and ionized intermediate configurations, created in the solid, is obtained from the analysis of their radiative and non-radiative decay processes. In the case where the creation of the intermediate configuration is independent of the recombination process, this configuration is perfectly defined. It intervenes only by its lifetime and the transitions involved give the description of the final state. This is the *final state principle*, applicable in X-ray emission and Auger spectroscopy. For photoabsorption, photoemission, energy loss spectroscopy, the creation of the excited and ionized configurations is observed during the transitions and these configurations are present at the final states. Their energy can be measured but no information on their decay dynamics can be deduced.

A change of the number of the valence electrons results in a shift of the binding energies of the core levels. This is the *chemical shift*. The core level energies are observed to be shifted in the compounds with respect to those in the metal. The shift increases with the ionicity of the bonds. This characteristic makes the measurement of the transitions concerning core levels a direct method to identify the fractional number of the charges of the analyzed ions and consequently their oxidation states. Among the various possible measurements, one can mention the energy of the core level related peaks seen in photoemission, the energy of the X-ray emissions between core levels, that of satellite emissions seen in photoemission and in X-ray emission. All these data are very useful for chemical applications. Other particularities are observed in the compound spectra: there are electron transitions, which can take place between electron distributions associated with different atoms. They are named *inter-atomic transitions*.

In a solid, a strong Coulomb interaction exists between a core hole and the quasi-localized electrons of an unfilled sub shell. The interaction parameters can be deduced from spectroscopic observations. In the case of the rare-earths and the actinides, the presence of the open nf sub shell determines the final configuration.

The relaxation accompanying the creation of the core hole can be strong enough to push down the empty nf levels which are then mixed with the energy levels of valence electrons [26, 27]. Such redistribution has been widely considered. However, from experimental results obtained for lanthanum and cerium, the process is slightly probable and it is observable only in a few cases. Interactions between sub shells have an important role on the spectral characteristics. They are clearly stronger between nl and nl' sub shells than between nl and $n'l'$, or nl and $n'l$ ones. Thus, interactions 4d–4f are very strong and dominate completely the 4d spectra. Interactions 3d–4f or 4d–5p are much weaker and interactions 3d–4f are stronger than interactions 3d–5p. In the 4d range, the spectra are dominated by the 4d–4f electrostatic interactions while in the 3d range, the spectra are well described with the help of j–j coupling and the 3d spin–orbit interaction is the dominant one.

Several experimental parameters need to be considered in the choice of the spectroscopic analysis method. The analyzed depth must be mentioned first. It depends on the method and on the conditions of the experiment and increases with the energy of the incident particles. The range of the electrons in the material is much shorter than that of photons of same incident energy. In photoemission, the observations depend on the mean free path of the escaping photoelectrons. Its value is a function of the incident photon energies. The penetration depth of the UV photons in the sample is small and surface effects can affect the spectra stimulated by VUV. It is important to work under experimental conditions that reduce these surface effects. The transition probability is also an important parameter. It intervenes differently for each type of transition and must be taken into account to deduce the density of states from the spectra. It is generally estimated theoretically. Another consideration is an eventual saturation effect. This effect exists in the presence of a very large intensity variation; it reduces the spectral resolution and modifies the shape of the spectral lines.

3.2 Resonance Effects in the 3d Range

The 4f sub shells of the rare-earths were studied by X-ray absorption spectroscopy (XAS), i.e. by analyzing the continuous radiation transmitted through an absorbing screen of convenient thickness [28]. Strong absorption lines involving dipolar transitions from the internal atomic $3d_{3/2}$ and $3d_{5/2}$ sub shells to the partially unoccupied 4f sub shells were identified for metals and compounds [29] revealing the presence of narrow 4f distributions in the solid.

Resonant emissions located at the same energy as the absorption lines were known to be present in gas spectra. They are expected in the same energy range as the absorptions every time that unoccupied localized levels are present. In solids, such emissions were thought to be absent since the unoccupied levels were considered as forming extended bands. However, very structured wide emissions spreading over all the range of the absorptions had been observed in electron-stimulated 3d X-ray emission spectra (EXES) of some rare-earths [30, 31]. Emission lines in exact energy coincidence with the absorption lines were observed for the first time in the 3d spectrum of Gadolinium induced by electron beam; they were called *resonance lines* [32, 33]. These observations have been extended to all rare-earths [34]. Resonance lines have also been observed in the rare-earth 4d spectra [35] and some actinide nd spectra [36]. Other emissions are present on each side of the resonance lines, making the nd emission spectra more complex than the absorption spectra, in particular for configurations near the half-filled 4f shell.

Radiative decays of excited or ionized configurations are always accompanied by non-radiative processes, or Auger processes. Auger transitions associated with resonance lines were observed for the first time in the Gadolinium 3d spectra and designated as *resonant Auger transitions* [37]. Often incorrectly named "resonant photoemission", the resonant Auger transitions have been widely observed [38] but unlike the case of X-ray emissions no systematic study has been performed for the series of rare-earths. The observation of X-ray and Auger emissions from excited configurations with a core hole are two reliable methods to prove the localized character of a partially filled sub shell.

Lanthanum, as well as the preceding element barium, is of particularly interest because no 4f electron is present in their ground configuration. It appears only in the nd^{9}4f^1 excited configurations and calculations of its transitions to the excited as well as the ionized configurations were performed. The spectroscopic results obtained for lanthanum in metal and compounds are described in detail in this chapter, first in the 3d range, then in the 4d range. Comparison between the experimental and calculated results is discussed in each case. The excellent agreement between these results is underlined.

3.2.1 *X-ray Photoabsorption*

In a photoabsorption process a photon of energy $h\nu$ loses all its energy during its interaction with another particle. In the course of this interaction, various other processes are possible. The photon can be absorbed, all its energy being transferred to the absorbing system, or scattered elastically or inelastically. During an X-ray photoabsorption process, an electron present in a filled sub shell nlj of a Z atom is transferred to an unoccupied level by absorbing a photon. This final state may be either an excited state of the neutral atom (photoexcitation) or any state of the ion (photoionization).

The transition probability $w_{i,f}$ between an initial state i and a final state f of an atom Z is given, according to the Fermi golden rule (W. Heitler, *The quantum theory of radiation*, Oxford University Press, 1965), by

$$w_{i,f} = 2\pi/\hbar \left[\langle f|H_{\mathrm{int}}|i\rangle \right]^2 \delta\left(h\nu - E_f - E_i\right)$$

where $\langle f|H_{\text{int}}|i\rangle$ is the matrix element of the interaction Hamiltonian between the i and f states. For an interaction treated as a time-dependent perturbation in the electric dipole approximation, the transition probability is, in the one-electron approximation, proportional to the square of the dipolar matrix element $[\langle \psi_f|er|\psi_i\rangle]$, where ψ_i and ψ_f are the initial and final wave functions of the electron that makes the transition. In this case, only the electron that absorbed a photon is taken into account and all the correlations with the other electrons are neglected. The matrix element determines the electric dipole selection rules.

The photoabsorption cross section σ for all transitions starting from the initial state i is given by the sum over all the final states f of the transition probability $w_{i,f}$.

$$\sigma(v) = 4\pi^2 \alpha h v \sum (\langle \psi_f|er|\psi_i\rangle)^2 \delta(hv - E_f - E_i)$$

where α is the fine structure constant and er is the electric dipole operator.

In a free atom, photoexcitation corresponds to electric dipolar transitions of an electron from the ground state of a neutral atom to various J levels of an excited configuration. In the excited configuration, a nlj hole is associated with an $n'(l \pm 1)j'$ electron. The excited configuration is noted $(nl)^{-1} (n'(l \pm 1))^{+1}$. The number of transitions is equal to the number of the final levels accessible by the electric dipolar selection rules. These levels are arranged in terms called multiplets. The separation between them is called *multiplet splitting*. Multiplet splitting is characteristic of an atomic process. X-ray photoabsorption is the single direct probe of the orbital angular momentum of the unoccupied level distributions. Experimentally, it has been observed that the angular momentum increases in most transitions.

Each transition is a Lorentz curve and transitions to levels with $n' = n + 1, n + 2, n + 3, \ldots$ are present with decreasing intensities. If a *discrete* level of the excited configuration is within the energy range of the *continuum*, absorption is allowed to take place to both the discrete level and the continuum, provided that the selection rule is satisfied. In the absence of the discrete–continuum interaction, the excitation line is not modified and the spectrum consists of both discrete and continuum transitions. In the presence of interaction, the discrete level acquires some of the properties of the continuum and the excitation line has an energy distribution described by a Breit-Wigner-Fano curve, $F(\varepsilon)$ [39]. Contrary to a symmetric Lorentz curve, $F(\varepsilon)$ is asymmetric and this asymmetry becomes more pronounced as the interaction between the discrete level and the continuum is stronger.

Photoionization corresponds to the transition of an electron from a sub shell nl of a Z atom to a continuum. This transition, noted $nlj \rightarrow \varepsilon(l \pm 1)$, takes place between the ground state of the neutral atom and the continuum of the ion with the nlj hole. One names nlj *state*, or X *state*, the configuration with an nlj hole and an electron of kinetic energy equal to zero in the continuum and also the energy necessary to create this state. The level of kinetic energy zero is labelled the level of the vacuum. The spectrum is the convolution of the distribution of the unoccupied levels of the ion with an nlj hole by the energy distribution of the nlj level, weighed by the ionization probability of $nlj \rightarrow \varepsilon(l \pm 1)$. Let us note that in photoabsorption, the transitions from l to $(l + 1)$ are generally two orders of magnitude more probable than the transitions from l to $(l - 1)$.

The creation of a core hole in an atom Z changes its electron distribution. The energies of all the nlj levels are modified with respect to that in the neutral atom. The levels are all pulled down by roughly the same amount According to the *equivalent core approximation*, a Z atom with an inner hole is equivalent to the $Z + 1$ atom [40].

In a free atom, an X absorption transition from a nlj sub shell to a continuum state is a Lorentz line of half-height width equal to the Γ_{nlj} (or Γ_X) width of the nlj (or X) level. The entire array of the transitions from the ground state of an atom to all the states of the continuum of the ion, associated with the X hole in the atom, is a sum of Lorentz curves of equal width. If the continuum has a distribution of constant density N, formed by states of energy ε having the same symmetry, all the ionization transitions have the same probability and the total ionization is described by an arctangent curve. The abscissa of its inflexion point indicates the position of the ionization *threshold* and its ordinate corresponds to the absorption edge. Γ_X is equal to the width measured between the quarter and the three-quarter of the absorption edge height.

In a solid, a core hole remains localized on the same atom during its lifetime. The unoccupied level distribution is extended and forms the conduction band. An absorption transition takes place when a core electron of l symmetry is excited to the $(l \pm 1)$ continuum of conduction states of density $N_c(\varepsilon)$, characteristic of the solid. The absorption spectrum is the convolution of the Lorentz distribution of the X level by the density $N_c(\varepsilon)$ of the unoccupied levels in the conduction band, if the perturbations due to the presence of the core hole can be neglected. The shape of the absorption curves reveals the shape of the unoccupied level distributions. If this distribution has a quasi-uniform density, the absorption is nearly an arctangent curve, whose inflexion point defines the position of the threshold. Observed features reveal variations in the density of the $(l \pm 1)$ unoccupied levels.

The energy of the threshold is the Fermi energy in the case of a metal and the energy of the bottom of the conduction band in the case of an insulator or a semiconductor. A shift of the threshold energy reveals a change in the oxidation state of the element. In the solid the core hole is partially screened by the valence electrons and this induces a decrease of the core-hole binding energy with respect to the binding energy in the free atom. The energy of the ionized configuration is predicted to be lower than the value calculated from an atomic model. On the other hand, the electronic structure obtained from a first principles-based method does not necessarily reproduce the details of an X-ray absorption spectrum since it has been observed that Kohn–Sham eigenvalues may not necessarily be compared to excited states of materials. Indeed, because the presence of the core hole, the absorption spectrum does not give the exact position of the unoccupied levels.

Strong maxima can be present in the absorption spectra of solids. These maxima are usually so-called *white lines* because they appeared in this way on a photographic plate. They are associated with transitions towards large maxima of densities of unoccupied states in the solid or towards quasi-bound levels involving the presence of impurities, defects, excitons,…, all absent in the free ion. Excitation lines, well described by an atomic model and characteristic of the multiplet splitting of the initial and final configurations, were also observed in spectra of solids but only for elements having incomplete 4f or 5f sub shells. These transitions take place to unoccupied discrete levels located just above the Fermi level in a conductor or present in the band gap of a non-conductor material. When a bound electron is excited into a discrete level, it partially screens the hole and repels the valence electrons. Consequently, in the presence of the excited electron, energy perturbations due to the presence of the core hole are reduced and thus satisfactory agreement is obtained between experimental and calculated energies of the excited configuration.

At the nd absorption threshold of the rare-earths, the open 4f sub shell manifests a large spread of the energy levels that determines the structure of the $nd^{-1}4f^{m+1}$ configuration. The J levels of this configuration can be spread to an extent of several electronvolts depending on the nd sub shell and on the number m. The splitting of energy levels of the electronic configuration associated with each excited ion is large compared to the perturbations produced by the interaction of this ion with its neighbours and the spectra are characteristic of transitions of localized electrons. Indeed, the 4f electrons can be described as localized whenever sharp and clear peaks are seen by X-ray photoabsorption at the energies corresponding to the excited configuration. If itinerant nf electrons are present, they partially screen the core hole and the Coulomb attraction between the hole and the photoelectron is reduced. No excited configuration is present. This is the case in the tetravalent cerium compounds.

When intense absorption lines are present in the spectrum, the thickness of the absorbent becomes an important experimental parameter. Indeed, it must be such as to keep the experimental response linear at the maximum of absorption. Otherwise, the absorption lines are strongly deformed. This is named the *thickness effect* [13]. The very strong absorption lines necessitate very small thicknesses. Experimental results concerning the shape of this type of lines can hardly be valid if the thickness is not known.

The nd–n'f transitions between atomic-type levels have Lorentzian shapes. In the nd spectra of various rare-earth and actinide materials, one expects the presence of absorption lines of the same shape. The energy and the width of the lines vary with the chemical state and enable its characterization. Because the chemical shift is small with respect to the experimental resolution in the energy range concerned, the spectra must be obtained in high conditions. Specially, the thickness of the absorbent must be compatible with the experimental conditions, i.e. sufficiently small. As an example, a chemical shift has been established between U and UO_2 from the measured energy difference between the U 3d absorption lines. This chemical shift reveals a change of the oxidation state of uranium between the two solids. This result underlines the interest of the nd absorption spectra for the study of these materials, in contrast to what is sometimes mentioned.

In summary, photoabsorption allows to select each element in a compound material and collect information about its oxidation state and the nature of the chemical bonds of the absorbing atoms in the solid and about its unoccupied levels, such as their symmetry, the orbital localized character, the spin state.

3.2.1.1 Lanthanum 3d Absorptions

For lanthanum, the effective potential in the atom core region becomes sufficiently strong to induce the collapse of the 4f wave function. In contrast, the 5f, 6f, wave functions remain largely outside the core region, preventing the observation of transitions to higher series terms. Thus, one expects the oscillator strength to be concentrated in the 3d–4f transitions.

In the solid, the lanthanum 3d absorption spectrum is completely dominated by intense narrow lines characteristic of transitions to the excited 4f configuration. Indeed, the conduction levels have essentially s and d character and only very little p character levels are present. From the electric dipole selection rules, absorptions to the conduction states are expected to be the weak d–p transitions. The relaxation caused by the inner hole is strong enough to pull down the empty 4f levels below the Fermi energy of the metal or in the band gap of the insulator compounds. A 3d electron has a high probability to be excited to the 4f levels and the $3d^9 4f^1$ excited configuration is formed by direct excitation process. Since no 4f electron is present at the ground state, the singly ionized configuration is $3d^9$.

In the metal, the J levels associated with the excited $3d^9 4f^1$ configuration are about -1.8 eV below the Fermi level; they are mixed with the valence levels. If an electron of the valence band participates in the creation of this excited configuration, the energy of the system is minimized. The formation of the excited $3d^9 4f^1$ configuration by ionization followed by electron relaxation is thus energetically possible in the metal but its probability is low as will be shown in the next paragraphs. The excited configuration is created preferentially by direct excitation of a 3d electron into the 4f level.

In the insulator compounds, the levels associated with the $3d^9 4f^1$ configuration are localized in the band gap; the formation of a $3d^9$ ion followed by relaxation to the $3d^9 4f^1$ configuration requires the excitation of an electron from the valence band. The formation of this configuration would need additional energy with respect to that of the ionization process. That is confirmed by compiling together the results of X-ray emission and photoemission spectra obtained for metal and compounds (cf. Sects. 3.2.2 and 3.5.1). The excited configuration is, thus, created by direct excitation.

Experimentally, for a detector with linear response, only two absorption lines are observed for lanthanum metal and its compounds. These lines were denoted $3d_{5/2}$ and $3d_{3/2}$ [41]. Their position is independent of the material considered, within the experimental error of ± 0.1 eV. The $3d_{3/2}$ line is slightly stronger than the $3d_{5/2}$ one. By increasing the absorbing thickness an additional weak line is observed towards the lower energy of the $3d_{5/2}$ line. When the detector observes the three lines

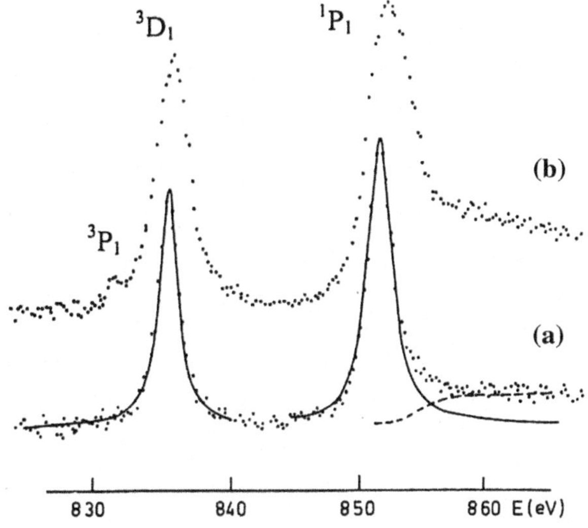

Fig. 3.1 Lanthanum 3d photoabsorption in La$_2$O$_3$ for samples of thickness equal to 280 Å (*curve a*) and 1350 Å (*curve b*). The two main peaks correspond to ^{3}D$_1$ and ^{1}P$_1$. The ^{3}P$_1$ is observable only at large thickness. *Dots* denote experimental data. *Solid line* denotes lorentzian curves. *Dashed line* denotes arctangent curve [41]

simultaneously, its response is already not linear in the range of the two stronger lines. Additional structures are also observed towards the higher energies and introduce an asymmetry of the 3d$_{3/2}$ line [42]. An analysis of the 3d$_{5/2}$ and 3d$_{3/2}$ absorption line shapes has shown them to be Lorentz curves of full widths at half-maximum, FWHM, equal to 1.2 and 1.8 eV respectively (Fig. 3.1) [41]. As indicated above, such line shape is characteristic of a non-interacting level decaying exponentially with a lifetime τ equal to $\hbar$/FWHM.

Independently, it was observed that in the metal the transition from the ground state to the discrete level 3d$_{5/2}^{-1}$4f^1 ^{3}D$_1$ has effectively a lorentzian profile of FWHM 1.1 eV, while the transition to 3d$_{3/2}^{-1}$4f^1 ^{1}P$_1$ was observed to have a Fano profile with FWHM of 1.7 eV [43]. It was suggested that the Fano profile of the ^{1}P$_1$ line was due to the interaction between the excited level and the ionized levels associated with the 3d$_{5/2}$ threshold. It must be underlined that, if the absorption is observed under non-linear response conditions, a 3d$_{3/2}$ absorption edge appears quite clearly and a very faint edge is detected at the 3d$_{5/2}$ threshold. For La$_2$O$_3$, an edge is also clearly observed at about 2.2 eV off the 3d$_{3/2}$ line towards higher energy. It corresponds to a direct ionization to the solid continuum states. From the observed ratio of the strengths of the 3d$_{3/2}$ line and the 3d$_{3/2}$ edge in the metal, the probability of a direct ionization from 3d$_{3/2}$ was estimated to be approximately 9 % of that of the direct excitation while from 3d$_{5/2}$ it is negligible. This ionization explains the weak anisotropy of the 3d$_{3/2}$ emission. On the other hand, the singlet level ^{1}P$_1$ is energetically above the 3d$_{5/2}$ threshold. The non-radiative *Coster–Kronig* decay 3d$_{3/2}^{-1}$4f^1 ^{1}P$_1$–3d$_{5/2}^{-1}$ can take place and it is known to have a high transition probability (cf. Sect. 3.2.3). This decay process of the ^{1}P$_1$ level results in a broadening of the ^{1}P$_1$ line and a decrease in the height of its maximum with respect to the ^{3}D$_1$ line. This presence explains why the intensity ratio of the two lines is smaller than the theoretical value. We shall return to this point in the comparison between the absorption and emission spectra.

The La 3d absorption spectrum has been calculated with the help of an atomic calculation programme (cf. Sect. 3.2.4). For xenon-like triply ionized La^{3+}, photoexcitation can take place from the ground state ($^1S_0$) to two J levels ($^3P_1$ and $^3D_1$) of the $nd_{5/2}^{-1}4f^1$ configuration and to one J level ($^1P_1$) of the $nd_{3/2}^{-1}4f^1$ configuration. Three lines are expected but generally only two strong lines are observed, the more intense one being the ($^1S_0$)–($^1P_1$) line. Indeed, the transition probability of the $^1S_0$–$^3P_1$ line is very low. This line was observed at the predicted energy by using a thick absorbing screen [44]. Under these experimental conditions, the response is not linear and no measurement of the absorption coefficient is possible. In summary, the 3d–4f spectra are well described in intermediate coupling and the designations $3d_{5/2}$ (M_5) and $3d_{3/2}$ (M_4) are only approximate. However, for lanthanum, the spin–orbit is the dominant interaction and $\Delta j = +1$ is the dominant channel. Agreement is good between the experimental and calculated spectra.

Calculations have also been made for La^+ with the external configuration $6s^2$. The two 6s electrons are valence electrons in the solid. As expected, the calculated energies of La^+ are lower than the corresponding ones in La^{3+} and in a better agreement with the experimental values. The difference between the observed and ab initio calculated energies is of the order of one electronvolt, i.e. lower than one percent.

Comparison between the ratio of the calculated probabilities and the ratio of the experimental line heights must take into account the widths of the lines. Since the initial state of the absorption process has a negligible width, the FWHM of each absorption line is equal to the FWHM of the excited state, noted here as $\Gamma_{3/2}$, or $\Gamma_{5/2}$. Let us take two lorentzian curves of area unity; their heights are inversely proportional to their FWHM. Consequently, the calculated transition probabilities of the $nd_{5/2}$ and $nd_{3/2}$ absorption lines must be adjusted by taking into account the $\Gamma_{3/2}/\Gamma_{5/2}$ ratio before they can be compared with the experimental lines. This is done in a next paragraph.

3.2.2   *Electron-Stimulated X-ray Emission (EXES)*

Unlike the photon–electron interaction, an electron loses only a part of its initial energy during a collision with another particle. Each electron can participate in a large number of *inelastic collisions*, each with different energy transfer, before it slows down completely [45]. The collision distribution is determined from statistical laws. Because electrons undergo a large number of inelastic collisions in a material, their range is small. It is clearly smaller than the range of photons having the same energy.

When an electron beam hits a target, the emitted X-ray spectrum is composed of a continuous radiation due to the electron slowing down in the target, labelled bremsstrahlung, and of discrete lines, characteristic of the target elements. The discrete lines are emitted after the creation of a core hole by direct collisional interaction between the incident electrons and the core electrons of the atoms constituting the target. The probability of creating a core hole under electron bombardment is a function of the incident electron energy. This energy must be equal to or higher than the threshold energy necessary to create the hole. Consequently, the creation of a core hole is controlled by the electron slowing down laws through the thickness of the target. The *effective thickness* is defined as the thickness in which the energy of the electron is reduced to the limit energy necessary to create the hole. Numerous formulas giving the line intensity in the X-ray spectra produced by an electron beam have been established [46].

An important characteristic of the interaction of the incident electron beam with the target is that both excited and ionized configurations with a core hole can be created. The rates of excitation and ionization are a function of the incident energy: the excitation processes are predominant in the vicinity of the threshold. In contrast, the excitation diminishes rapidly above the threshold, where the ionization predominates. The electron excitation probability is proportional to the photoexcitation probability in the electron energy range where the Bethe–Born approximation is valid. This approximation holds if the energy of the incident electron is equal or superior to twice the threshold energy. In this energy range, the incident electrons can excite atomic electrons according to the dipole selection rules. For incident electron energies near the excitation threshold, the generalized oscillator strengths characteristic of the electron–electron collisions are not proportional to the optical oscillator strengths and all the J-levels of the excited configuration can be populated but the probability to populate the J-levels accessible by photoexcitation stays predominant.

As the width of the incident electron energy distribution is large with respect to that of the excited states, the excitation and ionization processes are very fast and of the collisional type. Consequently, they can be considered as first-order processes and the radiative and non-radiative recombination processes of the excited, or ionized, state can be decoupled from the initial process and studied separately. The timescale associated with the excitation, or the ionization, of a core hole is of the order of the information time, i.e. 10^{-17}–10^{-18} s; it is much shorter than the time of the decay of the core hole. A fundamental question is to know the relaxation time of excited or ionized configurations. They can relax on the timescale of core-hole decay [47]. This supposes that, after the ejection of the core electron, the system relaxes towards the ground level of the configuration. Only after it is fully relaxed can it be considered as the precursor state of the X-ray and electron emissions. This "two-step model" can be used to predict the spectra and to calculate the width associated with the core-hole lifetime. In this model, the core-hole decay is considered to be independent of the creation of the initial core-hole state. As an example, if a secondary excitation process (shake-up process) occurs together with the core-hole creation, this process is decoupled from the core-hole decay. Each decay process takes place independently and their probabilities are added to determine the lifetime of the core hole. The two-step model holds generally in the soft X-ray region but might not be valid for the emissions at very high X-ray energies. For those, the timescale of the core-hole annihilation can be shorter than that of the relaxation. It becomes, thus, necessary to treat this annihilation in a one-step model. Such a model takes into account the interferences between the different core-hole decay channels. The problem now is to distinguish between two cases where intense resonant lines are expected: a) The radiative reorganization takes place in a very short time, thus the excited configuration has not enough time to relax; or b) The decay process is slower than the time of the J levels statistical reorganization and all the J levels of the configuration can contribute to the emission. This will be easily illustrated from results concerning lanthanum.

The relaxation of a discrete excited level that involves the valence orbitals depends on the solid considered. This is the case if the relaxation is due to an energy transfer between the discrete level and a valence orbital; this process is dependent on the energy of the various levels. The relaxation is faster for a metal than for an insulator. In a compound, the process was compared to an inter-atomic charge transfer.

The ionization of the core electron creates a positive charge localized at the position of the hole, causing a change of potential in the core region. All the electrons respond to this potential change and the system relaxes, resulting in a redistribution of the electrons around the core hole. This change is large for localized electrons and can induce their additional excitation or ionization (shake-up or shake-off processes); it is small for delocalized valence electrons. In the case of an initial excitation, the screening is due mainly to the excited electron and the change of other electrons is small. Consequently, X-ray emission spectral profiles vary with the incident electrons energy above the threshold. They depend on the way the configuration with the core hole is prepared.

The intensity of the nlj–$n'l'j'$ emission line depends on three factors, the number of initial configurations with one nlj core hole created per unit time, the probability of the nlj–$n'l'j'$ radiative transition and the self-absorption of the radiation in the emitter. The electric dipole transitions are governed by the dipole selection rules. Various types of lines are present in the spectrum, depending on their initial and final states, the initial state always including a hole in a core level. These lines can be regrouped in normal lines, resonance lines and satellites.

An X-ray *normal line* is emitted following an electron transition between two inner levels of a singly ionized Z atom (Fig. 3.2). The initial configuration is that of a singly ionized atom with a hole in an nlj sub shell. The final configuration is obtained after transfer of an electron from a less energetic $n'(l \pm 1)j'$ sub shell to the nlj hole. The energy of the normal line is equal to the difference between the energies of the nlj and $n'(l \pm 1)j'$ levels, $E(nlj)$–$E(n'(l \pm 1)j')$. If the energy of the nlj level is known, the energies of other levels of the Z element may be deduced from the normal X-ray emission energies. This property was used to determine the energy levels of elements. Conversely, the energies of normal X-ray emissions can be determined from the energies of levels established by photoemission.

The energy distribution of a line emitted between two discrete levels, nlj and $n'l'j'$, is the convolution of the energy distributions of these two levels. It is a Lorentz curve

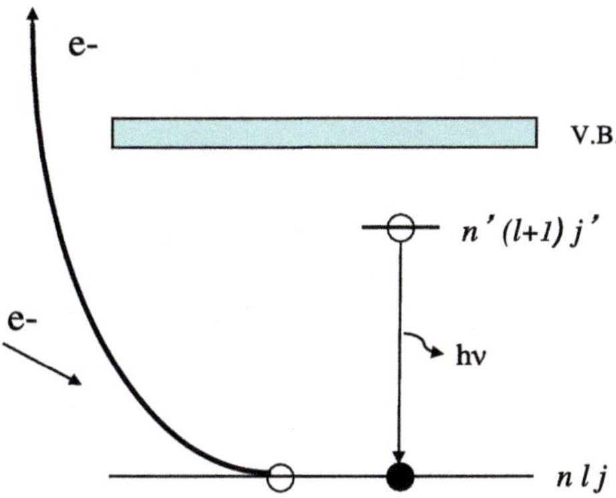

Fig. 3.2 Schematic representation of X-ray normal line

with a full width at half-maximum Γ equal to the sum of the width of the two levels. As already indicated, a lorentzian curve is characteristic of the exponential decay of a discrete atomic state with the lifetime $\hbar/\Gamma$. When the transition takes place in an atom having an open sub shell, two or more holes can be present simultaneously in localized sub shells and each configuration may undergo a splitting that can reach 10 eV. The experimental resolution may not be enough to distinguish between the various sub-levels. The observed emission is then one broad asymmetric peak.

If a normal transition takes place from a level belonging to the valence band of a solid to a core level, its energy distribution is the convolution of the Lorentz distribution of the nlj inner level by the density $N(\varepsilon)$ of the valence states having the symmetry $(l \pm 1)$. It is named *emission band* (Fig. 3.3). These transitions are observed in the soft X-ray range where the energy resolution is better. From emission bands, it is possible to deduce the distribution of the valence states of predetermined symmetry, characteristic of each of the elements present in the solid. The emission bands are generally interpreted successfully in terms of one-electron

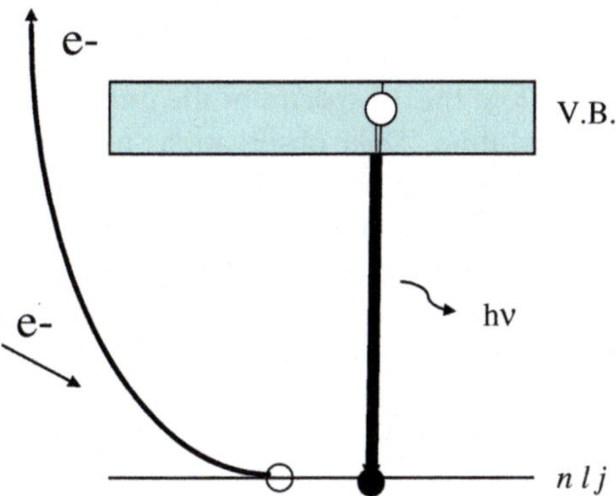

Fig. 3.3 Schematic representation of X-ray emission band

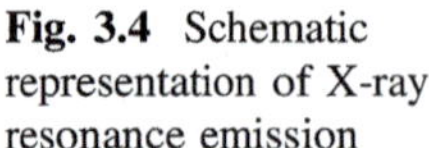

Fig. 3.4 Schematic representation of X-ray resonance emission

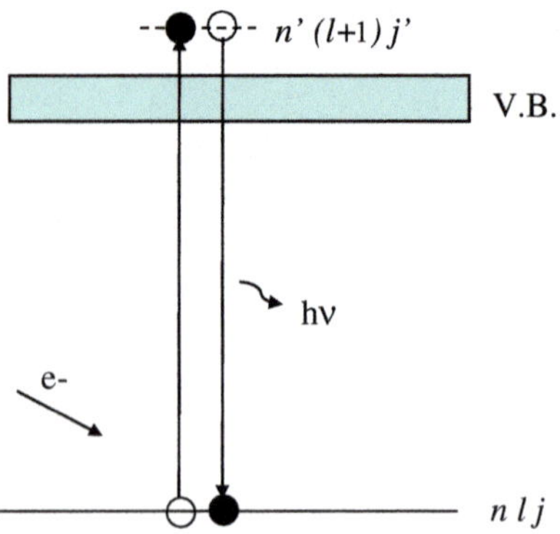

band structure calculations and by considering the two-step model, often named the final-state rule. They provide a measure of the angular-momentum-selected partial density of valence states.

A *resonance line* is emitted during the direct recombination of an excited electron with its hole (Fig. 3.4). The initial configuration is an excited X state with an electron in the discrete levels $n'(l + 1)(j, j \pm 1)$ and a nlj hole. The final state is the ground state. This emission is the inverse process of the photoexcitation: indeed, the resonance lines have the same energy, the same shape and the same width as their counterparts in photoexcitation. A resonance line is observed only if the excited electron is localized in the same atom with the same spin during the lifetime of the excited X state. There are transitions that take place between quasi-atomic discrete states for which configuration interactions, particularly with the continuum states, are negligible. Consequently, their decay may be described with the help of the Weisskopf–Wigner independent particle model.

Initially, this type of lines was considered observable only in spectra of free atoms. In solids, the excited states are generally extended states and it was thought that such lines could not be observed. However, around 1970, X-ray resonance lines from the 3d spectra of the rare-earth metals and compounds were reported for the first time. Their observations received little attention. In these experiments, the initial excited X state $3d^9 4f^{m+1}$ was created by bombardment with electrons having energies between the threshold and twice the threshold of the excited state. Subsequently, resonance lines were also observed in the 3d spectra of the actinides but with lower intensity since the interactions between the localized 5f electrons and the other valence electrons compete with the Coulomb interactions between the 5f electrons and the core hole.

Let us note that the usual practice is to name "resonance lines" only the emissions stimulated by photons. The electron-stimulated $3d^9 4f^{m+1}$–$3d^{10} 4f^m$ emissions had been designated as resonance lines [32] because they were observed in an exact energy coincidence with the corresponding photoexcitation lines. This energy coincidence was confirmed by observations made in strong radiation self-absorption conditions. Indeed, since the observation of the inversion of the sodium D lines by Gustav Kirchhoff in 1859, which has revealed for the first time the existence of the absorption process, it has been known that an excitation line can be reabsorbed and appear as an absorption line, which consequently was named *white line*. When an incident electron beam of energy several times the excitation threshold is used, the resonance line is emitted inside a noticeable thickness of the target and is strongly self-absorbed. Under such conditions, an absorption line was observed at the position of the emission line, thus confirming that the emission and absorption lines between the same pair of levels have the same energy. The observed coincidence of an emission line with its corresponding intense absorption line is a direct proof of the *localized character* of the concerned excited electrons in the solid.

Identification of the resonance lines is possible in the following way: the spectrum is induced with electrons having incident energy E_0 between only several eV above the threshold energy and 1.6 times this energy. For each E_0 value, the intensity of the emission from the excited state is compared to that of an emission from the ionized state, labelled normal atomic emission. With E_0 increasing, the intensity of the resonance line decreases while that of the normal line increases considerably. The change of the intensity with E_0 serves to identify these two emissions. This method does not require the observation of the photoexcitation spectrum and it can easily be generalized to complex materials.

It was suggested that the resonant emission process could be regarded as a resonant scattering with temporary negative ion formation and subsequent radiative decay [48]. But, in this hypothesis, no relaxation of the excited configuration could take place and the lines in coincidence with the absorption lines should be the only emitted lines following the excitation, in contrast with the observation of other lines emitted from the excited configurations. Another suggestion already mentioned was that the excited X state, which is the initial state of the resonance lines, could be obtained by ionization following by relaxation. From energy and intensity considerations, it will be shown for lanthanum, then for the rare-earths, that only direct radiative recombination processes can account for the high relative intensity of resonance lines observed in the neighbourhood of the nd thresholds.

Other emissions can be observed from an excited configuration; they are *normal lines in the presence of a spectator electron* (Fig. 3.5). These emissions are radiative transitions equivalent to those observed from the ionized configuration. The excited electron is not involved in the process and remains "spectator". These transitions take place between excited initial and final configurations and, consequently, their energy is different from that of the normal lines. As already underlined, under bombardment with electrons of incident energies near the threshold, all the J levels of the excited configuration can be populated and the energy of the configuration is that of its barycenter. This energy E_S is the sum of the energies E_J of all the J levels weighed by their statistical weights, divided by the sum of the statistical weights

Fig. 3.5 Schematic representation of X-ray normal emission in the presence of a spectator electron

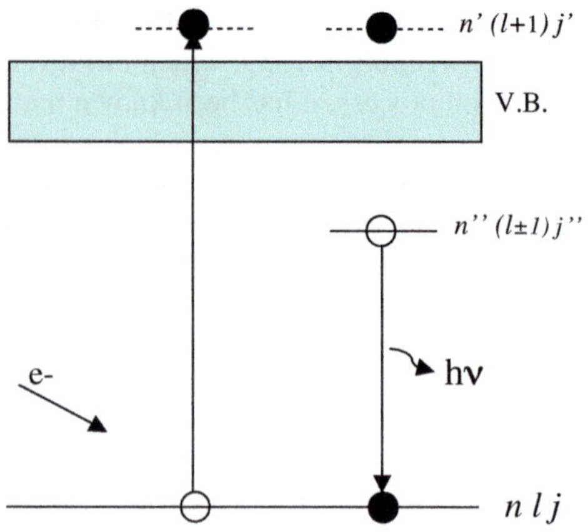

$$E_s = \frac{\sum_J (2J+1)E_J}{\sum_J (2J+1)}$$

A normal line in the presence of a spectator electron is the transition between levels of two such excited configurations. In contrast, each resonance line originates from one of the J levels of the excited X configuration, which can decay to the ground state by the dipole selection rules. Consequently, the initial configuration of the normal lines in the presence of the spectator electron and the initial configuration of the resonance lines have different energies and the same holds for the emission lines. Normal lines in the presence of a spectator electron were observed in solids only when elements with f electrons are present. In the rare-earth 3d spectra, they are located towards the lower energies of the corresponding normal lines. Their observation reveals that a sufficient number of electrons remain localized in the vicinity of the core hole throughout all the duration of the X-ray emission. *Satellite lines* are emissions in multiply excited or ionized systems. The most probable additional excitations or ionizations are due to *shake-up* and *shake-off* processes accompanying the sudden creation of the inner hole (Fig. 3.6).

These processes are caused by changes in the effective charge that accompany electron ejection. A shake-up process corresponds to the excitation of a second electron by ($\Delta l = 0$) transition during the ionization of the core; in the final state, the atom is doubly excited and ionized. In a shake-off process, the atom is doubly ionized. The satellite intensity is low and increases with the incident electron energy. These emissions are observed on the higher energy side of the normal or resonance lines. These lines are generally difficult to interpret because various quantum states with similar energies can be created concurrently. These satellites are present in the spectra of free atoms as well as solids. In the solid compounds, other satellites can be present with quite noticeable intensity: an inter-band transition can take place from ligand valence levels to unoccupied levels of the metal ion simultaneously with the normal transitions.

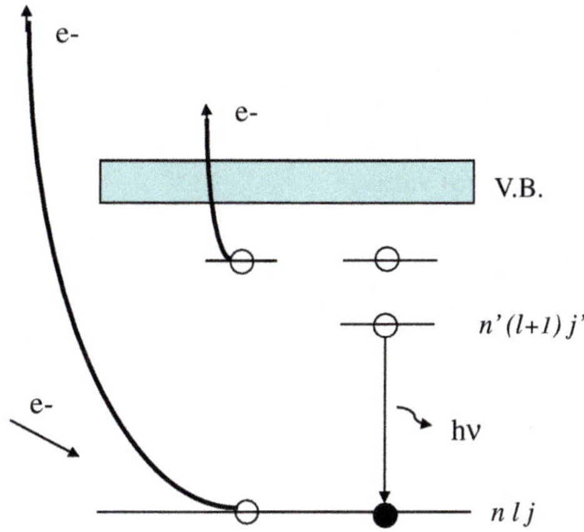

Fig. 3.6 Schematic representation of X-ray shake-up satellite

3.2.2.1   Lanthanum 3d Emission

Two intense lines are observed in the electron-stimulated La 3d emission spectrum; they coincide in energy with the two intense 3d absorption lines. These lines are due to the inverse process of absorption, i.e. to the direct radiative emission from the two levels $^3D_1$ and $^1P_1$ of the singly excited configuration $3d^9 4f^1$ to the $3d^{10}\ ^1S_0$ ground state. They are resonance lines. The experimental and calculated energies are given in Table 3.1 along with their calculated transition probabilities. Other transitions appear in the La 3d emission spectrum, making this spectrum more complex than the 3d absorption one (Fig. 3.7). These transitions are the $3d^9$–$5p^5$ normal emissions in the ion, the $3d^9 4f^1$–$5p^5 4f^1$ normal emissions in the presence of a 4f spectator electron, the $3d^9 4f^1$–$3d^{10}$ emissions from the non-resonant levels of the excited configuration to the ground state and emission bands from valence electron distributions. The appearance of normal lines in the presence of a 4f

Table 3.1 Energies (eV) and transition probabilities of lanthanum $3d^9 4f^1$–$3d^{10}$ radiative recombination [41]

	Calculated values		Experimental values	
	Probability	Energy	Energy	
			Metal	La_2O_3
$^3P_1$	0.0009×10^{13}	830.3		
$^3D_1$	0.900×10^{13}	834.6	834.9	834.9
$M3d_{5/2}$		831.8		
$^1P_1$	1.847×10^{13}	850.9	851.5	851.5
$M3d_{3/2}$		848.5		

$M3d_{5/2}$ and $M3d_{3/2}$ are the barycenters of the configurations $3d_{5/2}^{-1}4f^1$ and $3d_{3/2}^{-1}4f^1$. The intensities of the emissions from $^3D_1$ and $^1P_1$ depend on the transition probabilities but also on several other parameters such as the probability to create the excited states under electron irradiation and the self-absorption

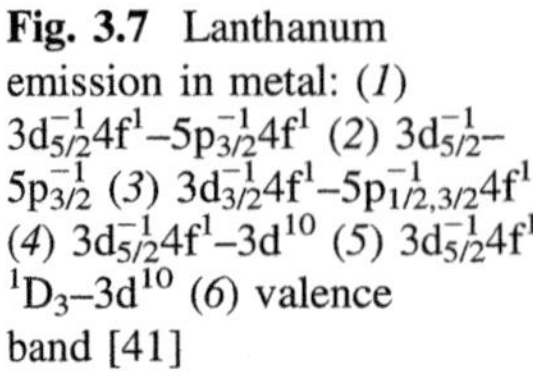

Fig. 3.7 Lanthanum emission in metal: (*1*) $3d_{5/2}^{-1}4f^1$–$5p_{3/2}^{-1}4f^1$ (*2*) $3d_{5/2}^{-1}$–$5p_{3/2}^{-1}$ (*3*) $3d_{3/2}^{-1}4f^1$–$5p_{1/2,3/2}^{-1}4f^1$ (*4*) $3d_{5/2}^{-1}4f^1$–$3d^{10}$ (*5*) $3d_{5/2}^{-1}4f^1$ $^1D_3$–$3d^{10}$ (*6*) valence band [41]

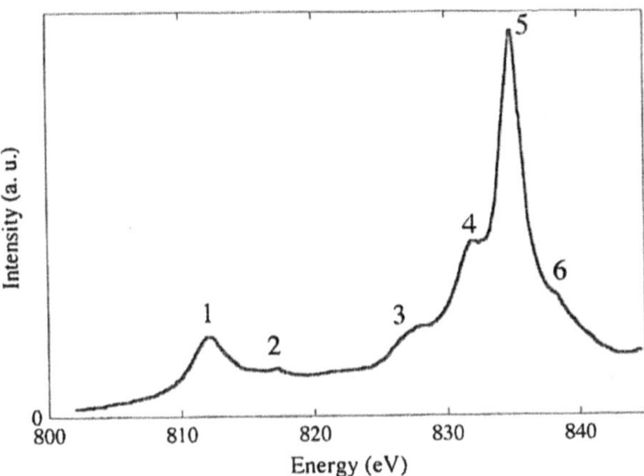

spectator electron confirms the localized character of the 4f electron. Transitions with spectator electron, as well as resonance lines, have never been observed in the spectra of solids that do not possess an f electron.

The classification of the various X-ray emissions is based on the energies of the X levels and those of the excited configuration obtained from the absorption spectra. As an example, let us consider the La $3d_{5/2}$–$5p_{3/2}$ normal line in the metal. Its energy is equal to the difference between the $3d_{5/2}$ and $5p_{3/2}$ energy levels. The energy of each level can be determined by photoemission. On the other hand, the energy of the $3d_{5/2}$ level is equal to the difference between the energies of the $2p_{3/2}$ (L_{III}) absorption and the $2p_{3/2}$–$3d_{5/2}$ ($L\alpha_1$) emission. No excitation process is present in the $2p_{3/2}$ absorption and the measured energy corresponds to the energy of the $3d_{5/2}$ level with respect to E_F. From these values, one deduces that the $3d_{5/2}$–$5p_{3/2}$ line is located at 820 eV (Table 3.2), in agreement with the weak line observed at 819 eV. The line at the lower energy, 812.4 eV, is not interpretable from the 3d photoemission peaks. It is not due to a transition in the ion with a 3d hole. It has been identified as the transition $3d_{5/2}^{-1}4f^1$–$5p_{3/2}^{-1}4f^1$ from a theoretical calculation taking into account all the dipolar transitions possible between the J levels of both excited configurations (cf. Sect. 3.2.4). These results easily prove that the 5p–3d transitions in the excited configuration are located towards the lower energies of the corresponding transitions in the ionized configuration. Moreover, this line is observed at the same energy in metal and compounds and this confirms that the initial configuration is obtained by direct excitation and not following an interaction with the valence band, contrary to what was suggested by several authors [49].

The doubly excited configuration $3d^94f^2$ can be created in lanthanum according to the process $3d^{10} + e^- \rightarrow 3d^94f^2$, i.e. by elastic collisions with electrons having energy just equal to the 3d–4f excitation energy. The number of such incident electrons is small in the beam and so is the number of the elastic collisions, but the emissions $3d^94f^2$–$3d^{10}4f^1$ are not reabsorbed, in contrast with the resonance lines. Structures have effectively been observed around the $3d_{5/2}$ and $3d_{3/2}$ excitation

Table 3.2 Lanthanum 5p–3d lines [41]: the energies (eV) of the emissions in the ion are predicted from the difference between the level energies

	Deduced from levels	Experimental	
		La	La_2O_3
$3d_{5/2}^{-1}4f^1$–$5p_{3/2}^{-1}4f^1$		812.4	812.4
$5p_{3/2}$–$3d_{5/2}$	818.8	819.0	
$3d_{3/2}^{-1}4f^1$–$5p_{1/2}^{-1}4f^1$		828.6	828.6
$5p_{1/2}$–$3d_{3/2}$	833.1		
$^3D_1$	834.6	834.9	

The other lines correspond to the emissions in the excited atom, $3d_{5/2}^{-1}4f^1$–$5p_{3/2}^{-1}4f^1$ and $3d_{3/2}^{-1}4f^1$–$5p_{1/2}^{-1}4f^1$. They are located towards the lower energies of the normal lines. The normal line to $5p_{1/2}$ is not resolved from the $^3D_1$ intense line

thresholds of the lanthanum towards the lower energies of the main peaks and interpreted as due to the scattering of incident electrons into vacant 4f levels [48].[1] Subsequently, these structures were treated theoretically as a resonance in the bremsstrahlung emission at the 3d threshold [50]. But their observation using incident electrons of energy relatively far above the 3d threshold contradicted these interpretations and other processes had to be considered. These lower energy structures can be identified with the decay towards the ground state of all the J levels of the excited configuration $3d^9 4f^1$ other than the levels $^3D_1$ and $^1P_1$. Indeed, agreement exists between the energies of the observed structures and the average energies of the $3d^9 4f^1$ configurations in the lanthanum $3d_{5/2}$ and $3d_{3/2}$ ranges. Moreover, their intensity varies with the energy of the incident electrons in agreement with the variation of cross sections. In the vicinity of the threshold, the electron *generalized oscillator strengths* do not follow the dipolar selection rules. The excitation is the dominant process and takes place to all the J levels of the excited configuration. The energy of the excited configuration can be represented by its barycentre and the population of its levels is statistical. The above structures are more intense at the threshold than the resonance lines. When the energy of the incident electrons increases, dipolar excitation increases. The intensity of these emissions decreases and there are weak structures located towards the lower energies of each $3d_{5/2}$ and $3d_{3/2}$ intense resonance lines.

The intensities of the lines emitted from $3d^9 4f^1$ are slightly lower in the metal than in the compounds. On the other hand, the emission line $3d_{5/2}$–$5p_{3/2}$ in the $3d^9$ ion at 819 eV is observed in the metal but not in La_2O_3 and LaF_3. These differences reveal the presence of a small percentage of ionized $3d^9$ configurations in the metal, while in the compounds the excited configurations dominate completely. As deduced from photoabsorption, excited $3d^9 4f^1$ levels other than $^3D_1$ and $^1P_1$ are energetically mixed with the valence level distribution of the metal with a 3d hole.

---

[1] In Ref. [48], main peaks observed at the position of the excitation energies $3d^9 4f^1$ $^3D_1$ and $^1P_1$ were designated as being $M\alpha$ and $M\beta$ whereas no 4f electron is present in lanthanum and such emissions are not observable.

Consequently, these levels can interact with the solid continuum during the time of core-hole decay. The interaction leads to a very partial delocalization of the 4f electrons in the metal with respect to the compounds, making the intensity of the lines from the excited state weaker in the metal. However, this intensity variation is small, as is the intensity of the normal line $5p_{3/2}$–$3d_{5/2}$. In spite of an energy mixing of the 4f levels and valence levels, the excited 4f electron remains definitely localized in the presence of a $3d_{3/2}$, or $3d_{5/2}$, core hole in the metal.

Various results based on energy considerations confirm that the precursor configurations of resonance lines do not result from ionization followed by relaxation but from the direct excitation of the ground state. First, the variation of resonance emission intensity with the energy of the incident electrons is different from that of normal X-ray emissions. Indeed, the intensity of the resonance emission decreases with increasing incident electron energy, while the inverse is true for lines that were emitted after the ionization of a core level took place. Secondly, compilation of the energies of the normal X-ray lines and of the photoemission peaks confirms that it is not possible to explain the presence of the emissions from the excited X states. Thirdly, from atomic calculations, the 3d transitions from the excited and ionized configurations are predicted at energies in agreement with the energies of the observed lines for metal and compounds. Lastly, the relaxation processes following an initial ionization are different in the metal and in the compounds and, consequently, the emissions from such relaxed configurations do not have the same energy. The same holds for the transitions of the excitonic type, which are different for a conductor and a non-conductor. In contrast, emissions observed during the direct recombination from the excited state to the ground state are expected to have the same energy for lanthanum metal and compounds because they are the inverse process of the photoexcitation and the photoexcitation transitions are observed to have the same energy in the various materials.

Only the two levels $^3D_1$ and $^1P_1$ of the La $3d^9 4f^1$ excited configuration have a large enough probability to be photoexcited and to disappear by direct recombination before the relaxation, resulting in the emission of the two intense resonance lines. In contrast, all the J levels are involved in the emission of the 5p–3d lines, whose energy is equivalent to the energy difference of the barycentres of $3d_{5/2}^{-1}4f^1$ and $3d_{3/2}^{-1}4f^1$ excited configurations, located several eV below the $^3D_1$ and $^1P_1$ levels. If relaxation took place before the direct radiative recombination to the ground state, i.e. before the emission of the resonance lines, the probability for populating the $^3D_1$ and $^1P_1$ levels should be only 3.6 and 5.3 % respectively of the direct excitation probability. The ratio of the radiative transition probabilities of the 4f–3d and 5p–3d emissions is about 20 to one but the 4f–3d resonance lines are strongly self-absorbed. By taking into account the three parameters, excited level population, radiative probability and self-absorption, the $^3D_1$ resonance line should be less intense than the 5p–3d emission. This is contrary to the observations. Indeed, for all the incident electron energies, the resonance emissions are much more intense than the 5p–3d ones. This shows that the intense lines in coincidence with the absorption lines are emitted before the relaxation takes place and are indeed resonance lines as described above.

Let us recall that it was suggested to describe the La M_4 ($^1P_1$ line) absorption by a Fano curve and the La M_5 ($^3D_1$ line) absorption by a Lorentz curve. Since the 4f–3d emissions are the reverse transitions of the absorptions, the same shape difference should be expected between the $^1P_1$ and $^3D_1$ emission lines. However, no such shape difference was observed [51].

The intensity ratios of the peaks $^3D_1$ and $^1P_1$ observed in X-ray emission and in photoabsorption are different. The intensity ratios of the peaks observed in photoabsorption and in EELS are different, while they are the same for the peaks observed in X-ray emission and EELS. In the latter case, the primary process of the core-hole creation is the same, while it is different in the two other cases. By taking these results into account, it is possible to explain entirely the intensity difference between X-ray emission and absorption and to conclude that the two processes, absorption from the ground state to the excited state and recombination from the excited to the ground state, are, indeed, inverse processes.

In summary, from the study of the 3d emission spectra observed for lanthanum in metal and compounds, it is shown that the characteristics of the nf electrons in any material can be deduced from their spectral analysis. Indeed, if the nf electrons are localized, two types of nf–3d emissions associated with the localized nf electrons are expected using incident electrons: (a) the resonance lines associated with the initial excited configuration nf^{n+1} which correspond to the transition $nf^{n+1} \rightarrow nf^n$; (b) the normal emissions from the 3d ionized configuration which correspond to the transition $nf^n \rightarrow nf^{n-1}$. The resonance lines are located towards the lowest emission energies of the normal lines and their intensity varies with the energy of the incident electrons inversely to the normal lines. They have particular characteristics: they have the same energy for the same element present in different compounds; they have a large probability to be emitted before relaxation processes have time to intervene; they are very strongly reabsorbed in the target; they largely dominate the spectrum; independently of the considered compound, their energy is in good agreement with the energies calculated for the transitions in the free ion. Moreover, the direct decay is more probable if the atomic character of the excited state is more pronounced and the interactions with valence or conduction electrons are weaker. However, very small differences can exist in a metallic material because some levels interact and relax before they decay but their number remains small. Consequently, from the presence or absence of resonance lines, it is possible to deduce what is the localized character of the nf electrons and, by comparison with the calculations made for various configurations, to deduce the numbers of nf localized electrons.

In the case where itinerant nf electrons are present, they can be more or less mixed with the dsp valence electrons and a narrow emission band is present, only partially mixed with the dsp band. This is the case for cerium under pressure and in its tetravalent compounds (cf. Chap. 4). In light actinides, itinerant and localized 5f electrons are present and more complex spectra are predicted.

3.2.3 *Electron-Stimulated Auger Transitions*

The Auger transitions are the non-radiative transitions competitive with the X-ray emissions because they take place from the same initial quantum states. Consequently, there exist normal Auger peaks taking place from singly ionized configurations, resonant Auger peaks and normal Auger peaks with a spectator electron taking place from mono-excited states, satellite Auger peaks taking place from multiply excited or ionized states (Figs. 3.8, 3.9 and 3.10).

In a normal Auger transition of a Z element, the initial quantum state is a core-ionized atom with the ionized electron in the continuum. An electron jumps from a higher $n'l'j'$ sub shell to the initial nlj core hole and transmits its energy to an electron $n''l''j''$ which is emitted with the kinetic energy E_c

$$E_c = E(nlj) - E(n'l'j') - E(n''l''j'', n'l'j').$$

$E(nlj)$ and $E(n'l'j')$ are the energy levels of the singly ionized configurations with a nlj, or $n'l'j'$, hole. $E(n''l''j'', n'l'j')$ is the energy of the $n''l''j''$ configuration in the presence of the $n'l'j'$ hole; this energy can be approximated by the energy of the $n''l''$ j'' level in the element of $Z + 1$ atomic number. The Auger line shape reflects the convolution of the initial and final configurations, i.e. an atom with an nlj hole and a doubly ionized atom. It therefore depends on the width of the two configurations.

The energy gained by the system following the electron jump out of the initial nlj level can be transmitted either to an electron of the same atom (non-radiative transition) or to a photon (radiative transition). The *fluorescence yield* of an (nlj) level, ω_{nlj}, is defined by

$$\omega_{nlj} = \Sigma P_R / (\Sigma P_R + \Sigma P_{NR})$$

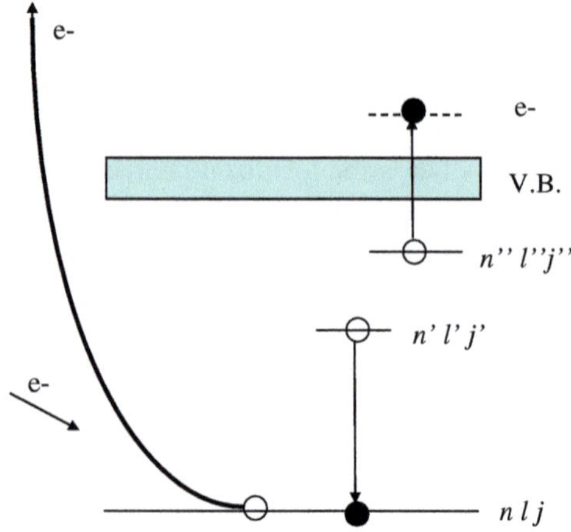

Fig. 3.8 Schematic representation of Auger normal transition

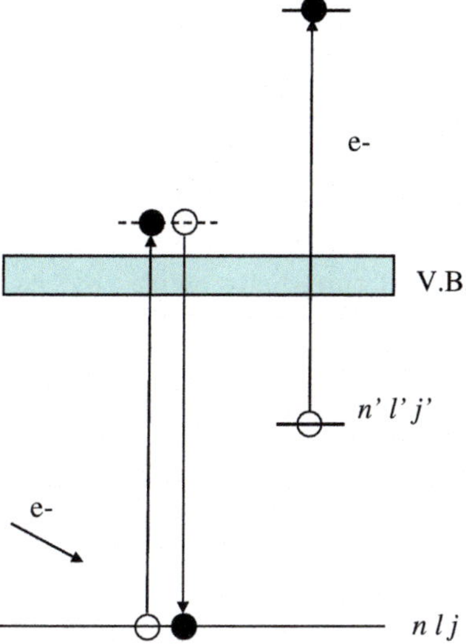

Fig. 3.9 Schematic representation of Auger resonant transition

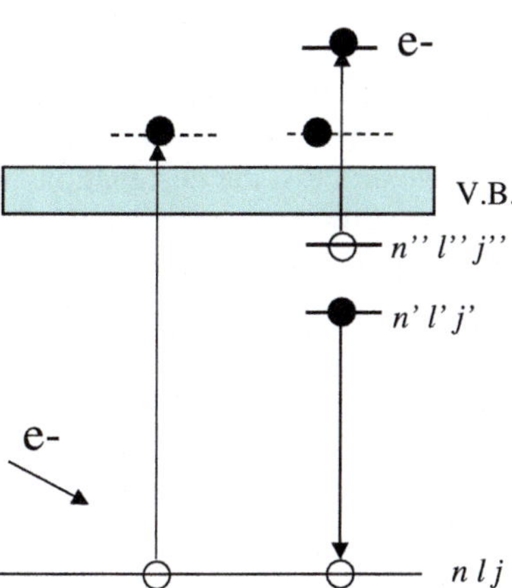

Fig. 3.10 Schematic representation of Auger normal transition in the presence of a spectator electron

ΣP_R and ΣP_{NR} are the sums of the radiative and non-radiative decay probabilities of the nlj core hole. The fluorescence yield ω_{nlj} of a Z atom decreases with n and l increasing. It varies approximately as Z^4. It follows the variation of the radiative probabilities, which increase practically as Z^4 while the non-radiative probabilities decrease slowly according the relation $a/Z^4/(1 + a/Z^4)$, where a is a constant depending on the considered sub shell. Contrary to radiative transitions for which interaction of electromagnetic origin is present, no symmetric selection rule exists in Auger spectroscopy because the interactions are of the Coulomb inter-electronic type. This type of interaction is dominant for low-Z elements, leading to a fluorescence yield in the order of about 10^{-4} for first row atoms. The Auger spectra are more complex than the emission spectra because of the absence of selection rules. This makes their interpretation difficult.

When two levels belong to the same shell the Auger transitions are named Coster–Kronig transitions. They are very intense and take place when the energy difference between the two levels is bigger than the energy necessary to eject an outer electron. Their probability varies with Z as that of the Auger transitions. The three types of Coster–Kronig transitions, normal, with a spectator electron and resonant, were observed but sometimes misidentified. As an example, unnecessary complicated explanations have been proposed to interpret transitions between the two 3d sub shells, while, indeed, these transitions are simply the well known Coster–Kronig $3d_{3/2}^{-1}4f^{m+1}$–$3d_{5/2}^{-1}4f^m\varepsilon f$, always normally present as soon as they are energetically possible.

In a resonant Auger transition, the initial energy level is an excited level of energy E_i that belongs to the singly excited configuration $(nlj)^{-1}(n'l'j')^{+1}$ of energy $E[(nlj)^{-1}(n'l'j')^{+1}]$. The energy received by the system during the direct recombination of the $n'l'j'$ excited electron into the nlj core hole is transmitted to an electron $n''l''j''$ which is emitted with the kinetic energy E_c.

$$E_c = E\left[(nlj)^{-1}(n'l'j')^{+1}\right] - E(n''l''j'')$$

In contrast with a normal Auger process, the initial state is a bound state of energy equal to the energy lost by the incident particle, electron or photon. The final state is a singly ionized state while it is doubly ionized in a normal process. Like for the radiative process, the designation "resonant Auger transition" is justified by the fact that the initial state of the transition has energy equal to E_i. This process is also called "direct recombination channel".

In a normal Auger transition with a spectator electron, the initial state is the same as that of the resonant Auger process and the final state is an excited and doubly ionized configuration. The kinetic energy E_c of the emitted electron is

$$E_c = E\left[(nlj)^{-1}(n'l'j')^{+1}\right] - E\left[(n'l'j')^{+1}(n''l''j'')^{-1}(n'''l'''j''')^{-1}\right]$$

Let us note that the resonant Auger transitions are sometimes designated as auto-ionization processes. An excited X state is mixed with continuum levels of the same energy, i.e. with levels associated with sub shells of lower energies than the X state. This mixing induces an interaction between the excited X level and the continuum levels of the same energy. Only when this configuration interaction is strong, auto-ionization is possible. The decay process is represented by an energy distribution described by a Breit-Wigner-Fano curve, $F(\varepsilon)$. Contrary to a symmetric Lorentz curve, $F(\varepsilon)$ is asymmetric and this asymmetry becomes more pronounced as the configuration interaction between the discrete and continuum levels becomes stronger. Auto-ionization has to be taken into account in the UV region where it is as fast as the radiative and Auger processes. However, in the X-ray region, where the radiative and Auger processes are much faster, the auto-ionization is negligible.

As for X-ray emission, a two-step model is adequate for describing the essential processes of the Auger-like decay. In the first step, a core hole is created and in the second step, an Auger electron or a photon is emitted. The two steps are treated as consecutive and independent, i.e. as incoherent events. Moreover, no interference exists between the radiative and Auger transitions.

3.2.4 Theoretical Calculations of the X-ray and Auger Spectra

Until this review, the decay of excited configurations has not been the subject of systematic study for the rare-earths while the decay of the ionized quantum states has been studied theoretically in a systematic manner for all the elements. In order to compare the dynamics of nd excited and ionized configurations, the radiative and Auger decay rates from La^{3+} nd^9 and $nd^9 4f^1$ configurations have been calculated. The calculations are presented in this paragraph. When the spectra are stimulated by electrons or by photons of energy higher than the threshold energy, the core hole creation and its subsequent decay by X-ray or Auger emission are treated as two independent processes. The case of the core hole creation by photons of energy lower than or very near the threshold is considered in a following paragraph. Electron–phonon coupling can be neglected in the X-UV energy range, making the matrix elements independent of the arrangement of the atoms.

Only transitions concerning localized electrons have been considered. Wave functions and energies were computed with a multiconfiguration Dirac–Fock (MCDF) program including Breit interaction and QED corrections [52]. Corrections due to Breit interactions increase with the configuration energy. In the 3d range, they vary between -1.6 and -2.1 eV for La^{3+} and between -3.7 and -4.7 eV for Yb^{3+}. The QED correction increases also along the rare-earth series but remains below $+0.05$ eV. The calculation of the quantum states was made with the help of the extended average level (EAL) extension of the MCDF method. The wave functions of the initial and final states of each transition were built from the same set of orbitals. To minimize the relaxation contributions, the transition state defined by Slater [53] was used, giving equal weight to the initial and final states. Thus, at the first order of the development, the relaxation corrections cancel.

The energies and the oscillator strengths of the photoexcitation transitions from the ground state to the excited configurations with an nd hole and an additional excited 4f electron have been calculated for all the rare-earths. Intermediate coupling was used. Only electric dipole transitions were taken into account. The velocity form of the transition matrix elements was used in the X-ray region, i.e. above about 500 eV, and the length form was employed for transitions of lower energy. The calculations were made by using the golden rule. The photoexcitation spectra were obtained by summing up all the electric dipole allowed lines from the ground level of the initial configuration. Each line was fitted by a Lorentz curve whose surface is equal to the excitation cross section and whose width results from the lifetime of the excited state. The average energy of each one of the configurations is the barycenter of all its J levels.

For calculating each X-ray emission it was assumed that all the J levels of the initial configuration are populated statistically. The energy and the probability of all the allowed electric dipolar lines from these J levels to all the J' levels of the final configuration were included. Each J–J' line was obtained by folding the theoretical probability with a lorentzian broadening function whose width results from the sum of the lifetimes of the J and J' levels. The total emission was determined as the weighed sum of all the lines participating in this emission. This model gives the energies and the probabilities of all the transitions from the excited and ionized configurations with a 3d hole to the corresponding $3d^{10}$ final configurations with one 4f or 5p electron missing [34]. The calculations were made for all the triply ionized rare-earths. From these results, the absorption and emission spectra have been simulated. As discussed in Chap. 4, they reproduce well the main features of the experimental spectra.

The probabilities of transitions between an initial and final sub shells increase with the overlap of their wave functions. Overlap of the nd (n = 3, 4) and 4f sub shells increases in the presence of an nd core hole and due to its localized form the 4f wave function leads to a large dipole matrix element. Indeed the nd–4f observed transitions are very strong in the solid rare-earth metals and compounds.

As already mentioned, it had been suggested that the excited $3d^9 4f^1$ state of lanthanum could be created by an ionization followed by a relaxation. In such a process, the twelve J levels of the $3d_{5/2}^{-1}4f^1$ configuration and the eight J levels of the $3d_{3/2}^{-1}4f^1$ configuration must be taken into account with a probability which depends on the statistical weight of each level. The weight of the $J = 1$ (5/2, 7/2) level is only 3.5 % of the total contribution of the twelve J levels and the weight of the $J = 1$ (3/2, 5/2) level is only 5.5 % of the total contribution of the eight J levels. Owing to the ratio of ionization and 3d–4f excitation cross sections as well as the contribution of the $J = 1$ levels in the relaxation, the ionization followed by a relaxation is negligible with respect to direct excitation.

Concerning lanthanum, the radiative transitions have been calculated for Xe-like triply ionized La^{3+}, for singly ionized La^+ whose external configuration is $6s^2$ and for neutral lanthanum with an external configuration $5d6s^2$. The radiative decay probability of the levels La^{3+} $3d^{-1}4f^1$ $^3D_1$ and $^1P_1$ is much greater than that of the ionized states $3d^{-1}$.

The Auger decay of La^{3+} with one additional 3d hole can proceed in several channels: the normal transitions in the ionized system $3d^9$–$3d^{10}X^{-1}Y^{-1}$; the normal transitions in the presence of a spectator excited electron $3d^9 4f^1$–$3d^{10}4f^1 X^{-1}Y^{-1}$; the resonant transitions $3d^9 4f^1$–$3d^{10}X^{-1}$, in which a single hole is present in the final state. The matrix elements are calculated together for all levels involved in the Auger decay in order to take into account the correlations between the various configurations. The probabilities of the normal Auger transitions in the $3d^9$ ion have also been calculated for La^{3+}.

The excited La^{3+} $3d^9 4f^1$ configuration has twenty levels. All the levels of the excited configuration contribute to each Auger transition. These transitions take place from the excited configuration to all the possible $3d^{10}4f^1 X^{-1}Y^{-1}$ and $3d^{10}X^{-1}$ configurations, with X and Y being the 4s, 4p, 4d, 5s or 5p levels. There are fifteen

$3d^{10}4f^1X^{-1}Y^{-1}$ configurations and five $3d^{10}X^{-1}$ configurations, i.e. 632 final levels. The probabilities of the processes $3d^94f^1$–$3d^{10}4f^1X^{-1}Y^{-1}$ were calculated to all the 632 final levels but only from two $3d_{5/2}^{-1}4f^1$ levels and two $3d_{3/2}^{-1}4f^1$ levels of the initial configuration. The probabilities obtained for each of the two $3d_{5/2}^{-1}4f^1$ levels are practically equal to those calculated for the $3d_{5/2}^{-1}$ ion (Table 3.3). The same is true for the transitions from $3d_{3/2}^{-1}4f^1$, showing that the presence of the 4f spectator electron, while it modifies the energies of the normal transitions, affects only very slightly their probabilities.

All the direct recombination processes, or resonant Auger, $3d^94f^1$–$3d^{10}X^{-1}$, have been calculated. The sum of the resonant Auger probabilities is greater for the $^1P_1$ and $^3D_1$ levels, i.e. for the levels involved in the resonant radiative emission, than for the other levels of the $3d^94f^1$ configuration. Indeed, it is equal to 1.628×10^{14} for $^3D_1$ and 2.815×10^{14} for $^1P_1$, whereas it is only of the order of 2–3×10^{13} for each of the other 18 levels of the excited configuration. All the levels of the $3d^94f^1$ configuration must be taken into account statistically in an Auger recombination process. Consequently, the probability of a $3d_{5/2}$ resonant Auger process is the

Table 3.3 Probabilities of radiative, Auger and Coster–Kronig transitions from La^{3+} 3d^9 and La^{3+} 3d^{9}4f^1 and widths of the 3d levels

Recombination from La^{3+} ion		
	$3d_{5/2}^{-1}$	$3d_{3/2}^{-1}$
Σ Radiative probabilities	0.061×10^{13}	0.073×10^{13}
Σ Auger probabilities	127.50×10^{13}	124.00×10^{13}
Σ Coster Kronig probabilities		2.0×10^{13}
Σ Probabilities	127.561×10^{13}	126.073×10^{13}
Level width (eV)	0.84	0.83
Recombination from La^{3+} excited ion		
	$3d_{5/2}^{-1}4f^1$	$3d_{3/2}^{-1}4f^1$
With 4f spectator electron		
Σ Radiative probabilities	0.061×10^{13}	0.074×10^{13}
Σ Auger probabilities	127.40×10^{13}	124.00×10^{13}
Coster-Kronig probability		60.0×10^{13}
With 4f recombination		
Σ Radiative probabilities	0.039×10^{13}	0.108×10^{13}
Σ Auger probabilities	3.20×10^{13}	4.00×10^{13}
Σ Probabilities	130.700×10^{13}	188.182×10^{13}
Level width (eV)	0.87	1.25
Recombination from $^3D_1$ and $^1P_1$ of La^{3+} excited ion		
	$3d_{5/2}^{-1}4f^1\ ^3D_1$	$3d_{3/2}^{-1}4f^1\ ^1P_1$
Resonant transitions		
Σ Radiative probabilities	1.093×10^{13}	1.946×10^{13}
Σ Auger probabilities	16.28×10^{13}	28.152×10^{13}
Σ Probabilities	17.373×10^{13}	30.098×10^{13}

weighed average of the probabilities of all the processes from the 12 levels of the configuration $3d_{5/2}^{-1}4f^1$. These values are much smaller than the sum of the probabilities of normal Auger processes. The same result is obtained for the eight levels of the configuration $3d_{3/2}^{-1}4f^1$. Indeed, the direct recombination is about fifty times weaker than the normal recombination. The situation is the reverse of that found for the radiative transitions.

Coster–Kronig transitions $3d_{3/2}$–$3d_{5/2}X^{-1}$ are energetically possible only in neutral and singly ionized lanthanum. They have been calculated for La^+ $(6s^2)$. The normal transitions are weak while the resonant transitions are intense. Their presence explains the intensity anomalies observed between the $3d_{3/2}$ and $3d_{5/2}$ transitions.

Because of the very large number of the Auger transitions, the calculation of the entire transition array is practical only when a single f electron or hole is present. These calculations can thus be available only for La^{3+} and Yb^{3+}. For Yb^{3+}, the Auger transitions in the excited or ionized systems have equivalent probabilities. Indeed, for a full or almost full 4f sub shell, change of the screening effect is not sufficient to introduce a change of the Auger probabilities.

From the calculation of the radiative and Auger decay processes for La^{3+} $3d^9$ and $3d^94f^1$ configurations, it was possible to deduce important data such as the lifetime of both configurations. Another important result was obtained: the energies of transitions between levels of the excited configurations are smaller than their counterparts in the ionized configurations. These essential characteristics are discussed in the next paragraph.

3.2.5 *Lanthanum 3d Excited Configurations*

Each quantum state is characterized by its creation and annihilation processes and by two parameters, lifetime τ and width Γ, connected together by the uncertainty principle and related to the sum of rates of all the annihilation processes.

For lanthanum, the widths of the $3d_{5/2}^{-1}$ and $3d_{5/2}^{-1}4f^1$ configurations were deduced directly from the transition probabilities given in Table 3.3. The same was done for the $3d_{3/2}^{-1}$ and $3d_{3/2}^{-1}4f^1$ configurations. For the latter, the width can be enlarged because of the eventual presence of $3d_{3/2}^{-1}$–$3d_{5/2}^{-1}X^{-1}$ Coster–Kronig transitions. No Coster–Kronig transition is energetically possible in the La^{3+} free ion. In the solid, Coster–Kronig transitions from the valence electrons are possible. Estimation of their probabilities was made for the lanthanum neutral atom with $6s^25d^1$ ground configuration. From that, the contribution of the Coster–Kronig transitions to the normal Auger transitions in the presence or absence of a spectator electron is found negligible. In contrast, probabilities of about 60×10^{13} s^{-1} were obtained for the Coster–Kronig transitions from the excited configuration $3d_{3/2}^{-1}4f^1$ $(^1P_1)$ $6s^2$ $5d^1$. As a result, the level $3d_{3/2}^{-1}4f^1$ $(^1P_1)$ is wider than the $3d_{5/2}^{-1}4f^1$ $(^3D_1)$ and the calculated ratio is found to be 1.44. This ratio is expected for the widths of the experimental $3d_{3/2}$ and $3d_{5/2}$ resonance lines in the solid. Indeed, a resonance line is the transition from an excited discrete level to the ground level, which has no width. The width of the resonance line, therefore, gives directly the width of the excited level. In the case of La_2O_3 irradiated by an electron beam accelerated under 1 keV, the experimental FWHMs of the resonance lines from $^1P_1$ and $^3D_1$ are in the ratio 1.8/1.2, i.e. in the ratio 1.50, in agreement with the calculated ratio.

As already underlined, the absorption lines correspond to transitions between the ground state and a discrete excited state; they are lorentzian curves. Their intensities are in the ratio of their transition probabilities corrected by a factor taking into account their respective widths. The calculated intensity ratio of the $3d_{3/2}$ and $3d_{5/2}$ absorption lines is 1.23. This value agrees with the experimental curve observed in linear response conditions, i.e. with an absorbing sample having a thickness compatible with the response of the detector.

Thus, the experimental results are in good agreement with the predictions of the atomic model. This shows that the excited configurations have an atomic character and that the interaction of the 4f electron with the valence electrons is very weak, or negligible, in the metal and in the compounds.

The fluorescence yields are larger for the excited configurations than for the ionized configurations because of the presence of terms resulting from the recombination of the 4f excited electron. Indeed, in the presence of these terms only, the partial fluorescence yield of the $3d_{5/2}^{-1}4f^1$ configuration should be 0.039/ $(0.039 + 3.20) = 0.012$, i.e. much larger than the total yield, 0.00088, calculated by taking into account all the recombination processes. In order to characterize the excited levels $^1P_1$ and $^3D_1$ alone, the ratios ω_R of the radiative resonant transitions to the sum of radiative and non-radiative resonant transitions has been determined for these two levels, taken independently. The corresponding values are indicated in Table 3.4. These values are remarkable because they are approximately two orders of magnitude bigger than the fluorescence yields of $3d_{5/2}^{-1}$ and $3d_{3/2}^{-1}$ levels in the ion La^{3+}. They do not concern the excited configuration in its totality, i.e. the levels from which resonant transitions and transitions in the presence of the spectator electron are emitted, but only each of the levels from which a resonant transition takes place directly to the ground state; they are named *resonant levels*. The big fluorescence yields appear as a particularity of these resonant levels. This explains the strong intensity of the resonant radiative transitions and their relatively small widths. This unexpected characteristic of the transitions originating from the $3d^{-1}4f^{n+1}$ configuration in the X-ray range has already been well known for the 4f–5d transitions in the optical range.

The large values of ω_R are due to two factors: first, the presence of the intense 4f–3d radiative transitions in excited $La^{3+}3d^94f^1$; second, the very low probabilities

Table 3.4 Fluorescence yield of La^{3+}

$3d_{5/2}^{-1}$	0.00048	$3d_{3/2}^{-1}$	0.00056
$3d_{5/2}^{-1}4f^{1a}$	0.00088	$3d_{3/2}^{-1}4f^{1a}$	0.00097
$3d_{5/2}^{-1}4f^{1b}\ ^3D_1$	0.067	$3d_{3/2}^{-1}4f^{1b}\ ^1P_1$	0.069

[a]All J levels are statistically populated
[b]Only the $J = 1$ levels are populated

of the Auger resonant processes. Indeed, whereas the probabilities of normal Auger transitions, with or without a spectator 4f electron, are very similar, the probabilities of resonant Auger transitions are much smaller than those of the normal transitions. This decrease of the Auger probabilities causes the lifetime broadening of the emission and absorption transitions $La^{3+}3d^{10}$–$3d^{9}4f^{1}$ to be smaller than generally expected in this energy range. Similar calculations have been made for the iso-electronic Ba^{2+} and Ce^{4+} in the $3d^{9}$ ionized and $3d^{9}4f^{1}$ excited configurations. Results obtained for both these elements are analogous to those obtained for La^{3+}. As is well known, the fluorescence yield ω increases with the atomic number of the core ion, i.e. along the isoelectronic series from Ba^{2+} to Ce^{4+}. The resonant yield ω_R varies in the same way but its value is about a hundred of times larger than that of ω for the three ions (Table 3.5). These data are important in research areas that look for high fluorescence yields.

During the direct non-radiative transition of the excited electron to its initial level, called here resonant Auger transition, the bound 4f electron fills the core hole and, simultaneously, a new vacancy is produced in a level higher than the 3d level. This process was sometimes suggested as being comparable to the recombination of core excitons. A net narrowing of the spectral width is observed in passing from the recombination of the core hole to that of the core exciton. It was explained by a reduction of the probability of the Auger recombination processes. But it was suggested that the probabilities of the normal Auger transitions decreased, while the contribution of the resonant Auger transitions was noticeable [54], contrary to that which is seen for lanthanum. An exciton is not created by direct transfer of a core electron but during the relaxation of the system with a previously created core hole. In contrast, as discussed above, the La $3d^{9}4f^{1}$ excited configuration is created by direct transfer of a 3d electron to the empty 4f level. The two processes are different. Excitation following the relaxation of an ionized state was observed in the 3d photoemission of metal lanthanum but with a very small probability. This process, discussed in photoemission, is associated with a wide peak, which does not have the characteristics of the excitonic transition. It is, moreover, absent in the lanthanum insulator compounds.

Table 3.5 Probabilities of radiative and Auger transitions and fluorescence yields of the $3d^{9}4f^{1}$ $^{3}D_{1}$ and $^{1}P_{1}$ levels of Ba^{2+}, La^{3+} and Ce^{4+}

$^{3}D_{1}$	Ba^{2+}	La^{3+}	Ce^{4+}
Resonant emission	0.643×10^{13}	1.093×10^{13}	1.169×10^{13}
Resonant Auger	13.721×10^{13}	16.280×10^{13}	18.574×10^{13}
Fluorescence yield	0.047	0.067	0.063
$^{1}P_{1}$			
Resonant emission	0.913×10^{13}	1.946×10^{13}	2.600×10^{13}
Resonant Auger	21.005×10^{13}	28.152×10^{13}	34.902×10^{13}
Fluorescence yield	0.043	0.069	0.074

Here no variation of the normal Auger processes is expected in passing from the ionized system to the excited one. Indeed, the dynamic screening effects due to the presence of one strongly localized 4f electron are comparable in the initial and final levels of normal Auger processes with 4f spectator electron. Consequently, the differences between the transition probabilities in the presence or absence of an additional 4f electron are negligible. In contrast, during a resonant process, the electron orbiting at shorter range from the core hole has a large probability to recombine with it. This reduces the probability of the Auger recombination processes since incomplete electron relaxation could, then, be present.

It was mentioned that in La^{3+} the auto-ionization from $3d_{3/2}^{-1}4f^1$ to the $3d_{5/2}^{-1}$ continuum states was energetically possible [55]. If this process is present, it would introduce changes in the shape of the $3d_{3/2}$ absorption transition. Experimentally, the La 3d absorption and emission lines have their maxima at the same energy and their shapes are identical on the low energy side. On the high energy side, the $3d_{3/2}$ absorption line is accompanied by an edge of weak intensity, shifted to the higher energies. After removal of the edge, the $3d_{3/2}$ absorption line is well approximated by a lorentzian shape [41]. One concludes that the interaction between the La^{3+} $3d^94f^1$ excited states and the La^{4+} ion continuum states is very weak. Indeed, the hybridization strength between discrete and continuum states is strongly reduced because the part of the 4f wave functions close to the nucleus increases in the presence of a 3d core hole and becomes very large. That is different in the 4d energy range, discussed in the next part, where interactions between discrete excited states and continuum states are strong.

In summary, X-ray and Auger emission spectra induced by electron bombardment are very well adapted to investigate core level excited configurations because these configurations can be created directly by electron impact. They can be identified by comparing absorption and emission spectra. Their decay dynamics can be investigated on a timescale defined by the lifetime of the core hole. Indeed, the measurement of the widths of the absorption and emission resonance lines gives a direct determination of the decay rate of the excited states. All the competitive decay processes can be taken into account and compared to the decay processes from the ionized states, deduced from the analysis of the normal emissions.

3.3 Resonance Effects in the 4d Range

Excitations from the 4d inner sub shell into the 4f open sub shell of the rare-earths have attracted much attention of both atomic and solid state physicists. The various decay channels of the $4d^94f^{m+1}$ excited configurations were widely analyzed [56]. As opposed to the 3d case, where the dominant interaction is the 3d spin–orbit interaction and j–j coupling is expected, in the 4d range, it is the electrostatic interaction, and particularly its exchange part, that dominates and L–S coupling is present. Moreover, the large overlap between the 4d and 4f wave functions results in a large multiplet splitting of the open shell configurations. Consequently, the distribution of the oscillator strengths extends over a broad energy range. These new coupling conditions introduce large changes in the spectral characteristics of the 4d photoabsorption and of the 4d radiative and non-radiative decay processes with respect to the 3d ones. Emphasis is put here on these particularities and on the 4f electron characteristics which can be deduced from the observed spectra. The interaction between the 4f orbitals and the valence band in the presence of a 4d core hole is also mentioned. Results for lanthanum are considered as an example. Generalization to rare-earths and to 5d spectra of the actinides is presented in the next chapters.

3.3.1   4d Photoabsorption

The rare-earth 4d absorptions are located between 100 and 200 eV, i.e. in the ultrasoft X-ray region. The longer the radiation wavelength, the thinner must be absorbing material and more difficult is the production of chemically pure homogeneous screens. All the measured quantities are proportional to the absorption coefficient of the incident photons and attention must be paid to eliminate the saturation effects that show up for very large values of this coefficient.

The presence of a strong exchange electrostatic interaction in this energy range causes the 4d photoabsorption to be governed by the characteristics of the spectral terms. The strong 4f–4d exchange electrostatic interaction drives the energy of some multiplets above the 4d ionization threshold. Consequently, some 4d–4f intense excitation lines, referred to sometimes as *giant resonances* [57], are observed above the ionization threshold in the absorption spectra of the rare-earth free atoms as well as in the solid spectra. The photoabsorption process involving the $4d^9 4f^{m+1}$ levels, that are pushed above the 4d ionization limit by the large 4d–4f multiplet splitting, is the outcome of a competition between 4d–4f photoexcitation and 4d-εf photoionization. Both these processes are strongly connected by the configuration interaction $\langle 4d^9 4f^{m+1} | 1/r | 4d^9 4f^m \varepsilon f \rangle$. In this energy range, the excited electrons are also coupled to $4f^{-1}$ and $5(s,p)^{-1}$ ionization continua by the $4d^9 4f^{m+1}$–$4f^{m-1}$ and $4d^9 4f^{m+1}$–$4f^m 5(s,p)^{-1}$ Coster-Kronig decay channels.

The absorption transitions in the presence of an energy overlap of a discrete level with continuum levels can present various characteristics [39]. First, in the absence of interaction between the discrete level and the continuum, the transitions to the discrete excited level are superimposed upon the transitions to the continuum, their shape is lorentzian and their width depends on the lifetime of the discrete level. The absorption is the sum of a Lorentz curve and a continuum edge. Secondly, in the presence of an interaction between the discrete level and the continuum, the transition shape is still a Lorentz curve if the absorption takes place only to the discrete level but the width of the lorentzian is related to the matrix element of the interaction. Lastly, when the absorption takes place to both the discrete level and the continuum that interact, the discrete energy level and a continuum interfere. The total wave function is a linear combination of the wave functions of the unperturbed discrete level and of the continuum. The transition amplitudes are added, or subtracted, and the line shape results from the fact that the absorption to the discrete level and to the continuum adds, or subtracts, on each side of the energy of the perturbed discrete level. As a result, the discrete level acquires to some extent the possibility to ionize spontaneously. Therefore, this effect induces an *auto-ionization* of the excited configuration by radiationless transition to the continuum. Auto-ionization is thus a consequence of the presence of electrostatic matrix elements connecting a discrete level above the ionization energy with the continuum of same energy. Its probability is nearly independent of the atomic number. This process greatly changes the shape of the absorption line, which has a Breit-Wigner-Fano shape and is broadened. Its width arises from the lifetime of the discrete level modified by the interaction with the continuum.

As indicated in Sect. 1.2, the centrifugal potential barrier prevents the penetration of nl orbitals with $n > 4$ near the core and the 4d wave function hardly overlaps the wave functions of higher energy f orbits. Then, the probability to excite a 4d electron to the higher energy f, or p, orbitals is very small. This explains the absence of 4d–nf ($n > 4$) transitions and of absorption edges in the photoabsorption spectra.

The lanthanum 4f wave function is confined within the centrifugal well of the attractive potential due to the 4d inner hole. The 4d and 4f wave functions overlap strongly in coordinate space. This large 4d–4f overlap causes a strong admixture between the various levels of the excited $4d^9 4f^1$ configuration. Consequently, the average spatial charge distribution is unaffected by the 4d–4f transition and the relaxation involving the 4d hole is negligible. The 4d and 4f spin–orbit interactions are small compared to the strong 4d–4f electrostatic interactions, which are responsible for the distribution of levels over a broad energy range. Thus, for La^{3+}, and also Ba^{2+} and Ce^{4+}, the $^1P_1$ level, with almost a pure (L–S) designation, is observed at about 15 eV higher than the other levels of $4d^9 4f^1$ and is located above the 4d ionization thresholds. The energy difference between the $^3D_1$ and $^1P_1$ terms is almost one order of magnitude higher than the difference between the $4d_{5/2}$ and $4d_{3/2}$ levels, estimated to be 2.5–2.8 eV from the X-ray 4d–2p emissions.

The 4d–4f excitation probability is high owing to the strong 4d–4f overlap. The oscillator strengths are determined essentially by LS coupling selection rules. Therefore the line connecting $^1P_1$ of $4d^9 4f^1$–$^1S_0$ of $4d^{10}$ dominates the absorption spectrum. This $^1S_0$–$^1P_1$ peak, located above the threshold, forms the 4d–4f giant dipole resonance because of its very strong intensity. The transitions to $^3P_1$ and $^3D_1$ have $\Delta S = 1$ or $\Delta L = 2$ and are forbidden by the L–S selection rules. Their weak transitions are due to their admixture with the $^1P_1$ level owing to the spin–orbit interaction. The transitions from 4d to the continuum have very small probability at the threshold and no absorption jump is observed. However, transitions to the conduction levels are present beyond the threshold and their probabilities vary with the energy. The $^1P_1$ discrete level is situated about 10 eV above the threshold. Transitions to the conduction levels are superposed on the $^1P_1$ absorption peak and modify its shape (Fig. 3.11).

An attempt was made to explain the high energy of the line $4d^{10}$–$4d^9 4f^1$ $^1P_1$ and its high intensity by strong configuration interaction effects, including several

Fig. 3.11 **a** Lanthanum 4d absorption in LaF$_3$. The vertical line represents $^1P_1$ absorption line calculated for the free La^{3+} ion [60]. **b** Lanthanum 4d absorption in La$_2$O$_3$ [61]. **c** Lanthanum 4d absorption in LaB$_6$ measured by photoemission partial yield method [62]

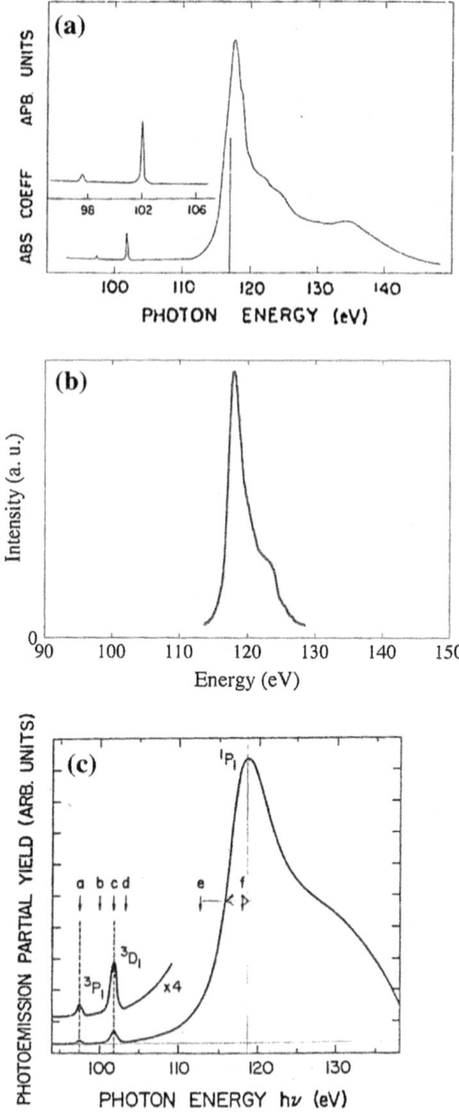

configurations of the type $4d^9nf^1$ (n = 5–7), extending into the continuum [58]. In this model, configuration interaction effects between levels belonging to the same Rydberg series would produce wide spread redistribution of the oscillator strength from one spectral region to another. It has been verified that, contrary to results claimed in precedent reference, the introduction of this configuration interaction explains neither the position nor the intensity of the transition $4d^{10}$–$4d^94f^1$ $^1P_1$. In fact, this transition retains always the same energy and intensity, independently of the considered configuration number. This proves that the reason for the high position of the $^1P_1$ line is found in the $4d^9 4f^1$ configuration itself: it is due to the high value of the G_1 exchange electrostatic interaction parameter that causes what Racah calls a "Russell-Saundersation" of this energy level.

Competition between discrete and continuum excitations (4d–4f,εf) has been widely discussed for the Xe isoelectronic ions. As already indicated, no lanthanum 4d edge is observed, proving that the ionization process is weak in the 4d photoabsorption. On the other hand, the energy of $4d^9 4f^1$ $^1P_1$, obtained from a calculation analogous to those made in the 3d range, i.e. in intermediate coupling, exceeds by several eV its experimental energy measured in the solid and the line $^1P_1$ is highly asymmetrical on the higher energy side [59]. This shows that the level $4d^9 4f^1$ $^1P_1$ interacts with the continuum according to $4d^9 4f^1$ $^1P_1$–$4d^9 \varepsilon l$. The probability of this auto-ionization process is expected to depend on the configuration interaction between the discrete excited $^1P_1$ level and the energy distribution of the continuum close to it. This auto-ionization process must remain weak in the solid because the density of the conduction levels is essentially of 5d, 6s symmetry. However, photoionization of the 4d electrons to the 6s, 6p conduction levels is present in this energy range.

The conduction levels change according to the material considered and the shape of the intense line 4d–4f $^1P_1$ varies accordingly, but it remains strongly asymmetrical towards the higher energies in all materials. In a wide band gap insulator, as for example LaF_3 [60], the $^1P_1$ absorption peak has a lorentzian shape on the low energy side (Fig. 3.11a). Its experimental half width agrees with that calculated from the atomic model taking into account all the radiative and non-radiative decay processes of the excited configuration $4d^9 4f^1$ and disregarding the configuration interaction with the continuum states (cf. Sect. 3.2.4). Towards the higher energies, the absorption line is strongly enlarged and presents weak broad features, characteristic of the density of the conduction states. In a solid conductor, the shape of the $^1P_1$ line remains the same but its broadening is larger, in particular on the lower energy side. These results suggest that, in LaF_3, the $^1P_1$ level is located just at the bottom of the conduction band and the mixing of the orbital 4f with the conduction states is very small.

In summary, due to the presence of the centrifugal potential barrier and to the weak spin–orbit interaction, the predominant interaction between 4d and 4f is electrostatic and the most probable process is the excitation of a 4d electron to a 4f orbital. The main features of the spectrum are well reproduced by atomic calculations and the three expected lines from $^1S_0$ to $^3P_1$, $^3D_1$ and $^1P_1$ are observed with energies and relative intensities in agreement with those calculated from the atomic model taking into account all the radiative and non-radiative decay processes of the excited configuration $4d^9 4f^1$. As shown in the next paragraphs, the $4d^9 4f^1$ $^1P_1$ level preserves the characteristics of a discrete state although it is above the 4d ionization threshold. Indeed, it decays essentially by direct de-excitation, by recombination in the presence of a spectator electron or by discrete Coster–Kronig process to the excited level $4d^9 4f^1$ $^3D_1$. But these various processes explaining the decay of $^1P_1$ of 4d have already been encountered in the case of 3d. Therefore no extra explanation is needed, contrary to that which had been suggested. The La 4d photoabsorption appears thus as a 4d–4f photoexcitation, although the excited electron has a non-vanishing probability to escape to continuum f orbits or to the conduction levels of the solid and the shape of the intense $^1P_1$ line depends on the mixing of the orbital 4f with the conduction levels.

It must be underlined that the shape of the $^1P_1$ absorption line depends also strongly on the experimental conditions. The choice of the absorbing sample thickness is particularly critical in this low energy range. In this highly absorbing energy range the probability to saturate an intense absorption line is high. An example of non-saturated absorption spectrum, obtained with sufficiently thin screens, is seen in Fig. 3.11b for lanthanum oxide La_2O_3 [61]. The experimental 4d photoabsorption spectrum has the general characteristics of the calculated spectrum for the La^{3+} free ion except for the $^1P_1$ line, which is strongly deformed because the self-absorption effect is very large; it is asymmetric towards the higher energies because the transitions to the continuum of the solid are dominant in this range. A more pronounced asymmetrical shape was observed for LaB_6 by means of the photoemission partial yield method [62, 63] (Fig. 3.11c).

3.3.2 Electron-Stimulated X-ray Emission

By analogy with the observations made in the 3d energy range, one expects that, under electronic bombardment, the relative intensities of emissions connected with excited or ionized systems will be functions of the ratio of excitation to ionization cross sections. Near the threshold, the electron ionization cross section is small and the excitation cross section largely predominates. The probability to create the excited configuration $4d^9 4f^{m+1}$ by direct jump of a 4d electron to the unpopulated 4f levels is high. When the incident electron energy E_0 increases, the ionization cross section increases strongly with respect to the excitation one, making the relative intensities of the lines in the ion increase with respect to that in the excited atom. Let us recall that the core-hole creation by excitation or ionization is a process much more rapid than its decay and the excited and ionized configurations are able to reach a fully relaxed core-hole quantum state on the timescale of the order of the hole decay. However, the relaxation probability must be of the same order, or higher, than the radiative probability in order for the relaxation to all the J levels of the excited configuration to take place prior to the emission.

As in the 3d energy range, the $4d^9 4f^{m+1}$ excited configuration can decay by two types of the radiative processes: the direct de-excitation, or resonant emission, which corresponds to the return of the excited f electron to the 4d hole, and the transitions in the presence of a spectator electron, i.e. the $4d^9 4f^{m+1}$–$5p^5 4f^{m+1}$ emission. Energies and probabilities of the lines can be calculated by the same method as described in Sect. 3.2.4. From these calculations, the lines in the presence of a spectator electron have lower energies than those of the corresponding normal lines. Indeed, the 4f–nd electrostatic interactions are stronger than the weak 4f–5p interactions. This partially explains the difference in value of the energies of the 5p–nd lines in the presence or absence of a spectator electron.

In the La^{4+} ion with 4d^9 ground configuration, three lines belonging to the 5p^{6}4d^9–5p^{5}4d^{10} transition are present. These lines, $4d_{5/2}^{-1}$–$5p_{3/2}^{-1}$, $4d_{3/2}^{-1}$–$5p_{3/2}^{-1}$ and $4d_{3/2}^{-1}$–$5p_{1/2}^{-1}$, are expected around 87.3 eV and are generally not resolved. The sum of their transition probabilities is 6×10^{10} s^{-1}.

From the La 4d^{9}4f^1 excited configuration, the resonant emissions, 4d^{9}4f^1–4d^{10}, and the 5p–4d emissions in the presence of the 4f spectator electron, 4d^{9}4f^1–5p^{5}4f^1 have been calculated [59]. Concerning the direct recombination, as in the 3d range, three resonance lines are expected from the three $J = 1$ levels of the 4d^{9}4f^1 configuration to the ground configuration 4d^{10} ^{1}S$_0$. But here the configurations manifest a predominant L–S coupling structure. Consequently, the transition probability of the ^{1}P$_1$–^{1}S$_0$ line is predicted to be of the order of 10^{12} s^{-1} while the probabilities of the ^{3}D$_1$–^{1}S$_0$ and ^{3}P$_1$–^{1}S$_0$ lines are only of order of 10^9 and 10^8 s^{-1} respectively. As indicated for the photoabsorption, the discrete level 4d^{9}4f^1 ^{1}P$_1$ is separated from the two other levels by about 15 eV and it is located above the 4d ionization threshold. The very strongly self-absorbed ^{1}P$_1$ resonant line has been observed in coincidence with the giant absorption peak (Fig. 3.12) [64].

The twenty J levels associated with the 4d^{9}4f^1 configuration contribute to the La 4d^{9}4f^1–5p^{5}4f^1 emission. Among the 130 lines of this transition array, 4 lines associated with the ^{1}P$_1$ level are located around 95 eV; they are about 15 eV higher than the other 126 lines, associated with the 19 other J levels of the configuration and located around 80.0 eV. Owing to the small 4d spin–orbit splitting, these 126 5p–4d lines are confined inside only 4–5 eV. The sum of the probabilities of the four 5p–4d lines associated with ^{1}P$_1$ level is 0.08×10^{10} s^{-1} while the sum of those associated with all the other 5p–4d lines is 3×10^{10} s^{-1} [59].

In analogy with the 5p–3d transitions, the 5p–4d lines in lanthanum with the excited configuration 4d^{9}4f^1 have lower energies than the corresponding lines in the

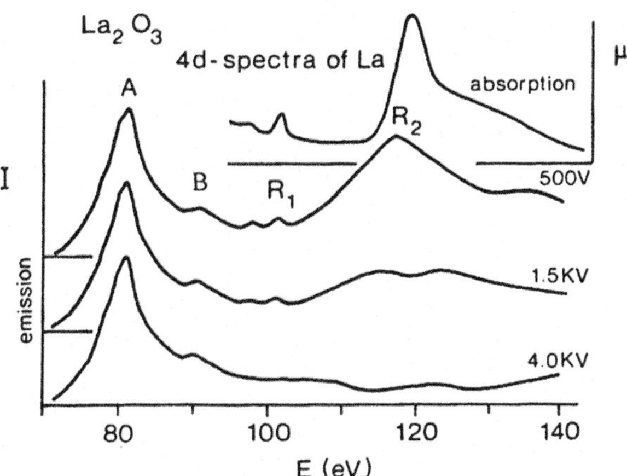

Fig. 3.12 Lanthanum emission in metal: (*A*) 4d^{9}4f^1–5p^{5}4f^1 (*B*) 4d^9–5p^5 (*R$_1$*) 4d^{9}4f^1 ^{1}P$_1$–5p^{5}4f^1 ^{1}P$_1$ (*R$_2$*) 4d^{9}4f^1 ^{1}P$_1$–4d^{10}. The transition 4d^{9}4f^1 ^{1}D$_3$–4d^{10} is not detected [62]

ion $4d^9$ [65]. However, exception exists for the four $4d^9 4f^1$ $^1P_1$–$5p^5 4f^1$ lines, located above the threshold [59]. Emission spectrum was observed under irradiation by incident photons of energies higher than the ionization threshold [66]. Lower energy lines, whose position was found to be independent of the incident photon energy, were assigned as transitions belonging to the $4d^9$ ion and two lines observed towards the higher energy as transitions in the presence of a spectator 4f electron, their energies varying with the incident photon energy. In fact, these two lines belong to the unresolved group of four $4d^9 4f^1$ $^1P_1$–$5p^5 4f^1$ lines; they are observed around 10 eV above the low energy lines, which are part of the other 126 lines predicted in the excited atom, in agreement with the theoretical predictions.

The potential barrier associated with the 4d hole and the 4f sub shell increases the excitation cross sections of the 4d–4f transitions with respect to the 4d ionization cross sections. The 5p–4d emission probabilities have the same order of magnitude in excited atom and ion. Consequently, the lines in the excited atom are more intense than the corresponding lines in the ion. Thus the 5p–4d emissions from La $4d^9 4f^1$ are more intense than the 5p–4d ones from La $4d^9$. It has already been seen in the 3d range that it is easier to excite a lanthanum atom than to ionize it. On the other hand, from theoretical radiative probabilities, one expects the 4d emission spectrum to be dominated by the $^1P_1$–$^1S_0$ resonance line. Indeed, this line is intense when it is excited by incident electrons of energy near the threshold. However, the $^1P_1$ line depends on various other factors, which may modify it, for instance the non-radiative recombination processes. Among those, the Coster–Kronig and Auger transitions with direct recombination of the excited 4f electron to the 4d hole and ejection of an *nl* electron to the continuum are the most probable (cf. next section). These very intense transitions broaden strongly the line $4d^9 4f^1$ $^1P_1$–$4d^{10}$ $^1S_0$ and its observed height decreases. This broadening is lorentzian and does not change the shape of the peak. The $^1P_1$ line depends also on the interaction of the discrete level $^1P_1$ with the continuum. Since the energy distributions of the continuum depend strongly on the solid considered, so does the shape of the $^1P_1$ emission line, as discussed in the case of the absorption line.

Another process perturbs the $^1P_1$ resonance line, the self-absorption. When the incident electron energy E_0 is near the 4d excitation threshold, self-absorption is weak and the line $^1P_1$–$^1S_0$ is intense. At incident energies E_0 equal to several times the threshold, a strong self-absorption perturbs the emission. An intensity minimum appears at the position of the peak maximum, confirming the coincidence between resonance and absorption lines in the solid. The emission is distorted and appears as asymmetric. If the self-absorption is sufficiently large, the 5p–4d emission in the presence of a spectator electron appears as the stronger emission.

Two important results can be deduced from these spectra. First, the line corresponding to the decay of the excited level $4d^9 4f^1$ $^1P_1$ to the ground state is highly predominant and it is in coincidence with the corresponding absorption line. Consequently, no electron transfer is present during the decay of the excited configuration. Secondly, the four $4d^9 4f^1$ $^1P_1$–$5p^5 4f^1$ lines are observed about 15 eV above the other lines involving the $4d^9 4f^1$ configuration and their observed energies and intensities agree with the theoretical predictions made for the atom. The observation of such lines proves that the $^1P_1$ level decays by the same processes as the other discrete levels of the La $4d^9 4f^1$ configuration. This is an important result because it shows that this level is subject to exactly the same processes as all the other levels of this configuration despite the fact that it is located above the ionization threshold.

3.3.3 Electron-Stimulated 4d Auger Spectra

As described in the previous paragraph for the radiative transitions, the non-radiative Auger transitions take place both from excited and ionized configurations. A single type of transitions is present from the ionized configuration $4d^9 4f^m$, leading to a final state with two holes shared among the 5s, 5p, 4f or valence levels and one Auger electron present in the continuum. Starting from the $4d^9 4f^{m+1}$ excited configuration, electrons can make two types of transitions: - first, a direct recombination of the 4f excited electron to the 4d hole with ejection of a 5s, 5p or valence electron; this is called a *resonant Auger process*; - second, they can make the transitions in the presence of a 4f spectator electron, leading to final configurations of the type $nl^{-1}n'l'^{-1}4f^{m+1} + e^-$, with two core holes, an excited 4f electron and one Auger electron. The direct recombination transitions appear on the high kinetic energy side of the other Auger peaks. Because of the large extension of the various configurations, the transitions from $4d_{3/2}^{-1}$ and $4d_{5/2}^{-1}$ are not resolved except for the heavier rare-earths. The transition probabilities of the Auger processes involving the localized sub shells 5s and 5p are much higher than those of the radiative transitions and they govern the lifetime of the 4d levels. Transitions to $5s^1 V^{-1}$ and $5p^5 V^{-1}$, where V denotes the valence electrons, are of low intensity and difficult to observe.

From the La $4d^9$ ionized configuration, only two Auger processes to the final configurations $5s^1 5p^5$ and $5p^4$ are observed because the transitions from $4d_{3/2}^{-1}$ and $4d_{5/2}^{-1}$ are not resolved. Their probabilities are about 10^3 times higher than those of the radiative transitions. The transitions to $5s^1 V^{-1}$ and $5p^5 V^{-1}$ were not observed.

From the La $4d^9 4f^1$ excited configuration, there are two transitions with direct recombination of the 4f excited electron to the 4d hole followed by the ejection of a 5s or 5p electron and three transitions in the presence of the 4f spectator electron to the final configurations $5s^0 4f^1$, $5s^1 5p^5 4f^1$ and $5p^4 4f^1$. In total 59 final J levels combine with each of the 20 levels of the $4d^9 4f^1$ initial configuration, giving a maximum of 1180 lines. Each transition is, thus, strongly split (Fig. 3.13) [67].

As for the radiative transitions, the group of higher energy peaks, located at 98 eV, corresponds to transitions from the excited level $4d^9 4f^1$ $^1P_1$; their final configuration is $5p^5$. In this process, the 4f electron returns to the 4d core hole and the energy gained is transferred to a 5p electron. This is *resonant Auger transition*. The resonant transition from $4d^9 4f^1$ $^1P_1$ to $5s^1$ falls among the resonant transitions from the 19 other levels of $4d^9 4f^1$ to the $5p^5$ final configuration. They are located between 70 and 80 eV and form the more intense structure. The transitions in the

Fig. 3.13 Auger emission
probability of 4d lanthanum:
a simulated spectra from
$4d^9 4f^1$; **b** from $4d^9$, compared
c to experimental
spectrum [67]

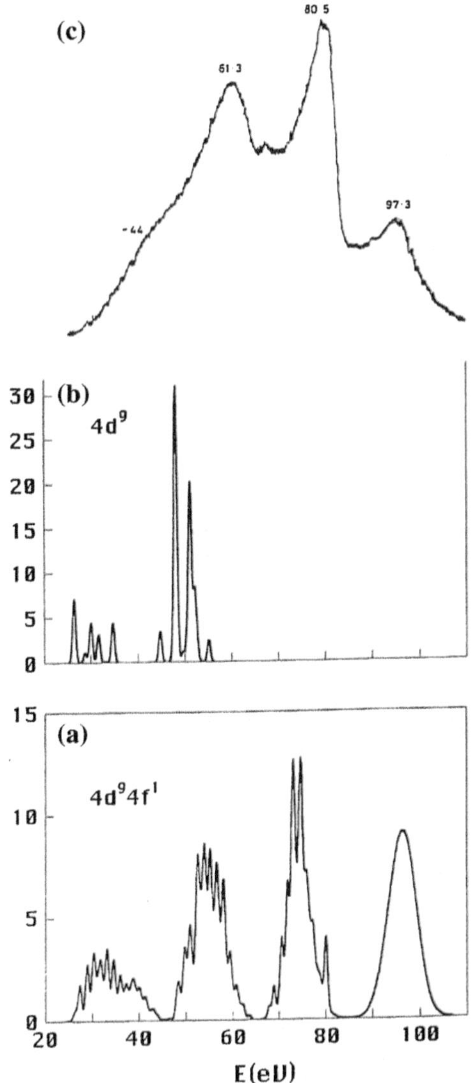

presence of the 4f spectator electron are distributed in two large peaks: the large
peak located between 50 and 60 eV corresponds to $4d^9 4f^1\ ^1P_1-5s^1 5p^5 4f^1$ and
$4d^9 4f^1-5p^4 4f^1$ from the 19 other levels; the large peak between 30 and 40 eV
corresponds to $4d^9 4f^1\ ^1P_1-5s^0 4f^1$ and $4d^9 4f^1-5s^1 5p^5 4f^1$ from the 19 other levels.
The sum of the probabilities associated with the resonant transitions $4d^9 4f^1\ ^1P_1-$
$(5l)^{-1}$ is about $4.5 \times 10^{15}\ s^{-1}$. These transitions are consecutive to the direct
recombination $^1P_1-^1S_0$, which is the dominant process of all three ions La^{+3}, Ba^{+2}
and Ce^{+4} owing to the strong 4d–4f interaction. In contrast, the sum of the

probabilities of all the lines from each of the 19 other J levels of $4d^9 4f^1$ is of the order of 10^{14} s^{-1}, i.e. clearly lower.

Another Auger transition with direct recombination of the excited 4f electron is energetically possible, the transition $4d^9 4f^1$ $^1P_1$–$4d^9$ + e^-. In this process, of the Coster-Kronig type, the 4f electron returns from the $^1P_1$ level to the 4d hole and the energy gained is used to ionize a 4d electron to the continuum. Its probability is smaller than that of the resonant Auger to $(5l)^{-1}$. This process is often described as an auto-ionization from the $4d^9 4f^1$ $^1P_1$ discrete level into the adjacent continuum leaving a free electron and the $4d^9$ positive ion, i.e. taking place mainly through the exchange interaction $(4d4f\,|V|\,\varepsilon f4d)$. However, the decay rate of a discrete excited level into an auto-ionization channel is known to be smaller than the decay rate to another discrete level. The interactions among $4d^9 nl^{m\pm1}$ discrete levels are, thus, stronger than the interaction between the discrete level $4d^9 4f^1$ $^1P_1$ and the continuum in which it is embedded.

Since the decay probabilities from excited and ionized configurations are different, the lifetime of a core hole is not the same in the excited atom and in the ion. Indeed, the decay from an excited atom includes the transitions in the presence of the 4f spectator electron, whose probabilities are of the same order of magnitude as those of the transitions in the ion, and also the resonant processes to single-hole final configurations. The decay probabilities from the various levels of the excited configuration are different too. The high probabilities of the Auger transitions from $4d^9 4f^1$ $^1P_1$ explain the large broadening of the La $^1P_1$ radiative transition. The Auger broadening is symmetrical. It can modify strongly the width of the lines emitted from the different J levels of the $4d^9 4f^1$ configuration and thus their relative heights. This must be taken into account when the radiative probabilities are derived from the measurements of the emission lines.

The rather large number of levels in each of the configurations and consequently the very large number of transitions, explains the presence of unresolved broad peaks that appear in the Auger spectra. Similar to the case of the radiative transitions, the Auger transitions from the excited atom are more intense than those from the ion. These results confirm the presence of a configuration with a localized excited 4f electron in lanthanum.

3.3.4   Theoretical 4d Spectra

The 4d photoexcitation of La^{3+} and the radiative and Auger decays of the La^{4+} ion with a 4d hole and of the La^{3+} $4d^9 4f^1$ excited ion were calculated in order to compare the dynamics of the excited and ionized configurations [59]. The model previously described in Sect. 3.2.4 was used. The energies and the probabilities of all the radiative transitions were obtained from a calculation that took into account the various interactions. Similar calculations were also made for all the Auger transitions of the ion and of the excited configuration.

Good agreement exists between the theoretical and experimental energies of all the 5p–4d emissions in the presence or absence of a 4f spectator electron, including the transitions of the type $4d^9 4f^1\ {}^1P_1$–$5p^5 4f^1$; in this case the 4f electron is present both in the initial and final states. For all the Auger transitions, the same good agreement is obtained. The spectrum calculated by including the sum of the transitions from $4d^9$ and $4d^9 4f^1$ reproduces well the experimental spectrum. Indeed, the main observed characteristics of the La 4d photoabsorption and 4f–4d emission, such as intensity, width and energy of the lines, are in agreement with the theoretical calculations, except for the energy of the transitions involving the ${}^1P_1$ level.

Experimentally, the ${}^1P_1$ level is located above the 4d ionization threshold and about 15 eV higher than the ${}^3D_1$ level. It is thus clearly separated from the others 19 levels of the excited $4d^9 4f^1$ configuration. Theoretically, this separation of the ${}^1P_1$ level results mainly from large value of the radial electrostatic exchange parameter G^1 that appears only in the energy expression of the ${}^1P$ and not in that of the other levels. Consequently, due to this large exchange interaction, the $4d^9 4f^1\ {}^1P_1$ level is pushed to higher energies and is shifted above the ionization threshold. The calculated energy of the ${}^1P_1$ level is several electron volts higher than the experimental value while the agreement is good for the other 19 levels of the configuration, which are all regrouped on an interval of about 8 eV. The agreement is good in particular for the other $J = 1$ levels, $4d^9 4f^1\ {}^3P_1$ and ${}^3D_1$. It was verified that the ${}^1P_1$ level does not interact with levels of the multiply ionized atom that are present in the continuum in this energy range, such as $5p^4$ (109 eV) or $5s^0 4f^1$ (111.6 eV). It is suggested that the high energy predicted for ${}^1P_1$ is due to the calculation mode: indeed the wave functions obtained from a MCDF EAL calculation minimize the contribution of this particular level because its statistical weight is only 3 while the total statistical weight of the other 19 levels is 137. On the other hand, the difference between the experimental and theoretical energies of the ${}^1P_1$ level [68, 69] was explained away as resulting from its strong interaction with the continuum. Its energy calculated in this model is in agreement with the experiment.

Intense peaks are associated with the ${}^1P_1$ level. These peaks have a 99.8 % LS character and completely dominate the 4d emission and absorption spectra. Indeed the theoretical radiative transition probabilities from or to the ${}^1P_1$ level are three orders of magnitude higher than the probabilities from or to the ${}^3D_1$ and ${}^3P_1$ levels. They are of the order of $10^{12}\ s^{-1}$ for ${}^1P_1$ while they are only of order of 10^9 and $10^8\ s^{-1}$ for ${}^3D_1$ and ${}^3P_1$. Two orders of magnitude higher are found for ${}^1P_1$ with respect to each of the other 19 levels when all the radiative and no radiative transitions are taken into account. These differences explain why the 4d spectral characteristics differ strongly from those observed for the 3d spectra though the processes are the same.

The calculated lifetimes of the 19 J levels other than ${}^1P_1$ are practically the same, leading to widths of the order of 0.1 eV, while the life-time of ${}^1P_1$ being two orders of magnitude shorter, leads to a width of 3 eV. This very short lifetime is due to the large probabilities of all the transitions from the ${}^1P_1$ level. Among them, the discrete Auger transitions from the level $4d^9 4f^1\ {}^1P_1$ to the $5p^5$ and $5s^1$ configurations have the highest probabilities and their sum explains the very large experimental width. As already underlined, differences between the $^1P_1$ level and the other levels of the excited configuration had been attributed to its energy position above the ionization threshold, which should induce strong interactions with the continuum, of the type $4d^94f^1\ ^1P_1$–$4d^9\varepsilon f$. However, the characteristics of the $4d^94f^1\ ^1P_1$ level can be totally explained by considering the same decay processes to discrete levels as for the other levels. In this model, intensity and width of the experimental transitions from the $^1P_1$ level are in agreement with the theoretical predictions. This result shows that the $^1P_1$ level is of the same type as the 3d levels. Only the exchange interaction of the 4f electron is much stronger with the 4d electrons than with the 3d ones.

From calculations for the neutral atom of $5d^16s^2$ configuration, it was shown that the presence of the three valence electrons can change very little the 4d spectral characteristics and great similarity would exist between the spectra of La^{3+} and neutral lanthanum. From this result, one deduces that the 4d–4f spectra are almost independent of the lanthanum configuration. This confirms that the valence electrons interact very little with the 4f electron.

3.3.5 Characteristics of 4d Excited Configuration

While excitations from La 3d levels lead to a 4f wave function localized within the inner well of the double-well potential, excitations from 4d levels lead to a 4f wave function residing in the inner potential well for 19 of the $4d^94f^1$ levels but not for the $4d^94f^1\ ^1P_1$ level. The excited electron wave function was expected to have the $(4,\varepsilon)f$ character. It was suggested that the dominant decay channel of the configuration $4d^94f^1\ ^1P_1$ was the auto-ionization to $4d^9$ with an f electron in the continuum. This process of $4d^9\varepsilon(d,g)$ final configuration, comparable to a tunnelling, should be in a competition with the direct photoionization $4d^{10}$–$4d^9\varepsilon(f)$. In this framework, the X-ray and Auger spectra should be dominated by the transitions from $4d^9$. The $4d^9$–$5p^5$ X-ray emissions located around 90 eV and the Auger transitions located between 45 and 55 eV should be the most intense transitions while direct recombinations from $4d^94f^1\ ^1P_1$ should be weak or unobservable. This is in disagreement with the bulk of the observations.

In fact, all the direct decay processes expected theoretically from a discrete level are observed from the $4d^94f^1\ ^1P_1$ level in both X-ray and Auger emissions. As an example, the two $4d^94f^1\ ^1P_1$–$5p^54f^1$ lines are observed at 99.4 and 101.5 eV, confirming the theoretical predictions. The more intense $4d^9$–$5p^5$ transitions are those from the $4d^94f^1$ excited configuration. However, there is a difference if the level of the $4d^94f^1$ excited configuration is below or above the $4d^9$ ionization limit. For the levels that lied below the ionization limit, the decay takes place mainly through the 4f spectator Auger processes and the width of each level is of the order of 0.1 eV. The decay channels from the level $4d^94f^1\ ^1P_1$, positioned above the $4d^9$ threshold, are completely dominated by direct recombination. Among these processes, the most probable has the final configuration $5p^5$ and predicts a level width as large as 3 eV, in agreement with the experimental width. The observation of the resonant X-ray and Auger emissions from the $4d^9 4f^1$ $^1P_1$ level, along with their predicted intensities with respect to the other transitions, shows that the direct de-excitation takes place before the excited electron interacts with the continuum or the extended levels in the solid.

The decays of the excited and ionized configurations are different, and so are their lifetimes. In the energy range of the $^1P_1$ excitation line where the excitation is the most intense process, the ionization is weak and its probability is small. The $4d^9 4f^1$ $^1P_1$ level is analogous to a discrete level with its discrete resonant transitions, which make its lifetime noticeably shorter than that of levels of the $4d^9$ configuration. By comparing the calculated and experimental spectra, it is possible to deduce the relative importance of the initial configurations that contribute to the observed spectra and, consequently, to determine the relative proportion of the excited and ionized configurations, all of which are essential data in the characterization of the 4f and the 5f electrons in solids.

The energies and the probabilities of the transitions from or to the $4d^9 4f^1$ excited configuration do not vary with the chemical composition. Their interpretation is in terms of atomic transitions and agrees with observed spectra of different solid materials. The ionization energies, in contrast, vary with the charge of the ion in the solid. Consequently, the relative position of the $4d^9 4f^1$ and $4d^9$ configurations varies, leading to the mixing of the $4d^9 4f^1$ $^1P_1$ level with various continua but still without a change in the observed spectrum, confirming that the interactions with the continuum are clearly smaller than the 4d–4f interactions.

3.4   Different Characterizations of the Unoccupied Levels

Various techniques available to probe the unoccupied levels have been developed in which the incident radiation does not cross the sample but is reflected by its surface. Let us recall that, in XAS, the absorbing thickness must be such that linearity of the measurement is preserved at the absorption maxima. The lower the radiation energy, the thinner must be the thickness of the absorbing material. For strong absorption lines this condition imposes the use of very thin absorbing screens. This can become quite a restricting condition because producing chemically pure homogeneous screens becomes difficult. In the other techniques, the incoming and outgoing particles are either different or of same type. Electrons or photons are used as incident radiation. The measured parameters are either the energy lost by the particle in the material or the energy of the radiation created during the interaction of the incident particles with the material. Important methods employed are detailed in the following.

3.4.1 L_{III} Photoabsorption

As already underlined, the absorption spectra are related to the configuration of a final state with a core hole. Following this process, the number of electrons screening the nuclear charge diminishes and the energies of all the levels increase. Based on this characteristic, the L_{III} absorption spectra have been used to detect an eventual variation of the number of the localized electrons in the rare-earths. The L_{III} spectra describe the transfer of a 2p electron into the unoccupied 5d6s levels. They present a large peak of which the inflexion point of lower energy gives the position of the Fermi level. The energy of the absorption peak depends on the number of the localized electrons. This energy increases with a decreasing in the number of localized electrons that screen the nucleus charge. If the same rare-earth is present with two different valences, two absorption peaks of different energy and intensity are simultaneously present in the L_{III} spectrum. The measurement of their relative intensities has been used to determine the valence. However, this determination is not very accurate since the peaks are broadened by two factors, the intrinsic width of the L_{III} levels, about 3–4 eV for the rare-earths, and the spectral resolution, of the order of 1 eV in the L_{III} energy range.

Actually, intense X-ray synchrotron radiation is used and the absorption spectrum is recorded in the *photoyield mode*, also named photoemission yield spectroscopy [70]. Incident continuous radiation interacts with the surface of the sample, and the photoemission yield of electrons of fixed energy is measured. The absorption length of the incident photons is larger than the average mean free path, or escape depth, of the photoelectrons. Consequently, the latter determine the analyzed effective depth and make this method surface sensitive. However, if photoelectrons of only a few electron volts are considered, they move in the conduction band and have relatively long escape depth. Bulk properties can then be determined. The measured quantity is proportional to the absorption coefficient of the incident photons. Therefore, as in the measurements by transmission, attention must be taken to eliminate the saturation effects existing for very large values of this coefficient. Saturation effects are known to be present when the photon penetration depth is comparable to the electron escape depth.

Other detection mode was suggested, the fluorescence yield. The escape depth of the lower energy fluorescence emission is different from the penetration depth of the incident radiation. Moreover, it was shown in the case of the rare-earth M_V absorption that the fluorescence yield can be distinctly different from the absorption cross section and the photoyield [71]. Indeed, at the final state of the absorption process, only the excited configurations which can be obtained from the ground state by the dipolar selection rules are present. Relaxation of the excited configurations takes place prior to the fluorescence emission and, consequently, absorption and fluorescence spectra are different. No absorption obtained by fluorescence yield is considered here.

3.4.2 *Electron Energy Loss Spectroscopy (EELS)*

In EELS experiments, the energy of an electron beam is analyzed after its inter-action with the target material. During this interaction, the incident electrons excite or ionize various atomic shells of the elements present in the sample. One measures the kinetic energy of electrons after their inelastic scattering by the core electrons. Generally, energetic incident electrons are necessary to obtain a sufficient intensity. The dipole selection rules are obeyed only if the incident electron energy is larger than the ionization threshold energy. In that case, the spectra give information comparable to that of X-ray absorption. In contrast, if the incident electron energy is only a few hundredths of electronvolt above the threshold, all the levels of the excited configuration are populated as discussed in Sect. 2.2.2. Different specific regions of the sample can be studied by attaching the spectrometer to an electronic microscope.

For Lanthanum, at incident electron energy very near the 3d thresholds, double peaks were observed in each energy range $3d_{5/2}$ and $3d_{3/2}$. The main peaks are at 834 and at 851 eV. Their energies match those of the two $^3D_1$ and $^1P_1$ intense lines observed in photoabsorption. They correspond to excitations from the ground state ($J = 0$) towards the excited levels $3d^9 4f^1$ $J = 1$. A structure is observed on the lower energy side of each main peak and its intensity decreases with the increase of the incident electron energy. Initially, these secondary peaks were interpreted by involving the decay of the intermediate $3d^9 4f^2$ configuration with emission of the characteristic radiation $h\nu$ according to the process $3d^{10} + e^-(E_0) \rightarrow 3d^9 4f^2 \rightarrow 3d^{10} 4f^1 + h\nu \rightarrow 3d^{10} + e^-(E_0 - h\nu)$ [72, 73]. In the $3d_{5/2}$ range, the structure is located at 831.3 eV, i.e. about 3 eV lower than the peak $^3D_1$. The same energy interval exists in the $3d_{3/2}$ range between the structure observed at 848 eV and the peak $^1P_1$. As a matter of fact, these structures correspond to the transitions towards the excited levels $3d^9 4f^1$ with $J = 2-6$ [74], in agreement with the calculated energies for these transitions (cf. Sect. 3.2.4). Their intensity decreases with increasing energy of the incident electrons (Fig. 3.14) [75].

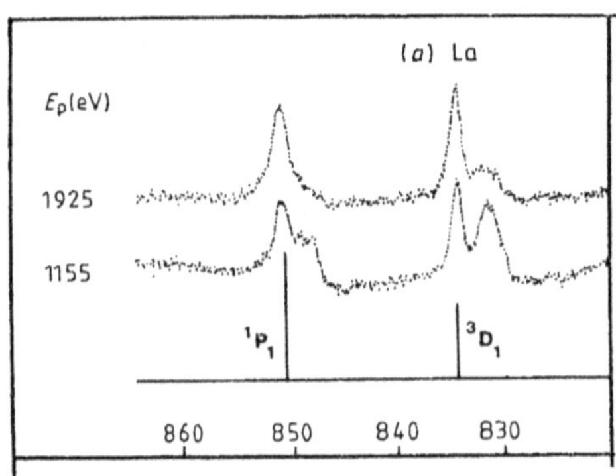

Fig. 3.14 Electron energy loss spectra of 3d lanthanum [75]

Indeed, under irradiation by electrons, all the J levels are populated in the vicinity of the excitation threshold while at higher energy the intensity of the non-dipolar transitions decrease, the electron cross sections of the optically allowed transitions increase and the levels corresponding to the dipolar transitions become preferentially populated. Concerning the $^3D_1$ and $^1P_1$ intensities, the more intense peak is $^3D_1$ while in photoabsorption the more intense peak is $^1P_1$. This inversion of the relative intensity between the $^3D_1$ and $^1P_1$ peaks is due to the difference in the primary process of the core-hole creation. For incident electrons of energy such that the dipolar selection rules are not obeyed, the intensity ratio of the lines is that of the statistical weights of the $d_{5/2}$ and $d_{3/2}$ sub shells. This accounts for the observed inversion of the intensity ratio.

The energies lost by mono-kinetic incident electrons after they excite or ionize a core 4d electron were also measured and the 4d–4f excitation energies deduced from these measurements. As already underlined, all the J levels of the $4d^9 4f^{m+1}$ configuration can be excited by incident electrons of energy near the threshold. In the case of lanthanum (Fig. 3.15) [75], the spectrum consists of the three excitation lines to the $J = 1$ levels, identical to the three photoabsorption lines. However, towards the lower energy of the ^3D or ^1P line, other lines are observed that are characteristic of the excitations to the other seventeen J levels of the excited configuration $4d^9 4f^1$. These lines are all located within an interval of about 10 eV below the $4d_{3/2}$ ionization edge. Only the intense line ^1P is situated above the threshold at about 12 eV of the other lines. The relative intensities of the various lines vary with the energy of the incident electrons and the number of well-resolved lines increases when the incident electron energy decreases. Near the threshold, the probability for interacting electrons to populate all the J levels increases because deviation from dipole selection rules becomes possible in this energy range. It appears that among the possible transitions, those with no spin exchange and

Fig. 3.15 Electron energy loss spectra of 4d lanthanum [75]

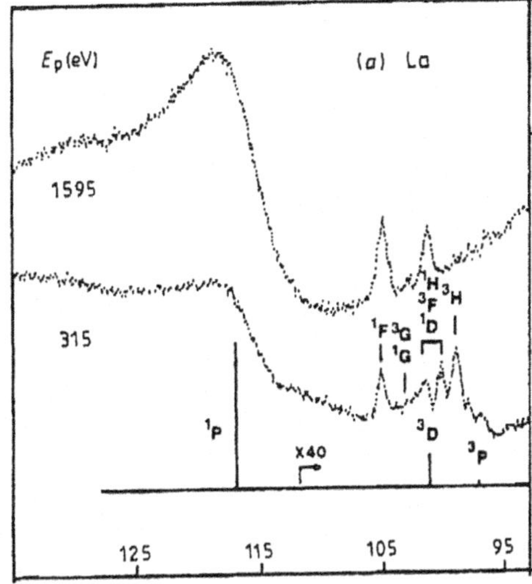

$\Delta L = 1$, 3 or 5 dominate. Indeed, the ^{1}F and ^{1}H ($\Delta S = 0$) transitions have intensities greater than those of the ^{3}D$_1$ or ^{3}P$_1$ transitions. But other transitions are also present, the more intense multiplet being ^{3}H, and a complex structure is observed. As discussed for the 4d absorption spectrum, the broad ^{1}P$_1$ resonance involves correlation effects with the continuum of the solid but it retains its dominant atomic-like character.

The NIXS, or NRIXS, method has been developed in parallel to EELS. This method is connected with EELS but high energy X-ray radiation is substituted to the high energy electron beam. While in XAS an incident photon is absorbed and its energy transferred to a core electron during a dipolar transition, in NIXS, as in Compton scattering, only a small part of the energy of the incident photon is transferred to an electron of the material. The matrix element of this scattering process is proportional to $\langle \psi_f | e^{iqr} | \psi_0 \rangle$. When the momentum transfer q is low, the matrix element tends to the dipolar matrix element and the spectrum is equivalent to the EELS spectrum. But at high q value, transitions with large momentum transfer can take place. High-order (up to 32) multipole transitions are possible to localized levels connected with high orbital angular momentum electrons and give complementary information on the nf electron distributions [76].

3.4.3   Isochromat Spectroscopy

In isochromat spectroscopy, one measures the intensity of the photons emitted at a fixed energy as a function of the incident electron energy, contrary to EXES, in which the spectral distribution of the emitted photons is analyzed during the irradiation by an electron beam of constant incident energy. This method makes possible the intensity measurement of the radiation emitted in a narrow spectral band in the vicinity of the appearance threshold. It differs from the appearance potential spectroscopy (APS) in which the measurement involves the emitted X-ray intensity in the entire spectral range. The spectral resolution depends on the energy width of the incident electron beam; it is different from that of EXES. In order to compare the emitted intensities at different photon energies, the isochromat spectra are analyzed with the help of a spectral band of constant width. Two distinct methods exist, the BIS and the CIS (Figs. 3.16 and 3.17).

In BIS, electrons of about 1–2 keV induce the emission of X-rays via a Bremsstrahlung process [25]. Bremsstrahlung is the continuous radiation emitted by electrons during their slowing down in the nuclei field of a solid. The radiation energy spreads from the maximum energy of the electrons down to their minimum energy. In BIS, the intensity of the emitted X-rays is recorded at a fixed energy as a function of the energy of the impinging electrons. The spectrum represents the density of empty levels weighed by the transition probabilities.

In a more detailed manner, at the appearance threshold of the radiation of energy $hv = E_S$, an incident electron of kinetic energy $E_C = E_S$ is completely stopped in the target. Its entire initial kinetic energy is transformed into radiative energy and the

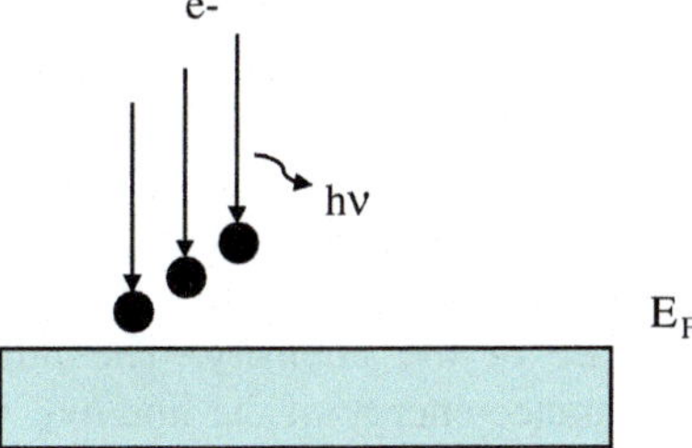

Fig. 3.16 Schematic representation of bremsstrahlung isochromat transition

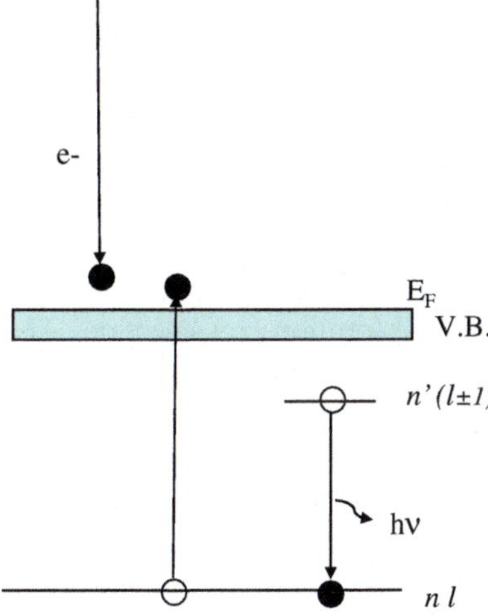

Fig. 3.17 Schematic representation of characteristic isochromat transition

electron is present in the continuum with zero energy. The kinetic energy E_C of the incident electrons is increased continuously. After the collision resulting in the emission of a photon of energy $hv = E_S$, these electrons are in one of the conduction levels of the solid of kinetic energy E_C–E_S. The probability of this process is proportional to the density of these empty levels. For insulator samples, charge effects are present, resulting in level shifts. These shifts are taken into account by recording the spectrum at different electron beam currents and by extrapolating the absorption edge to zero current.

From BIS it is possible to determine the density of all the conduction levels of a solid in its ground state. No core hole is present. Consequently, no perturbation due to a hole potential exists. The establishment of the levels is independent of their parity and of dipole selection rules. In contrast, the photoabsorption depends on the dipole selection rules and gives the distribution of the levels of symmetry $(l \pm 1)$ in the presence of an nl core hole in the Z atoms. The two methodologies are thus complementary.

CIS is regarded as an inverse XPS measurement. The incident electrons induce a characteristic X-ray emission. If the energy of the incident electrons is superior or equal to the energy of the level nlj, E_{nlj}, needed to create the initial core hole nlj, a photon of energy $h\nu_X$ is emitted. E_{nlj} is always higher than $h\nu_X$ for the normal emissions. The isochromat centred at $h\nu_X$ shows at first a threshold, which corresponds to the appearance of the bremsstrahlung of energy $h\nu_X$, then a sudden increase of the intensity. The inflexion point of this curve corresponds to the appearance energy of the line $h\nu_X$. CIS gives directly the energy of the initial level of the each characteristic emission and makes possible its identification. The spectral resolution depends on the energy width of the incident electron beam.

At the appearance threshold of a line, two electrons of kinetic energy zero are present, the incident electron and an atomic electron. When the energy of the incident electrons is increased, the isochromat describes the self-convolution of the distribution of the unoccupied levels, i.e. the self-convolution of the curve observed by BIS. It is necessary to take into account the probability of the process. This probability depends on the electron–electron collision cross section. Beyond twice the threshold energy, the conditions of the Bethe–Born approximation are satisfied and the collision cross section depends on the optical oscillator strength, thus on the photoabsorption cross section. At the threshold, the Bethe–Born approximation is not valid and the collision cross sections are generally not known. However, it is commonly accepted that the isochromat is the product of the photoabsorption cross section by the self-convolution of the unoccupied level distribution in the solid.

Characteristic isochromat of the resonance lines has also to be considered. The appearance energy of the line is equal to the energy of the emitted photon $h\nu_R$. Consequently, the isochromat of a resonance line is superposed on that of the bremsstrahlung. The incident electron loses a discrete amount of energy equal to the energy of the excited state. The ejected electron is localized in a discrete level and a single electron moves to an unoccupied level of the solid. Characteristic isochromat of a resonance line is the product of the probability of forming an excited level under electron bombardment by the density of the unoccupied levels in the excited system.

3.4.3.1   Lanthanum 3d Isochromat

Lanthanum bremsstrahlung isochromat was observed in the metal and LaF_3. Its general shape was comparable in both materials. The intensity increase at the threshold energy $E_S \sim h\nu$ was attributed to the scattering of the incident electrons into the lowest empty level of the conduction band. A strong peak observed above the energy of this level was due to scattering into the unoccupied 4f levels. These levels are situated at 5.5 eV above the threshold in the metal and 6.9 eV in LaF_3. A feature was seen at 15 eV above the threshold.

In characteristic isochromat of lanthanum metal, two resonance enhancements were observed in each of the $3d_{5/2}$ and $3d_{3/2}$ ranges. In the $3d_{5/2}$ range, one is to

831.1 eV and the other to 835.4 eV [77]. Initially, interpretation was made according to the suggestions of reference [48]. In fact, the enhancement at 835.4 eV, labelled R_V CIS, corresponds to the resonance line $^3D_1$ $3d_{5/2}^{-1}4f^1$-ground state while the other corresponds to transitions from the other eleven levels of the $3d_{5/2}^{-1}4f^1$ excited configuration towards the ground state. Independently, a characteristic isochromat was observed at 812 eV, i.e. at the energy of the $3d_{5/2}$–$5p_{3/2}$ line in the presence of a spectator electron. The threshold energy of the $3d_{5/2}$–$5p_{3/2}$ line was shifted down by about -2 eV with respect to the threshold energy of the R_V line. Indeed, all the J levels of the configuration $3d_{5/2}^{-1}4f^1$ contribute to the $3d$–$5p$ transitions while only the $^3D_1$ level contributes to the R_V line. The mean energy associated with the configuration $3d_{5/2}^{-1}4f^1$ is predicted about 2 eV below the $^3D_1$ level, making the threshold energy of the $3d_{5/2}$–$5p_{3/2}$ line lower than the threshold of the R_V line. This observation confirms that the emission observed at 812 eV does not occur in the ion since the energy of the $3d_{5/2}$ ionization threshold is higher than the excitation energy of $3d_{5/2}^{-1}4f^1$.

When the incident electron energy increases, a feature analogous to that observed by BIS appears, followed by a more intense feature labelled CK, near the $3d_{3/2}$ threshold. The feature CK is seen at the same energy in the $3d_{5/2}$–$5p_{3/2}$ and R_V characteristic isochromats. Its presence reveals the possibility for a system with a $3d_{3/2}$ hole to decay very rapidly to a system with the $3d_{5/2}^{-1}4f^1$ configuration. Various processes can transfer a $3d_{3/2}$ hole into a $3d_{5/2}$ hole. As an example, the Coster–Kronig transition $3d_{3/2}^{-1}4f^1$–$3d_{5/2}^{-1}$ has a very high probability but it leads to an ionized state and does not contribute to the feature present in the R_V and $3d_{5/2}$–$5p_{3/2}$ with spectator electron CIS. Only the processes leading to the precursor state of the R_V line are to be taken into account. The Coster-Kronig transition $3d_{3/2}^{-1}4f^2$–$3d_{5/2}^{-1}4f^1$ can contribute to these emissions because all the J levels of the initial and final configurations participate in this process. Then, the probability to obtain the configuration $3d_{5/2}^{-1}4f^1$ depends on the probability to create the initial configuration $3d_{3/2}^{-1}4f^2$ and its decay. Two processes can lead to the initial state $3d_{3/2}^{-1}4f^2$. If, after excitation of the $3d^94f^1$ configuration, the energy of the incident electrons is comparable to that of the excited electron, then the two electrons remain correlated until a dissociation process of this complex takes place. The number of electrons having the required energy is large in the vicinity of the threshold. The condition can be satisfied after slowing down of the incident electrons. The initial configuration $3d_{3/2}^{-1}4f^2$ could also be created directly by inelastic scattering of an incident electron into a 4f level accompanied by the excitation of an atomic $3d_{3/2}$ electron into this level. The transition probabilities of the Coster-Kronig type transitions from $3d_{3/2}^{-1}$ to $3d_{5/2}^{-1}$ are always very high when they are energetically accessible and, in both cases under consideration, the final state of the process is the $3d_{5/2}^{-1}4f^1$ excited configuration. In summary, the Coster-Kronig feature found in the R_V CIS contributes to enhance the intensity of the R_V line with respect to that of the resonance line at the $3d_{3/2}$ threshold (R_{IV}) and beyond it.

3.4.4 Inverse Photoemission Spectroscopy (IPS)

When the emitted radiation is in the UV energy range, the isochromat method is named IPS. The sample is impinged upon with electrons that first populate the unoccupied levels and subsequently undergo a radiative decay [78]. In IPS, the transition takes place between an initial unoccupied level, located above the vacuum level, and a final unoccupied level above the Fermi level. The energy dependence of cross sections in IPS was shown to be similar to that of the UV photoemission experiments. IPS investigates the levels between the Fermi level and the vacuum level and gives the unoccupied level distribution in the absence of a core hole. The emitted radiation is between a few volts to 30 eV. The sensitivity of the method is low compared to the photoemission. The process is governed by polarization dependent dipole selection rules [79]. Therefore the orientation of the polarization vector of the outgoing photon has to be taken into account. However, dipole selection rules still apply.

In IPS, as in XAS, the theoretical calculations assume the dipole selection rules and the one-electron approximation. But in IPS, in the initial state, the incident electron wave function is taken as fully symmetric, or even, with respect to the mirror plane, i.e. the plane that contains the surface normal and detection direction. This difference plays an important role in the case of light polarization measurements. Another difference refers to the reciprocal space position probed in the two techniques. In IPS, one can probe only one k-point in the Brillouin zone for each incidence angle. In XAS, the signal is given by the charge carriers collected throughout the entire Brillouin zone. Other differences exist between IPS and XAS techniques. XAS provides data on the unoccupied levels related to each type of atom separately. This is the unique direct probe of the orbital angular momentum of the unoccupied electronic states. Information is also collected by XAS on the spin state, the oxidation state of each atomic site and on the nature of chemical bonds for the absorbing atom. IPS is not site selective and comparison between IPS and XAS can reveal valuable information on the interactions with the core hole. IPS measures the unoccupied electronic structure of all the atoms located in the region of the sample defined by the range of the impinging electrons. As UPS, this is a surface analysis method

Spin-polarized inverse photoemission has been considered to obtain the spin-dependent density of unoccupied levels in magnetic samples. But its use has remained limited because the complexity of the measurement and the difficulty to obtain a source of spin-polarized low energy electrons. Alternative technique, the spin-resolved X-ray absorption, has been used to study the ferromagnetic oxide EuO [80]. A magnetized sample, irradiated by synchrotron radiation, was analyzed by photoyield using an analyzer coupled to a spin polarimeter. A spin-up and spin-down splitting, as large as 0.6 eV, was observed near the bottom of the conduction band.

3.4.5   X-ray Anomalous Scattering

From optical theorem, the photoabsorption cross section σ (ω) of a photon $\hbar$ ω is a function of the imaginary part of the scattering factor, $\mathrm{Im}\, f(\omega)$. The scattering factor is related to the optical constants, among them the complex index of refraction, n $(\omega) = n - i\beta$ by the relations

$$n = 2\pi(c/\omega)^2 r_0 \sum_i N_i \mathrm{Re} f(\omega_i)$$

$$\beta = 2\pi(c/\omega)^2 r_0 \sum_i N_i \mathrm{Im} f(\omega_i)$$

where N_i is the number of scattering atoms i per volume unity; r_0 is the classical radius of the electron. Consequently, it is possible to determine $\sigma(\omega)$ experimentally from reflectivity measurements made under the condition of total reflectivity, named *specular reflectivity*. In the X-ray range, the total reflection angles are very small and the experiments must be made under grazing incidence.

Measurements of reflectivity over a range of energy around an absorption discontinuity, labelled *anomalous region*, enable the evaluation of the photoabsorption cross section in this region. This method is often used in regions where the absorption measurements by transmission necessitate the use of very thin absorbing screens. It permits the determination of the unoccupied level distributions in the material. A theoretical model was developed to account for the observed variations of the specular reflectivity in anomalous regions [81]. It was shown that the spectral distribution of the specular reflection changes noticeably according to whether the absorption takes place towards a continuum of unoccupied levels or towards quasi-localized levels. The first case predicts a small dip in the reflection spectrum almost at the same energy of the absorption minimum. The second case predicts a narrow intensity maximum slightly shifted towards the higher energy side of the absorption line.

Experiment was carried for a mirror containing 50 wt.% La_2O_3 in the vicinity of the La $3d_{5/2}$ absorption edge [81]. The spectral distribution of the intensity reflected by the mirror is localized in a slightly asymmetric line, shifted +1 eV with respect to the absorption line. This result is in good agreement with the theoretical prediction and confirms that the photoabsorption involves final discrete levels.

3.5   Different Characterizations of Occupied Levels

3.5.1   Photoemission

In this process, a hole is created in a core or valence orbital by photon irradiation of energy hv and the kinetic energy E_c of the emitted photoelectron is measured. Given hv and E_c, it is possible to deduce the ionization energy of the photoelectron,

$E_X = hv - E_c$. Photoemission depends on the ionization probability of the various electronic shells, or ionization cross section, which are a function of the energy of the incident radiation. In X-ray photoemission spectroscopy (XPS), an intense characteristic line, very often Al Kα (1486.7 eV) or Mg Kα (1253.6 eV) is used for irradiating the target. In ultraviolet photoemission spectroscopy (UPS), the incident photons are generally those of He I (21.2 eV) or He II (40.8 eV). When a tunable light source is used, it is possible to choose photons of incident energy so as to favour a selected electron orbital. The combined use of synchrotron radiation and apparatus with a very high resolution improves the precision of the energy measurement to the order of 0.05 eV.

XPS is a powerful tool for determining core electron binding energies in materials. The binding energies of the core electrons depend strongly on the chemical surrounding. The presence of different valence states is manifested by a shift of the core levels due to the different degrees of screening of the nuclear potential by the valence electrons. Indeed, changes in the oxidation number, in the coordination number of the atoms, in the ligands, produce a *chemical shift* of the binding energies. The timescale, 10^{-16} s, of the analysis, enables the detection of alternating valence by the observation of distinct peaks in the spectra involving core levels. The precision of the energy measurement depends on the precision of the zero reference energy determination. The energy of an X-ray normal emission is the difference between the energies of the two X levels concerned. Comparison between X-ray emission and photoemission spectra can be used to determine the energies of normal X-ray lines and, conversely, to identify the various peaks present in a photoemission spectrum.

XPS, as XAS, is governed by the absorption cross section of the incident radiation. The variation of the photoionization cross section with the energy of the incident radiation can be used to identify the parity of the orbitals. As an example, the photoionization cross section of the 5f orbitals increases with the radiation energy more rapidly than that of the other orbitals. While the 5f cross section is considerably lower than the one of the 6d sub shell at the excitation energy of the He I line, the two cross sections are nearly equal at the excitation energy of the He II line. This rapid increase has made the 5f electron distribution easily identifiable and constitutes one merit of this analysis method.

Relaxation processes are noticeable in the ion but they are very weak in the presence of an excited electron because its screening of the core hole is then more efficient. In XPS, the photoelectron quits the atom with a kinetic energy that increases with the photon energy and transfers it into the continuum. Consequently, the screening of the core hole remains weak. It is due to itinerant valence electrons. The shape and the intensity of the peaks strongly depend on the surrounding of the atom with the core hole. The one-electron picture generally breaks down and multi-electron structures have to be introduced. In contrast, in XAS, the photoelectron is excited into empty levels just above the Fermi level where the electron is able to screen the core hole. If the excitation takes place into localized unoccupied levels, the excited electron stays in a sub shell of the same ion, the screening is practically complete and the photoexcitation does not depends on the surrounding atoms. Consequently, relaxation effects are present in photoemission while they are negligible in photoabsorption owing to the presence of the excited electron that screens partially the core hole. Satellites due to the rearrangement of the electronic charge after the ejection of the photoelectron are present in photoemission and absent in photoabsorption. As an example, the shake-up processes, which are negligible near the threshold energy of the core-hole creation in photoabsorption, have an important role in photoemission. These processes depend strongly on the surrounding atoms in the solid. In a compound, they depend on the ligands. Numerous models have been developed to treat the photoemission of correlated electrons in solids [82–84].

Another difference exists between XPS and XAS because the detected particles are not the same. In XPS, the thickness of the sample contributing to the spectrum is conditioned by the range of the photoelectrons, which is clearly shorter than that of photons of same energy. Due to the short electron probing depth, XPS is surface dependent. Moreover, charge effects can make difficult the characterization of the insulators by XPS. Surface sensitivity can be modified by varying the incident photon energy $h\nu_0$ or the analysis angle. Indeed, the inelastic scattering length of the photoelectron decreases with $h\nu_0$, increasing the contribution of the surface atoms. All the same, if the analyzed electrons are emitted at grazing angle, the analysis concerns a superficial thickness. XPS is often used as an analysis method of the surfaces. The coordination number of the atoms varies from the surface to the bulk. Consequently, the potential is different and a shift of the energy levels of the atoms occurs going from the surface to the bulk. Changes observed for 4f photoemission excited with photons of 100 eV or X-ray photons are discussed below [85].

The shape of the photoemission peaks is characteristic of the final state. The energy variation across an emission band is almost negligible and the spectral density is considered as describing the distribution of a hole in the unperturbed electron band. Consequently, XPS and UPS have been widely used in order to obtain the valence electron distributions in the bulk. If the incident energy is well above the ionization potential of the core electrons, the sudden approximation might be used. In this approximation, the electronic transitions are considered as having a one-electron character. Nonetheless, secondary effects can exist and lead to a breakdown of the one-electron picture. The core level spectra of metallic materials show asymmetric line shapes due to the excitation of electrons present at the Fermi level. Two models were used to describe the process. The formation of electron–hole pairs in the immediate vicinity of E_F, consecutive to the creation of the core hole, was considered in a many body model and the line shapes of the core level spectra was formulated in this model [86]. Other explanation was based on the existence of a shake-up mechanism in which the sudden occurrence of a core hole causes simultaneous excitation of electrons from occupied levels just below E_F to just above E_F [87]. This phenomenon requires the presence of a high density of states at E_F and a more asymmetric line shape is expected in the materials with a similar high density of states at E_F. It can contribute up to 0.5 eV broadening on the high binding energy side of all the observed core level spectra while the tail on the high binding energy side appears to decrease exponentially with increasing binding

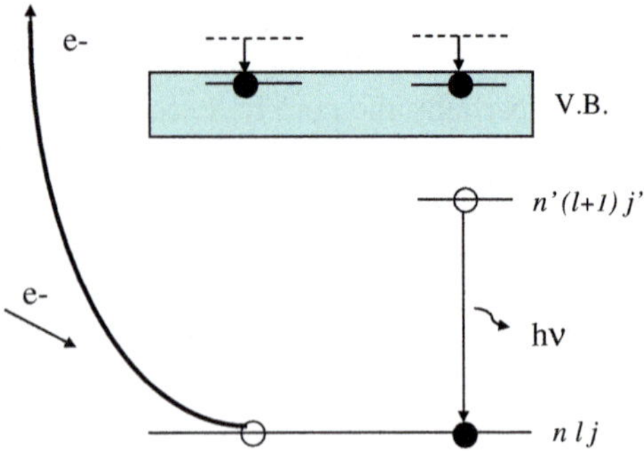

Fig. 3.18 Schematic representation of shake-down satellite

energy. This broadening mechanism is removed when no electron is present at E_F. Thus in insulators the core level peaks are found to be symmetric.

Comparison between photoemission and X-ray emission spectra is useful to identify the photoemission spectra when several peaks of comparable intensity are simultaneously present. Thus considerations based on the results of EXES have made possible the identification of the *shake-down* satellites, also named *well-screened* peaks (Fig. 3.18).

A shake-down process is an extra-atomic relaxation in which a valence electron is transferred to an unoccupied localized orbital in the presence of a core hole. Indeed, the extra positive nuclear charge due to the presence of the core hole pulls down the filled and unfilled localized-like orbitals making their energies higher in the ion than in the neutral system. This energy shift follows that of the core levels. A similar effect but weaker exists for the partially delocalized valence orbitals. The energy shift of an empty localized-like orbital can be large enough to pull the orbital below the top of the valence band and an electron can be transferred from this band to the orbital nl leading to the $c^{-1}(nl)^{m+1}V^{-1}$ configuration where c^{-1} and V^{-1} designate a hole in the core sub shell and the valence band respectively. Consequently, a well-screened configuration is created in the presence of the core hole when a localized orbital and the valence band overlap energetically. Thus, the transfer of an electron from the valence band to the localized orbital require no energy. Probability of a shake-down process is weak because overlap integral between initial and final states is small.

Shake-down satellites were observed in lanthanum and some light rare-earth spectra but not in barium spectrum (Fig. 3.19) [88]. The well-screened final configuration was obtained by ionization of the 3d core level, followed by the localization of a valence electron in an unoccupied localized 4f orbital. This relaxation process contributes to screen the charge due to the 3d hole and stabilizes the system. The energy of this well-screened configuration is lower than that of the ionized state. It gives rise to a shake-down satellite, or "well-screened" photoemission peak, located towards the lower energies of the main peak The main peak is, then, named the "poorly screened" peak [89]. The presence of the satellite depends on the energy of the empty nf orbital with respect to the valence band in the presence of an 3d

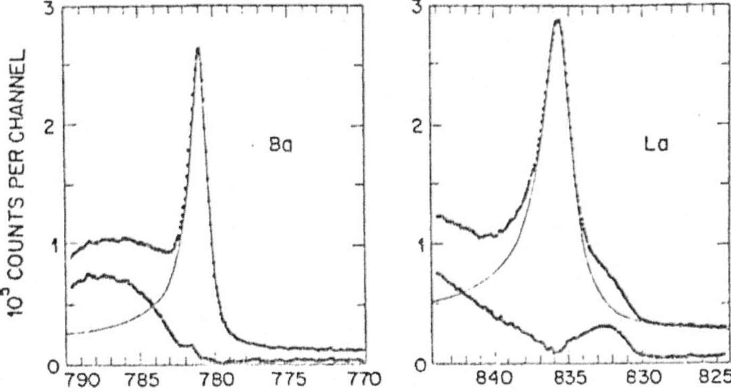

Fig. 3.19 $3d_{5/2}$ photoemission of barium and lanthanum [88]

hole. This relaxation process leads to the $3d^{-1}(4f)^{m+1}V^{-1}$ configuration and is not governed by dipole selection rules: all the J levels of the well-screened configuration can be populated with probabilities equal to their statistical weights. The energy of this configuration is represented by its barycentre.

In the rare-earth compounds, inter-atomic charge transfers from the ligand to the 4f orbitals are possible during the creation of a core hole in the rare-earth atom. This transfer is made easier by the stabilization of the 4f levels, i.e. the increase of their binding energies in the presence of the core hole. According to the respective energies of the ligand orbitals and the rare-earth 4f orbitals, the secondary emissions associated with these charge transfers are shake-up or shake-down satellites. If the band gap is large, the 4f states are located inside the band gap in the presence of the core hole and the charge transfer requires energy, resulting in the appearance high energy satellites, or shake-up satellites. In the compounds with a narrow band gap, the 4f levels can be mixed with the valence levels in the presence of the core hole and shake-down satellites can be observed towards lower binding energies. Intra-atomic ligand–ligand charge transfer can also be present.

Lastly, interference can exist between a photoemission process and the de-excitation of a core hole, leading to the same final state. This corresponds to an Auger transition. When the photon energy is varied across the region of the core excitation threshold it causes a strong variation in the photoemission cross section. This phenomenon forms the basis of *resonant photoemission*, which is discussed conjointly with the Auger resonant process.

Photoemission is not bound by selection rules. In the general case, it gives a measure of the energy of the photoelectrons emitted by all atoms located in the analyzed region, without directional restriction. Particular experimental methods were developed with the aim to limit the analysis to that of photoelectrons of particular characteristic. Thus, the angle-resolved photoemission spectroscopy (ARPES) makes possible the measurement of the kinetic energy and the angular distribution of the photoelectrons emitted from a valence band. This is equivalent to an analysis in the k-space. In the constant final state photoemission spectroscopy

(CFS), photoelectrons of chosen kinetic energy are collected as a function of the continuously varied incident photon energy. This method gives information directly related to the initial density of states (DOS) independently of the possible final state DOS modulation. It requires incident photons of variable energy, which is best provided by synchrotron radiation.

3.5.1.1   Lanthanum nd Photoemission

Experimental and calculated energies of the photoemission peaks are discussed here for lanthanum and compared to data of the X-ray emission spectroscopy. This discussion is generalized to other rare-earths in Chap. 4. Let us consider La $3d_{5/2}$ photoemission in the metal: the energy of the main peak is 835.6 eV. This value corresponds to the binding energy of the $3d_{5/2}$ sub shell [88]. A weaker peak is observed at 832.4 eV. These two clearly separated peaks have approximately the same energy and the same shape as the two peaks seen by EELS (cf. Sect. 3.4.2). The peak at 832.4 eV has been interpreted successfully as a shake-down satellite of the $3d^{-1}4f^1V^{-1}$ final configuration. In the presence of the core hole, an empty 4f orbital, situated above the Fermi level, relaxes in the core-hole field and mixes with the valence levels, making possible the creation of the $3d_{5/2}^{-1}4f^1V^{-1}$ excited configuration by transfer of a valence electron into the 4f orbital. This excited configuration is more stable than the $3d_{5/2}^{-1}$ ionized one because the 4f electron partially screens the positive charge due to the 3d hole. Its energy is lower than that of the dipolar transition to $3d_{5/2}^{-1}4f^1$ $^3D_1$. It is close to that of the barycentre of the $3d_{5/2}^{-1}4f^1$ configuration. The screening effect is possible only in the metal because the 4f and valence electrons are mixed in the presence of the core hole and the large mobility of valence electrons favours the population of the 4f level. The intensity of the shake-down satellite depends on the coupling energy between valence band and atomic-like 4f electrons. This satellite is weak; its intensity amounts to only a few percent of that of the main peak and decreases with an increase of the 4f electron localization. This shake-down satellite is named also "well-screened peak". The two names are used here. The well-screened peak corresponds to transitions from the ground state to an excited atomic configuration. Its energy is therefore independent of the considered solid.

In order to confirm this interpretation, the energies of the twelve J levels belonging to the configuration $3d_{5/2}^{-1}4f^1$ and the energy of the barycentre of this configuration were calculated [34]. The barycentre is at 831.8 eV, i.e. at the energy of the shake-down peak with a precision better than one per thousand. The agreement between the experimental and calculated energies of the shake-down peak confirms the localization of the final configuration. The intensity of the peak is only 5–6 % of that of the main peak. It depends on the interaction between the 4f orbital and the valence band. The presence of a low intensity peak demonstrates that this interaction exists but is weak in metallic lanthanum.

The energy of the photoemission peaks is approximately the same in lanthanum metal and its intermetallic compounds. In contrast, it varies strongly from one chemical compound to another [90, 91]. In the $3d_{5/2}$ photoemission of lanthanum insulator compounds, such as La_2O_3 and LaF_3, no peak is seen in the vicinity of 832 eV. No process having as final configuration the excited configuration $3d_{5/2}^{-1}4f^1$ is observed, implying that there is no relaxation of the ionized system with a core hole and no presence of a well-screened configuration. In La_2O_3, two peaks were observed at 836.0 and 840.2 eV. The binding energy of the $3d_{5/2}$ singly ionized level can be deduced from the L_{III} absorption and $3d_{5/2}$–$2p_{3/2}$ emission. It is predicted at 837.7 eV and corresponds to the main peak located at 836.0 eV. The peak at 840.2 eV was attributed to an inter-atomic shake-up process. In this process, the creation of the $3d_{5/2}$ hole is accompanied by an electron excitation from the valence band to an empty 4f orbital, usually designated as an electron charge transfer from the ligand to the rare-earth. Although the stabilization of the empty 4f orbital increases in the presence of the core hole, this orbital still remains situated in the band gap of the insulator and the transfer of an electron from the valence band to the 4f levels requires an additional energy. Analogous interpretation was given for 3d photoemission of $La(OH)_3$ [92].

Information can also be obtained from the La 3d–5p emissions, by comparing the experimental energy of the lines to the difference between the 3d and 5p energy levels. For La_2O_3, the La $5p_{3/2}$ and $5p_{1/2}$ peaks are observed at 18.0 and 19.9 eV. No satellite is present in the 5p photoemission spectrum, as opposed to the presence of satellites in the 3d, 4d, 4s and 5s photoemission spectra. Based on the $5p_{3/2}$ energy at 18.0 eV and the energies of the two $3d_{5/2}$ peaks, 836.0 and 840.2 eV, the emission $3d_{5/2}$–$5p_{3/2}$ is expected at either 818 or 822.2 eV. No X-ray emission is detected in the vicinity of 822 eV. A low intensity emission is observed at 817.3 eV; it corresponds to the lower binding energy La $3d_{5/2}$ peak located at 836.0 eV. Moreover, a strong emission is observed towards the lower energies, at 812.2 eV. This line is not expected from the photoemission spectrum of the oxide La_2O_3. As underlined in Sect. 3.2.2, this lower energy line is the $3d_{5/2}$–$5p_{3/2}$ emission in singly excited lanthanum in the presence of the 4f spectator electron. Its energy depends on that of the excited configuration $3d_{5/2}^{-1}4f^1$, i.e. on the energy of the lower peak seen at 832 eV in the metal. It is predicted at about 814 eV, in agreement with the experimental value. This result confirms that the photoemission peak observed at 836.0 eV for the oxide does not correspond to the lower peak seen at 832 eV in the metal, i.e. to a well-screened excited configuration in the compound, as had been wrongly suggested [89]. The peak, at 836.0 eV, corresponds to the singly ionized configuration. The higher energy peak is a shake-up peak. It results from the excitation of a valence electron into a higher 4f orbital in the presence of the $3d_{5/2}$ core hole. Its presence accounts for the perturbation undergone by the system during the creation of the core hole.

From these observations, the main results can be described as follows: In metal lanthanum $3d_{5/2}$ photoemission, a low energy peak is present; it corresponds to the $3d_{5/2}^{-1}4f^1V^{-1}$ final configuration, where V is the valence band. This configuration is created by the relaxation of the ionized $3d_{5/2}^{-1}4f^0$ configuration and the peak is a shake-down satellite. As a result of the creation of a $3d_{5/2}$ hole, an empty 4f level is pulled down below the Fermi level and is occupied by an electron from the valence band with stabilization of energy. The energy of the barycentre of the configuration $3d_{5/2}^{-1}4f^1V^{-1}$ is very close to that of the barycentre of the excited configuration $3d_{5/2}^{-1}4f^1$ that was created by direct excitation of a $3d_{5/2}$ electron into the empty 4f orbital. However, the energies necessary to create these two configurations are not the same because the $3d_{5/2}$ sub shell must be ionized to obtain a relaxed configuration $3d_{5/2}^{-1}4f^1V^{-1}$ and the ionization energy is higher than the excitation energy. The intensity of this shake-down satellite, or well-screened peak, characteristic of the $3d_{5/2}^{-1}4f^1V^{-1}$ relaxed configuration, is low because the probability of the relaxation process is small. This peak is observable only if the 4f orbital is mixed with the valence states in the presence of the core hole. Its low intensity is due to the very weak 4f—valence state interaction, already underlined from EXES results. The same results are obtained in the $3d_{3/2}$ region. The two main peaks observed in the $3d_{5/2}$ and $3d_{3/2}$ photoemission of lanthanum metal have approximately the same energy and the same shape as the peaks observed by EELS.

In the lanthanum 3d photoemission of compounds as LaF_3 and La_2O_3, in contrast to the metal, no well-screened peaks, or shake-down satellites, are observed and the low energy peaks are observed above the higher energy peaks of the metal. It had been predicted that in the rare-earth compounds a 4f orbital could be populated by the relaxation of a core hole ionized state $(nlj)^{-1}4f^m$ to the $(nlj)^{-1}4f^{m+1}V^{-1}$ well-screened configuration. The lower energy La 3d photoemission peaks were interpreted with the help of this process. On the other hand, when all the energy data are considered, it becomes clear that these peaks correspond to the singly ionized $3d_{5/2}^{-1}4f^0$ and $3d_{3/2}^{-1}4f^0$ configurations. Satellite peaks are observed towards the higher energies of the main peaks. In these insulator compounds, in the presence of the core hole, the 4f orbitals are localized in the band gap above the valence band and an additional energy is needed to transfer an electron from the valence band into a 4f level. Therefore, the high binding energy satellites are due to an excitation from the valence band to the empty 4f levels taking place during the creation of the 3d inner hole; these are shake-up satellites. No relaxation process exists between the 4f and valence levels in these compounds.

The 3d photoabsorption is mostly dominated by the excitation of electrons to the 4f discrete levels and this process does not vary with the electronic structure of the material because the excited 4f electron strongly screens the core hole. On the other hand, the 3d photoemission depends on the studied material. It represents the ionization and is very sensitive to perturbations accompanying the formation of the core hole. These perturbations originate from the band structure of the material. The screening effect leading to a well-screened configuration, i.e. to a shake-down transition such as described above for lanthanum metal, is absent in La_2O_3 and LaF_3. In contrast, for both these insulator compounds, and also for the other La insulator compounds, energy is needed to transfer a valence electron into a 4f orbital, i.e. for a charge transfer. Shake-up satellites involving the valence electrons are observed towards the higher energies of the main peak.

In the lanthanum 4d energy range, X-ray photoemission is expected to show the spin–orbit splitting in the 4d sub shell. Indeed, the 4d spectrum of lanthanum metal exhibits a well-defined doublet due to $4d_{3/2}$–$4d_{5/2}$ spin–orbit splitting. The experimental splitting is equal to 2.85 eV [93], in agreement with the value of 2.9 eV calculated for the La^{4+} ion. Although the 4d spin–orbit interaction is weaker than the 3d one, the $4d_{5/2}$ and $4d_{3/2}$ peaks are still well resolved. Let us note that only the calculated interval between the peaks is in agreement with the experience. The theoretical absolute energies are too big because the valence electrons were not taken into account in the calculation. The $4d_{3/2}$ peak is the stronger of the two and the density of the $4f_{5/2}$ states is predominant at the threshold. As already underlined, the relaxation accompanying the presence of a 4d hole is weak and no shake-down satellite is observed as opposed to what was seen in the 3d metal spectrum.

In contrast with the metal, the 4d and 5p photoelectron lines observed for atomic lanthanum do not show a well-resolved spin–orbit splitting [94]. This result was interpreted by assuming that the 4f orbital is occupied in the final state. Theoretical models were developed to predict this experimental result and explain the 4f orbital occupation by a *collapse* of the 4f orbital in the atom with a 4d core hole.

The 4d photoemission of lanthanum compounds are analogous to that seen for the metal. Only ionized configurations are involved and no resonance effect is present. The energy and relative intensity of the $4d_{3/2}$ and $4d_{5/2}$ peaks vary in the various components. These variations are connected with the chemical changes. An interesting study was made for lanthanum in LaB_6. The photoemission was observed at various incident photon energies in the energy range of the 4d photoabsorption giant resonance [63]. In this incident energy range, the photoemission partial cross sections, 5s, 5p and 5d, show a strong increase. An explanation was given as followed: at these incident energies, a number of excited levels of the configuration $4d^9 4f^1$ are created. These levels have a large probability to decay by resonant Auger process with emission of a 5p or 5s electron. The energy transferred to the 5nl electron being equivalent to the energy of the initial photon, it is not possible to differentiate the photoelectron from the Auger electron and the emission probability of the 5nl photoelectron appears strongly increased.

3.5.2   Inelastic X-ray Scattering

In this section, we consider the scattering of X-ray photons by bound electrons. Photon sources are either characteristic lines or the continuous radiation produced by the electron slowing down in an X-ray tube, or the continuum radiation emitted by relativistic electrons in rotation in a synchrotron. High brightness third generation synchrotron sources are tunable photon sources especially in the soft X-ray and X-UV ranges. This bright tunable photon source has a large interest since it enables the control of the photon energy and polarization. According to the energy of the incident photons with respect to the threshold, the observed secondary emission can be attributed to either a Raman scattering process or a fluorescence emission. Big progress has been made in the field of inelastic scattering using quasi-monochromatic tunable incident radiation [95]. As for EXES, a non-radiative

Auger process, named Auger Raman scattering, is associated with the radiative scattering process.

From a general point of view [96], scattering is a second order process compared to absorption or emission, which are first order processes, because the number of quanta changes by two during scattering. Consequently, the inelastic scattering probability is generally small. X-ray scattering can be described as the absorption of a primary photon $h\nu_{in}$ by the scattering system combined with the emission of a secondary photon $h\nu_{out}$. In the inelastic case ($h\nu_{out} \neq h\nu_{in}$), the scattering system is left in a quantum state different from its initial state. If the energy $h\nu_{in}$ of the primary photon is large with respect to the binding energies of the atomic electrons, an electron that absorbed a photon is ionized and can be considered as free; this process is *Compton scattering*. The interaction of the radiation field with the electron system may be obtained from the A^2-term, where A is the vector potential of the field. If the energy $h\nu_{in}$ is of the same order of magnitude as the binding energies of the electrons, the electrons are excited; the process is *Raman scattering*. The contribution of the (p.A)-term, where p is the electron momentum, becomes preponderant [97, 98]. At the resonance, a bound electron has just enough energy to be ejected.

Raman scattering is a powerful tool for probing elementary excitations in materials, including crystal field excitations, phonons, magnons [99]. It was initially used in the infrared range. The scattering particles are then the phonons and the transitions take place between vibrational levels. The resulting features consist of two different contributions, one with absorption followed by emission (Stokes lines) and the other with emission followed by absorption (anti-Stokes lines). The anti-Stokes lines are present only if the vibrational levels of the ground state are populated. These transitions are much stronger in the infrared than in the X-ray region.

In the X-ray region, transitions involving two photons take place through an intermediate state and, as described above, two possible intermediate states exist, one with $h\nu_{in}$ absorbed and no photon present, the other with $h\nu_{out}$ emitted and two photons present. According to the energy conservation law, $h\nu_{in} - h\nu_{out} = E_{0,F}$, where $E_{0,F}$ is the energy difference between the final state F and the initial state 0 of the scattering system. The initial state is the ground state; the final state is a low-energy electronic excitation state. $E_{0,F}$ is called Raman–Stokes shift of the scattered radiation. The lifetime broadening of the final state is generally smaller than the lifetime broadening of the intermediate state, which is a core-excited state in the X-ray region. Energy conservation is satisfied with an accuracy defined by the inverse of the lifetime of the final state. The scattered radiation is red shifted and this shift is limited by the time-energy uncertainty principle. The theoretical description uses a one-step model with the absorption and emission treated as a single, non-separable, event. In a one-step picture, several paths exist leading from the initial state to the same final state. In this model, the intermediate state is virtual and the order of occurrence of the two processes of incident photon absorption and of final photon emission is indeterminate.

The differential cross section for the scattering of the incident X-ray photon $h\nu_{\text{in}}$ into an elementary solid angle $d\Omega$ is proportional to

$$d\Omega\left[\Sigma\left(\frac{\langle f[H_{\text{int}}]I\rangle\langle I[H_{\text{int}}]0\rangle}{E_0 - E_I + h\nu_{\text{in}}} + \frac{[\langle 0[H_{\text{int}}]I\rangle\langle I[H_{\text{int}}]f\rangle]}{E_0 - E_I - h\nu_{\text{out}}}\right) + \delta(h\nu - E_f + E_0)\cos\theta\right]^2$$

where H_{int} is the interaction Hamiltonian, $h\nu_{\text{in}}$ and $h\nu_{\text{out}}$ are the energies of the incident and scattered photons, 0, I and f designate the initial, intermediate and final quantum states and θ the angle between the directions of polarization of the incident and scattered photons.

An important intermediate state is one with a vanishing denominator. This condition holds for $h\nu_{\text{in}} = E_f - E_0$, that is to say the photon $h\nu_{\text{in}}$ coincides with a resonance frequency of the atom. In this case, only one intermediate state is to be taken into account, where the atom is excited and one photon $h\nu_0$ is absorbed. In the final state, the atom is again in the state E_0 and a photon $h\nu_{\text{out}}$ is emitted. The photon $h\nu_{\text{out}}$ can differ from $h\nu_{\text{in}}$ only by an amount of the order of magnitude of the natural line width. This process is the *resonance fluorescence*. The ordinary time sequence, i.e. absorption before emission, is satisfied. This is a two-step process equivalent to two independent processes, absorption and subsequent emission. Its intensity is comparable with that of a first order process. It can be described by a lorentzian curve with a FWHM equal to the sum of the intermediate and final states widths. Only resonant processes contribute to the spectrum in the vicinity of a strong X-ray photoabsorption line and the non-resonant ones can be neglected.

In the case of monochromatic incident radiation having a narrow energy distribution compared to the natural emission line, the emitted line is expected to have the same width as the primary radiation; it is therefore sharper than the natural line and its shape is different from that which is emitted spontaneously. Emission and absorption cannot be regarded as two independent processes but are to be considered as a single quantum process. However, that is true only if the system is unperturbed. As soon as the quantum state of the system is determined, the process is composed of two independent processes, absorption and emission, and the emitted line has the natural shape.

Let us consider now the case where the energy of the scattered photon is equal to the energy of the primary photon $h\nu_{\text{in}}$, i.e. the energy conservation law is strictly fulfilled: that is the *elastic scattering* or *Rayleigh scattering*. Only the direction of the photon changes in this process. The phase ϕ of the scattered wave is the same as the phase of the primary beam. The absorption and emission steps are coherent except if $h\nu_{\text{in}}$ is equal to a characteristic energy of the system, i.e. at the resonance. Indeed, in the general case, the phase is only determined if the number of photons N is undetermined, corresponding to the uncertainty equation $\Delta N \cdot \Delta\phi \geq 1$. In the case of the scattering of a single photon, the two phases of the incident and scattered waves are undetermined. On the other hand, for the scattering of a photon by two atoms, the phase difference of the two waves scattered by each atom can have a definite value, even if the total number of photons is determined. This corresponds to the case of the Bragg reflection: if the incident photon arrives on a crystal under the Bragg angle, it is scattered symmetrically with respect to the incident direction and this process is the *coherent elastic scattering* or *diffraction by a crystal*. Consequently, the elastic scattering exhibits a strong angular dependence, as does also the Compton scattering, while it was shown experimentally that the Raman X-ray scattering is isotropic.

Lastly, when the energy of the incident monochromatic photons exceeds the core ionization threshold energy, no excited state is created. The core levels are ionized and two-step case applies. The X-ray emission probability is independent of the incident photon energy. The only dependence of the emission spectral profile on the incident photon energy is due to the self-absorption of the emitted radiation and this is noticeable only for emissions located near an absorption threshold. The X-ray emission lines are treated as discrete transitions between ionized core levels by using a two-step model with the emission decoupled from the initial core hole creation process. This is the *X-ray fluorescence emission*, well known and largely used for quantitative chemical analysis. The fluorescence emissions are identical to the normal lines observed by EXES.

From this brief summary, it is clear that the characteristics of the scattered radiation depend strongly on the energy $h\nu_{\text{in}}$ of the incident X-ray photons. When $h\nu_{\text{in}}$ is tuned just below an absorption threshold, at a distance of the order of the threshold width, core excitation can take place, promoting the system to a neutral core-excited intermediate state. This intermediate state can decay radiatively to a final state with a hole, an excited electron and an emitted photon (Fig. 3.20). This process is called *Resonant Inelastic X-ray Scattering* (RIXS) or *X-ray Raman scattering* [97, 100–104]. The linear energy dispersion of the final state, characteristic of the RIXS structures, is referred to as the Raman dispersion law [105]. This process is present only if the final state is discrete. It breaks down when the final state is in the continuum, i.e. when the excitation energy exceeds the core ionization threshold. The scattering is designated as resonant because it involves the creation and the disappearing of a core hole. The energy difference of the incoming and outgoing photons is taken up by the electronic system. The spectrum is continuous because the scattered photon and the excited electron share the available

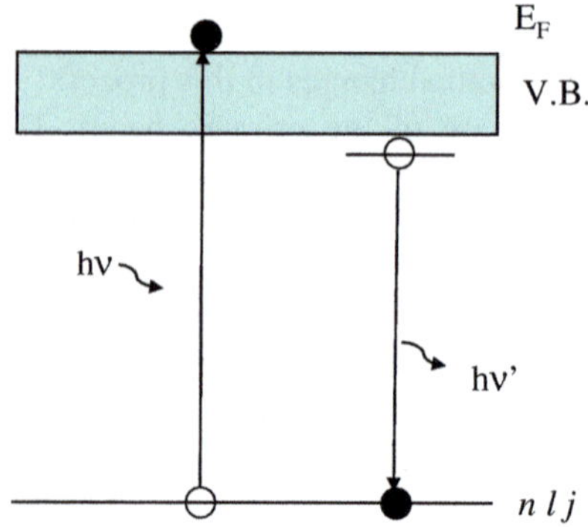

Fig. 3.20 Schematic representation of Resonant Inelastic X-ray Scattering (RIXS)

energy, $hv_{in}-\varepsilon_h$, where ε_h denotes the energy of the level with the final hole. The amplitude of the non-resonant scattering is usually negligible compared to the resonant amplitude. Contrary to the fluorescence, which is emitted at fixed energies regardless of the excitation energy, the resonant scattering occurs at given available energy, or constant energy loss, and therefore is present over an energy range larger than the excitation width.

In RIXS, energy conservation between the initial and final quantum states is independent of the intermediate state. Consequently, it was deduced that the natural width of the intermediate state, i.e. its lifetime, should not limit the widths of the spectral features [102, 106]. In this process, the number of electrons remains unchanged because the system is excited but still neutral. However, in solids, the potential change due to the presence of the hole induces a relaxation of the excited level, thus a variation of its energy ε_e. If ε_h is well defined, to a variation $\delta\varepsilon_e$ there corresponds a variation δhv_{out}. The uncertainty in ε_e introduces, therefore, a broadening of hv_{out}. It was shown that this broadening induces an asymmetric low-energy tail with decreasing hv_{in} [107].

As the incident energy hv_{in} approaches the absorption threshold energy E_{nl}, the scattered intensity increases approximately as $(E_{nl} - hv_{in})^{-1}$ [108]. In a first approximation, the scattering cross section is independent of the scattering angle and the polarization state of the incident beam. The inelastic X-ray scattering reaches its maximum intensity for the resonance fluorescence. It is very weak off the photoabsorption threshold and increases in a narrow energy range around the resonant region. In this region, the scattered spectra consist of three different subcomponents [109, 110]:

- the elastic peak; it is at the incident photon energy, except in the presence of losses by phonon excitation. This is the most intense peak up to the excitation threshold. Its intensity decreases with the increase of incident photon energy above the threshold.
- the normal X-ray emission lines; they appear at fixed X-ray emission energies. They are the fluorescence lines, originating from ionized intermediate states.
- the RIXS structures; these are radiative transitions appearing at constant energy loss and dependent on low energy electron excitations, as in optical Raman scattering.

The RIXS structures are generally complex. For atoms or molecules, the decay from an excited configuration towards the ground configuration involves all the transitions allowed by the selection rules. Thus, it takes place to the ground state and also to low-lying excited states of this configuration, so that the multiplet structure of the ground state configuration can be established. Furthermore, low energy dipole-forbidden transitions, whose intensity is enhanced strongly at certain energies, can also be observed. These are, for example, the d-d excitations in the transition elements [111, 112], the f–f excitations in the rare-earths [113] and in the actinides [111]. These excitations are always dispersed linearly with the incident photon energy and show up as energy loss peaks. Since they are normally forbidden by the dipole selection rules, their observation by RIXS is, thus, significant.

RIXS is related to the absorption process as shown by the strong dependence of the scattered intensity on the incident energy in the vicinity of the core hole excitation. In the case of solids, this dependence can been used to establish the band structure of complex materials from momentum-resolved experiments. In principle, the momentum conservation should apply only to electrons present in systems with translational symmetry. In fact, this law holds also for Compton diffusion by bound electron. On the other hand, perturbations of the intermediate state by its environment are possible [114]. These perturbations are due to electronic or nuclear relaxations like, for example, the scattering of a non-ionized electron or a loss of structural symmetry. They may destroy the energy and momentum conservation relations.

Like all spectroscopies involving a core hole, RIXS presents characteristics specific to each element, enabling its observation in the different sites of the solid. The technique is bulk sensitive, in contrast to electronic spectroscopy, such as photoelectron spectroscopy (XPS) and EELS, which are surface sensitive. It involves charge-neutral excited states whereas XPS (PES) and fluorescence involve ionization. In principle, the RIXS structure widths are considered as independent of the energy distribution of the intermediate state, and thus the core hole lifetime is not a limit on the resolution of the method. Among the potential applications of the RIXS, the observation of transitions between levels belonging to different atoms is generally mentioned. These transitions, often named charge transfer transitions, result from inter-ionic excitations in systems with strong metal-ligand hybridization. Another application is the study of selected vibrational states [115, 116]. But the most interesting possibility is the ability to probe low-energy intra-ionic electron excitations in systems with weak hybridization effects, such as the d–d excitations in the transition elements and the f–f excitations in the rare-earths. These excitations are not expected from the pure Hund's rule ground state and are generally unobserved by optical spectroscopy and EELS.

The RIXS structures appear towards the lower energy of the elastic peak at constant energy loss. For such structures to be observed, the energy loss between incident and emitted photons has to remain within the energy extension of the considered process. For example, for the structures due to charge-transfer excitations between the metal ions and a ligand in a compound, the energy extension is defined by the width of the ligand band from which charge transfer can take place. Therefore, correlations must exist between incident and emitted photons within this energy range. However, in some cases, this condition is not satisfied; the structures are observed at constant photon energy and appear as normal emission lines. On the other hand, the presence of a spectator electron or vibrational excitations, not taken into account in this model, may lead to non conservation of energy.

As already underlined, the intensity of the scattered electrons is low and depends strongly on the energy of the incident photons around the threshold. Features resulting from different excitation energies can be distinguished by utilizing the tunability of the synchrotron radiation to create the inner-shell vacancies. The energy distribution of the incident photon beam can be used to govern the spectral function of the scattering process. Thus, a photon beam width narrower than the lifetime width of the core level is often used. Consequently, the width of the emitted lines is smaller than the natural width of the core-excited level. Nevertheless, as already mentioned, some of the spectral lines may be broadened and asymmetric.

In summary, for incident energies below the ionization threshold, the spectra obtained consist of the elastic peak and of structures due to low-energy valence excitations, extending over about 10 eV. These structures appear at constant energy loss and are radiative transitions induced by low energy electron excitations, as in optical Raman scattering. The effect of the wings of the lorentzian distribution gives rise to the highly asymmetric line profiles. On the other hand, it has been often claimed that the absorption maxima can be probed by scattering with a better resolution than by photoabsorption. However, the fact that the absorption spectra were generally observed from very thick absorption screens [117], i.e. in unfavourable experimental conditions, casts doubt about this claim. Above the ionization threshold, a fluorescence spectrum is observed. It is independent of the excitation energy since the continuum reached by core ionization is degenerate. It is easy to distinguish between scattering peaks and normal fluorescence lines because the energy of the scattering peaks increases with the excitation energy while the fluorescence lines have constant energies. The distribution of the valence states in solids is determined only from fluorescence emissions observed in an energy range sufficiently above the ionization threshold. In this case, the fluorescence spectra and EXES are identical.

3.5.2.1 RIXS and Momentum Conservation

RIXS structures were observed as a function of the energy of the incident photons in the neighbourhood of the absorption thresholds for various solids. The incident energy dependence was attributed to the variations of the density of the unoccupied levels present just above the absorption edge and, consequently, to the variations of the absorption probability of the incident photons. The effect is strong only for solids with wide valence bands. It is related to the momentum of the photoelectron and of the valence hole present in the final state. As shown below, the law of crystal momentum conservation has to be taken into account in the inelastic scattering process [118–120]. The energies of the RIXS structures, defined in the momentum conservation framework, may vary with respect to the elastic peak only within the energy range covered by the occupied part of the valence band. This law, always satisfied in the case of radiation interaction with free electrons, is known to be valid also for the interaction with weakly bound electrons.

Let us recall that in a solid the energy variation of occupied and unoccupied quantum states as a function of the wave vector, called band structure, presents features at the high symmetry points in the solid, each of which is characterized by a particular value of the electron momentum (cf. Chap. 1). The same features appear

in the curves of the density of states as a function of the energy. The photoabsorption from a core level reveals the presence of these symmetry points in the
conduction band. When the energy of the incident photon is varied, the excitation of
a core electron into quantum states of different symmetry in the conduction band
takes place. In this process, the initial state is the ground state; the intermediate state
consists of a core hole and a photoelectron in the conduction band; the final state
has the photoelectron in the conduction band and a hole in the valence band. Energy
and momentum are conserved, the energy lost by the incident photon being used to
create the pair electron–valence hole. Consequently, momentum conservation law
applies to the scattering and makes it a one-step process as already underlined.
From this conservation law, the momentum difference of the electron–hole pair (q_e
and q_h) is equal to the momentum change of the incident and emitted photons (p_{in}
and p_{out}). The conservation relation for the overall scattering process is thus

$$p_{in} - p_{out} = q_e - q_h = G$$

where G is a vector in the momentum space, or wave vector space, called reciprocal
lattice. By definition, $G = 2\pi/d$ where d is the inter-reticular distance of the crystal
real space (cf. Chap. 1).

The energy of the scattered photon is

$$h\nu_{out} = h\nu_{in} - (E_{nl} - \varepsilon_e) + (E_{nl} - \varepsilon_h)$$
$$= h\nu_{in} + \varepsilon_e - \varepsilon_h$$

$h\nu_{in}$ is the energy of the incident photon; E_{nl} is the ionization energy of the core
hole; ε_e and ε_h are the energies of the valence electron and hole with respect to the
ionization threshold. By tuning the incident photon energy so that the photoelectron
is excited into a specific critical point in the conduction band determined by
$h\nu_{in} = (E_{nl} - \varepsilon_e)$, the absorption process selects a particular momentum p_e.

In the soft X-ray range and lower energies, the momentum of the photons is
small compared to the momentum of the valence electrons in the solid and can be
neglected in comparison with q_h and q_e. As a result of the negligible momentum
transfer from the photons, the final state of the solid has equal momentum for the
photoelectron and the valence hole. These momentums are correlated in the final
state and all the transitions are vertical in the momentum space. In the momentum
conservation framework, the valence state involved in the emission process has the
same crystal momentum as the conduction state to which the excitation takes place,
up to one reciprocal lattice vector. This momentum selectivity assumes that the core
hole lifetime is shorter than the phonon relaxation time and electron–phonon
coupling can be neglected. This approximation is generally adequate for non-polar
and non-metallic materials. However, the restrictions imposed by the momentum
conservation law can be relaxed by transfer of momentum from other particles,
phonons for example, not taken into account in the interpretation of the experiment.
No vertical transitions are, then, present.

Large intensity changes observed in the spectra resulting from merely a small variation of the incident energy can be due to modulations of the density of states and of the crystal momentum. Emission enhancements are observed if the excited photoelectron is present at symmetry points. Comparison between the emissions of amorphous and crystalline materials supplies a direct demonstration of the role of the band structure and of the long-range order. As an example, while the single-crystal emission manifests strong variations with the incident energy, the amorphous silicon emission is completely independent of it [121]. Thus, the incident energy is tuned around the excitation threshold in order to excite the core electron into unoccupied states of the conduction band with a well-defined momentum. In contrast, far above the excitation threshold, no dependence on the incident energy is observed [118].

The momentum conservation law does not hold in a few cases: For solids with narrow energy bands, as, for example, transition metals with an almost full d band [107]. In the presence of levels with an atomic or excitonic character; the photoelectron remains then localized near the core hole as a spectator electron; it can influence the emission process through its interaction with the valence electrons and the spectrum is dominated by the transition involving this localized electron. For an intermediate state relaxing in a timescale shorter than or equal to the core hole lifetime. Last, for systems whose interaction with the surrounding is strong during a time scale of the order of the core hole lifetime.

3.5.2.2 RIXS of Lanthanum

Observations of lanthanum aluminate $LaAlO_3$ in the La 3d range are presented to illustrate the RIXS method (Fig. 3.21) [122]. Only incident radiation of energy below the ionization threshold is considered. However, for lanthanum, particular cases exist in this range. They correspond to incident photons of energy equal to the three excited levels, $3d_{5/2}^{-1}4f^1$ $^3P_1$, $^3D_1$ and $3d_{3/2}^{-1}4f^1$ $^1P_1$ and are discussed more in details in the next paragraph. For incident radiation of energy inferior to $^1P_1$, the spectrum is complex and is dominated by inelastic scattering. The observed transitions are the 5p–3d emissions. Complexity is due to the presence of the two thresholds $3d_{3/2}$ and $3d_{5/2}$, to the highly probable $3d_{3/2}^{-1}$–$3d_{5/2}^{-1}X^{-1}$ Coster-Kronig transitions and to the valence-4f charge transfer, enabling 5p–3d transitions from the different initial configurations, $3d_{5/2}^{-1}$, $3d_{5/2}^{-1}X^{-1}$, $3d_{5/2}^{-1}4f^1$, $3d_{3/2}^{-1}$ and $3d_{3/2}^{-1}4f^1$ in this energy range. As in 3d absorption, a single excited level, $^1P_1$ or $^3D_1$, dominates the $3d_{3/2}^{-1}4f^1$, or $3d_{5/2}^{-1}4f^1$, scattering process, respectively. Owing to the strong 3d spin–orbit interaction the $^1P_1$ and $^3D_1$ levels are partially mixed. This explains the strengthening of the transitions from $^3D_1$ when the incident energy is that of $^1P_1$. A strengthening of the $3d_{5/2}4f^1$–$5p_{3/2}4f^1$ emission is also observed when the incident energy is that of $^3D_1$.

RIXS process was observed also in the lanthanum 4d range. Between about 116–140 eV, excitation takes place to the final configuration $4d^94f^1$ with the localized 4f electron. The process is independent of the geometry of the experiment

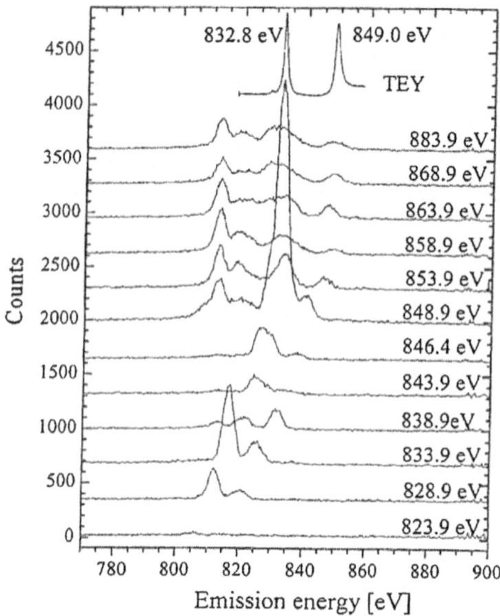

Fig. 3.21 RIXS of LaAlO$_3$ in the lanthanum 3d region. The excitation energy is given for each spectrum. From an excitation energy of 858.9 eV, the spectra are the same as fluorescence emission spectrum [122]

and obeys the energy conservation law. Above 140 eV, the 4d electron is ejected into the continuum and the final configuration is 4d^9. The process becomes a Compton process; it depends of the incidence and diffusion angles and it obeys the momentum and energy conservation laws. Separation is then observed between the curve corresponding to low energy process and the one corresponding to high energies. Emissions 4d^9–5p^5 in the presence of a 4f spectator electron originate from the excited configuration 4d^{9}4f^1 and are not expected to shift if the incident energy varies within the energy range of the excited 4d^{9}4f^1 configuration. In contrast, the emissions 4d^9–5p^5 in the ion are induced by photons of incident energy higher than the 4d ionization threshold. Inelastic scattering turns into fluorescence and the emissions are fluorescence normal lines.

Resonant inelastic scattering spectra are, thus, totally different from the EXES and vary strongly with the incident photon energy. The above experimental data confirm the essential results obtained from other spectroscopies for the spatial localization and the energy distribution of the 4f electrons in the lanthanum insulator compounds. Thus, transitions in the presence of a spectator 4f electron are observed and energies of these spectral features are in agreement with their values calculated by atomic models.

3.5.2.3   X-ray Fluorescence

Fluorescence is a two photons process, resulting from an ionization followed by an emission. The ionization is due to the absorption of a single photon. It is followed by an electronic recombination and the spontaneous emission of a single photon.

The two processes, ionization and emission, are independent. Fluorescence is observed if the energy of the incident photons is higher than the ionization threshold energy. This is an incoherent two-step process, each step being a first order process. As EXES, fluorescence obeys the dipole selection rules and is characteristic of each element. All the same, the width of the emissions is a function of the decay rate of the ionized state, i.e. of the core hole. But the fluorescence intensity is lower than the intensity of the electron-stimulated emissions since, beyond the resonance, the ionization probability of a core electron by photon interaction is much smaller than by electron interaction.

Thus, lanthanum 3d fluorescence emissions follow the radiative reorganization of the 3d hole ionized configurations and their energy is independent of the incident photon energy. No transition from excited configuration is expected. Indeed, in order to excite a configuration, the energy of the incident photons must be exactly equal to the energy of this configuration. Consequently, the 3d fluorescence spectra of the rare-earths describe the occupied 4f distributions in the ion with a 3d-hole. However, owing to the presence of secondary radiations, particularly of numerous secondary electrons, the above excitation is possible but remains weak. This has been shown for the Yb^{3+} ion of the configuration $4f^{13}$ [123]: with electrons accelerated under a voltage of 11 kV, the emission $3d_{5/2}^{-1}4f^{13}$–$4f^{12}$ ($M\alpha$) was observed in the M_V spectrum of the oxide Yb_2O_3, towards the higher energy of the M_V absorption line, and a weak structure was observed towards the lower energy; it corresponds to the $3d_{5/2}$ emission taking place from the excited configuration $3d_{5/2}^{-1}4f^{14}$ (cf. Chap. 4).

All the same, transitions involving excited states were observed in the 3d fluorescence spectrum of lanthanum metal induced by a rhodium target X-ray tube operating at 40 kV (Fig. 3.22) [124]. In a conventional high energy X-ray tube, intense bremsstrahlung radiation and numerous secondary electrons are created in the target, producing photons and electrons of low and middle energies. The creation of the excited configuration $3d^9 4f^1$ becomes possible, specially since the 3d excitation cross sections are at least three orders of magnitude higher than the 3d ionization cross sections. It was suggested that the $3d^9 4f^1$ excited configuration was created by ionization accompanied by relaxation, i.e. by a hole-induced shake-down transition analogous to the one seen in photoemission. The intensity of the shake-down photoemission peak is only a few percents of that of the main peak; thus the 5p–3d emissions should clearly be weaker in the excited atom than in the ion while the inverse was observed. In lanthanum metal, the excited configuration is, thus, created essentially by irradiation of particles having energy of the order of the La 3d–4f excitation energy, this process being favoured by its very high excitation cross section.

Fluorescence spectra of lanthanum in aluminate $LaAlO_3$ was also analyzed in the 4d energy range [125]. The transitions $4d^9$–$5p^5$ and $4d^9 4f^1$–$5p^5 4f^1$ were observed with 1.5 keV beam energy and were found nearly identical to the spectra excited by 1 keV incident electrons. Independently, energies and relative intensities of the electron induced lines were found in agreement with the calculated spectrum [59]. In this simulation, the initial state $4d^9 4f^1$ of the resonance lines and the lines in the

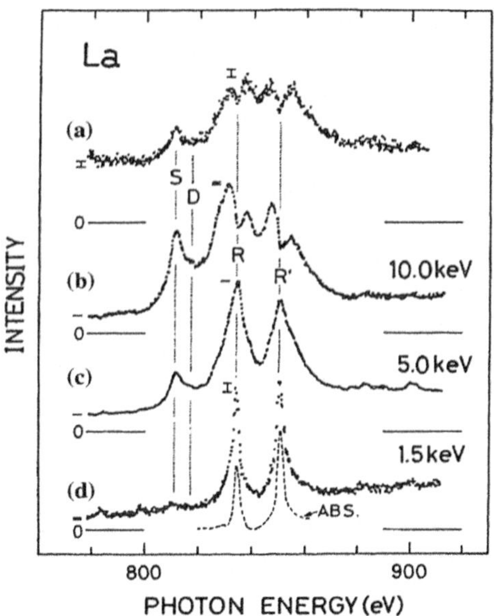

Fig. 3.22 3d spectrum of lanthanum: **a** fluorescent spectrum; **b–d** electron excited spectra with electron beam energies of 10, 5 and 1.5 keV. The dashed line is the 3d absorption spectrum by [51]. It is important to note the strong reabsorption of the lines R and R', which appear as absorption lines, or white lines, in the fluorescence spectrum and in that excited by electrons accelerated below 10 kV [124]

presence of a spectator electron were obtained by 4d–4f direct excitation. As the spectra observed by electron excitation and by fluorescence are identical, in particular the ratio between the emissions $4d^9 4f^1 - 5p^5 4f^1$ and $4d^9 4f^0 - 5p^5 4f^0$, one can deduce that the processes are the same in the two cases. Indeed, the energy of the 1.5 keV photons is widely superior to the 4d excitation energies and in this energy range numerous secondary electrons are created in the medium and can excite the 4d level. It was suggested that the $4d^9 4f^1$ configuration could be excited by electron transfer from ligand orbitals to 4f empty levels in the presence of the core hole. But transitions from the $4d^9 4f^1$ excited configuration are seen in metal and compound at nearly the same energy. Relative energy and intensity of the transitions from $4d^9$ and $4d^9 4f^1$ are nearly the same in the metal or the compound. This suggests that the presence of the excited configuration is due to the same process in the two cases and the ligands do not intervene in the spectrum of the compound.

When excitation takes place with photons of energy just equal to the excited level energy, the emission is expected at the same energy as the absorption line; this is the resonant fluorescence. Energy transfer is the same in these two opposite transitions and resonant fluorescence is at a single atomic site. Resonant excitation and de-excitation probabilities are equal and higher than the non-resonant probabilities. But the radiative recombination is strongly decreased due to the presence of highly competitive non-radiative processes, such as Auger and Coster-Kronig.

Moreover, self-absorption is strong and strictly limits the observations if attention is not taken in the choice of the experimental conditions. The geometry of the experiment must be such that the path of the incident and emitted photons in the sample is the shortest possible. Indeed, resonance fluorescence is observable only if the self-absorption is weak. In the case where a strong self-absorption is present, the resonance line may become a "white line", observed as an absorption in the place of the emission.

The energy distribution of the incident photon beam can be used to govern the characteristics of the resonance fluorescence. As an example, the energy distribution of the exciting radiation can be adjusted to select only one intermediate level of given parity among the different levels of the excited configuration. In this case, the energy width of the emission at the resonance is smaller than that observed under the fluorescence regime, where all the levels of the excited configuration contribute to the emission. On the other hand, if the energy distribution of the exciting radiation is more extended than that of the core hole, it is the lifetime of the hole that limits the width of the resonant photon distribution.

One expects that only ionized 3d configuration be created by photons of energy higher than the M_{IV} absorption threshold. However, fluorescence from an excited configuration of the type $3d_{3/2}^{-1}4f^1$ was observed under irradiation by photons of energy about 8 eV higher than the $3d_{3/2}$ absorption threshold [122]. The presence of this excited configuration was explained by electron excitation from the valence band to a 4f level, taking place during the creation of the 3d core hole. Such a transfer necessitates an energy supplement of several electronvolts, in agreement with the photoemission spectra (cf. Sect. 3.5.1.1): for lanthanum oxygen insulator compounds, as for example La_2O_3, shake-up (not shake-down) satellites due to the transfer of a valence electron into an unoccupied 4f level are observed towards the higher energies of each $3d_{3/2}$ or $3d_{5/2}$ main peak. Indeed, in the insulator compounds, the 4f levels are in the band gap, several electronvolts above the valence band. When the 3d inner hole is created by photons of energy exceeding by only several electronvolts the ionization energy, a simultaneous valence-4f excitation becomes energetically possible, the configuration $3d_{3/2}^{-1}4f^1V^{-1}$ can be formed and fluorescence from it can be observed. The configuration $3d_{3/2}^{-1}4f^1V^{-1}$ can also decrease by the highly probable $3d_{3/2}^{-1}$–$3d_{5/2}^{-1}X^{-1}$ Coster-Kronig transitions to $3d_{5/2}^{-1}4f^1V^{-2}$ or $3d_{5/2}^{-1}V^{-1}$ and all these configurations decay by radiative and non-radiative transitions of the type 4f–3d and 5p–3d in the presence or absence of the 4f spectator electron.

3.5.3 Auger Raman Scattering

Auger Raman scattering is a non-radiative process created under irradiation by incident photons of energy $h\nu_0$ close to the core ionization threshold [126–128]. As already underlined, several non-radiative processes are expected in parallel with the radiative processes, making a core hole disappear predominantly through Auger

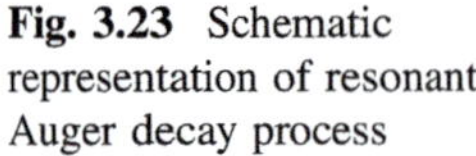

Fig. 3.23 Schematic representation of resonant Auger decay process

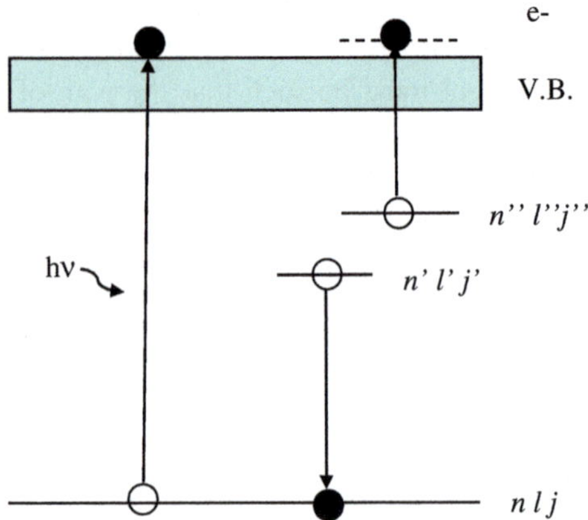

processes. In the energy range just below the threshold, these processes are usually designated as a non-radiative resonant Raman scattering, or *non-radiative RRS* (Fig. 3.23). It must be mentioned that this process is often named "resonant photoemission" but this terminology is not used here because it might be misleading: indeed, photoemission is the process associated with the ionization by a photon, while non-radiative RRS takes place during the reorganization that follows the primary ionization process. RIXS and non-radiative RRS have the same intermediate state but they are treated as independent.

Non-radiative RRS is interpreted as the virtual excitation of a core electron, followed by the filling of the core hole and, simultaneously, the excitation, or the ionization, of one weaker bound electron [129]. The intermediate state consists of a core hole and an excited electron. If E_{nl} is the ionization energy of the core hole, $h\nu_{in} = E_{nl} + \varepsilon_e$ where ε_e is negative for the RRS regime and positive above the threshold. The final state consists of two holes and two electrons in excited bound levels or the continuum. One is an excited electron with the energy ε_e and the other is an Auger electron of kinetic energy ε_A. Energy conservation for the initial and final states is

$$h\nu_{in} = E_{2h} - \varepsilon_e + \varepsilon_A$$

where E_{2h} is the energy of the configuration with two final holes. One deduces that ε_A depends linearly on $h\nu_{in}$ if ε_e is fixed. Consequently, if the energy of the excited electron is fixed, the energy of the Auger RRS peaks follows that of the incident photon by remaining at constant energy loss, as observed for the RIXS structures. The Auger RRS transition thus exhibits an energy dispersion that is a linear function of the incident photon energy. On the other hand, if the spectral distribution of the incident photon beam has a width narrower than the lifetime width of the intermediate state, no broadening due to this state is present in the process. This *narrowing* effect, analogous to that observed for RIXS, has been verified experimentally for free atoms.

As hv_{in} is tuned through the energy of the intermediate state, the process becomes resonant for e_e close to zero, and the resonant scattering channel dominates the weak non-resonant channel, as for the radiative components. Then the intensity of the Auger transition increases sharply and this non-radiative process becomes analogous to the resonance fluorescence. It is characteristic of the formation of two-hole configurations in the solid.

Above the threshold, the two emitted electrons are in the continuum and may interact with each other, one electron tending to be slowed down and the other speeded up. This effect is denoted as the post-collision interaction (PCI). The post-collision effect is maximal just beyond the threshold and gradually vanishes as hv_{in} increases. Towards higher incident energies, photoionization and Auger electron emission can be treated as separate sequential processes and Auger-like regime is present. The changing of the non-radiative RRS into Auger lines is analogous to that of the corresponding RIXS into fluorescence.

Two different channels are present in the non-radiative RRS regime, either Auger peaks which change into the normal Auger peaks above the threshold, or resonant Auger peaks, characterized by the presence of a photoelectron in the same initially unoccupied orbital during all the process. The observation of a resonant peak and its characteristics depend on the distribution of the unoccupied levels, thus on the shape of the absorption curve. A resonant peak becomes observable in the presence of a strong absorption maximum and its intensity varies in the same way as that of the maximum. The resonant peaks remain at constant binding energy. However, they shift linearly to higher kinetic energy when the incident photon energy increases, as observed also in photoemission lines.

It is important to underline that the observation of a resonant peak in the non-radiative RRS does not indicate the presence of localized electrons in the material. Indeed, resonant Auger peaks were observed in solids characterized by a distribution of delocalized unoccupied levels. These peaks reveal the trapping of a photoelectron with a zero kinetic energy in a high density unoccupied level characteristic of a non-uniform crystal field. Their intensity depends on the wave function of the excited electron and the mixing of the excited level with the continuum. Such peaks must not be mistaken for resonant Auger transitions that take place between levels of atomic character.

The Auger transition M_{45}–$N_{45}N_{45}$ ($3d^{-1}$–$4d^{-2}$) was studied in RRS regime for xenon-like ions in solids [127]. The presence of a resonance peak was investigated and its change from I^- to La^{3+} as a function of the degree of localization of the 4f orbital both in the initial and final configurations was established. The curves obtained from atomic calculations treating the Auger decay as a two-step process reproduce well the experimental spectra (Fig. 3.24). It appears thus clearly that the La Auger transition in LaF_3 is well described by an atomic model. The main structure observed is the Auger transition $3d^{-1}4f^1$–$4d^{-2}4f^1$, i.e. the transition La M_{45}–$N_{45}N_{45}$ in the presence of a spectator 4f electron. In addition, the resonant

Fig. 3.24 Resonant 3d–4d4d Auger spectra in LaF$_3$ (*thick lines*) compared to the calculated transitions (*thin lines*). (*1*): **a** 3d^{9}4f ^{3}D$_1$–4d^{-2}4f; **b** and **c** 3d^{9}4f ^{3}D$_1$–4p^{-1}; (2) **a** 3d^{9}4f ^{1}P$_1$–4d^{-2}4f; **b** and **c** 3d^{9}4f ^{1}P$_1$–4p^{-1}. The computed spectra are broadened with a Gaussian of 3 eV width [127]

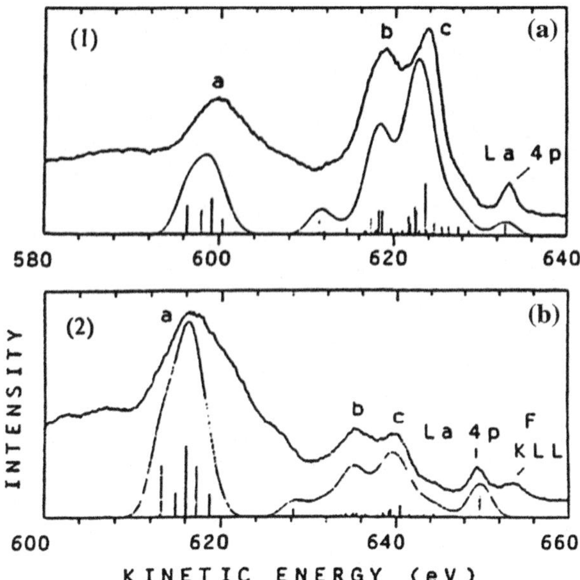

transition 3d^{-1}4f^1–4p^{-1} shows up with a much lower intensity. The Auger transition with a spectator electron is very sensitive to the degree of localization of the 4f orbital; it is observed also for barium but with a lower intensity. It is not observed for lighter elements.

3.6   Conclusion

Two kinds of spectral analysis have been described: the excitation and the ionization of a core electron by interaction with an incident particle, photon or electron; the decay processes of a core hole following its creation. The ratio of the excitation and ionization cross sections is different according to whether the incident particle is a photon or an electron and according to the energy of the particle. These differences play a decisive role in the observed processes. The decay analysis provides information on the dynamics of the excited and ionized quantum states in the solid. The various spectroscopic methods described are complementary and each interpretation can be confirmed by comparison with the results obtained from the other methods.

Creation and recombination of an excited configuration are first order independent processes. No interference exists between them and they are described in a two-step model. In contrast, in Raman scattering, energy conservation is applied to the whole excitation–emission process and no lifetime broadening of the intermediate state is observed, leading to an increase of the spectral resolution. That is why

RIXS experiments were considered as very attractive. However, this method gives information neither on the dynamics of the excited configurations in the solids nor on the resonant character of the associated levels. In RIXS, at the resonance, the energy taken from the incident photons is equal to the energy necessary to transfer the core electron with a zero velocity to an unoccupied level where it is trapped during the entire process. In contrast, in the radiative and non-radiative processes stimulated by electrons, excited intermediate configurations can be created and their decay observed in a timescale defined by the lifetime of the core hole. It is then possible to determine whether the excited electron remains localized on the same atom during the entire decay process.

Lanthanum has been chosen as an example because all the radiative and non-radiative transitions can be calculated. It was observed that the transitions from the $nd^9 4f^1$ excited configurations completely dominate the nd spectra. Such excited configurations were initially believed to be absent in solids. Thanks to EXES, they have been identified in the metal as well as in insulator compounds and their energies were found the same in the various materials. It is remarkable that the emission lines following the decay of the excited levels to the ground state coincide exactly with the corresponding absorption lines. This shows that the two processes are reciprocal and no energy is dissipated in the system during the excitation–recombination transitions.

An important question was to know how these $nd^9 4f^1$ levels are excited. The presence of a core hole is expected to induce perturbations of the valence distributions, which are very weak for conductor materials. But, among the unoccupied levels, several are modified by the core hole through its electrostatic interactions. For the rare-earths and the actinides, the electrostatic terms tends to pull the excited 4f, or 5f, levels down, while an exchange interaction drives the energy of some multiplets upwards. This last effect is particularly strong in the presence of a 4d, or 5d, hole, in the rare-earths and the actinides respectively. Excited configurations of the type $nd^{-1} n'f^{+1}$ are created either directly during the primary excitation process or by collapse of an unoccupied $n'f$ orbital into the field of the hole, during a shake-down process. These two processes are easily identifiable because their energies are different. For the first, the energies are those of the principal dipolar discrete lines and, for the second, the energy of the configuration barycentre.

When lanthanum interacts with an electron beam or with photons of energy just equal to the $nd^9 4f^1$ excitation energy, direct excitation of a 3d, or 4d, electron into the empty 4f orbital is possible. By EXES, it was observed that in the metal the emission $3d^9 4f^1 - 5p^5 4f^1$ is more intense than the normal emission $3d^9 - 5p^5$ and in the oxide it is the only one observable. In contrast, in photoemission, the irradiation by high energy photons makes the direct excitation of 4f impossible. The creation of the La $3d^9 4f^1$ excited configuration by a shake-down process following the formation of the $3d^9$ ionized configuration is highly improbable. Indeed, the shake-down peak is very weak compared to the normal peak in the metal and it is absent in the oxide. Consequently, the probability to create the excited configuration $3d^9 4f^1$ by direct transfer of a 3d electron to the empty 4f level is several orders of magnitude higher than by relaxation of the $3d^9$ ion. The observations of resonant transitions and transitions in the presence of the 4f spectator electron have revealed the localized character of the 4f electron in the $3d^9 4f^1$ excited configuration in lanthanum. The 4f orbitals are very similar to the core orbitals. Their radius is small; they are confined in the core region and are very sensitive to the attractive Coulomb potential of the inner hole. Auger spectroscopy together with EELS and CIS enable the identification of the excited X-ray configurations created under electron impact. These experiments confirm that the interaction between the 4f levels and the solid valence states is weak in the metal and the $3d^9 4f^1$ configuration retains a well-defined atomic character.

In summary, spectral analysis of the excited $nd^9 4f^1$ configurations in lanthanum showed that the 4f electron stays mostly localized on the same atom during the lifetime of the nd^9 core hole and the process conserves the spin of the core hole. From these, one deduces that the excitation is a *single-site process* and the excited states can be described with the help of an atomic model. From this model, the La 3d lines in the excited state are predicted to have lower energies than the corresponding lines in the ionized state, in agreement with the observations.

References

1. Y. Cauchois, J. Phys. Radium **13**, 113 (1952)
2. Y. Cauchois, J. Phys. Radium **16**, 253 (1955)
3. J.A. Bearden, A.F. Burr, At. Energ. Levels (1965)
4. F. Bechstedt, Phys. Stat. Sol. (b) **112**, 9 (1982)
5. A.J. Glick, P. Longe, Phys. Rev. Lett. **15**, 589 (1965)
6. P. Nozieres, C.T. DeDominicis, Phys. Rev. **178**, 1097 (1969)
7. B. Bergersen, F. Brouers,P. Longe, J. Phys. **F1**, 945 (1971)
8. B. Bergersen, F. Brouers, P. Longe, Phys. Rev. B **5**, 2385 (1972)
9. R.A. Ferrell, Rev. Mod. Phys. **28**, 308 (1956)
10. M. Cardona, L. Ley, Photoemiss. Solids I **26**, 1–104 (1978)
11. D.A. Shirley, Photoemiss. Solids I **26**, 165–195 (1978)
12. J. Farineau, Ann. dePhysique, **10**, 20 (1939)
13. C. Bonnelle, *Annual Report C* (Royal Society of Chemistry, London, 1987), p. 201
14. P. Auger, C.R. Seances, Acad. Sci. (France) **177**, 169 (1923)
15. M.P. Seah, Surf. Sci. **32**, 703 (1972)
16. C. Nordling, E. Sokolowski, K. Siegbahn, Phys. Rev. **105**, 1676 (1957)
17. A. Delobbe, A.-M. Dias, M. Finazzi, L. Stichauer, J.-P. Kappler, G. Krill, Europhys.Lett. **43**, 320 (1998)
18. R. Stumm von Bordwehr, Ann. Phys. Fr. **14**(4), 377 (1989)
19. C. Colliex, Ultramicroscopy **18**, 131 (1985)
20. J.C.H. Spencer, Rep. Prog. Phys. **69**, 725 (2006)
21. J.M. Andre, R. Barchewitz, Solid State Comm. **38**, 489 (1981)
22. D.C. Koningsberger, R. Prins, in *Chemical Analysis*, vol. 92 (Wiley, New York, 1988)
23. D.E. Sayers, E.A. Stern, F.W. Lytle, Phys. Rev. Lett. **27**, 1204 (1971)
24. D.E. Sayers, E.A. Stern, F.W. Lytle, Phys. Rev. Lett. **35**, 584 (1975)
25. K. Ulmer, in *Advances in X-ray Spectroscopy*, ed. by C. Bonnelle, C. Mande (Pergamon, New York, 1982), p. 451
26. M. Marabelli, P. Wachter, Phys. Rev. B **36**, 1238 (1987)

27. E. Wuilloud, Phys. Rev. Lett. **53** (1984)
28. H.F. Zandy, Phys. Rev. **162**, 1 (1967)
29. P.A. Lee, E.A. Stewardson, J.E. Wilson, Proc. Phys. Soc. A **65**, 668 (1952)
30. E.A. Stewardson, J.E. Wilson, Proc. Phys. Soc. London **A69**, 93 (1956)
31. D.W. Fischer, W.L. Baun, J. Appl. Phys. **38**, 4830 (1967)
32. C. Bonnelle, R.C. Karnatak, C.R. Seances, Acad. Sci. Ser. **A268**, 494 (1969)
33. C. Bonnelle, R.C. Karnatak, J. Phys. (Paris), Colloq. **32**, C4-230 (1971)
34. C. Bonnelle, P. Motais, Phys. Rev. A **73**, 042504 (2006)
35. T.M. Zimkina, V.A. Fomichev, S.A. Gribovskii, Sov. Phys. Solid State **15**, 1786 (1974)
36. C. Bonnelle, in *Theoretical Chemistry and Physics of Heavy and Superheavy Elements*, ed. by U. Kaldor, S. Wilson (Kluwer Academic Publishers, Berlin, 2003), pp. 115–170
37. G. Dufour, C. Bonnelle, J. Phys. Lett. (France) **35**, L255 (1974)
38. M. Richter, T. Prescher, M. Meyer, E. v. Raven, B. Sonntag, H.E. Wetzel, S. Aksela, Phys. Rev. B **38**, 1763 (1988)
39. U. Fano, Phys. Rev. **124**, 1866 (1961)
40. B. Johansson, N. Märtensson, Phys. Rev. B **21**, 4427 (1980)
41. P. Motais, E. Belin, C. Bonnelle, Phys. Rev. B **30**, 4399 (1984)
42. M. Elango, A. Maiste, R. Ruus, Phys. Lett. **72A**, 16 (1979)
43. R.C. Karnatak, J.M. Esteva, J.P. Connerade, J. Phys. B: At. Mol. Phys. **14**, 4747 (1981)
44. R. Hellmann, K. Ulmer, Z. Phys. B: Condens. Matter **58**, 259 (1985)
45. N.F. Mott, H.S.W. Massey, Theory At. Collis. (1965)
46. P.F. Staub, P. Jonnard, F. Vergand, J. Thirion, C. Bonnelle, X-Ray Spectrom. **27**, 58 (1998)
47. M. Ohno, J. Electron Spectr. Relat. Phenom. **143**, 13 (2005)
48. R.J. Liefeld, A.F. Burr, M.B. Chamberlain, Phys. Rev. A **9**, 316 (1974)
49. K. Jouda, S. Tanaka, K. Soda, O. Aita, J. Phys. Soc. Jpn, **64**, 192 (1995)
50. G. Wendin, K. Nuroh, Phys. Rev. Lett. **39**, 48 (1977)
51. J.M. Mariot, R.C. Karnatak, J. Phys. **F4**, L223 (1974)
52. J. Bruneau, J. Phys. B **16**, 4135 (1983)
53. S.J. Rose, N.D.C. Pyper, I.P. Grant, J. Phys. B **11**, 755 (1978)
54. G. Strinati, Phys. Rev. B **29**, 5718 (1984)
55. J.P. Connerade, R.C. Karnatak, J. Phys. F: Metal Phys. **11**, 1539 (1981)
56. A.S. Shulakov, A.P. Baiko, T.M. Zimkina, Sov. Phys. Solid State **23**, 1081 (1981)
57. J.P. Connerade, J.M. Esteva, R.C. Karnatak, *Giant Resonances in Atoms, Molecules and Solids* (Plenum, New-York, 1986)
58. J.-M. Bizau et al., Phys. Rev. Lett. **87**, 273002 (2001)
59. C. Bonnelle, G. Giorgi, J. Bruneau, Phys. Rev. B **50**, 16255 (1994)
60. C.G. Olson, D.W. Lynch, J. Opt. Soc. Am. **72**, 88 (1982)
61. M. Berland, C. Bonnelle, P. Dhez (1984 unpublished)
62. M. Aono, T.-C. Chiang, J.A. Knapp, T. Tanaka, D.E. Eastman, Phys. Rev. **21**, 2661 (1980)
63. M. Aono, T.-C. Chiang, F.J. Himpsel, D.E. Eastman, Solid State Comm. **37**, 471 (1981)
64. T.M. Simkina, A.S. Shulakov, V.A. Fomichev, A.P. Braiko, A.P. Stepanev, in *X-ray and Inner-Shell Processes in Atoms, Molecules and Solids*, ed. by A. Meisel, J. Finster (Karl-Marx-Universität, Leipzig, 1984), p. 263
65. A.S. Shulakov, T.M. Zimkina, A.P. Braiko, A.P. Stepanov, Opt. Spektrosk. **58**, 776 (1985)
66. J.-E. Rubenson, J. Lüning, S. Eisebitt, W. Eberhardt, Appl. Phys. A **65**, 91 (1997)
67. J.C. Riviere, F.P. Netzer, G. Rosina, G. Strasser, J.A.D. Matthews, J. Electron Spectr. Relat. Phenom. **36**, 331 (1985)
68. J. Sugar, Phys. Rev. B **5**, 1785 (1972)
69. J.L. Dehmer, A.F. Starace, Phys. Rev. B **5**, 1792 (1972)
70. M.O. Krause, in *Advances in X-ray Analysis*, ed. by L.S. Birks, C.S. Barrett, J.B. Newkirk, C.O. Ruud, vol. 16 (Plenum, New-York, 1973), p. 74
71. M. Pompa, A.M. Flank, P. Lagarde, J.C. Rife, I. Stekhin, M. Nakazawa, H. Ogasawara, A. Kotani, Phys. Rev. B **56**, 2267 (1997)
72. J. Kanski, P.O. Nilsson, Phys. Rev. Lett. **43**, 1185 (1979)

73. J. Kanski, G. Wendin, Phys. Rev. B **24**, 4977 (1981)
74. J.A.D. Matthew, G. Strasser, F.P. Netzer, Phys. Rev. B **27**, 5839 (1983)
75. G.Strasser, G.Rosina, J.A.D.Matthew, F.P.Netzer, J.Phys.F:Met.Phys.15, 739, 1985
76. R. Caciuffo, G. VanderLaan, L. Simonelli, T. Vitova, C. Mazzoli, M.A. Denecke, G.H. Lander, Phys. Rev. B **81**,195104 (2010)
77. K. Ulmer, Z. Phys. B: Condens Matter **43**, 101 (1981)
78. J.B. Pendry, J. Phys. C: Solid State Phys. **14**, 1381 (1981)
79. C.N. Borca, T. Komesu, P.A. Dowben, J. Electron Spectr. Relat. Phenom. **122**, 259 (2002)
80. P.G. Steeneken, L.H. Tjeng, I. Elfimov, G.A. Sawatzky, G. Ghiringhelli, N.B. Brookes, D.-J. Huang, Phys. Rev. Lett. **88**, 047201 (2002)
81. J.-M. André, R. Barchewitz, A. Maquet, R. Marmoret, Phys. Rev. B **29**, 6576 (1984)
82. O. Gunnarsson, K. Schönhammer, Phys. Rev. B **28**, 4315 (1983)
83. A. Fujimori et al., Phys. Rev. B **46**, 9841 (1992)
84. A. Kotani, H. Ogasawara, J. Electron Spectr. Relat Phenom. **60**, 257 (1992)
85. F. Gerken, A.S. Flodström, J. Barth, L.I. Johansson, C. Kunz, Phys. Scr. **32**, 43 (1985)
86. S. Doniach, J. Sunjic, J. Phys. **C3**, 285 (1970)
87. G.K. Wertheim, P.H. Citrin, *Optics in Applied Physics*, ed. by M. Cardona, L. Ley, vol. XXVI (Springer, New-York, 1978), p. 197
88. G. Crecelius, G.K. Wertheim, D.N.E. Buchanan, Phys. Rev. B **18**, 6519 (1978)
89. J.C. Fuggle, F.U. Hillebrecht, Z. Zolnierek, R. Lässer, Ch. Freiburg, O. Gunnarsson, K. Schönhammer, Phys. Rev. B **27**, 7330 (1983)
90. C.K. Jorgensen, H. Berthou, Chem. Phys. Lett. **13**, 186 (1972)
91. H. Berthou, C.K. Jorgensen, C. Bonnelle, Chem. Phys. Lett. **38**, 199 (1976)
92. D.F. Mullica, C.K.C. Lok, H.O. Perkins, V. Young, Phys. Rev. B **31**, 4039 (1985)
93. S.P. Kowalczyk, N. Edelstein, F.R. McFeely, L. Ley, D.A. Shirley, Chem. Phys. Lett. **29**, 491 (1974)
94. M. Richter, M. Meyer, M. Pahler, T. Prescher, E. v. Raven, B. Sonntag, H.E. Wetzel, Phys. Rev. A **39**, 5666 (1989)
95. W. Eberhardt (ed), Resonant inelastic soft X-ray scattering (RIXS). Special issue of Appl. Phys. A: Mater Sci. Process. **65** (1996)
96. W. Heitler, *The Quantum Theory of Radiation* (Oxford University Press, Oxford, 1965)
97. C.J. Sparks, Phys. Rev. Lett. **33**, 262 (1974)
98. P. Nozières, E. Abrahams, Phys. Rev. B **10**, 3099 (1974)
99. A. Kotani, S. Shin, Rev. Mod. Phys. **73**, 203 (2001)
100. Y.B. Bannett, I. Freund, Phys. Rev. Lett. **34**, 372 (1974)
101. D.L. Rousseau, G.D. Patterson, P.F. Williams, Phys. Rev. Lett. **34**, 1306 (1975)
102. P. Eisenberger, P.M. Platzman, H. Winick, Phys. Rev. Lett. **36**, 623 (1976)
103. T. Aberg, Phys. Scr. **21**, 495 (1980)
104. J.P. Briand, Phys. Rev. Lett. **46**, 1625 (1981)
105. F. Gel'mukhanov, H. Agren, Phys. Rep. **12**, 88 (1999)
106. J. Tulki, Phys. Rev. A **27**, 3375 (1983)
107. M. Magnuson, J.E. Rubensson, A. Föhlisch, N. Wassdahl, A. Nilsson, N. Martensson, Phys. Rev. B **68**, 045119 (2003)
108. S. Manninen, P. Suortti, M.J. Cooper, J. Chomilier, G. Loupias, Phys. Rev. B **34**, 8351 (1986)
109. J.J. Jia, T.A. Callcott, E.L. Shirley, J.A. Carlisle, L.J. Terminello, A. Asfaw, D.L. Ederer, F.J. Himpsel, R.C.C. Perera, Phys. Rev. Lett. **76**, 4054 (1996)
110. J.M. Platzman, E.D. Issacs, Phys. Rev. B **57**, 11107 (1998)
111. S.M. Butorin, J. Electron Spectr. Relat. Phenom. **110–111**, 213 (2000)
112. T. Schmitt, L.-C. Duda, M. Matsubara, A. Augustsson, F. Trif, J.H. Guo, L. Gridneva, T. Uozumi, A. Kotani, J. Nordgren, J. Alloys Compd, **362** (2004)
113. S.M. Butorin, D.-C. Mancini, J.-H. Guo, N. Wassdahl, J. Nordgren, J. Alloys Compd. **225**, 230 (1995)

114. O. Karis, A. Nilsson, M. Weinelt, T. Wiell, C. Puglia, N. Wassdahl, N. Märtensson, Phys. Rev. Lett. **76**, 1380 (1996)
115. M.Neeb et al., Chem. Phys. Lett. **212**, 205 (1993)
116. O. Björneholm et al., Phys. Rev. Lett. **73**, 2551 (1994)
117. K. Hämäläinen, D.P. Siddons, J.B. Hastings, L.E. Berman, Phys. Rev. Lett. **67**, 2850 (1991)
118. Y. Ma, N. Wassdahl, P. Skytt, J. Guo, J. Nordgren, P.D. Johnson, J-E. Rubensson, T. Boske, W. Eberhardt, S.D. Kevan, Phys. Rev. Lett. **69**, 2598 (1992)
119. Y. Ma, K.E. Miyano, P.L. Cowan, Y. Aglitzkiy, B.A. Karlin, Phys. Rev. Lett. **74**, 478 (1995)
120. J.A. Carliste, S.R. Blankenship, L.J. Terminello, J.J. Jia, T.A. Callcott, D.L. Ederer, R.C.C. Perera, F.J. Himpsel, J. Electron Spectr. Relat. Phenom. **110–111**, 323 (2000)
121. K.E. Miyano, D.L. Ederer, T.A. Callcott, W.L. O'Brien, J.J. Jia, L. Zhou, Q.-Y. Dong, Y. Ma, J.C. Woicik, D.R. Mueller, Phys. Rev. B **48**, 1918 (1993)
122. A. Moewes, S. Stadler, R.P. Winarski, D.L. Ederer, M.M. Grush, T.A. Callcott, Phys. Rev. **58**, R159521 (1998)
123. R.E. LaVilla, Phys. Rev. A **9**, 1801 (1974)
124. M. Okusawa, K. Ichikawa, O. Aita, K. Tsutsumi, Phys. Rev. B **35**, 478 (1987)
125. D.R. Mueller, C.W. Clark, D.L. Ederer, J.J. Jia, W.L. O'Brien, Q.Y. Dong, T.A. Callcott, Phys. Rev. A **52**, 4457 (1995)
126. G.S. Brown, M.H. Chen, B. Crasemann, G.E. Ice, Phys. Rev. Lett. **45**, 1937 (1980)
127. R. Ruus, A. Kikas, A. Maiste, E. Nommiste, A. Saar, M. Elango, J. Electron Spectr. Relat. Phenom. **68**, 277 (1994)
128. H. Aksela, S. Aksela, O.-P. Sairanen, A. Kivimäki, A. NavesdeBrito, E. Nömmiste, J. Tulkki, S. Svensson, A. Ausmees, S.J. Osborne, Phys. Rev. A **49**, R4269 (1994)
129. H. Wang, J.C. Woicik, T. Aberg, M.H. Chen, A. Herrera-Gomez, T. Kendelewicz, A. Mäntykentta, K.E. Miyano, S. Southworth, B. Crasemann, Phys. Rev. A **50**, 1359 (1994)

Chapter 4
Rare-Earth Spectroscopy

Abstract This chapter points out the importance of high energy spectroscopies, like X-ray emission and absorption, photoemission, Auger emission, electron energy loss spectroscopy, bremsstrahlung and characteristic isochromat spectroscopy, inelastic X-ray scattering, in the study of the electron distribution in the rare-earths and their compounds. Examples of materials characterized by valence fluctuations and their connections with the localized or delocalized character of the 4f electrons are given.

Keywords Localized 4f electrons · Rare-earth metal · Rare-earth intermetallics · Rare-earth compounds · Mixed-valence compounds

4.1 Introduction

High energy spectroscopy of the rare-earths has been the object of systematic experimental studies since 1960. The aim was to determine the characteristics of the 4f electrons by comparison with the other valence electrons such as 5d and 6s and also with the nd electrons of the transition elements. It has rapidly appeared that the rare-earth 4f and the transition metal nd electrons had very different characteristics. Resonant effects were observed in all the rare-earth nd radiative spectra whereas they were completely absent in the np radiative spectra of nickel and its compounds, which are considered as the most strongly correlated systems among transition element materials. The same observation can be made for the corresponding non-radiative transitions [1]. It was suggested [2] that for NiO a band model might exist for the ground state and a local model for the excited configurations. However, from the analysis of the 3d–2p self-absorbed emission, the absorption maximum was deduced to be present at higher energy than the emission peak [3]. No emission in coincidence with the absorption maximum, i.e. in resonance with the absorption process, has been observed in these materials in spite of intensive research. That demonstrated that there are no localized valence electrons in the excited configurations of the 3d compounds.

© Springer Science+Business Media Dordrecht 2015
C. Bonnelle and N. Spector, *Rare-Earths and Actinides*
in High Energy Spectroscopy, Progress in Theoretical Chemistry
and Physics 28, DOI 10.1007/978-90-481-2879-2_4

In spite of the fact that the 4f electrons remain localized in the vicinity of the core, they have energies near the Fermi level. Some properties of the rare-earths were then explained with the help of a 4f band model while others were explained using a 4f-core model. This suggested that the 4f electrons in the rare-earths could have a dual character. Some authors also argued that the unoccupied f levels interact with the s–d bands present at the Fermi level causing it to acquire a partial f character. However, the interaction between the 4f and s–d valence electrons remains negligible except in particular cases, as when the valence of the rare-earth changes from trivalent to tetravalent, discussed in Sect. 4.3. The energy of transitions in the solid is lower with respect to the energy in the free ion and the difference is approximately the same for all the levels of each trivalent lanthanide ion. Due to this difference, an excited $4f^{m-1}$ 5d level in a compound may be more bound than the ground state in the free atom. That can induce photon cascades of interest for various applications, such as scintillator, tunable laser,... The crystal field is an order of magnitude smaller than the spin–orbit interaction and can be treated as a perturbation. Consequently, the electron—phonon interactions are weak for the 4f electrons and are to be taken into account only in the treatment of second order effects such as magnetic effects.

Electron bombardment, often considered as complicating the analysis of the X-ray emission spectra because it produces multiple excitations, is very convenient in the case of the rare-earths because excited configurations with a core hole are created directly, without any secondary effects. In contrast, the irradiation by quasi-monochromatic photons causes large difficulties because only a single level of the excited configuration is reached and the entire spectrum cannot be obtained in a single experiment. All the electrons react to the creation of the core hole and screen it from the sudden change in the electrostatic potential [4]. If a core electron is ejected into the continuum in the solid, the relaxation increases with the kinetic energy of the ejected electron. Various relaxation processes take place, increase of the level energies, stabilization of the unoccupied localized orbitals and their occupation by a valence electron followed by the emission of a shake-down satellite, excitation or ionization of a secondary electron producing a shake-up, or shake-off, satellite. These last processes depend on the neighbouring atoms and give information on the chemical characteristics of the material. In contrast, if the core electron is transferred into an excited level the perturbation remains negligible because the excited electron now screens the effect of the hole and the screening is almost perfect. Thus, the effect of the hole on the neighbouring atoms can be neglected. The photoabsorption describes the multiplet splitting of the excited configuration. This is the case for the nd holes in the rare-earths.

The study of the decay processes following the creation of a core hole sheds light on the dynamics of the excited and ionized electrons. X-ray emission and Auger spectroscopy, X-ray resonant and Auger resonant emissions can be used for this purpose. If the width of the incident energy distribution, electrons or photons, is large with respect to the width of the excited or ionized core levels, the excitation process is very rapid and of the collisional type. Consequently, the excitation can be considered as a first order process and the radiative and non-radiative parts of the process can be separated from the excitation part making the study of excited states by XES and AES possible. In contrast, if the incident energy distribution is narrower than the width of the core level, the excitation process is a second order one. That is valid only for incident photons and corresponds to the resonant scattering process, whose interpretation is complex.

Decay probabilities of atoms, or weakly charged ions, present in excited X levels, i.e. in excited configurations with a hole in an inner sub shell, had received little attention. However, the excited X levels play an important role in the absorption and radiative and non-radiative nd emission spectra of rare-earths stimulated by electrons. The observation of emission lines whose energy coincides with intense excitation lines is a direct proof of the localized character of the excited configurations $3d^9 4f^{m+1}$ in the lanthanides. In order to identify the initial configuration of each X-ray emission line, the 3d spectra have been analyzed, comparatively to the photoabsorption, for several values of the incident electron energy from threshold to several times the threshold value. As underlined in the previous chapter, EXES is a powerful tool to identify the presence of excited X-ray levels populated by electron impact and to investigate their decay dynamics on a time scale defined by the lifetime of the core hole. The same applies to the Auger spectroscopy initiated by electron impact. Interpretation of the spectra is made in conjunction with calculations of the energies and the probabilities of all the radiative and non-radiative transitions taking place from the excited configurations to the ground state.

Let us recall the main results obtained for lanthanum, detailed in the previous chapter. The excited levels can be populated directly by interaction between an incident particle and an nd electron in the metal as well as in the compounds. The 3d–4f transitions are at the same energy in the trivalent metal and compounds such as La_2O_3 and LaF_3. The 3d–4f lines coincide in absorption and in emission. Good agreement was found between experimental measurements and theoretical predictions based on an atomic model for the X-ray nd–4f transitions. Agreement of the same type was obtained for the nd Auger transitions. All the energies obtained from various spectroscopies are in agreement. The experimental energies of the $3d^9 4f^1$ configuration obtained by direct excitation of a core electron to an empty 4f level or by relaxation of the ionized configuration $3d^9$ are not the same. Finally, the transitions in the excited configuration were predicted to have lower energies than the corresponding transitions in the ionized configuration. All these results confirm the high stability of configurations with an excited 4f electron in lanthanum. It will be shown here that the same observations hold for all trivalent rare-earths.

The choice of the experimental methods and of the analysis conditions is guided by the sample type and the wish to obtain properties characteristic of the sample volume or surface. Sharp contrast exists between the surface sensitivity of electron or X-ray spectroscopy. Indeed, the penetration range of the electrons is much smaller than that of the photons of the same energy. For the same initial energy, electron spectroscopy is much more sensitive at the surface than photon spectroscopy.

In photoemission, the observations depend on the mean free path of the escaping photoelectrons. This path is small under the generally used experimental conditions, making the photoemission extremely sensitive to surface contaminations. Contamination by oxygen can be detected from the presence of the O2p line, located at 6 eV, which is a very sensitive indicator to small traces of oxidation.

Among spectroscopic methods giving a direct insight into the electronic structure, soft X-ray emission spectroscopy is known to be one of the better adapted techniques to determine the distribution of the valence states in solids because of its selectivity with respect to element and level parity. It is not very sensitive to the surface. It can be initiated by photons or electrons. It competes with valence electron photoemission that has been the most used method although it is sensitive at the surface and not selective. In fact, photoemission of the valence electrons is generally considered as a direct and detailed tool for experimentally determining the electronic structure of a material. In the same way, photoabsorption and BIS are competitors. On the other hand, photoemission and BIS are often the only two methods to be taken into account when discussing the electronic properties of the rare-earth compounds, in spite of the fact that these two processes are based on different principles and are not complementary. Observations of the complete series of the rare-earths were made in the 3d and 4d ranges by photoabsorption, X-ray emission, Auger spectroscopy, photoemission and electron energy loss spectroscopy. Systematic observations of the ensemble of the rare-earths are cited preferentially because they allow a better assignment of the spectral features and their progress along the series.

It was shown experimentally from X-ray emission spectroscopy that, under standard conditions, the 4f electrons are localized in all the rare-earth metals and their trivalent compounds. The number of 4f electrons is thus integral on each site, in agreement with the magnetic properties and in disagreement with the band models. Europium and ytterbium metals are listed among the trivalent elements in spite of their bivalence. This particularity is due to the exceptional stability of the half shell-$4f^7$- and closed shell-$4f^{14}$- which makes the divalent configurations very stable and competitive with the $4f^6$ and $4f^{13}$ trivalent configurations. From spectroscopic investigations, the degree of 4f localization was found similar in the light and heavy rare-earths. However, differences exist between them because the heavy rare-earths are insensitive to the environment whereas the light ones can be sensitive and undergo a valence change. Whereas the number of valence electrons of the trivalent rare-earths is constant, the number of the s or d electrons can change according to the compound. The cohesive energies as well as the energies of the levels can also change along the series. A change in the distribution of the valence electrons can cause an energy shift of the 4f sub shell as can also an increase of the atomic number Z. The spectra can be interpreted simply in analogy with those described for lanthanum that represents the typical case of a trivalent rare-earth. The trivalent behaviour is treated in the first part of this chapter for all the rare-earths.

Some rare-earths have several valences. This valence change can be attributed to the variation of the radial term of the wave function. Changes from trivalent to divalent are observed. But the more frequent changes are from trivalent to tetravalent. In this case, the number of the localized 4f electrons decreases. One of the 4f electrons becomes strongly correlated with the other valence electrons and the transitions associated with this electron are fast compared to the fluctuations involving all the 4f electrons. The typical case is that of α-cerium. The second part of this chapter deals with tetravalent cerium, other tetravalent or divalent rare-earths and intermediate valence compounds. Cerium is the most studied rare-earth because of its interesting solid-state properties. Its valence is very sensitive to the local environment. Its chemical reactivity is particularly high and makes the experiments difficult. This reactivity is responsible for the simultaneous presence of Ce^{3+} and Ce^{4+} in numerous compounds in standard conditions. The existence of ions in the two states of oxidation can perturb the observation of the physical phenomena.

Both for the rare-earth oxides and for the pnictides or other compounds, it is difficult to obtain impurity free single crystals with ideal chemical composition and structural lattice. Instability of valence at the surface of the samples is also present and can foil the spectroscopic observations. The contamination of crystal surfaces by oxygen from the surrounding air is rapid. Rough surfaces are often present. To circumvent these problems, cleaving under ultra-high vacuum at low temperature is desirable.

4.2 Trivalent Rare-Earths

In trivalent rare-earths, particularly the heavy ones, the radial extent of the 4f wavefunctions can be approximated by that of the core wavefunctions. Consequently, the f–d hybridization between 4f and 5d valence electron wavefunctions is negligible. Very weak interaction exists between a discrete excited 4f level and the continuum neighbouring in energy. Thus no autoionization channel exists. The unoccupied 4f levels are predicted just above the Fermi level, as seen from spectroscopic analysis at room temperature [5].

As already underlined, all the levels of the excited configuration $3d^9 4f^{m+1}$ are modified by the presence of the additional 4f electron and their energies are reduced with respect to those of the ionized configuration $3d^9 4f^m$. The energy of the excited configuration is thus lower than that of the corresponding ionized one and this difference is a function of the 3d–4f correlation energy. It varies with the number of the 4f electrons. On the other hand, after the creation of a core hole, the energies of all the levels increase. Calculation by Kotani and Toyozawa [6] showed that the consequence of the presence of a core hole in a metal is to shift down the unoccupied levels with narrow widths and to make possible their overlap with the valence band. This well-known shake-down effect is discussed in the photoemission paragraph.

4.2.1   Density of Unoccupied States

4.2.1.1   X-ray Absorption

Dipolar transitions completely dominate the photoabsorption spectra in the edge region, i.e. in an energy range of about 15 eV above the threshold. Consequently, absorptions are observed from the nd core levels to the unoccupied 4f levels and from the np levels to the nd conduction state distributions. The effect due to the potential barrier term, $-l(l + 1)/r^2$ where r is the electron coordinate, has to be taken into account for the 4f electrons because of the large value of their angular momentum l. In a qualitative way, it can be argued that this large l value implies that a large escape energy is required for the 4f electron to overcome the potential barrier while it is not needed for the 5d and especially 6s valence electrons because of their low value of l.

The $2p_{3/2}$ (L_{III}) absorption spectrum of the trivalent rare-earths shows a broad maximum, which corresponds to the transitions to the unoccupied 5d6s conduction levels and is highly widened by the Lorentzian shape of the $2p_{3/2}$ level and weighed by the transition probabilities. The $2p_{3/2}$ level width increases from 3.4 eV for lanthanum to 4.6 eV for ytterbium [7]. A similar shape was observed for the $3p_{3/2}$ (M_{III}) absorption spectrum [8]. But the M_{III} core-hole lifetime is longer than the L_{III} one, making the $3p_{3/2}$ level only 1–1.5 eV wide for the rare-earths and increasing the spectral resolution in this energy range. The Fermi level energy E_F corresponds to the escape energy of the 5d and 6s valence electrons. By definition, the energy of the nd levels is equal to the threshold energy of the np discontinuities determined with respect to E_F. From this energy and that of adequate X-ray lines, the energies of all the other levels can be determined. Spectral features might be obscured in the L_{III} range by the large broadening due to the core hole. An alternative determination is to measure the fluorescence intensity excited by photons of incident energy that varies along the L_{III} absorption. By this method better resolution is obtained for the absorption spectrum by using a spectrometer of spectral band width narrower than the natural line width [9]. From this measurement method, low energy structures were observed in the dysprosium L_{III} edge and attributed to 2p–4f quadrupole transitions. However, the very small values of the quadrupole transition probabilities make the transitions to pure 4f levels negligible.

Two separate peaks of similar shape were observed in the L_{III} absorption spectra of intermediate valence compounds [10]. In the absence of interaction between the 4f levels and the valence band, these peaks are characteristic of the configurations $4f^m(5d6s)^q$ and $4f^{m+1}(5d6s)^{q-1}$ whose 4f levels are atomic-like. The measurement of their respective intensities can, then, be used to determine the valence. This determination necessitates an accurate analysis of the shape of the peaks. Such measurements were made, for example, for various samarium compounds [11]. In contrast, in the presence of 4f and band level interaction, spectra with peaks of very different intensity were observed. In Sect. 4.3 they are discussed in more detail for the case of the tetravalent compounds of light rare-earths and particularly of cerium.

The 3d absorption spectra of rare-earths had been observed for a long time [12–14] along with the theoretical interpretation of their fine structures [15, 16]. The observation of pronounced narrow 3d absorption peaks reveals transitions to localized 4f levels. These transitions are intense due to large dipole matrix elements. Indeed, a large overlap exists between the $4f^{m+1}$ orbital and the 3d sub shell. The excited 4f electron screens the perturbations due to the core hole almost completely and no secondary effect has to be taken into account. That underlines the importance of the 3d spectral region. In contrast, owing to the delocalized nature of excited electron in the 5d6s levels, screening of the L_{III} core hole is incomplete and secondary effects due to this incomplete screening can be present in this range.

The 3d absorptions form two groups corresponding to the $3d_{5/2}$ and $3d_{3/2}$ inner hole states and each group shows strong sharp peaks well described in terms of $3d^{10}4f^m$–$3d^9 4f^{m+1}$ absorption lines in the triply, or doubly, ionized rare-earth atom. The ratio between the intensities of the two groups, $3d_{5/2}/3d_{3/2}$, is observed to increase with increasing 4f orbital occupancy, showing that the $4f_{5/2}$ orbitals are filled up preferentially. The width of the $3d_{3/2}$ peak is larger than that of $3d_{5/2}$ peak because additional transitions are possible in the $3d_{3/2}$ range. Among these, the resonant Coster–Kronig transitions $3d_{3/2}^{-1}4f^{m+1}$–$3d_{5/2}^{-1}4f^m V^{-1}$ are the most probable when they are energetically possible.

Initially, the photoabsorption spectra of the metals were obtained by analyzing the radiation transmitted across an absorbing screen. In the region of the rare-earth 3d spectra, the absorption probabilities 3d–4f are exceptionally high and so are the absorption peaks. The absorbing sample must be very thin: the choice of the thickness is critical in order to avoid saturation in the absorption spectrum. The difficulty of obtaining sufficiently thin films of high quality limits the direct absorption measurements. To surmount this difficulty, measurements of the secondary particles created during the absorption process were developed. Several possibilities exist. The measured parameter is either the fluorescence radiation, giving photoyield method, or fluorescence yield; or particular Auger electrons, giving the partial electron yield method; or all the emitted electrons, primary, Auger and secondary; this is the total electron yield method [17, 18].

Analysis of 3d absorption for all the elements of the series was made by photon yield method using the intense X-ray synchrotron radiation (Fig. 4.1) [19]. Attention was taken to avoid saturation effects, present also in this experimental method. Calculations of the dipolar transitions from the ground state of the configuration $4f^m$ to the excited configuration $3d^9 4f^{m+1}$ were made in intermediate coupling by using Cowan's atomic Hartree–Fock program with relativistic corrections. The electrostatic parameters were taken equal to 80 % of their Hartree-Fock valuesobtained from Cowan's program. The 3d spin–orbit parameter was deduced from the energy interval measured between the $3d_{5/2}$ and $3d_{3/2}$ levels. Only the average energies of the ground state and excited configurations were considered in these calculations. The lifetime broadening was introduced by convoluting each line with Lorentz shape in the $3d_{5/2}$ range and with Fano shape in the $3d_{3/2}$ range. This broadening increases across the series. The resulting spectrum was convoluted by a Gauss curve to take into account the experimental broadening.

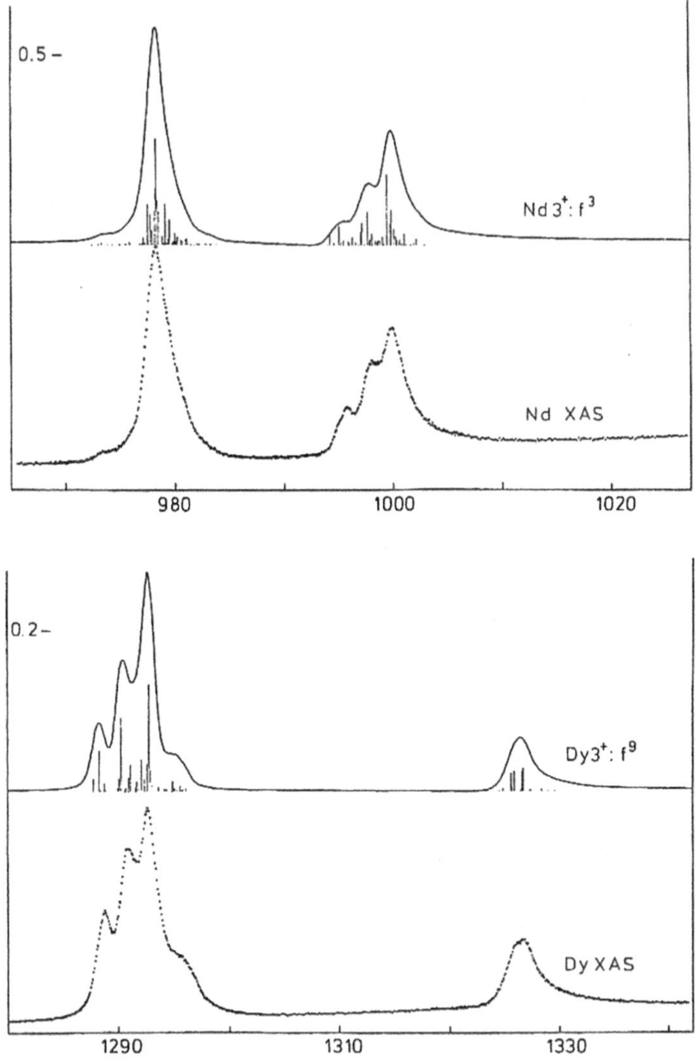

Fig. 4.1 3d absorption spectra of neodymium and dysprosium metals (*points*) compared to theoretical $4f^n$–$3d^9 4f^{n+1}$ dipole excitation spectra [19]

Good agreement was found between the experimental and calculated spectra. The 3d absorptions are, therefore, well described as line spectra and no edge towards the continuum is observed. Indeed, edges characteristic of 3d absorption transitions towards the empty extended states in the solid are predicted to be weak with respect to the 3d–4f transitions because they concern transitions to the unoccupied 6s levels whose density is low. The position of the empty 4f orbitals with respect to the Fermi level cannot be determined from the 3d absorptions. It is necessary to use the level energies such as determined from the level table, i.e. from the energies of the L discontinuities and emission lines. This method takes partially into account the perturbations due to the presence of the core holes and can be generalized to any material. More precise determination can be made from the M spectra because the resolution is better in this range. The M_{III} absorptions describe the transition of 3p electrons to the 5d and 6s unoccupied levels and the M_{III} threshold gives the position of the first unoccupied state, e.g. of the Fermi level in the metal. As expected for this type of transitions, the M_{III} absorption edges show almost an arctangent form, whose inflexion point corresponds to the position of E_F. The M_{III} absorption thresholds can be located with an accuracy of about 0.3 eV. The M_{III}–M_{IV} and M_{III}–M_V intervals are determined from X-ray emission lines within 0.5 eV. The positions of the M_{IV} and M_V thresholds, i.e. the energies of the ionized $3d^9 4f^m$ configurations, are obtained with an accuracy of less than an electronvolt. From these determinations, the ionized $3d^9 4f^m$ configurations are predicted around several electron volts above the excited levels $3d^9 4f^{m+1}$.

Due to the great number of the final levels $3d^9 4f^{m+1}$ $J, J \pm 1$, reached from the ground state $3d^{10} 4f^m$ J [20], it is not possible to resolve the multiplets except for La^{3+}, Tm^{3+} and Yb^{3+} for which the line numbers are respectively, 3, 4 and 1. As an example, the number of the levels associated with the configuration $d_{5/2}^{-1} f^7$ is equal to 1832. Remarks must be made concerning the simulation of the spectra. Discrete absorption line shapes are expected to be well approximated by Lorentz curves. That was verified for the $3d_{5/2}$ absorption lines. In contrast, the $3d_{3/2}$ absorption lines are represented by Fano curves that take into account their observed asymmetry [19]. This asymmetry, distinctly observed for the light rare-earths, progressively disappears into an increasing continuous background of the $3d_{5/2}$ and $3d_{3/2}$ transitions for the heavy rare-earths. It is explained by the presence of transitions to the continuum, located to the higher energies of the $3d_{5/2}$ and $3d_{3/2}$ discrete transitions. These transitions induce a non-uniform intensity distribution in this energy range. It was suggested that interaction exists between the excited levels and the continuum and leads to a shape change of each line of the $3d_{3/2}$ multiplet. The absence of interaction and simple superposition of discrete transitions and transitions to the continuum can also be considered. No $3d_{5/2}$ and $3d_{3/2}$ absorption jumps are observed. It is difficult to predict their positions and to evaluate the energy beyond which the line spectrum is perturbed. However, by taking into account the effective Coulomb interaction between the 3d hole and the 4f electrons, each $3d_{5/2}$, or $3d_{3/2}$, edge was estimated to be present 3–5eV above the discrete 3d–4f transitions and the 6s conduction levels are expected further above.

The 4d–4f dipole matrix elements are large because of the large overlap between the 4d and 4f wave functions, which increases by the contraction of the 4f wave functions in the presence of the 4d core hole. Consequently, the 4d absorption spectra of the rare-earths are dominated by excitation of a 4d electron to the unoccupied 4f levels. The levels of each core-excited $4d_{5/2}^{-1} 4f^{m+1}$ and $4d_{3/2}^{-1} 4f^{m+1}$ configuration are strongly mixed and the oscillator strengths are distributed over a large energy extend due to the multiplet splitting in the open shells [20]. The strong

4d–4f Coulomb interaction energy largely dominates the 4d spin–orbit coupling energy inducing LS coupling structures.

The 4d absorption spectra differ significantly from the usual X-ray absorption spectra of metals [21–23]. They consist of a very structured array of narrow peaks extended over 8–10 eV and an intense broad band, generally designated as *giant resonance*, extending over an energy range of about 10 electronvolts. The absorption coefficient at the giant resonance was an order of magnitude greater than at the peaks of the fine structure. But response linearity was not verified in these experiments and a larger ratio, of the order of 100, was later suggested. The number of lines increases greatly up to gadolinium then decreases [24]. It agrees with the calculated number of transitions to the $4d^9 4f^{m+1}$ configuration. Thus, at the $4d_{5/2}$ threshold, the number of possible lines is two for lanthanum, 30 for cerium, three for thulium and one for ytterbium. At the $4d_{3/2}$ threshold, it is one for lanthanum, cerium and thulium. In agreement with these predictions, three peaks were observed at the Tm $4d_{5/2}$ threshold, one at the Tm $4d_{3/2}$ threshold and one at the Yb $4d_{5/2}$ threshold [25]. For lanthanum and cerium, the fine structure was resolved and the measured width at half-height of each line is equal to 0.5 eV. For the rare-earths following cerium, the fine structure was seen as superimposed on the beginning of the giant resonance. The strength of the giant resonance decreases with increasing Z. For the heavy rare-earths, it varies proportionally to 14-m, where m is the number of the 4f electrons [26] and vanishes just when the 4f sub shell is filled. In ytterbium metal, no giant resonance is present. Agreement was found between the 4d absorption spectrum in the vapour phase and in the solid for the rare-earths having the same number of 4f electrons in the two physical states [27, 28]. For the other rare-earths, the number of the 4f electrons decreases by a unity in the solid with respect to the atom and close resemblance was found between the Z free atom and the (Z + 1) solid. Analogy between metal and free atom cerium has also been observed from the measurement of partial 4d cross sections [29].

In all the rare-earth metals, the giant resonance is due to the presence of intense discrete excitations. Absorption spectra were found nearly identical for the oxides and the trifluorides but with narrower and better resolved structures [30]. The excitation energies vary with respect to ionization energies along the series. In the heavy rare-earths, energies of the multiplets associated with the configuration $4d^9 4f^{m+1}$ are lower than the ionization threshold. The spectral characteristics are interpreted as 4d–4f resonant excitations to the multiplet terms of the $4d^9 4f^1$ excited configuration and no influence of the environment is present. In the light rare-earths, part of the $4d^9 4f^{m+1}$ multiplet structure is located above the threshold. A competition is expected between 4d $\rightarrow$ 4f excitation and 4d $\rightarrow$ εf ionization and these two processes could be connected by configuration interaction. Photoabsorption spectra, obtained in the total photoelectron yield mode, were calculated for all the rare-earth trivalent ions by taking into account all the multiplets [31]. These calculations reproduce well the experimental observations. They confirm that the spectra are mainly characteristic of the number of 4f electrons in the initial state and not of the chemical environment. However, the position of the threshold has strong influence on the shape of the giant resonance whose high energy part can be sensitive to the environment in the light rare-earths.

No 4d absorption edges corresponding to electron transitions towards the levels in the conduction band of the solid were observed. As in the 3d range, it is not possible to place the 4f levels with respect to the Fermi level from 4d absorption spectra. This placement must be made by comparing the energy of the 4d–4f transitions to the 4d energy level such as given in the energy level tables. Precise placement of the 4d levels with respect to the Fermi level in the metal can also be obtained from the measurement of M_{III} absorptions and the M_{III}–N_{IV} and M_{III}–N_V emissions. Absorption edges are often located in the fine structure range and multiplets can be found on both sides of the edge.

In summary, from comparison between experimental and theoretical results, it appears clearly that consideration of the radial parts of 4f wave functions alone is not enough to interpret the experimental absorption spectra. The angular parts also play an important role in explaining the multiplet structures present of these spectra. In the 3d energy range, the experimental 3d excitation energies are lower than the absorption energies and the multiplet energy range is approximately 5 eV. In the 4d energy range, owing to the strong 4d–4f exchange energy of the configuration $4d^9 4f^m$, the total energy range of the J-levels is about 20 eV and 4d excitation energies are above the threshold energies for the light rare-earths. For the heavy rare-earths, the effect of the exchange interaction along the series is more mitigated and the levels are found mostly below the ionization threshold.

4.2.1.2 Electron Energy Loss Spectroscopy

As already underlined, electron energy loss spectroscopy can be observed in a transmission electron microscope. It has the advantage of being a high spatial resolution technique and, consequently, is widely used in the investigation of materials. The rare-earth metal spectra were initially observed in transmission, in two energy ranges, the 4d and 5p excitations. An incident electron beam of 75 keV was used. The inelastic scattering cross sections are proportional to generalized oscillator strengths that converge towards the dipole oscillator strengths for incident electrons of energy much larger than the threshold and for small scattering angles. The spectra obey, then, the dipole selection rules and furnish information directly comparable with X-ray absorption. Indeed, in the 4d range [32, 33], the results obtained for the entire rare-earths series from praseodymium to ytterbium agree with those obtained by photoabsorption. Thus, the width of the "giant" peak above the Fermi level is smaller for the heavy rare-earths than for the light ones and its distance from the absorption edge decreases with Z as also does its intensity. The energy of this maximum was found almost identical in the metals and in their fluoride. For trivalent ytterbium, the 4d electron is excited into the 4f shell and the spectrum is located at energies just under the N_V threshold. As an example, the

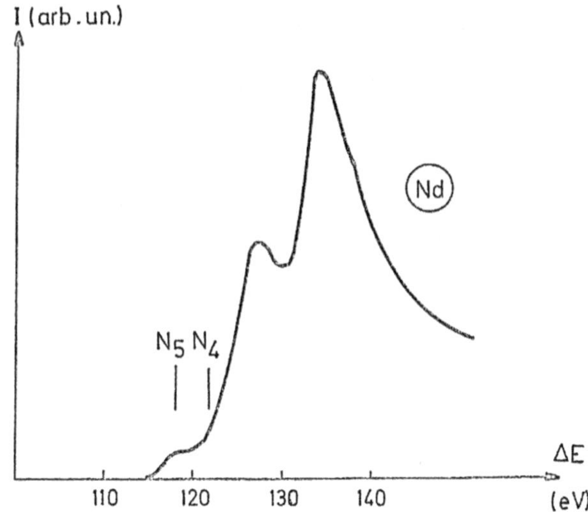

Fig. 4.2 Electron energy loss spectrum of neodymium metal observed at 75 keV [32]

spectrum obtained for neodymium is compared with the photoabsorption one (Fig. 4.2) [32]. The presence of a background prevents the detection of the weak discrete structures located under the threshold but good agreement is observed for the more intense peaks. As expected for a core hole, the 5p peak is observed to move towards higher energies with Z increasing and does not vary noticeably with the chemical state of the sample [34].

Spectra of γ and β-cerium have been observed by EELS in the 4d absorption range [35]. No change of the pure 4d–4f transitions located towards the lower energy of the giant peak was observed. In contrast, the shape of the 4d giant peak varied towards the higher energies, in the range where the extended levels are dominant. This result illustrates the influence of the crystalline structure on the electronic distributions.

As observed in absorption, experiments can also be made in reflection mode. The difference between transmission and reflection modes is related to the penetration depth in the sample and also depends on the incident energy. In reflection mode, the incident energy is chosen near the threshold, the dipole selection rules break down and all the J-levels of the excited configuration are populated. Peaks of full multiplets are expected. Indeed additional multiplets were observed towards the low energy in the 3d and 4d spectra [36, 37]. Their corresponding peaks are broader than the main peaks because they involve transitions to all the J value levels while the main peaks correspond only to transitions from the ground level J to levels J, $J \pm 1$. Consequently, by using incident electrons of energy near the threshold, it is possible to induce non-dipole transitions that are inaccessible with photons. 3d–4f and 4d–4f excitations were studied for all the rare-earths [38]. The spectral characteristics were observed for two different initial energies. At higher initial energies, the EELS spectra and the X-ray absorption spectra give the same informations for both 3d and 4d range. At low initial energy, multiplets gain strongly in intensity. In transmission, the spectra vary with the number of the 4f electrons. In the light rare-earths, the main peaks appear above the threshold energy whereas for the heavy ones they are located approximately at the threshold. Interesting characteristics of the core level excitation by an electron beam were obtained in cerium experiments. The intensity of the transitions is strongly enhanced with decreasing primary energy of the incident electrons [39].

Recently, non-resonant inelastic X-ray scattering NIXS, or NRIXS and EELS of the trivalent metals cerium, praseodymium and neodymium were observed in the 3d and 4d energy ranges [40]. In the two ranges, NIXS was in agreement with the EXES results observed at low incident electron energy. Good agreement was also obtained with atomic-like calculations. Dependence of the spectra on the momentum transfer in the excitation process was analyzed. These results merely confirm the ability of the atomic model already employed to describe with great precision the 4f electrons in these metals.

4.2.1.3 BIS and CIS

The energy of the 4f levels above the Fermi level was determined by BIS for all rare-earth metals [41, 42, 43]. Emitted radiation of energy equal to 1486.6 eV (AlKα) due to the deceleration of mono-energetic incident electrons in the sample was observed. The analyzed depth is determined by the mean free path of the incident electrons in the solid. In the considered energy range, depth of about ten Angströms was analyzed and the method has a surface sensitivity of the same order as XPS. The experimental resolution was lower than for the analysis of the occupied distributions by XPS. Transitions from $4f^m$ to $4f^{m+1}$ were observed for all the series. They were interpreted as atomic-like excitations. That showed that in both $4f^{m+1}$ final and $4f^m$ initial configurations, the electrons are localized. Transitions to the continuum levels were also observed but their contribution remains weak. Similarity was predicted between the photoemission of a $4f^m$ element and the BIS of a $4f^{14-m}$ element.

Partial 5p and 4f photoionization cross sections were obtained in the region of the 4d giant resonance by using CIS spectroscopy [44]. The line shapes of all the elements are similar except for cerium whose 4f and 5d6s valence electrons show characteristics different from those of the other rare-earths. This particularity of cerium can be explained by the presence of partially delocalized 4f electrons (cf. Sect. 4.3).

In metallic γ phase cerium, the $4f^2$ configuration in the absence of a core hole is known from BIS to be contained in the region of 3–7 eV above E_F. In contrast, for core-ionized cerium the situation is like in praseodymium where the $4f^2$ level lies 3.3 eV below E_F. In cerium the $4f^2$ levels interact poorly with the other valence levels.

4.2.2   Density of Occupied States

4.2.2.1   X-ray Emission

Large multiplet splitting of the dipolar L emissions was the first anomaly observed in rare-earth X-ray emission spectra. This splitting was attributed to the presence of the unfilled 4f sub shell [45].

Then, emission lines in coincidence with intense absorption lines in the electron-stimulated 3d spectra of a metal rare-earth were observed and this result was unexpected. The more evident interpretation was to consider these emissions as *resonance* lines, i.e. as the radiative recombination of X excited states [14, 46]. But this interpretation was doubted by solid-state physicists. Subsequently, it was shown that excited configuration $3d^9 4f^{m+1}$ was created by direct excitation and not by some secondary effect and that this configuration de-excited by direct recombination of the excited 4f electron towards its initial level [47]. The observation of the $3d_{5/2}$ and $3d_{3/2}$ resonance lines was the first direct evidence for the existence of localized electrons in the excited configuration $3d^9 4f^{m+1}$ of the lanthanides. The systematic observation of the resonant emissions of the type 4f–3d is actually one of the simplest experimental methods to determine the number of localized electrons in an excited configuration with a core hole. The same analysis could be made from the 4d emissions. But the splitting due to the 4d spin–orbit interaction is small compared to the corresponding widths of the $4d_{3/2}^{-1} 4f^{m+1}$ and $4d_{5/2}^{-1} 4f^{m+1}$ energy distributions. Consequently, these two distributions become mixed and the identification of the transitions is difficult.

The 5d valence distributions can be derived from the observation of the $5p_{3/2}$ emissions. In observation made for the light rare-earth metals, the 5d valence band was found quasi-symmetrical and its width increased gradually from 3.2 eV for lanthanum to 7 eV for samarium [48].

4.2.2.2   3d Emissions

Initially, the 3d emission spectra were observed with incident electron energies E_0 of the order of twice to three times the 3d threshold energy. Since the absorption spectra are situated in the same energy range as the emission spectra, strong self-absorption was present. Emissions were observed at both lower and higher energies around the position of the absorption lines [13]. The low energy emissions were erroneously considered as the Mα (4f–$3d_{5/2}$) and Mβ (4f–$3d_{3/2}$) normal emissions. The high energy emissions were attributed to satellites and received little attention.

Large differences between the spectra of lanthanum and the other rare-earths were expected because of the transitions from the 4f occupied levels to the 3d core hole. These transitions take place from the configuration $4f^m$ of the ion. There are the $3d^9 4f^m$–$3d^{10} 4f^{m-1}$ Mα and Mβ lines, present in the $3d_{5/2}$ and $3d_{3/2}$ spectra respectively. They are designated here as ionization, or *normal*, emissions. Since the configurations involving the open 4f sub shell are very large, the above emissions are widely spread but only slightly so by the broadening due to the level lifetime. Then, in the presence of unoccupied 4f levels, electron bombardment produces simultaneously 3d excitation to the 4f levels and 3d ionization to the continuum. Both these processes are present together in all the rare-earth series and simultaneous decay from the excited and ionized configurations is expected. The spectra are complex because of the presence of the following emissions: resonant emissions, non-resonant emissions from the excited configurations, emissions in the presence of a spectator excited electron, normal 4f–3d and 5p–3d emissions and satellite emissions. Moreover, self-absorption associated with the 4f–3d resonance lines can introduce a large asymmetry of the line shapes which makes the interpretation difficult.

Emissions induced by photons of energy equal to twice or more the threshold energy were also observed. These spectra, labelled fluorescence spectra, show only the normal emissions and eventually also satellite emissions. Indeed, each incident photon imparts its full energy to a bound 3d electron, which is ejected to the continuum by creating the ion $3d^9$. The emissions take place from this ion. A typical example of 3d fluorescence spectrum was observed from Gd_2O_3. Comparison with the electron-induced spectra confirmed the existence of resonance lines but did not give information on the position of the normal lines and their asymmetric shapes are due to strong self-absorption (cf. below).

To identify the initial configuration associated with each X-ray emission and to deduce the various decay processes of this configuration, it appeared necessary to compare the 3d spectra in emission and in photoabsorption. It is known that one of the advantages of excitation by electron impact is that it is possible to minimize considerably the self-absorption by reducing the energy of the incident electrons. This is difficult to achieve with excitation by X-ray photons. Indeed, the penetration of the electrons into the target is a function of their incident energy E_0. If E_0 is near the threshold, the penetration is small; the spectrum is emitted by a peripheral thickness and the self-absorption is small. If the electron beam energy increases, the penetration of the electrons into the target increases and with it the self-absorption of the emitted radiation. At sufficiently high incident electron energies, the resonant emissions are totally self-absorbed and an intensity minimum is observed at the position of the emission maximum. Under such conditions, the structures present in the emission spectra are a consequence of the self-absorption of the emitted radiation in the target. Moreover, the ionization probability increases with E_0 as $(E_0)^p$ $(1.5 < p < 2)$ while the excitation probability decreases. Thus, the resonant emissions are observable only with incident energies E_0 near the threshold.

The levels of the excited configuration $3d^9 4f^{m+1}$ are populated by collisions between the incident electrons and the rare-earth 3d electrons. The interaction duration between an incident electron and the irradiated atom depends on the energy distribution of the incident electrons, which is a few eV's wide. The lifetime broadening of the 3d core hole is of the order of 1 eV for the rare-earth series. Consequently, the incident electron-atom interaction takes place in a time shorter than the decay time of the 3d core hole, enabling a reorganization of the level population while the hole is still present. On the other hand, when the energy E_0 is equal or superior to about 1.5 times the threshold energy, the excitation process is governed by dipole selection rules. Closer to the threshold energy, electronic relaxation exists, populating all the J-levels of the excited and ionized configurations but with very different probabilities (cf. Sect. 3.2.4). The J, $J \pm 1$ levels populated by electric dipolar transition from the ground level have the highest probability to be excited and to decay to the ground J-level. If the direct recombination of the excited electrons from the J, $J \pm 1$ levels to the ground J-level is more rapid than the statistical reorganization, intense resonant transitions directly emitted from the excited J, $J \pm 1$ levels to the ground state are observed. As already underlined, statistical redistribution of the selectively excited levels among all the J-levels of the excited configuration can take place during the lifetime of the configuration. All the same, all the J-levels of the ionized configuration are statistically populated during its lifetime. In both excited and ionized cases, the population of the J-levels is proportional to their statistical weight. All the levels decay radiatively and contribute to the total emission spectrum.

The 3d emissions of all the rare-earths were observed by varying the energy E_0 from about 100 eV above the 3d ionization threshold up to three times this threshold [49]. If E_0 is near the 3d excitation threshold, intense 3d emission lines are seen at the same energy as the absorption lines. These are resonant emissions. Following the results described for lanthanum, they are emitted during the direct decay from the $3d^9 4f^{m+1}$ J, $J \pm 1$ excited levels to the ground level $3d^{10} 4f^m J$. They correspond to the inverse process of the 3d–4f photoexcitation and are identified easily by comparing the energies of the absorption and emission lines. Resonance lines are observed if excited atoms are present and if the additional 4f electrons remain localized on the same atom without change of spin orientation during a time less than or equal to the radiative lifetime of the excited configuration. As already underlined, the presence of an additional 4f electron that screens the nuclear charge, leads to a decrease of all atomic levels energies and the energy of the excited $3d^9 4f^{m+1}$ configuration is several electronvolts below the energy of the ionized $3d^9 4f^m$ configuration.

The remaining observed 3d emissions were also interpreted based on calculated spectra. As seen above, except for the resonance lines, the emissions take place from statistically relaxed configurations, i.e. after statistical reorganization of all their levels. The observation of such a level reorganization confirms that the X-ray emissions are two-step processes. Ab initio calculations of energy and transition probability of each emission were made for each rare-earth with the help of a multi-configurational Dirac-Fock (MCDF) program as described in Sect. 3.2.5. All the J-levels of the ground and excited configurations were included. The excited and ionized configurations, $3d^9 4f^{m+1}$ and $3d^9 4f^m$, are largely spread. Whereas the La^{3+} spectrum is simple, with only three excitation lines, which are resonance lines, and no ionization line, the number of lines increases very rapidly with m along the series. It is maximum for the half-filled 4f sub shell ($m = 7$) and decreases until the shell is filled. Thus, each $3d_{5/2}$ or $3d_{3/2}$ array of excitation and ionization transitions consists of a large number of lines and spread over an energy range of about 10 of electronvolts, thus clearly broader than the lifetime width of each line. All the lines were taken into account in the calculation of the spectra. Differences of the order of one or two percents between the ab initio calculated energies and the experimental energies were obtained.

All the levels of the $3d^9 4f^{m+1}$ configuration can be reached from the ground $3d^{10} 4f^m$ by irradiation with a probe electron beam of energy near the threshold. However, the electron–electron excitation cross sections corresponding to the dipolar electron transitions from the ground level $3d^{10} 4f^m$ ($^{2S+1}L_J$) to the excited levels $3d^9 4f^{m+1}$ $J, J \pm 1$ are among the highest, and they increase for the heavier rare-earths [38]. The $J, J \pm 1$ excited levels decay by dipolar electron transitions to all the allowed levels of the $3d^{10} 4f^m$ configuration while the resonance lines correspond only to transitions towards the ground level $^{2S+1}L_J$ of the configuration $3d^{10} 4f^m$. Thus, the resonance lines are only a part of the allowed emissions from the excited levels $J, J \pm 1$, to the various levels of the ground configuration. The case of lanthanum is unique since its ground configuration $3d^{10}$ $^1S_0$ has a single level. Additional peaks appear on the low energy side of each resonance line. They were labeled "$J, J \pm 1$ emissions". Simultaneously, the excited $3d^9 4f^{m+1}$ configuration relaxes making all the J-levels statistically populated. The ensemble of lines emitted from all the levels of $3d^9 4f^{m+1}$ to all the levels of $3d^{10} 4f^m$ were labeled "excitation emissions". The theoretical excitation spectra comprise all these lines. $\Gamma_{5/2}$ and $\Gamma_{3/2}$ widths of 0.8 and 1.3 eV for the light rare-earths and 1 and 1.5 eV for the heavy rare-earths, respectively, were used in the calculations [49].

When E_0 exceeds twice the threshold energy, the ionization of the core electrons is favoured. The ionization is completed in a time shorter than the lifetime of the core hole. Consequently, all the J-levels of the ionized configuration $3d^9 4f^m$ are populated statistically before the core-hole decays and transitions from all these levels contribute to the emission spectrum; they were labeled "ionization emissions" or normal emissions. The ionization spectra describe the distribution of the unperturbed occupied 4f levels. The energies and the probabilities of the all electric dipolar 4f–3d and 5p–3d transitions were calculated for each trivalent rare-earth ion from La^{3+} to Yb^{3+}. Energies of the maximum are given in Table 4.1 for the 4f–$3d_{5/2}$ excitation and ionization emissions and for the 5p–$3d_{5/2}$ emissions in the presence of a 4f spectator electron and in the ion. For all the rare-earths, the 4f–3d, and also 5p–3d, emissions from the $3d^9 4f^{m+1}$ excited configurations are located towards the lower energies of the emissions from the $3d^9 4f^m$ ionized configurations. The average energy of the $3d^9 4f^{m+1}$ configurations is lower than that of the $3d^9 4f^m$ ionized configurations because the additional 4f electron of the excited configuration partially screens the excess of positive nuclear charge induced by the ejection of the 3d core electron. On the other hand, the electrostatic interaction between the 4f electrons and the 3d hole is larger than that between the 4f electrons and a 5p or valence hole. Both these facts account for the respective energies of the various emissions. The calculated probabilities of the 4f–3d excitation emissions are always higher than those of the ionization emissions. In contrast, the probabilities, calculated for the normal $5p_{3/2}$–$3d_{5/2}$ atomic lines in the presence, or absence, of a spectator electron, are very similar. The 5p–3d emissions are not modified by self-absorption. Consequently, their relative intensities give direct information on the ratio of the excitation and ionization probabilities.

Table 4.1 Calculated energies (eV) of the 4f–3d$_{5/2}$ and 5p–3d$_{5/2}$ emissions: by excitation, columns (1) and (3); by ionization, columns (2) and (4)

	4f–3d$_{5/2}$		5p–3d$_{5/2}$	
	(1)	(2)	(3)	(4)
La	836.6		813.0	821.2
Ce	882.8	887.6	860.0	867.8
Pr	931.1	934.4	907.7	915.4
Nd	980.3	983.3	956.2	963.8
Sm	1080.0	1083.3	1054.1	1061.5
Eu	1131.9	1135.3	1105.8	1113.0
Gd	1184.6	1187.7	1158.2	1165.4
Tb	1238.3	1241.4	1211.7	1218.7
Dy	1293.2	1295.8	1267.1	1279.0
Ho	1348.7	1353.1	1323.2	1329.3
Er	1405.1	1407.9	1378.9	1386.5
Tm	1461.0	1465.2	1437.6	1442.9
Yb	1520.3	1521.9	1495.4	1500.2

The ratios of the transition probabilities 4f–3d$_{5/2}$/4f–3d$_{3/2}$ were predicted for the excitation and ionization emissions of the entire rare-earth series. The calculations were made for the 3d$_{5/2}$ and 3d$_{3/2}$ transitions taken either together or separately. The 3d$_{5/2}$/3d$_{3/2}$ transitions ratios were approximately equal to 1.5 for all rare-earths when the transitions to 3d$_{5/2}$ and 3d$_{3/2}$ are calculated separately, as in the previous theoretical works. In contrast, when both emission groups were calculated together, the 3d$_{5/2}$/3d$_{3/2}$ ratio increases with the number of 4f electrons [49]. It is equal to approximately 0.5 for lanthanum, 1.0 for the ions having the 4f sub shell half-filled and tends to the value 1.5 characteristic of the complete sub shell for ytterbium. Indeed, the relative probabilities cannot be deduced from the ratio of the statistical weights if one of the sub shells is not filled. This result is related to the various couplings. Among them, the 3d–4f electrostatic interaction plays a predominant role. However, due to their 4f spin–orbit interaction the 4f electrons have the tendency to present preferentially a 5/2 character. Consequently, the transition probabilities from 4f to 3d$_{3/2}$ are higher than those to 3d$_{5/2}$ for the light rare-earths and this explains the variation of the 3d$_{5/2}$/3d$_{3/2}$ ratio. The same calculations were made for the resonance lines and the $J, J \pm 1$ emissions. For the resonance lines, the probability ratio becomes greater than 1.0 from neodymium and it increases very rapidly along the series. The same variation exists for the absorption probabilities from the ground state J to the excited states $J, J \pm 1$ because this process is the inverse of the resonance lines.

Comparison between the various ab initio calculated 4f–3d emissions, resonance lines, $J, J \pm 1$ lines, excitation emissions and ionization emissions and the 3d EXES spectra was made for all the trivalent rare-earths, metal or oxide. The energy range of each type of emission is wide and it extends over a great number of allowed lines that might blend, introducing some inaccuracy on the energy measurements (Fig. 4.3). However, the difference between the calculated and observed energies does not exceed 1–2 eV, i.e. about 1–2 % of the energy range.

Fig. 4.3 Calculated and experimental emissions: *a* resonance lines; *b* J, $J \pm 1$ emissions; *c* 4f–3d and 5p–3d excitations emissions; *d* 4f–3d and 5p–3d ionization emissions; **A** for praseodymium observed at *e* 1.1 kV and *f* 3.0 kV; **B** for gadolinium observed at *e* 1.4 kV and *f* 2.25 kV [49]

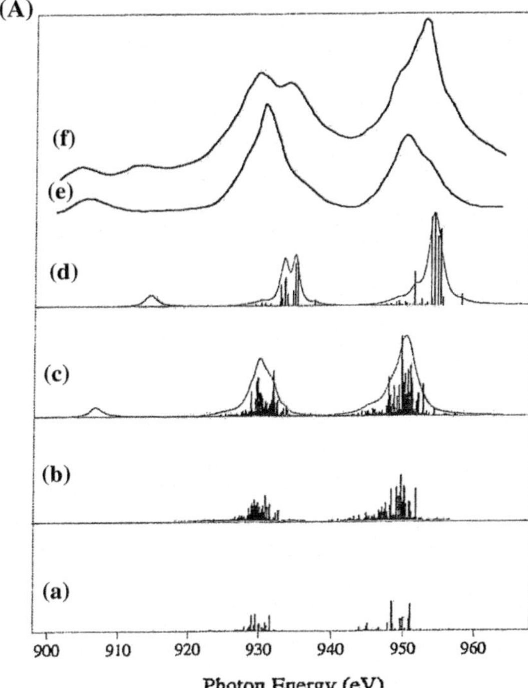

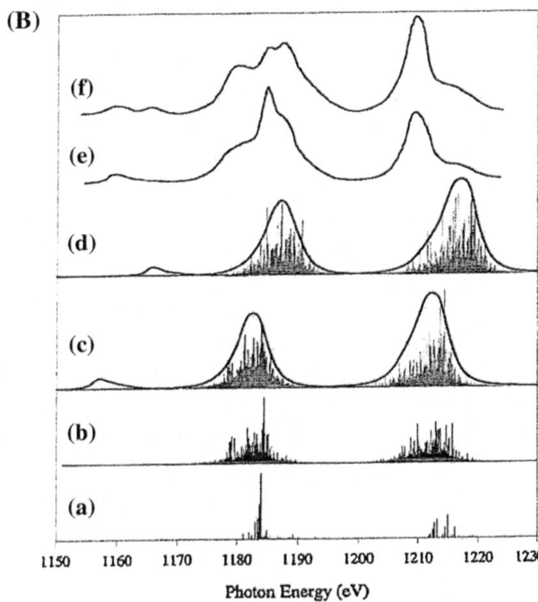

The calculated values of the emission energies are systematically higher than the observed values. This is in part due to the partial screening of the core hole by the valence electrons that is not taken into account in the calculations. All the observed features were interpreted and it appeared that the trivalent rare-earths were well described by atomic-type calculations. At an incident energy E_0 near the threshold, remarkable coincidence of some of the emission peaks with the absorption maxima is observed for the light rare-earths; these are the resonance lines. The 4f–3d excitation and ionization emissions are observed towards the lower and higher energies of the resonant lines, respectively. The spectra of the mid-series rare-earths are very complex because of the high number of levels associated with the half-full 4f sub shell. The $3d_{5/2}$ resonance lines are spread on a wide energy range while the $3d_{3/2}$ resonance lines are located clearly towards the higher energies of the excitation emissions. The energy interval between the excitation and ionization emissions decreases gradually along the series, making the resolution of the various types of transitions difficult. In general, the ionization emissions appear as weak features towards the higher energies of the excitation lines. For the heavy rare-earths, the number of lines decreases. In the $3d_{5/2}$ range, three resonance lines are seen for Tm^{3+} and only one for Yb^{3+}, in agreement with the predictions.

A good agreement is also found between experimental and calculated energies for the $5p_{3/2}$–$3d_{5/2}$ emissions in the presence or absence of a spectator electron. These lines are observed towards the lower energy of the $4f_{7/2,5/2}$–$3d_{5/2}$ emissions. Their intensity varies with E_0 as seen for the 4f–3d emissions. Thus, for E_0 near the threshold, the more intense lines are those in the presence of a spectator electron. The 5p–3d emissions are weak with respect to the 4f–3d emissions. For the light rare-earths, they are about an order of magnitude weaker than the 4f–3d emissions. Their intensity decreases for the heavy rare-earths and they become too weak to be observed.

Strong spectral intensity variations accompany a change of E_0. These variations concern mainly the resonance lines, which are the strongest emissions when E_0 is near the threshold. But for E_0 equal to two or three times the threshold there appears a central minimum in the strong resonant lines. As an example, at 5 keV, for Sm^{3+}, the main part of the $3d_{5/2}$ excitation emissions is strongly self-absorbed as shown by the presence of a minimum at the position of the more intense resonance emissions (Fig. 4.4). In contrast, the intensity strongly increases towards the lower energies of the resonance lines because the self-absorption is weak. Strong intensity increase is also observed for the weakly self-absorbed Mα emission. Under the same conditions, the main part of the $3d_{3/2}$ excitation emission varies little for the mid-series elements because the self-absorption is weak in this domain. Detailed analysis of all these changes is complex. However, it appears clearly that the excitation emissions located towards the lower energies of the resonance lines are more intense in the $3d_{3/2}$ than in the $3d_{5/2}$ region for ions like Sm^{3+}. Differences between the $3d_{5/2}$- and $3d_{3/2}$-based spectra confirm that the $4f_{5/2}$ levels are filled in first.

The intensity ratio of the $3d_{5/2}$ to $3d_{3/2}$ resonance lines can be measured only for lanthanum. For the other rare-earths, the resonance lines overlap the excitation lines

Fig. 4.4 4f-3d absorption (**a**) and emissions of samarium metal at 1.2 (**b**) and 5.0 keV (**c**). At 5.0 keV, a *"white line"* is observed at the position of the maximum of the resonance line. This is explained by the self-absorption in the target

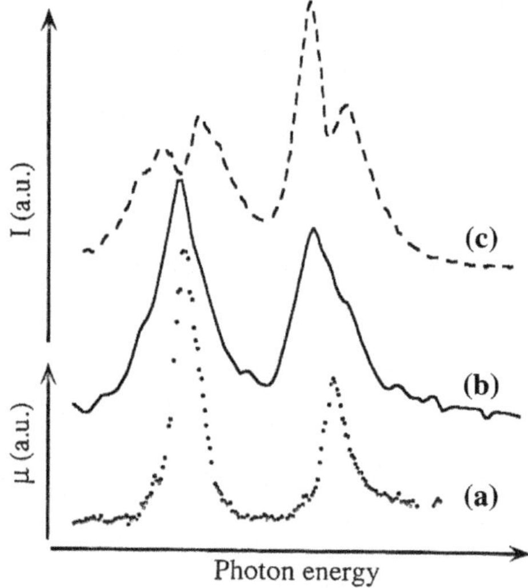

and this makes the experimental determination of this ratio difficult. Moreover, the intensity of each resonance line is a function of the probability of populating its initial excited J-level. This intensity can undergo large changes with the self-absorption because the most intense lines are the most self-absorbed. As already underlined, the $3d_{3/2}$ resonance lines are prominent in the beginning of the series; their transition probability diminishes considerably beyond gadolinium where the $3d_{5/2}$ resonance lines become prominent.

This paragraph concerns the $3d_{5/2}/3d_{3/2}$ ratio of all the excitation lines. Their initial configuration is in a statistically populated relaxed state. Each level is populated proportionally to its statistical weight and this is taken into account in the calculation. The self-absorption causes a relative decrease of the $3d_{3/2}$ emissions with respect to the $3d_{5/2}$ emissions for the lighter rare-earths and the inverse effect for the heavy rare-earths but these changes are not predominant. The predominant factor modifying the experimentally measured $3d_{5/2}/3d_{3/2}$ ratios with respect to calculated ratios for the emission lines is the presence of resonant Coster–Kronig transitions from $3d_{3/2}^{-1}4f^{m+1}$ to $3d_{5/2}^{-1}4f^{m}$. The presence of these non-radiative transitions has as consequence the decrease of the lifetime of the $3d_{3/2}$ level with respect to that of the $3d_{5/2}$ level and consequently the increase by about 0.5 eV of the width of the $3d_{3/2}$ level with respect to the width of the $3d_{5/2}$ level for the entire series of rare-earths. For La^{3+}, this effect increases the calculated $4f–3d_{5/2}/4f–3d_{3/2}$ transition probability ratios of the excitation emissions by about 50 % (Table 3.3). The ratio of the excitation emissions is predicted to vary from 0.60 for Ce^{3+} to 1.43 for Yb^{3+} (Table 4.2). But these values must be increased by more than 60 % to take into account the non-radiative transitions in the light rare-earths. Indeed, from spectra observed at low incident energy for Ce^{3+} metal, the experimental value of this ratio is of order of unity. It increases progressively along the series and is about 2 for Tm^{3+}.

Table 4.2 Calculated probability ratios $3d_{5/2}/3d_{3/2}$

	R_r	R_e	R_i
La	0.49	0.49	
Ce	0.74	0.60	0.45
Pr	0.86	0.66	0.55
Nd	1.09	0.72	0.62
Sm	1.58	0.86	0.77
Eu	2.13	0.94	0.84
Gd	2.00	1.01	0.92
Tb	2.62	1.08	0.99
Dy	3.80	1.15	1.07
Ho	6.19	1.23	1.15
Er	9.64	1.30	1.22
Tm	20.4	1.37	1.30
Yb		1.43	1.37
Lu			1.43

R_r for resonant lines; R_e for excitation emissions; R_i for ionization emissions

For the resonant emissions, the $3d_{5/2}/3d_{3/2}$ ratio is predicted to increase rapidly with Z and to become one order of magnitude larger for the heavy rare-earths.

Let us consider the $4f$–$3d_{5/2}/4f$–$3d_{3/2}$ intensity ratio of the ionization emissions. It is predicted to vary from 0.45 for Ce^{3+} to 1.37 for Yb^{3+} and to be equal to unity for Tb^{3+}. The self-absorption is negligible in the energy range of these emissions and it is convenient to stimulate them with incident electrons whose energy E_0 is of the order of twice the threshold. Thus, the Pr^{3+} and Nd^{3+} $4f$–$3d_{3/2}$ ionization transitions are the most intense emissions at $E_0 = 3$ keV. The calculated intensity ratios vary approximately as the experimental ratios. For Pr^{3+} and Nd^{3+} at any E_0 the $4f$–$3d_{3/2}$ ionization emissions are more intense than the $4f$–$3d_{5/2}$ ones, in agreement with the calculations. For Pr^{3+}, the $4f$–$3d_{5/2}/4f$–$3d_{3/2}$ intensity ratio is about equal to one half; it is approximately unity for Gd^{3+} and Tb^{3+} and superior to unity from Ho^{3+}. As already underlined, the $4f_{5/2}$ states are populated first.

Let us consider now the respective probabilities for excitation and ionization. For Ce^{3+}, the excitation and ionization lines are well resolved because their number is small and the $4f$–$3d_{5/2}$ and $4f$–$3d_{3/2}$ excitation emissions are clearly more intense than the ionization emissions. An analogous ratio is seen for all the series at rather low incident electron energy, e.g. at E_0 exceeding the threshold by about 200 eV. All the same, only the $5p$–$3d$ emissions in the presence of a $4f$ spectator electron are observed at low E_0 while the $5p$–$3d$ normal emissions are observed with a similar intensity only from $E_0 = 3$ keV. Self-absorption and radiative probability are equivalent for the two types of $5p$–$3d$ emissions, making these lines adequate for the study of the primary process of the $3d$ core-hole creation. One deduces from their observations that the excitation and ionization probabilities are approximately equal at 3 keV, i.e. for E_0 of the order of twice the threshold, while the excitation probability largely dominates below this energy.

No energy shift of the transitions between the 4f and *nlj* core sub shells is observed when going from the metal to its compounds. It is known that the energies of all the core levels of an element are shifted by the same amount when the element changes from one compound to another. As already underlined, the 4f sub shell can be considered as a core sub shell. It was suggested than the excited configurations present in the oxides had the form $3d^9 4f^{m+1} V^{-1}$, where V designates the valence electrons, and were created by charge transfer from the oxygen valence band to the 4f levels in the presence of the 3d core hole [50]. According to this suggestion, the emissions from the $3d^9 4f^{m+1} V^{-1}$ excited configuration in the compound and the resonance lines in the metal should have different energies. However, according to the observations by EXES the resonance lines in the oxide fall in the same energy range as their corresponding lines in the metal. Therefore, the above suggestion is not validated.

Only metal and compounds with the same number of 4f electrons are considered here. For such cases, the resonance lines, as well as the absorption lines, are predicted to be identical in shape and in position in both metal and compounds. However, the intensity of the resonant emissions of the metal are somewhat attenuated. Owing to the mixing between the discrete and band levels, the number of excited levels decreases slightly in the metals by interaction with the continuum in the solid. For a non-conductor, the spectra are dominated by the excitation process. On the other hand, the normal emissions show a small change in shape and in position for the elements of the mid sub shell. Thus, dysprosium Mβ is shifted by about 0.6 eV towards the higher energies in the oxide with respect to the metal, gadolinium Mβ is shifted by less in the same direction. Their shape is asymmetrical in metal and symmetrical in oxide. These observations reveal a difference in the filling of the $4f_{5/2}$ states; this is more pronounced in the compound than in the metal, making the sub shell more localized in the compound. In contrast, the Nd^{3+} spectra in the metal and in the oxide differ only by the relative intensities of the excitation and ionization emissions. For the oxide, the intensity of the 4f–3d ionization emissions is low and the 5p–3d emission in the ion is not observed, showing that excitation rather the ionization is the dominant process.

Important data have been deduced from the detailed analysis of electron-stimulated excited X-ray configurations observed by EXES. In the first place, from the variation of various $4f–3d_{5/2}/4f–3d_{3/2}$ intensity ratios, data were obtained on the respective $4f_{7/2}$ and $4f_{5/2}$ electron densities along the series. Next, the good agreement between experimental and calculated spectra confirmed that the 4f–3d transitions can be treated like transitions between core levels, i.e. taking place between quasi-atomic-like levels in the solid. The high probability of exciting the configurations $3d^9 4f^{m+1}$ in all the metal rare-earths as compared to the low probability of their ionization is a confirmation of the very strong localization of the 4f electrons in the atomic core. Finally, the observation of resonance lines at the same energy in the metal and the oxide has shown that the number and the character of the 4f electrons are the same in both materials and the excited configurations are not influenced by chemical bonding. In the metal, for light rare-earths, the excited levels are energetically mixed with the distributions of the delocalized valence levels.

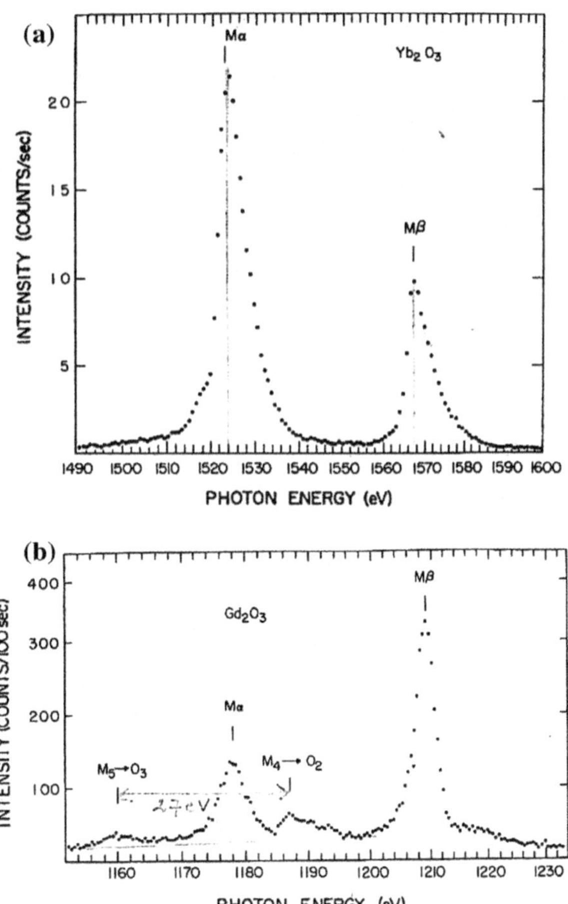

Fig. 4.5 3d emission spectra obtained by fluorescence: **a** from Yb$_2$O$_3$; **b** from Gd$_2$O$_3$ [51]

Tunneling of the excited 4f electrons to these distributions is then possible. However, the localized character of the 4f electrons is conserved and this confirms the stability of the excited structure.

Observation of the 4f–3d emissions by fluorescence was made for Gd$_2$O$_3$ and Yb$_2$O$_3$ [51]. The Yb$_2$O$_3$ 3d$_{5/2}$ spectrum is easily interpreted. Two resonance lines are possible, one intense and one very weak, clearly separated from the lines in the 3d^9 ion, located approximately 3–5 eV higher. In the fluorescence spectrum (Fig. 4.5a), the emission 4f–3d$_{5/2}$, denoted Mα, corresponds to all the 62 lines from the ion 3d$_{5/2}^{-1}$4f^{13} while the weak feature seen towards the lower energies of Mα corresponds to the main resonance line. As predicted, the ionization is predominant because the spectrum is produced under irradiation of high energy. The Gd$_2$O$_3$ 3d$_{5/2}$ spectrum is complex and shows a strong self-absorption (Fig. 4.5b). The resonance lines appear as a minimum of intensity. The emission, noted by error Mα in the figure corresponds to the excitation lines while the emissions in the ion, which correspond to $M\alpha$, are located at the limit of high energies of the resonance lines. They are partially reabsorbed and appear as a feature mixed with the very weak O_{II}–M_{IV} emission. This example shows the complexity of these spectra that require well-adapted experimental conditions. Careful comparisons between various experimental results are also necessary in order to prevent further erroneous interpretations.

4.2.2.3   4d Emission

Direct radiative recombination and transition in the presence of a 4f spectator electron, observed for the first time in the 3d range, were also observed in the 4d range. Indeed, the 4d emission spectra consist of a large number of lines characteristic of atomic-like 4f–4d transitions, analogous to transitions between core levels. Their presence confirms the highly localized character of the $4d^{-1}4f^{m+1}$ excited configurations but without theoretical computations it is not possible to identify each observed feature. Indeed, a detailed study of the emission spectra for the entire rare-earth series does not exist in the 4d range. Features related to the presence of a giant resonance appear in $4d_{3/2}$ absorption. The corresponding decay peaks are located above the $4d_{3/2}$ threshold. They are predicted as intense but are very strongly reabsorbed. Therefore, they are strong only when the excitation takes place in the vicinity of the threshold energy.

Electron-excited 4d spectra were observed for the elements from praseodymium to lutecium [52–54]. Main characteristics are the following [55]. The 4d emission spectra extend over about 50 eV including two groups of transitions, the 5p–4d and 4f–4d ones, which overlap towards the lower energies. 4f–4d transitions from the initial configurations $4d^9 4f^m$ and $4d^9 4f^{m+1}$ are present. The energy extension of each of these configurations is about 20 eV. Consequently, on the high energy side of the 5p–4d lines, overlap takes place between the 4f–4d emissions originating from the $4f^m$ and $4f^{m+1}$ configurations. Owing to the high number of levels associated with each of the $4d_{3/2}$ and $4d_{5/2}$ sub shells and the small 4d spin–orbit interaction that separates them, the two groups partially overlap in energy and the various transitions are not well resolved. The 4d emission spectra of rare-earth metals then correspond to the superposition of numerous multiplets and are very complex. Above the $4d_{3/2}$ threshold, $4d_{3/2}^{-1}4f^{m+1}$ high lying levels overlap the conduction levels and weak interaction exists among them. Consequently, the discrete transitions located above the $4d_{3/2}$ threshold are modified by interaction of the discrete levels with the continuum levels. In some rare-earths (cf. Sect. 4.3), this interaction can partially delocalize the 4f electrons and make the direct radiative recombination hardly probable. In insulator compounds, the mixing between the 4f and conduction levels is not possible and the intensity of the resonance lines increases with respect to that in the metal. As an example, let us mention the radiative decay of the $4d^9 4f^{m+1}$ excited configuration and the 4d photoabsorption observed in LaB_6 and CeB_6 [56]. The experimental conditions enabled a good comparison between the emission and absorption spectra. It was confirmed that the intense and broad peak observed in the emission coincides with the giant absorption peak. It must be the result of the radiative decay of the excited configuration as are also the multiplet structures observed under the 4d threshold.

It has been shown [55] that along the rare-earth series the probability of the 5p–4d transitions increases by a factor of 2 whereas the probability of the 4f–4d transitions increases by a factor of 20. The 5p–4d transitions are more prominent than the 4f–4d transitions in the beginning of the series but they decrease gradually and become barely detectable at lutecium while the transitions from 4f are predominant in this range. Radiative recombination between the 4f electrons and the 4d hole is expected even though autoionization channels are present.

4.2.2.4 Applications

In summary, owing to the localization of the 4f wave functions within the centrifugal well of the attractive potential due to the inner hole, the $nd^9 4f^{m+1}$ highly excited configurations in solid rare-earths can be treated as quasi-atomic. These configurations are time-localized during the time scale of the measurement. They differ from the optically excited configurations. Indeed, the lifetime of an inner hole is much shorter than that of a hole in an optical level. The decay of the excited X-ray configurations is rapid and tunneling has not enough time to take place. Due to the presence of the resonance lines, the nd emissions of the rare-earths depend strongly on the excitation energy.

The nd emission spectra are complex due to the presence of different recombination processes. The $nd^9 4f^{m+1}$ configurations can be de-excited through all the atomic transitions possible from a configuration with an inner-shell vacancy. Indeed, while in absorption, only transitions from the ground state $nd^{10} 4f^m\ ^{2S+1}L_J$ to the (J, $J \pm 1$) levels of the excited configuration are present, in emission all the J-levels of the configuration $nd^9 4f^{m+1}$ are populated and decay towards the different J-levels of the ground configuration. The emissions $nd^9 4f^{m+1}$–$nd^{10} 4f^m$ are thus richer in lines than the corresponding absorptions and clearly more spread in energy. Transitions in the ions are also present. These transitions are observed towards higher energies of the transitions in the excited configuration for all the rare-earths. The stronger ones are the resonance lines, the weaker are the lines in the ion.

The complexity of the spectra explains why erroneous interpretations have often been given to them and it is important to mention the following remarks. The resonance lines are sometimes designated as the Mα and Mβ emissions. In fact, the Mα and Mβ emissions are the transitions in the ion; they appear towards the highest energy of the resonance lines. The totality of the transitions from all the J-levels of the excited configuration are often not taken into account and only the emissions in resonance with the absorption lines are considered in the interpretations. When the energy of the incident electrons is varied, the contribution of the self-absorption process is often overlooked: indeed, the resonance lines undergo a very strong self-absorption and then they either are not observed or appear as an intensity minimum between two apparent peaks. This may introduce interpretation errors. Except for lanthanum, thullium and ytterbium, excitation emissions comprise a large number of lines whose relative intensities vary in the vicinity of the excitation threshold. Consequently, it is not possible to identify these emissions as a single peak.

The observation of intense lines from the excited configurations $3d^9 4f^{m+1}$ of open 4f-shell elements is a direct experimental proof of the highly localized character of the 4f electrons and shows that the interactions between the 4f and valence electrons are negligible. Inversely, when the excitation lines are weak or absent, one can conclude that no 4f localized electrons are present. From the shape of these emissions, the presence of interactions between the excited levels and the continuum levels can be deduced; it is negligible when the emissions are symmetrical and of lorentzian shape. Conversely, when the excitation emissions have a Fano shape, autoionization is present.

The 3d excitation emissions depend only on the valence of the rare-earth. As an example, they are the same for cerium in the metal and in the trivalent oxide. In contrast, they are absent in tetravalent cerium while 4f normal emissions are expected. Cerium is easily oxidized in Ce^{4+}. That must be taken into account in the analysis of the emission spectrum [57].

Let us mention that coincidence spectroscopy between $n'l'$ and nl X-ray emission and nl photoemission is now being developed for rare-earth compounds. Theoretical model has been developed for the 5p–3d transitions in these compounds [58].

4.2.2.5 Photoemission

In all the rare-earth metals, valence photoemission is governed by the 4f electron excitation. But, in contrast with XES, emissions from all the valence levels are observed simultaneously and a weak peak due to 5d6s valence electrons is then present in the spectra [59]. The 4f photoionization cross section is small at low energy and is known to increase rapidly with the energy E_{hv} of the incident photons. Thus, for an increase of the photon energy from 10 to 40 eV, 4f photoemission was observed to increase very strongly relatively to that of the valence band [60]. Increase was also observed from 32 to 80 eV (Fig. 4.6) [61]. In the X-ray radiation range, the 4f photoionization cross sections are high and this energy range is often used for obtaining photoemission.

For the metals, at low E_{hv}, valence electron distribution is observed as an asymmetric feature originating from the Fermi level E_F. Thus, for praseodymium, neodymium and samarium, levels with a 5d character are present around E_F and a peak is observed at 0.5–1 eV below E_F for the three elements [43]. It is equivalent to the peak observed at the same position in lanthanum and attributed to the 5d electron distribution. Occupied 5d valence levels were also found at E_F in f.c.c. divalent ytterbium metal [62].

After ionization of a 4f electron from the $4f^m$ ground state the rare-earth is in the highly spread $4f^{m-1}$ final configuration. The multiplet structure of this final state is complex and varies with the 4f occupation number. The spectrum becomes more complex the closer the 4f sub shell is to the half-full shell. This complexity of the

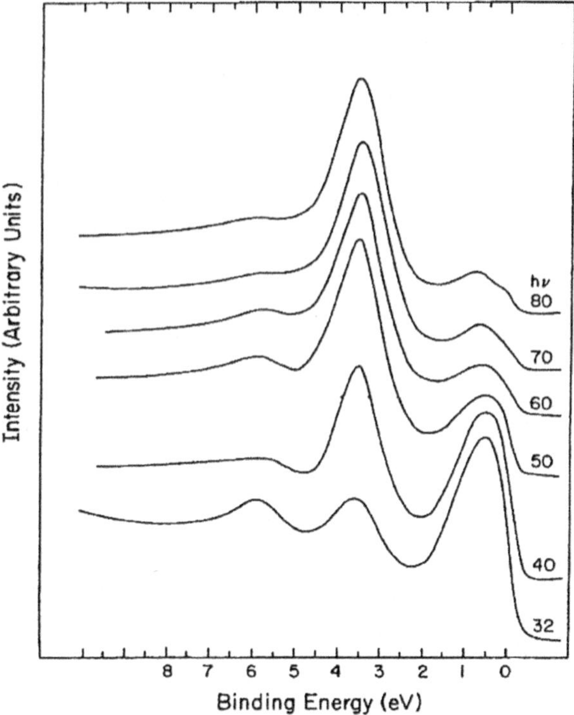

Fig. 4.6 Photoemission of praseodymium excited with photons between 32 and 80 eV [61]

rare-earth metal 4f spectra has been widely confirmed and, initially, the energy resolution was insufficient to resolve them [63, 64]. However, agreement has been found between the XPS observations for the bulk metal and the calculations made in intermediate coupling scheme [42, 43, 65]. In the beginning of the series, the energy interval between the 4f levels and the valence band is small. The interaction between 4f and valence electrons diminishes with Z increasing. The 4f spectrum extends over about 10 eV and generally exhibits two large structures. Their energies increase along the series and follow a regular variation with the atomic number (Fig. 4.7) [59]. As an example, for dysprosium, two peaks are observed in 4f photoemission, a highly structured peak at about −8 eV below the Fermi level and a peak at −4 eV. On the other hand, for the ytterbium $4f^{14}$ configuration, two narrow peaks with an energy separation of 1.3 eV were observed, in agreement with the spin-orbit splitting of the 4f levels in Yb $4f^{13}$ final configuration. Other exceptions were noted for europium and gadolinium, which have the $4f^7$ configuration in the metal. The spectra of these two metals are simple because of the spherically symmetric ground state of the half-filled $4f^7$ shell, which does not overlap in energy with the valence states. Only one 4f peak ($^7F_0$) is present. It is located at about −2 eV in europium while it is at −8 eV in gadolinium and can be considered as arising from a core level. A single peak is also observed for protactinium of $4f^1$ final configuration.

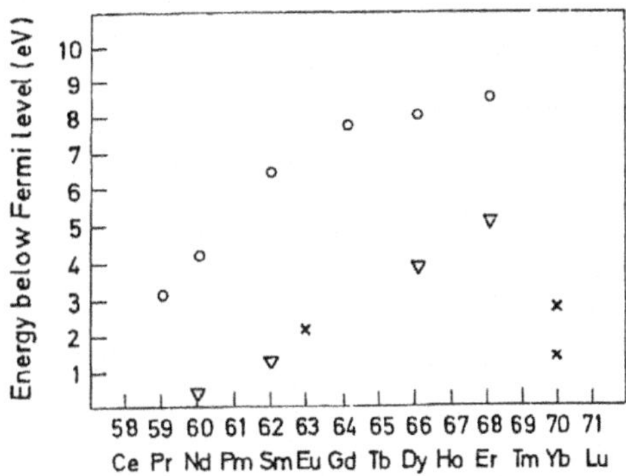

Fig. 4.7 Energies of the two observed valence photoemission peaks as a function of the atomic number [59]

More recently, the "Hubbard-I" approximation, or HIA, described in chap. 1, has been used to determine the excitation of the 4f shell in the light [66] and heavy [67] rare-earth metals in order to interpret the XPS and BIS experiments. From comparison of the calculated spectra with the experimental photoemission [42], it was shown that the main observed structures are reproduced by this theoretical approach except in the immediate vicinity of the Fermi level. In this range, structures due to valence 5d electrons are present, which were not taken into account in the calculated spectra. This study was made for praseodymium, neodymium and samarium (Fig. 4.8) in Ref. [66] and for terbium, dysprosium, holmium (Fig. 4.9) and erbium in Ref. [67] and confirm the atomic-like character of the 4f photoemission and the presence of multiplets in the spectra.

Differences were observed between the 4f multiplet structures excited with 100 eV or with X-ray photons [68]. The sensitivity to the surface increases when the photon energy decreases and the spectral changes observed with decreasing photon energy are due to the simultaneous ionization of surface and bulk atoms. Complexity of the 4f multiplets exists except for metal europium and ytterbium. For 4f photoemission of divalent metal Yb $4f^{14}$, two 4f peaks $^2F_{7/2}$ and $^2F_{5/2}$, characteristic of the final configuration $4f^{13}$, are observed in the XPS spectrum while for incident photons of 100 eV, four peaks are present, the two characteristic of the bulk and two extra peaks at 0.6 eV higher binding energy, characteristic of the surface atoms. For Eu $4f^7$, one 4f peak $^7F_0$ characteristic of the final configuration $4f^6$ in the bulk is observed in the XPS spectrum and an additional peak at 0.4 eV higher binding energy and characteristic of the surface atoms is observed when incident photons of 100 eV are employed. All the same, in $EuPd_5$ and $EuPt_5$ europium has the trivalent $4f^6$ configuration in the bulk and divalent $4f^7$ configuration at the surface [69]. For the other rare-earths, the observations are more difficult because only differences in relative intensities of the features appear. However, the 4f sub shell shift between the bulk and surface atoms was determined for all the elements of the series and it was found that it increases from about 0.4 to 0.7 eV going from trivalent cerium to thulium [68]. This is due to increasing Z. Another possibility was also considered.

Fig. 4.8 XPS (*squares*) and BIS (*triangles*) of neodymium and samarium metal compared to calculated spectral functions (*full line*). The Fermi level is at zero energy [66]

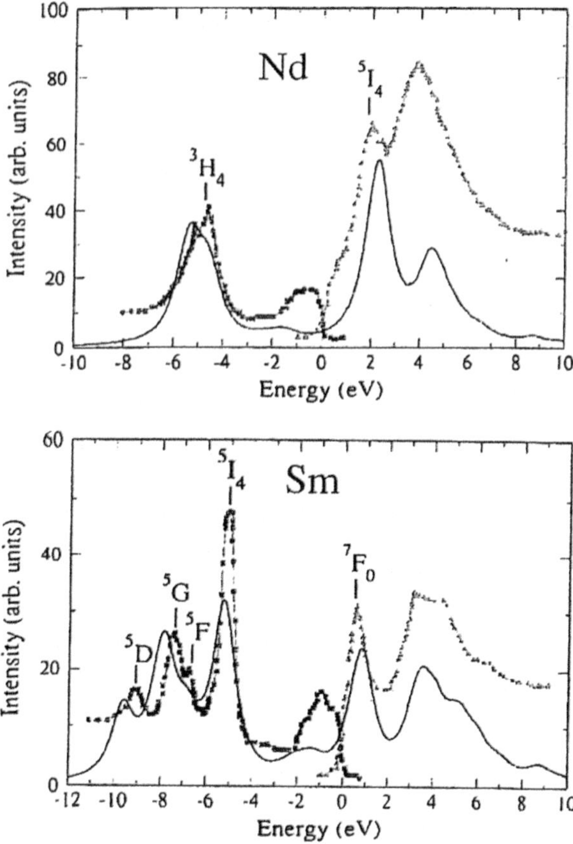

Fig. 4.9 XPS (*squares*) and BIS (*triangles*) of holmium metal compared to calculated spectral functions (*full line*). The Fermi level is at zero energy [67]

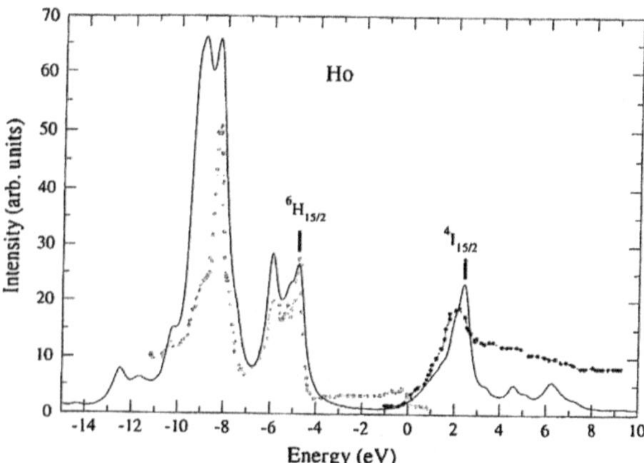

For the trivalent rare-earths the total number of the valence electrons remains constant but its division into electrons of each symmetry, s or d, can vary. This results in the variation of the cohesive energies and of the energies of the valence band. The 4f sub shell shift is, therefore, due to the change in the distribution of the valence electrons. Lastly, broadening of the 4f peaks characteristic of the surface was sometimes observed. This broadening was explained by different energy shifts characteristic of different single crystal grains in the surface. Broadening observed at low incident energies and absent in the XPS spectrum could also be due to small second layer surface shift.

Samarium is a particular case: it was shown that the surface ions have a 4f configuration different from that of the bulk ions [41, 70, 71]. In the bulk, samarium has the configuration $4f^5(5d6s)^3$ while at the surface its configuration is $4f^6(5d6s)^2$. The coordination number decreases at the surface, increasing the localization of the electrons and the electron distribution tends to fill the $4f_{5/2}$ sub shell. The valence change in samarium will be considered in the second part of this chapter.

Photoemission of the valence electrons was observed for the oxide RE_2O_3 series [72]. The valence bands are formed from the 2p-oxygen and 5d6s-rare-earth orbitals and their binding energies are larger in these insulator compounds that in the metals. For the trivalent heavy rare-earths, the 4f multiplet structures were only slightly modified by the chemical binding. This was a confirmation that the 5d6s valence electrons alone are involved in the bonding with ligand atoms [73]. The features characteristic of the 4f electrons were compared to the ionization probabilities of the trivalent $4f^m$ ions, deduced from the coefficients of fractional parentage, tabulated by Cox [74]. It was deduced that the 4f electrons were localized and core-like, resulting in relatively sharp 4f structures. The spectral resolution was limited in these experiments. The rare-earth 4f energies were found to increase along the series while the valence band energy remained constant. The energy necessary to transfer an oxygen 2p valence electron to the rare-earth 4f levels was found to decrease from lanthanum to europium, to jump to a larger value at gadolinium and to decrease again from terbium to ytterbium.

Valence photoemissions of the halides of lanthanum and cerium were observed and showed a large difference between the fluoride and the other compounds [75]. The binding energy of the fluoride np band is several eV higher than that of the chloride and the bromide. The energy interval between the np band and the cerium 4f level is thus clearly larger for the fluoride than for the two other halides. A big spectral difference was also observed in the photoemission of a core level indicating a chemical difference between the two types of compounds. This will be considered in a next paragraph.

Photoemission of the valence electrons of the cerium pnictides presents two peaks, a narrow near the Fermi level and a main peak about 2–3 eV below the first. For CeN, the main peak was shifted towards the higher binding energies with respect to the other pnictides while the intensity of the peak at E_F increases [76, 77]. This was attributed to the presence of a level of 4f symmetry, pinned at E_F, superimposed on a broad 5d band. When the anion atomic number increases, i.e. when the lattice constants increase, the 4f level is pulled toward the higher energies and gradually overlaps the main band. From these spectra, the characteristics of CeN were considered different from those of the other pnictides and it was deduced that the 4f electrons were delocalized in CeN while in the heavier pnictides they were localized and non-bonding. This could bring about some of the unusual properties of CeN, anomalously small lattice constant and gold-coloured metallic appearance (cf. Sect. 4.3).

The two peak structure present in cerium pnictides was also observed in praseodymium and neodymium pnictides but the peaks were further removed from the Fermi level [61, 78]. Consequently, the lower energy peaks cannot to be explained in a Kondo model since this model concerns only the features tied to E_F. Experimental results were compared to calculations where the 4f electrons were treated as either coupled with the valence electrons or localized and partial agreement was obtained with each of these theoretical models [79]. On the other hand, for the heavier cerium pnictides, it was not possible to reproduce the peak near E_F from the LDA/GGA + DMFT approach. The 4f electrons are localized and non-bonding in these materials. Indeed, it was possible to obtain a good experiment-theory agreement by treating the 4f electrons as core electrons, i.e. not hybridized with the other electrons and not hopping from lattice site to another site [80].

From valence photoemission and BIS, the occupied and unoccupied 4f levels of GdP, GdAs, GdSb, GdBi, were found 8–10 and 5–6 eV below and above E_F, respectively [81]. From X-ray photoemission, the 4f levels of GdN films were observed 7.8 eV below E_F [82]. Comparable results were obtained for all the pnictides of heavy rare-earths. The 4f electrons are thus localized core-like electrons with fixed spin values and do not contribute to chemical properties. They contribute only by their magnetic characteristics to particular properties of the compounds (cf. Chap. 2). These properties can change with the composition and the hybridization between the rare-earth 5d6s and pnictogen np electrons, with the crystalline structure and the inter-atomic distances. The rare-earth monopnictides have the simplest crystalline structure of the rare-earth compounds. But large changes of the physical properties can be obtained by a change of the constituents, for example by mixing different rare-earth pnictides. Numerous studies are now in progress on the realization of new materials [83].

Photoemission of various other compounds was observed, including intermetallic compounds of type $RE(M)_n$, $RE(Si, Ge)_2$, $RE(M)_2(Si, Ge)_2$, where M is a metal. For some rare-earths, the valence is mixed in these compounds. This is the case for most intermetallic compounds of cerium, in which a mixture of the $4f^0$ and $4f^1$ configurations is present at the ground state. These materials are discussed in Sect. 4.3.

Let us note that the Coulomb energy U can be derived from experiment by combining the results obtained by XPS and BIS or by X-ray emission and absorption. Experimental determination was made in the metals and values between 5 and 7 eV were found, in agreement with relativistic calculated values [84–86].

4.2.2.6 Resonant Auger Emission

In parallel with the resonant X-ray emissions, resonant Auger emissions were observed in the nd rare-earth spectra. These processes follow the mono-excitation $nd^9 4f^{m+1}$. Non-radiative decay takes place either through the Coster–Kronig transition

(a) $nd^9 4f^{m+1} \rightarrow nd^{10} 4f^{m-1} + \varepsilon d, g$

or through ionization of a core sub shell according to

(b) $nd^9 4f^{m+1} \rightarrow nd^{10} 4f^m 5l^{-1} (\text{or } 6s^{-1}) + \varepsilon(l \pm 1)$

or through double ionization of the excited configuration according to

(c) $nd^9 4f^{m+1} \rightarrow nd^{10} 4f^{m+1} 5l^{-1} \, 5l'^{-1} (\text{or } 6s^{-1}) + \varepsilon(l \pm 1) \text{ or } \varepsilon(l' \pm 1)$

Contrary to normal Auger emissions, (a) and (b) processes lead to a single-hole final configuration. The (c) process corresponds to an Auger emission in the presence of a spectator excited 4f electron. All the three types of energetically possible transitions are present simultaneously in the spectra, as was shown for Yb^{3+} [87]. Indeed, all the observed transitions in this spectrum have been identified by comparison with the various Auger decay channels calculated in an atomic model. As underlined in Chap. 3, the complete calculation of Auger transitions in the ionized and excited systems is only manageable when a single f electron, or a single f hole, is present, i.e. for La^{3+} and Yb^{3+}. In this case, the Auger transitions in the excited or ionized systems have equivalent probabilities and it is probable that a change of the screening effect will not be sufficient to introduce a change of the Auger probabilities.

The Coster–Kronig transition (a) is equivalent to the autoionization of the 4f level. It gives the same final state as the direct 4f ionization $4f^m \rightarrow 4f^{m-1} \, \varepsilon d, g$. It was supposed that autoionization and direct ionization could interfere, yielding asymmetric resonances with Beutler-Fano profiles. Transition of the (a) type was observed for the first time for gadolinium [88] and dysprosium (Fig. 4.10) [89] and later confirmed by numerous authors [90]. The final state of this transition is identical to that of the 4f photoelectron peak and, consequently, it appears as an enhancement of the photoemission peak when the incident photon energy is in a narrow band around the excitation energy $4d^9 4f^{m+1}$. It is thus generally named "resonant photoemission". But here, we keep the original name of *resonant Auger* in order to underline the analogy with the resonant radiative emissions. Indeed, each 4d–4f direct radiative recombination is in conjunction with a coupled non-radiative process. Resonant Auger is frequently used to obtain the 4f distributions because the intensity of this transition is clearly stronger than that of the normal 4f photoelectron peak. As already underlined (cf. Chap. 3), non-radiative recombinations between sub shells belonging to the same shell are always very fast and they dominate entirely the other processes. Let us note that since the decay processes from excited and ionized configurations are different, the lifetime of the core hole is not the same for the two configurations.

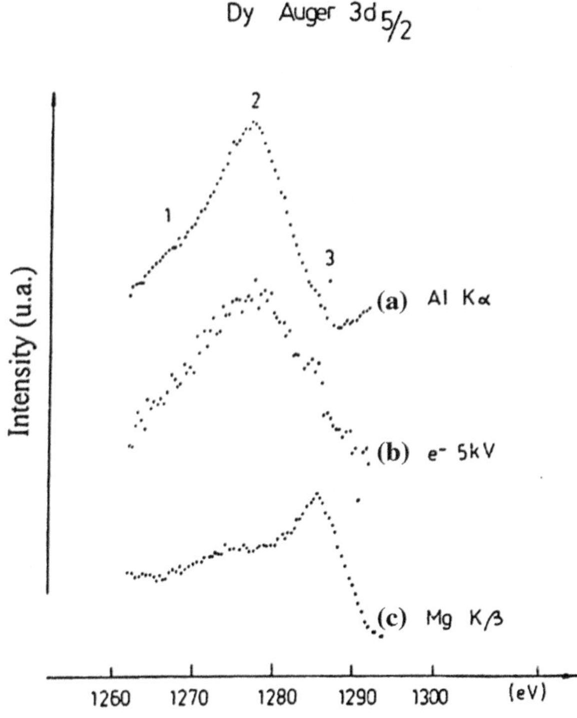

Fig. 4.10 Auger $3d_{5/2}$ spectrum of dysprosium induced by Al Kα, electrons of 5 keV and Mg Kβ [89]

A very complete study of the electron-excited 4d Auger spectra of the rare-earth metals from lanthanum to lutetium was published [91]. An advantage of the electron beam irradiation is that excited and ionized configurations are produced simultaneously and the entire non-radiative recombination spectrum was observed. Decay processes from the excited configurations $4d^9 4f^{m+1}$, named here "direct recombination", were observed to increase progressively in intensity along the series. They become more intense than the Auger and Coster-Kronig transitions from gadolinium. However, the use of an incident electron beam accelerated in the range 1600–2000 eV was not favorable to the 4d excitation because the excitation cross sections reach their maximum at the threshold energies, which are of the order of 100–350 eV along the series, but decrease rapidly with increasing energy. These results confirm the importance of the excitation with respect to ionization in the 4d range.

Comparisons between the rare-earth 4d and 3d spectra have shown that 4d resonant Auger depended on the surface while 3d resonant Auger was characteristic of the bulk [92]. However, most studies made in order to obtain the 4f distributions were in the 4d range. In fact, in this energy range, the transitions to 4f levels are strongly enhanced with respect to the transitions to valence band levels, which remain low. The method is used essentially to investigate the valence of the rare-earth (cf. Sect. 4.3). Some analysis of trivalent rare-earths are mentioned here as an example.

Following the first observation, 4d Auger spectra of gadolinium, stimulated by electrons and by photons were observed comparatively [93]. Resonant recombination from the $4d^9 4f^8$ excited configuration was identified in the electron-induced Auger spectrum as well as in the photoemission induced by photon energies in the region of the 4d–4f resonance. The presence of multiplet structure was later on identified [94].

Resonant Auger emission was observed for CeS, irradiated by photons of energies between 100 and 140 eV [95]. The 4f level was present at about -2.5 eV while the well-resolved sulfur p- and cerium d-bands were observed respectively at about -5 eV and near E_F. The cerium 4f peak was easily identified because its intensity was strongly enhanced for photons of about 120 eV. This peak is clearly separated from the valence bands, meaning that cerium is trivalent in this compound.

Resonant Auger emissions and Auger emissions in the presence of the spectator 4f electron were also observed in LaB_6 by excitation with photon energies in the range of the $4d^9 4f^1$ excited levels [96].

4.2.2.7   Resonant Inelastic X-ray Scattering (RIXS)

The study of RIXS in the gadolinium 4d range was among the first concerning rare-earths [97]. The incident photon energies were varied from the energy of 4d absorption maxima up to 30 eV above the ionization threshold. For incident photons of energy below the threshold, the $4d^9 4f^{m+1}$ excited configuration was created and the resonant 4f–4d recombination to the ground state was observed. At the same time, the 5p–4d transitions in the $4d^9 4f^{m+1}$ configuration, i.e. in the presence of the spectator 4f excited electron, were identified in the spectrum. These experiments confirmed the important role of the $4d^9 4f^{m+1}$ excited configurations as well as the presence of the various transitions predicted in this range. Other results concerning gadolinium can be cited. In $(Y, Gd)_2O_3$, calculations of the inelastic scattering associated with the excitation $4d^{10} 4f^7\ ^8S_{7/2}$–$4d^9 4f^8\ ^8D_{7/2},\ ^6D_{7/2},\ ^8P_J$ ($J = 5/2, 7/2, 9/2$) in the triply ionized Gd^{3+} agreed with the experimental observations [98]. The observed 4d–4f transitions of $(Y, Gd)_2O_3$ appeared nearly identical with those of Gd_2O_2S and in agreement with the atomic calculations. The same situation holds for the transitions $4d^{10} 5p^6 4f^7$–$4d^9 5p^6 4f^8$–$4d^{10} 5p^5 4f^7$. These results confirmed the localization of the 4f electrons in gadolinium [99].

Similar studies were made in the dysprosium 4d–4f range [100] In Dy_2O_3, triply ionized ions are present and transitions $Dy^{3+}\ 4d^{10} 5p^6 4f^9\ ^6H_{15/2}$–$4d^9 4f^{10}$ were predicted. Spectra were recorded for incident radiation below the threshold ionization and largely above ($h\nu = 208.4$ eV). In the latter case, fluorescence spectrum was observed. These results confirmed those obtained by EXES although the observations seem to be different. Indeed, the $4d^9 4f^{10}$ excited configuration is extended over more than ten electronvolts. In RIXS, the incident radiation is a very narrow energy band and only a few excitation energies of the 4d–4f excited configuration were selected and the corresponding spectra observed. Consequently, striking variations of the transitions and their intensity were observed depending upon which intermediate levels were excited. From such a method, it is thus possible to analyze separately only a few multiplets at a time of the excited configuration while the totality of the configuration is difficult to obtain. Two remarks can be made. 4d–5p transitions in the presence of a 4f spectator electron, predicted towards the lower energies of the corresponding transitions in the ion, were not clearly identified. The 4d ionization cross sections are small with respect to the 4d–4f excitation ones and particularly with respect to the giant resonance. That explains the low intensity of the emissions from ionized configurations.

Comparison between RIXS near the 3d and 4d thresholds of neodymium in Nd_2O_3 has confirmed the differences predicted between these two energy ranges [101]. Thus, different couplings must be used, intermediate coupling to treat the 3d–4f transitions, L–S coupling in the 4d–4f range. Coster–Kronig transitions of $3d_{3/2}$–$3d_{5/2}X$ type are highly probable in the 3d range and do not have an equivalent in the 4d range. In contrast, autoionization is strong in the 4d range. However, it is not clear why 5p–4f transitions are absent at the 4d threshold while they are clearly observed in EXES. RIXS excited in the region of the 3d–4f threshold is presented in Fig. 4.11 for Nd_2O_3. Transitions above the 3d–4f threshold were observed at the energies calculated for excited or ionized Nd^{3+} ion. These results confirmed the agreement between experiment and calculation in a model of 4f localized electrons.

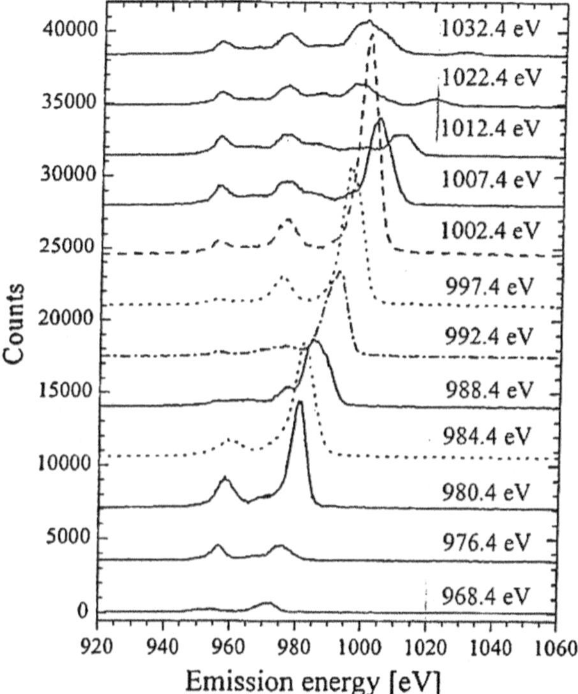

Fig. 4.11 RIXS of Nd_2O_3 excited at the 3d threshold. The excitation energy is indicated above each spectrum [101]

In the 2p range, RIXS of gadolinium in a $Gd_3Ga_5O_{12}$ garnet revealed the presence of the two multiplet families, attributed to the transitions from the $2p^54f^{m+1}$ and $2p^54f^m5d^1$ intermediate excited configurations to the $3d^94f^{m+1}$ and $3d^94f^m5d^1$ final configurations [102]. The presence of an initial $2p^64f^m \rightarrow 2p^54f^{m+1}$ quadrupolar excitation was then suggested to explain the observed low energy multiplet, in spite of its very weak probability and of its absence in the L_{III} absorption spectrum. Subsequently, experimental 2p–3d RIXS of samarium metal was compared with spectra calculated by a Hartree-Fock-type model with relativistic corrections to describe the 2p–4f excitation [103]. The observed transitions of low energy and weak intensity were associated with the $2p^54f^{m+1}$ excited configuration. Thus, the theoretical interpretation by quadrupolar excitation was not discarded.

The 2p–3d RIXS of the trivalent cerium compound CeF_3 was also analyzed [104]. Following the initial 2p excitation, the final configuration, Ce^{3+} $2p^63d^94f^1$, is equivalent to the final configuration of the 3d photoemission. As discussed in the following paragraph, the positioning of an initially unoccupied 4f level in the valence band in the presence of a 3d hole is possible in the metal. Indeed, a weak photoemission peak corresponding to the configuration $3d^94f^2V^{-1}$ was observed at about 878 eV in cerium and also in various intermetallic compounds but not in insulator compounds such as CeF_3. A very weak peak was detected in RIXS at this same energy. It was attributed to the $3d^94f^2V^{-1}$ final configuration. But, quadrupolar excitation followed by a dipolar transitions of the type $2p^63d^{10}4f^1 \rightarrow 2p^53d^{10}4f^2 \rightarrow 2p^63d^94f^2$ was supposed to explain the presence of this weak peak in cerium RIXS [105]. All the same, theoretical studies were made to justify the presence of quadrupolar excitation $2p^64f^m \rightarrow 2p^54f^{m+1}$ in Gd^{3+}, Dy^{3+}, Ho^{3+}, Er^{3+} RIXS [106]. This interpretation was based on the presence of a strong Coulomb interaction between the 4f electrons and the 2p core hole in the final state. However, in the case of cerium, the predominant interaction depends on which of the two above mentioned processes excited the configuration $3d^94f^2$. It was, therefore, surprising that both gave prediction of that weak peak at the same energy, 878 eV.

In the case of LaS, LaNi and La_2Ni_7, the low energy peaks observed in lanthanum 2p–3d RIXS were interpreted as a transition between the 4f orbitals and the valence band [107]. Indeed, it appears pertinent to associate these peaks to dipole transitions rather than to quadrupole transitions.

In summary, when a continuous radiation selected to match the $3d_{5/2}$, or $3d_{3/2}$, threshold is used, the spectral features must show resonance lines. Intensity of the photoexcited resonance lines seems to follow the intensity of the absorption lines. Above the ionization threshold, fluorescence is observed and is equivalent to EXES. Fluorescence is in a non-resonant energy regime and absorption and emission processes are independent. The ionized intermediate state can decrease by Auger processes and the low intensity of the fluorescence with respect to RIXS is due to the importance of the Auger decay. However, comparison between fluorescence stimulated by monochromatic photons just above the threshold and EXES must take into account the ratio of the photoionization and electroionization cross sections and the experimental conditions.

4.2.3   Core Levels

4.2.3.1   Photoemission

In photoemission of the core levels, the experimental resolution is limited by the broadening due to the lifetime of the core hole. Inelastic energy losses of the photoelectron in the solid also induce an asymmetry of the peaks towards the higher binding energies. In rare-earth metals and compounds, core photoemission is calculated to exhibit multiplet structure due to the presence of the open shell 4f electrons and to their coupling with the core hole. But these structures are only slightly resolved experimentally. However, the observation of a simple spin-orbit doublet, $3d_{3/2}$ and $3d_{5/2}$, in the 3d photoemission of metal ytterbium was a confirmation of its $4f^{14}$ configuration and of its divalent character. Various core levels of the rare-earths, 4s, 4p, 5s and 5p were determined, [73, 75]. Their energies were obtained with respect to the Fermi level in the metal.

Let us recall that the effects due to the core hole present at the final state in XPS are different from those in XAS. In XAS, perturbations due to the hole can be neglected at absorption threshold because the excited electron remains partially bound to the core hole and contributes strongly to the screening. In contrast, perturbations, sometimes named *final state effect*, must be taken into account in the core level photoemission. Thus, following the 3d ionization, the energy of all the levels increases, the partially filled 4f sub shell contracts, its radius decreases making the electronic charge in the limited volume of the atomic sphere higher in the $3d^9$ ion than in the neutral atom. Moreover, the perturbation due to the core hole is responsible for the double excitation or ionization.

Calculations of the configurations $3d^9 4f^m$ in intermediate coupling for praseodymium and neodymium were used for the interpretation of their corresponding 3d photoelectron spectra [108]. A width of about 10 eV was calculated for the spectra involving each of these configurations with the major part of the signal intensity within an interval of 3 eV. When the interval between the inner 3d levels was taken into account, the predictions were in agreement with the observations. Moreover, the intensity ratio of $3d_{5/2}$ and $3d_{3/2}$ photoelectron peaks is 6/4, which is the ratio of their statistical weights. This work shows that intermediate coupling scheme is convenient to describe the interaction between a core hole and an open valence sub shell. The relative intensity calculations based on the sum of squares of the Russell–Saunders amplitudes had to be supplemented by a ponderation factor P that took into account the coupling of the total J of the ground term with the j of the 3d hole, thus forming a Jj coupling.

Various types of satellites are present in photoemission. They have been widely studied because of their important role in the interpretation of the spectra. Low energy satellites were observed in the $3d_{5/2}$ and $3d_{3/2}$ spectra of metal lanthanum (cf. Chap. 3). Strong analogy exists between the 3d photoemission of lanthanum and that of the light rare-earth metals. For cerium metal, a main peak is seen at 883.8 eV and a weak peak at 878.9 eV. The main peak corresponds to the ionized configuration

$3d_{5/2}^{-1}4f^1$. The weak peak was successfully attributed to a shake-down satellite [109], also labeled "well-screened" peak. Its energy is close to the value calculated for the barycentre of the excited configuration $3d_{5/2}^{-1}4f^2$. This peak is obtained in the metal by transfer of a valence electron into a 4f orbital, initially above the Fermi level and stabilized by the presence of the core hole. The associated configuration is $3d_{5/2}^{-1}4f^2V^{-1}$. This electron transfer accounts for the perturbation undergone by the system during the creation of the hole. It depends on the distribution of unoccupied f levels above the Fermi level. The low intensity of this satellite shows that interaction between the 4f orbitals and the valence orbitals is weak in metal cerium. But it must be underlined that the decay probability of a core hole is very high and it increases with the hole energy. Consequently, the probability for the filling of the core hole to take place before the singly ionized system underwent a tunneling process is high, making the shake-down process hardly probable.

Analogous spectra were observed for praseodymium and neodymium. In these metals, well-screened $3d^{-1}4f^{m+1}V^{-1}$ configurations can also be created by relaxation of the $3d^{-1}4f^m$ ionized configurations because the energy mixing of unoccupied 4f levels with the valence levels is possible in the presence of the core hole. The "well-screened" configurations are close in energy to the $3d^{-1}4f^{m+1}$ excited configurations observed by X-ray absorption and EELS and calculated by the atomic model. They are more stable than the $3d^{-1}4f^m$ ionized configurations, sometimes named by contrast "poorly screened" configuration. The intensity of the shake-down satellites observed below each main $3d_{5/2,}$ or $3d_{3/2}$ peak for cerium, praseodymium and neodymium metals is only a few percents of the intensity of the main peak. Their intensity decreases with the increasing of the number of 4f electrons and no such peak is seen in the heavy rare-earths metals. This is due to the increase of the 4f electron localization and of the 4f binding energies with respect to the valence band.

Shake-down satellites were also predicted in intermetallic compounds of light rare-earths, such as $CePd_3$ [110]. In LaAu, $LaAu_2$ and their cerium analogues, shake-down peaks were observed about 4–6 eV below the main peak. In the light RE-Au alloys, distortion of the main peak was observed. It was suggested that, simultaneously to the creation of the rare-earth 3d hole, an Au 5d electron is excited, leading to the configuration RE $3d^94f^{m+1}$–Au $5d^9$ [111]. The energy of this transition is approximately equal to the energy of the shake-down peak increased by the average energy of the Au 5d band, i.e. by a few eV. The resulting peak is, therefore, partially mixed with the main peak causing its distorted observed shape. Shake-down satellites were also observed in cerium 3d photoemission of CeP, CeAs and CeSb (Fig. 4.12) [77] but their observation was not discussed.

In the rare-earth insulator compounds, fluorides and other ionic compounds, another type of satellite is present. The 4f orbitals of the light rare-earths are less bound than the ligand orbitals both in the neutral and ionized system. After the 3d ionization, the orbitals remain in the same energetic order and only the intervals between them decrease. An electron transfer from a ligand orbital or from the valence band to the rare-earth 4f levels is possible but necessitates energy. The 4f levels are located in the band gap and the corresponding peaks are towards the

Fig. 4.12 3d photoemission
of cerium pnictides.
A shake-down peak is present
towards the lower energies of
the main peak [77]

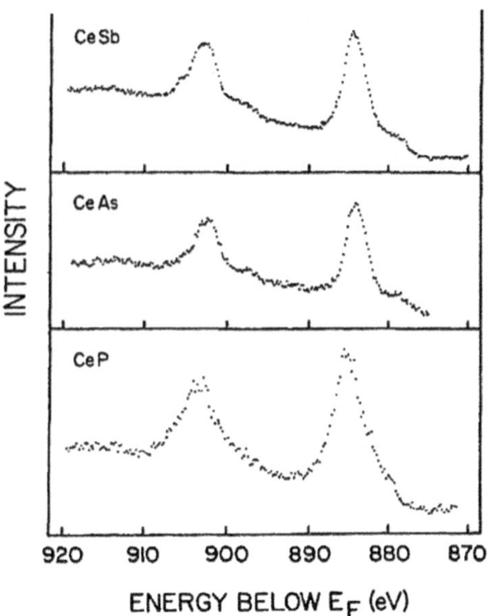

higher binding energies of the main peak; these are shake-up satellites. The binding energy of the 4f level increases with Z, making the energy interval between the satellite and the main peak decrease [71]. The energy interval between the top of the valence band and the 4f levels drops from 5 eV in LaF_3 down to practically zero for the europium compounds. In contrast, for YbF_3, the satellite is located at lower energy of the main peak. As the number of 4f electrons in the trivalent rare-earth fluorides increases, the satellites moves from higher to lower energy of the 3d main peak.

It appears that the spectral characteristics of the rare-earth 3d photoemission depend on the relative energies of 4f levels and valence band in the presence of the core hole. After the ionization of a 3d core electron, the $3d^{-1}4f^{m+1}$ excited configurations can be observed if the 4f levels mix and interact with the valence levels. Transfer of one valence electron into the 4f sub shell is then possible with no additional energy. As already mentioned, this condition is achieved in metal and metallic compounds of light rare-earths. But in their insulator compounds, there are unoccupied 4f levels within the band gap. If they remain there in the presence of the core hole, such a transfer is not possible and no shake-down satellite is to be expected. In contrast, a shake-up process becomes possible.

Absolute energies of the various peaks must be taken into account in the interpretation of the spectra. The normal peaks are predicted at higher energies in compounds than in metals but the energy difference is generally only 1–2 eV, except for the ionic compounds for which this difference can be of the order of 2–3 eV. As seen from X-ray absorption spectra, excited configurations $3d^{-1}4f^{m+1}$ have the same energy in the metal and the oxide if the rare-earth has the same valence in the two solids. In metals, shake-down satellites associated with these excited configurations are located about 5 eV below the normal peaks, associated with ionized configurations. Lastly, in compounds, peaks corresponding to simultaneous core ionization and valence excitation, or shake-up satellites, appear at energies higher than those of the normal peaks. But the multiple J-levels belonging to the $3d^9 4f^m$ final configuration make the spectra complex and must be taken into account [112, 113].

Let us consider cerium $3d_{5/2}$ photoemission in the insulator oxide Ce_2O_3. Two peaks are observed at 884.0 and 887.2 eV. The peak at 884 eV is assumed to correspond to the transition seen in the metal at 883.8 eV, namely the transition between the ground state and the configuration $3d_{5/2}^{-1}4f^1$ of the singly ionized cerium. No peak appears towards the lower energies in the oxide, unlike the metal for which a weak shake-down peak was observed at 878.8 eV, corresponding to the configuration $3d_{5/2}^{-1}4f^2V^{-1}$. From 3d photoabsorption, the $3d_{5/2}^{-1}4f^2$ excited configuration is observed at the same energy in the metal and the trivalent oxide. A similar statement can be made for the energy of the barycentre of the $3d_{5/2}^{-1}4f^2$ configuration. When the change in the band structure between metal and trivalent oxide is taken into account, the energy of the $3d_{5/2}^{-1}4f^2V^{-1}$ configuration is predicted to decrease slightly in the oxide with respect to that in the metal while no peak is observed below 884 eV in the oxide. No shake-down satellite is, then, present in Ce_2O_3. No energy transfer exists between the 4f and valence levels in the presence of the core hole in the insulator oxide. This interpretation is different from the one initially presented [110] where the photoemission peaks were compared by adjusting the peak of lower energy observed in Ce_2O_3 at 884 eV with the $3d^9 4f^2V^{-1}$ "well-screened peak" observed at 878.9 eV in the metal. All the same, the 3d main peaks observed for lanthanum and La_2O_3 at very different experimental energies were adjusted. From this adjustment the main peak of La_2O_3 was attributed to the $4f^1$ configuration, in disagreement with the usual interpretation [114].

Double peaks analogous to those described for Ce_2O_3 were observed for Pr_2O_3 and Nd_2O_3 and the stronger peak has the higher energy for the three oxides. Double peaks were also observed for lanthanum in La_2O_3 but in this case the peak of lowest energy is stronger. This confirms that it is the main peak and not a "well-screened peak" that is known to have a very low intensity. The peak located toward the higher energies in La_2O_3 was identified as a shake-up satellite, often named "charge transfer satellite". It corresponds to an electron transfer from the ligand to a rare-earth unoccupied 4f orbitals that takes place simultaneously with the perturbation caused by the creation of the 3d hole. The same interpretation is valid for the peak at 887.2 eV in Ce_2O_3, which is stronger than the peak at 884 eV. The increase of the excitation probability from cerium can be explained by the decrease of the energy interval between the top of the oxide valence band and the unoccupied 4f orbitals with the increase of the atomic number, already discussed in a preceding paragraph.

Some compounds have a complex crystalline structure. For example, two different La–O distances, 2.42 and 2.69 Å, i.e. two different atomic environments, are present in La_2O_3, making possible the existence of two $3d_{5/2}$ peaks of slightly different energies. The large variety of energies and shapes observed for the photoemission peaks of the light rare-earth compounds could then be attributed partly to differences of the latter's structural arrangements.

Other cerium trivalent compounds, CeF_3 and $CeCl_3$, were analyzed [115]. Double peaks were also present in each $3d_{5/2}$ or $3d_{3/2}$ core level spectrum at energies clearly higher than those of the excited configurations $3d_{5/2}^{-1}4f^2$ and $3d_{3/2}^{-1}4f^2$. A dramatic change in the intensity ratio of the double peaks was observed between these two compounds. For CeF_3 the more intense peak is the one of lower energy while for $CeCl_3$ and also $CeBr_3$, the reverse is observed. An analogous situation was observed in the spectra of LaF_3, $LaCl_3$ and $LaBr_3$ [116]. The lower energy peak observed for $CeCl_3$ around 884 eV corresponds to the one that appears at the same energy in the metal and in Ce_2O_3. As already underlined, this peak cannot be attributed to a shake-down satellite because such satellite is predicted at much lower energy. The second peak, with higher energy, was attributed to a shake-up transition. The intensity of this second peak is higher than that of the main peak. This anomaly appears in numerous compounds of light rare-earths and must be discussed.

In an insulating compound, a shake-up satellite is attributed to the excitation from the valence shell to levels in the band gap. The stronger the bond the more intense is the satellite. For lanthanum and cerium compounds, the intensity of this high energy second peak is observed to increase with the covalent character of the bonding by following an order comparable to that of the *nephelauxetic* series [117]. Let us recall that the nephelauxetic effect refers to the decrease of the inter-electronic repulsion of metallic ions present in a compound. As a result of this effect, the repulsion between two electrons present in the doubly occupied orbital of a metal ion in a compound is reduced with respect to that in the respective gaseous metal ion. The electron cloud expands, increasing the size of the orbital in the compound. This electron cloud expansion is a result of a decrease of the effective positive charge on the metal or an overlapping with ligand orbitals. Along the nephelauxetic series, the covalence of the bonds increases and also the expansion of the metal valence orbitals and the overlapping between metal and ligand valence orbitals. The valence electrons are distributed between the metal and the ligand. The inter-atomic transfers are facilitated and the intensity of the shake-up satellites increases.

In analogy to covalent rare-earth compounds, it appears that the expansion of the valence orbitals favours the excitation of the valence electrons into initially unoc-cupied f orbitals. The important data are thus the respective energies of the oxygen 2p and rare-earth 5d–6s valence levels and of the 4f unoccupied levels. Moreover, the energy interval between main and secondary peaks decreases with the increase of the covalence, i.e. with the increase of the bonding. In the presence of a core hole, the energy of the 4f levels involved is lowered owing to the attractive potential of the hole. The screening of the core-hole potential may occur by the occupation of additional f levels through electron transfer from the ligand levels depending on the local electronic structure. The diversity of the observed results is due to the diversity of the valence electron distributions depending on the considered solid. Theoretical studies of the 3d XPS of trivalent rare-earth compounds, in particular of the oxides Re_2O_3 (RE = La–Yb) were made [118]. They used the Anderson impurity Hamiltonian, which describes the excitation of rather localized electrons hybridized with delocalized valence electrons [119, 120]. The energy of the configurations is averaged over the multiplet terms, which are found to play a lesser role in the 3d photoemission of the trivalent compounds. This work concludes that two peaks are present. In the case of the light rare-earths, these peaks are interpreted by a mixing of the $4f^m$ and $4f^{m+1}V^{-1}$ configurations in the presence of a 3d hole. The relative intensity of the lower energy peak decreases with m increasing because its final state has a larger weight of the $4f^{m+1}V^{-1}$ configuration and this configuration is not connected with the ground state. This model included in the calculations parameters that were chosen arbitrarily to fit the theoretical and experimental spectra.

The Xα-molecular orbital method was also used to explain the double 3d photoemission peaks observed for La_2O_3, Ce_2O_3, Pr_2O_3, Nd_2O_3 and the single peaks observed for Eu_2O_3, Gd_2O_3, Dy_2O_3, [121]. First-principles calculations were performed with no arbitrary fitting parameters included. But the localized character of the 4f electrons is not taken into account. The double peaks characteristic of the light rare-earths were shown to correspond to the bonding and antibonding levels associated with the final configurations, $4f^{m+1}V^{-1}$ or $4f^m$, in the presence or absence of an additional rare-earth 4f electron. These peaks correspond then, respectively, to

$$a\left|3d^{-1}4f^m\right\rangle + b\left|3d^{-1}4f^{m+1}V^{-1}\right\rangle$$
$$b\left|3d^{-1}4f^m\right\rangle - a\left|3d^{-1}4f^{m+1}V^{-1}\right\rangle$$

The presence of the two peaks observed for the light rare-earths was ascribed to electron transfers from the ligand to the rare-earth 4f orbitals. The calculated intensity ratios of these peaks were found to follow the experimental ratios for the studied oxides.

The rare-earth 4d photoemission has received less attention than the 3d one because the $4d_{3/2}$ and $4d_{5/2}$ transitions are not resolved. However, any apparent anomalies observed are well explained in an atomic model that takes into account the interactions between discrete configurations and the continuum. The 4d spectrum of europium metal was one of the first studied in detail [122]. The final configuration was described as $4f^7(^8S)4d^9$, i.e. by considering only the ground level of $4f^7$. In this simple model, only two multiplets, ^9D and ^7D, are expected, in qualitative agreement with the experimental spectrum, which presents two main structured maxima. From comparison between metal and oxide spectra of the heavy rare-earths, terbium to lutetium, observed under ultra-high-vacuum, it was shown that the 4d peaks occur at different energies in metal and oxide [123]. The observations of the 4d photoemission for all RE_2O_3 oxides and their comparison to calculations in intermediate coupling scheme enabled the following of the evolution of the spectra along the series [124]. The electrostatic interaction between the core hole and the open 4f sub shell is large. Its effect is to introduce numerous unresolved components and to push a part of the levels far up above the ionization limit. This interaction is stronger in 4d than 3d photoemission because the 4d–4f repulsion terms are larger. For the heavy rare-earths, electrostatic interactions dominate completely the spectrum, which spans about 30 eV. However, secondary structures are also present. There are shake-up satellites due to the excitation of a ligand valence electron to the 4f unoccupied orbitals. They are less prominent in the 4d than in the 3d spectra because the change of the system is less sudden in the presence of 4d than 3d ionization and also owing to the large contribution of the electrostatic terms. They are clearly resolved for the light rare-earths up to terbium.

Strong enhancement of the photoemission involving the 5s, 5p and 4f peaks was observed by varying the energy of the incident photons through the region of the giant resonance $4d^{10}4f^m$–$4d^94f^{m+1}$ in numerous materials [125].

Lastly, it must be underlined that broadening of the valence peaks can be observed in the compounds. It is due to the decreasing of the lifetime of the core hole. This decreasing is due to the numerous valence electrons of the ligands contributing to inter-atomic Auger processes. This effect can be present also in EXES and photoabsorption.

4.2.3.2   Auger Emission

Different processes considered as equivalent photoemission processes have been observed in the 3d range of the rare-earths. When the incident photon energy is varied across the region of the core excitation threshold, two interfering transitions can lead to the same final state. For example, the $3d_{5/2}$ photoemission from the ground state to the core-ionized configuration $3d_{5/2}^{-1}4f^m$ and the de-excitation of the core-excited configuration, $3d_{3/2}^{-1}4f^{m+1}$, to the same ionized configuration, $3d_{5/2}^{-1}4f^m$. This last process is a Coster–Kronig transition. Interferences between these two processes can strongly increase the core excitation cross section when the incident photon energy is near the energy threshold. This phenomenon forms the basis of so-called *resonant photoemission*. Prerequisite for interference is the coherence of the processes involved. These processes will be considered along with the resonant Auger processes in a following paragraph.

4.3   Mixed-Valence Rare-Earths

Some rare-earths have a series of compounds, whose valence is different from that of the metal in its standard state. The ground state of these compounds may result from the hybridization of one 4f electron with the valence electrons. This is the case in tetravalent cerium compounds. In other compounds, rare-earths can have two different valences. These are "mixed-valence" compounds. The two $4f^m$ and $4f^{m-1}$ configurations are "frozen" on well defined sites and inhomogeneity is present in the material. The term mixed-valence then characterizes the fractional number of ions of each configuration [126]. Each rare-earth can have one or the other of the two $4f^m$ and $4f^{m-1}$ configurations and fluctuations between them can exist. This model was initially used to describe compounds of the heavier rare-earths. In order for the valence fluctuations to exist, the presence of one 4f level in the immediate vicinity of the Fermi level is necessary [127].

The time scale of the valence fluctuations, i.e. charge fluctuations, is predicted as 10^{-12}–10^{-13} s. It is of the same order, or longer than the time of the magnetic fluctuations and low enough to allow strong coupling between charge fluctuations and phonons. These fluctuations can cause a disorder of the nearest-neighbor positions. Properties associated with $4f^m$ and $4f^{m-1}$ configurations can depend strongly on the crystal structures. These characteristics render the compounds with valence fluctuations interesting physical properties. Rapid process techniques like X-UV and electron spectroscopies are necessary to study the fluctuations in mixed-valence materials.

Particular attention is given here to cerium and to its tetravalent compounds. The valence change of the cerium is due to a partial delocalization of the 4f electron and its admixture with the other valence electrons. Interaction between the cerium 4f electron and the valence electrons is larger for compounds in which the 4f electron does not exhibit normal magnetic properties. Indeed, magnetic momentum and localization decrease together. The trivalent $4f^2$ praseodymium can become also tetravalent in some cases. Among the other rare-earths having two different valences, samarium occupies a particular place because the change from the $4f^5$ configuration of trivalent samarium to divalent $4f^6$ samarium corresponds to the filling of the $4f_{5/2}$ sub shell and appears as associated with the relative strength of the spin–orbit interaction. Thus, the surface atoms of metallic samarium, initially considered of mixed-valence, were found to have a single valence and to be divalent though the bulk is trivalent [68, 71]. It is surprising that thulium, the corresponding heavy element, changes from $4f^{12}$ trivalent to $4f^{13}$ divalent in some compounds. Europium and ytterbium are known to have two valences. They are divalent in the metal and in some compounds because of the high stability of the $4f^7$ and $4f^{14}$ configurations but trivalent, like other rare-earths, in their stable oxide and numerous other compounds. This valence change does not induce modification of the localized character of the 4f electrons in these two elements. The same holds for trivalent terbium, which can be tetravalent because this valence change is favored by the great stability of the $4f^7$ configuration.

In a mixed-valence compound, the weight of the two final configurations can be different according to the characteristics of the employed experimental technique. This is because for each experimental technique secondary effects arise. These effects are different according to whether the measurements are obtained from occupied or unoccupied distributions or whether the analysis concerns excited, ionized or neutral configurations. Examples of experimental results obtained from the various experimental methods previously described are reported here.

4.3.1   Density of Unoccupied States

4.3.1.1   X-ray Absorption

The L_{III} absorption spectra of the trivalent rare-earths present a single large maximum characteristic of transitions into the 5d6s unoccupied levels. For various rare-earth compounds, these absorptions have complex shapes and two peaks of different intensities are present. These additional features reveal a change in the distribution of the valence electrons. As an example, let us compare the L_{III} absorption spectra of γ- and α-cerium metal. A large peak is located at the same energy in the two phases. An additional peak of low intensity is observed for α-cerium 9 eV above the main peak [128]. These observations are in agreement with the presence of 4f localized electrons in the two phases and with a change of part of them from localized in the γ-phase to delocalized in the α-phase without change in the 4f total occupancy. The spectrum of the α-phase appears as the superposition of two independent spectra, distinct in energy, each corresponding to a different valence configuration, thus to a different binding energy of the core level.

From a general point of view, based on the observation of two or more peaks in the L_{III} X-ray absorption, one can deduce the presence of two or more valence configurations in the analyzed rare-earth material. Indeed, the energy necessary for the transfer of an electron from the 2p sub shell to the unoccupied 5d6s levels depends on the number of the localized electrons. The valence may be deduced by comparing the relative intensities of the peaks. The method is elaborate because of limitations due to the energy resolution. These limitations are due to the intrinsic width of the 2p level, equal to several eV for the rare-earths, and to the large instrumental width in this energy range. The shape of the L_{III} absorption spectrum has to be taken into account in order to resolve the experimental curve. Moreover, the valence is generally determined by considering that the photoabsorption cross sections are the same for the various conduction levels, which is evidently not verified. In spite of these limitations, the energy shift between the two configurations is sufficiently large to be clearly observed and a large number of experiments were made using this method because it presents important advantages. Possible use of the synchrotron radiation makes the method very sensitive. It can be used for various samples, independently of temperature, pressure, concentration and surface state. This is a fast technique ($t < 10^{-16}$ s) with respect to charge fluctuations, which permits the simultaneous observation of the two configurations present in the compounds at intermediate valence. Consequently, it is very widely used for the study of the intermetallic and mixed-valence compounds.

From experiments concerning intermetallic compounds of cerium [129–137] it is possible to distinguish those for which the L_{III} absorption shows a single peak and cerium is trivalent, such as $CeCu_2Si_2$, $CeAl_2$, $CeIn_3$, $CeSn_3$, $CeCu_5$, Ce–Au, from those with two peaks in the L_{III} absorption, such as $CeFe_2$, $CeCo_2$, $CeNi_2$, $CeRh_2$, $CeIr_2$, $CeRh_3$, $CePd_3$, $CeCo_5$, $CeNi_5$, Ce_5Rh_4, more generally Ce–Co, Ce–Ni, Ce–Mo, Ce–Ru, Ce–Rh, ... alloys, where the cerium valence is superior to three.

Increase of the cerium valence is observed in the presence of nd (n = 3, 4, 5) transition elements. This electron transfer from the 4f orbital was found to take place until the cerium valence is around 3.3. This valence variation was correlated with the number of the valence electrons, characteristic of chemical environment, with which the cerium 4f electrons can hybridize. Consequently, the cerium valence increases in the presence of elements having a large density of states at the Fermi level, as, for example the transition elements of the end of the series, where the contribution of their d electrons to the cohesive energy is strong. This implies that the cerium valence is bigger in the presence of transition elements of the end of the series. Thus, the cerium valence deduced from lattice constant and neutron scattering is 3.4 for $CePd_3$ while it is only 3.07 for $CeAl_2$.

Numerous studies have been made for a large variety of cerium intermetallic compounds with three or four different atoms. Their aim was double, to study the response of the cerium 4f orbitals to a broad range of chemical environments and to determine the electronic configuration of compounds of particular physical properties. These researches necessitated the precise analysis of the near-edge X-ray absorption structures and also the elaboration of new stable materials having very different crystal structures. As an example, let us mention studies made for the $CeIr_{1-x}Rh_xIn_5$ series. From cerium L_{III} absorption, it has been shown that cerium was close to trivalent in these compounds and no measurable change of valence was present in the 20–300 K range [138]. This series exhibits interesting properties such as superconductivity and antiferromagnetism. The above results imply that the characteristic temperature is very small in these materials. Moreover, from observation of the iridium L_{III} edge as a function of the temperature, no disorder was observed around iridium and disorder-based models could not be applied to describe the low temperature properties. They were ascribed to strong spin fluctuations.

In L_{III} absorption of CeO_2, two peaks of practically equal intensities were observed. They are separated by only 6.7 eV. The spectrum of this insulator compound is clearly different from that observed in metal and intermetallics and this suggests the need to distinguish between the chemical compounds and metallic mixed-valence systems [132, 139, 140]. Many-body processes were suggested to interpret the second peak. In this model, the core hole induces a localization of the 4f orbitals, thus a change of the configuration in the final state. Other interpretation suggested the presence of 4f electrons of different degrees of localization which could be treated in terms of a one-electron model. The highest energy peak was attributed to the localized 4f electron and a valence around 3.33 for the ground state of CeO_2 was deduced from the relative intensities of the two peaks. [141]. However, the peak corresponding to the $4f^1$ configuration is predicted towards the lowest energy. In fact, in CeO_2, the cerium atoms are tetravalent and no atomic-like f level is occupied (cf. Chap. 1). The four outer electrons of cerium participate in the chemical bond and are involved in a covalent O2p–Ce5d6s and O2p–Ce4f admixture [142, 143]. Therefore, the interpretation of the two peaks observed in the L_{III} absorption of CeO_2 is the following: the highest energy peak is associated with transitions from the 2p shell of the ground state of the $4f^0$ configuration towards unoccupied levels of d–s character. The lowest energy peak corresponds to transitions towards the excited configuration created in the presence of the core hole by stabilization of a 4f electron in the potential of the hole. The existence of extended 4f levels was also pointed out for other insulating tetravalent cerium compounds such as CeF_4. The features, characteristic of the extended levels, were found to have a smaller spectral contribution for the fluoride than for the oxide because the ionicity of CeF_4 is higher, then its covalent character weaker. Close similarity was observed between the L_{III} absorptions of TbO_2 and TbF_4 and those of CeO_2 and CeF_4, respectively [144]. That gives clear evidence of an analogous 4f-ligand mixing in the tetravalent compounds of the heavy and light rare-earths.

Consequently, when a single peak is observed in the L_{III} absorption of a cerium compound, cerium is expected to be trivalent of $4f^1$ configuration. That is the case, for example, for CeTe [145]. However, controversy exist each time that a partial delocalization of the 4f electron in 5d levels appears possible. The number of 4f electrons may, then, become less than one. For example, in CeN, cerium is expected to be trivalent, in agreement with the presence of a single peak in the L_{III} absorption. However, thermodynamical properties of CeN as well as 3d and valence photoemission are characteristic of an intermediate valence compound. Indeed, the valence band spectrum shows a narrow 4f peak pinned at E_F, superimposed on a broader 5d band. From the simultaneous presence of occupied 4f and 5d levels at E_F, a partial delocalization of the 4f electrons into d levels was predicted leading to an intermediate valence character [146, 147].

Other rare-earths, samarium, thulium, trivalent in the metal, can be present in mixed-valence compounds. Valence fluctuations take values between 3 and 2 in these materials. Thus, the simultaneous presence of Sm^{3+} $4f^5$ and Sm^{2+} $4f^6$ was observed from experiments at the L_{III} absorption edge. A shift of about 7 eV towards lower energy is observed for the 2 valence with respect to the 3 valence, in agreement with the well known chemical shifts. As an example, let us cite SmB_6, which has been widely studied because these characteristics of mixed-valence compound were initially controversial. A samarium valence change from 2.60 at 300 K to 2.53 at 4.2 K was shown [148]. This change was explained by a variation with the temperature of the cubic lattice parameter. In the energy range between 10 and 25 eV above the L_{III} absorption line, labelled XANES, double structures were also observed in this compound. The energy separation, about 6.5 eV, was the same as that observed for the absorption line and confirmed the proposed interpretation (Fig. 4.13) [149]. Increase of the proportion of divalent ions with decreasing temperature was confirmed from the analysis of the L_{III} absorption between 300 and 10 K (Fig. 4.14) [150].

Two peaks were also observed at the L_{III} threshold of the samarium in the hexaboride solid solutions $Sm_{1-x}M_xB_6$ with M = Yb^{2+}, Sr^{2+}, La^{3+}, Y^{3+}, Th^{4+}, [148]. These peaks are characteristic of the simultaneous presence of Sm^{3+} $4f^5$ and Sm^{2+} $4f^6$. It was found that the samarium valence increases with an increase of the concentration of the Yb^{2+} and Sr^{2+} divalent ions whereas it decreases with an increase of the concentration of La^{3+}, Y^{3+}, Th^{4+}. These results agree with those

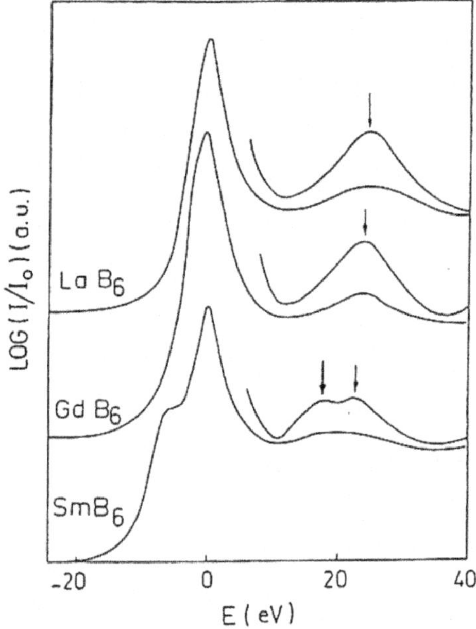

Fig. 4.13 L_{III} absorption of LaB_6, GdB_6 and SmB_6. The onset is magnified 2.5 times [149]

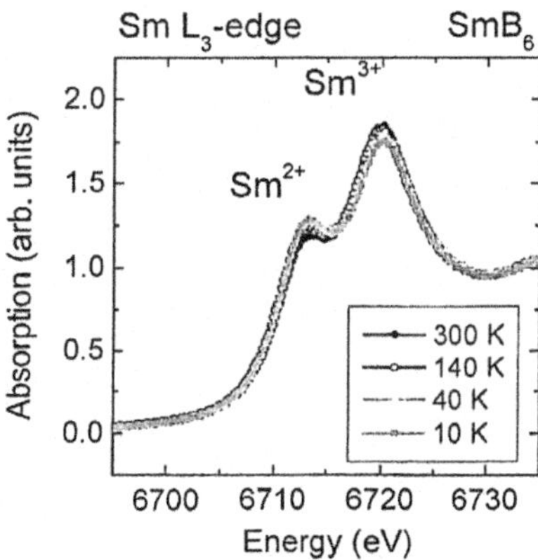

Fig. 4.14 Temperature dependence of L_{III} absorption of SmB_6 [150]

deduced from other physical measurements. Existence of a relation between the number of the 4f electrons and the nearest-neighbor distance was considered in mixed-valence materials. However, it was shown from the analysis of the extended L_{III} absorption fine structures that $Sm_{0.75}Y_{0.25}S$ is a mixed-valence compound and the Sm-S distance does not vary with the valence [151]. Each samarium ion adopts an average position and no observable dynamic electron-lattice correlation is found.

From L_{III} absorption of thulium, its valence is 3 in TmS and 2 in stoichiometric TmTe (Fig. 4.15) [152]. The absorption peak is shifted by nearly 8 eV towards lower energy in TmTe and its shape is different. Energy shift and shape difference are

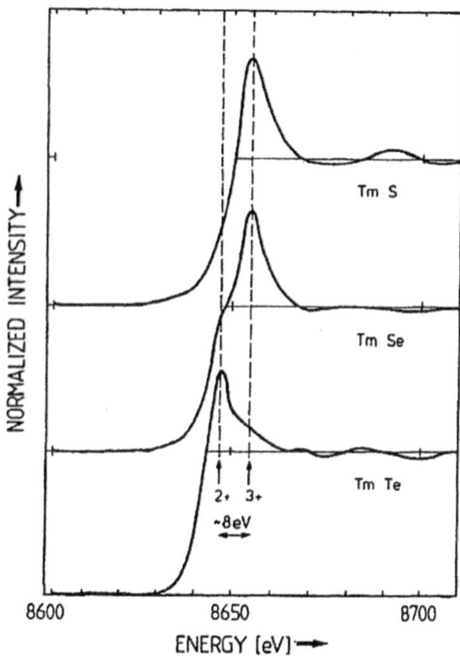

Fig. 4.15 L_{III} absorption of thulium in TmS (valence 3), TmSe (valence 2.58 ± 0.07) and TmTe valence 2) [152]

similar to those observed between Sm^{3+} and Sm^{2+} [153] and between Yb^{3+} and Yb^{2+}. TmSe is a mixed-valence compound, whose valence value has been debated because it is different according to whether its determination is from the lattice constant, from the magnetic neutron cross section or by X-ray spectroscopy. From L_{III} absorption, the main peak of TmSe was observed at the energy of the Tm^{3+} peak and a shoulder was present near the energy of the Tm^{2+} peak. Spectrum analysis gave a valence near 2.58, in agreement with the magnetically determined value. Values of 2.67, 2.65 and 2.70 were obtained for TmS, TmSe and TmTe, in closer agreement with the valence derived from the lattice parameter measurements [154].

Ytterbium L_{III} absorption of $YbInCu_4$ was measured above and below the valence transition temperature, equal to 42 K. Spectra in the partial fluorescence yield (PFY) and total fluorescence yield (TFY) are presented in Fig. 4.16 [155]. The resolution is considerably better in PFY because, in this case, the measured intensity is that of the Yb $L\alpha$ $2p^5 3d^{10}$–$2p^6 3d^9$ line. From these spectra, it is possible to follow the temperature dependence of the valence. However, better precision was obtained from RIXS (cf. Sect. 4.3.2.3).

When combined with EXAFS, which analyzes the crystal arrangement, the analysis of the L_{III} absorptions enables us, therefore, to determine the ground state of the mixed-valence compounds. This analysis can be made as a function of the temperature and the composition. Results deduced from the L_{III} absorption spectra

Fig. 4.16 L_{III} absorption of ytterbium in YbInCu4 observed above and below 42 K: *bottom*, in the total fluorescence yield (TFY); *top*, in partial ($L\alpha_1$) fluorescence yield (PFY) [155]

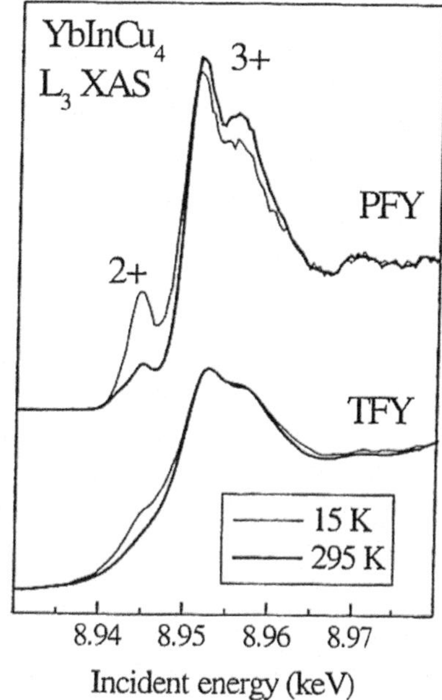

have been generally discussed in parallel with the results obtained from other techniques, XPS, BIS.

Now, let us consider results obtained from 3d photoabsorption. It was shown qualitatively that cerium metal absorption observed at 77 K was characteristic of a γ and α-phase mixing in proportions depending on the thermal history of the sample [156]. For CeO_2, which is often considered as the emblematic compound of the $4f^0$ configuration at the ground state, the 3d absorption is composed of two parts. The first part, also indicated by the predicted spectrum for the Ce^{4+} $3d^{10}4f^0$–$3d^94f^1$ transition (Fig. 4.17) [157] is analogous to the 3d absorption of lanthanum. It consists of two intense peaks at 883.4 and 901.0 eV, corresponding to the transitions to $3d^94f^1$ $^3D_1$ and $^1P_1$, and by a weak feature at 878.6 eV, due to the transition to $3d^94f^1$ $^3P_1$. The second part is characterized by two peaks at 888.6 and 906.3 eV. These peaks are located towards the higher energies of the two intense peaks $^3D_1$ and $^1P_1$ and are characteristic of transitions towards empty 4f levels hybridized with continuum levels. These transitions have probabilities lower than those between localized levels. The $3d_{3/2}$ peak at 906.3 eV is the stronger of the two secondary peaks. The empty conduction levels have, then, a predominant $f_{5/2}$ character, as expected [86, 157, 158]. No spectral characteristics of the presence of the $4f^1$ configuration at the ground state were observed in the 3d photoabsorption, contrary to various suggestions [159–161]. Consequently, no mixing of configurations exists in CeO_2. These results are in agreement with those obtained from L_{III} absorption. The same conclusions were obtained for PrO_2 and TbO_2 by comparing the 3d absorptions with those of the corresponding $Z - 1$ elements in trivalent compounds. In the two

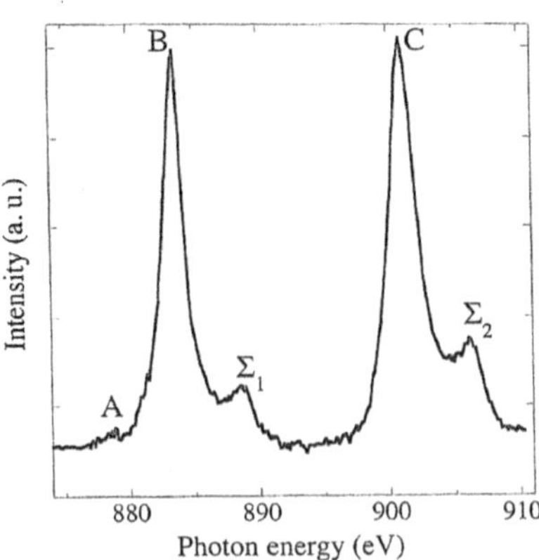

Fig. 4.17 Cerium 3d absorption in CeO_2: the peaks A, B and C correspond to the transitions to $3d^9 4f^1\ ^3P_1$, $^3D_1$ and $^1P_1$. The peaks Σ_1 and Σ_2 correspond to transitions towards empty 4f levels hybridized with continuum levels [157]

oxides, features are observed several eV above each main peak and attributed to transitions towards empty 4f levels mixed with ligand levels [86]. These observations show the analogy between the tetravalent insulator compounds of the light and the heavy rare-earths and their difference from mixed-valence compounds. The same observations were made for the CeF_4 and TbF_4 fluorides [144]. The transitions to non-localized 4f levels are weaker in the fluorides whose ionicity is high.

The advantage of the 3d absorption method is clear. From the 3d absorption, the number of localized f electrons present in a solid at the ground state can be deduced directly by comparing the experimental spectrum with the calculated atomic-like spectrum while in the L_{III} absorption, the presence or absence of localized electrons is deduced indirectly from the observation of a secondary effect.

From 3d absorption, thulium was found trivalent in $TmAl_2$, TmS, Tm_2O_3 while Tm^{3+} and Tm^{2+} were observed in TmTe and TmSe with mean valence values in agreement with those deduced from lattice constants [162].

In the intermetallic compounds, the 3d absorption spectra of cerium include, jointly, the peaks corresponding to the transition $3d^{10}4f^1\ ^2F_{5/2}$–$3d^9 4f^2$ in Ce^{3+} ions, the peaks corresponding to the transition $3d^{10}4f^0$–$3d^9 4f^1$ in Ce^{4+} ion, the features corresponding to transitions to 4f levels mixed in the conduction band and characteristic of the presence of Ce^{4+} ions. It was confirmed that the two Ce^{3+} and Ce^{4+} configurations were present in the $CeNi_2$ and $CePd_3$ mixed-valence compounds and, consequently, hybridization exists between the 4f and conduction levels [157]. The contribution of the configuration $3d^{10}4f^1$ was determined experimentally in each case. The average valence of cerium is 3.22 ± 0.06 for $CeNi_2$ and 3.23 ± 0.05 for $CePd_3$, in agreement with the values obtained from the L_{III} absorption [163]. The 3d absorption recorded by photo yield was used for the study of numerous other compounds of the γ- or α-cerium [164, 165].

The polarization dependence of the rare-earth 3d absorption spectra has recently been used to obtain the 4f charge distributions in the ground state [166]. The method was applied to $CePd_2Si_2$. The 3d absorption spectrum of Ce^{3+} $4f^1$ was calculated for different orientations of the polarization vector of the incident radiation with respect to the crystal. With the help of a set of calculated reference spectra, it was possible to determine the crystal-field ground state symmetry by comparison with the experiment. This method that complements neutron scattering is more sensitive and has the advantage to be specific of the element and its orbital symmetry.

Broadening of 4d–4f multiplet structures was also observed by photo yield at the γ-α phase transition of cerium metal [167]. It was interpreted as due to an increase of 4f hybridization, therefore, an f–d mixing, in agreement with the absence of local magnetic moment in the α-phase. No new structures appeared. This indicated that the number of the 4f electrons does not change.

Close correspondence was observed between the 4d photoabsorption spectra of cerium metal and CeF_3 in the region below the giant resonance. Nevertheless, fine structures were considerably sharper for the fluoride than for the metal because the bond is ionic in the compound and, consequently, the 4f electron is more localized. In contrast, only two peaks with no fine structure were observed for CeO_2. This spectrum is characteristic of the cerium $4f^0$ configuration in the ground state, as expected for a compound of tetravalent cerium. However, some very weak structures can be detected and this could suggest the presence of a very small number of localized 4f electrons [168]. From $4d_{5/2}$ photoabsorption observed by the total electron yield method, the spectrum of CeF_4 was found analogous to that of LaF_3, i.e. characteristic of the final configuration $4d^9 4f^1$, while it differed completely from that of CeF_3. [169]. This result confirms that cerium is tetravalent in the ground state of the fluoride. The same observation was made in the region of the $4d_{3/2}$ giant resonance.

4.3.1.2   X-ray Magnetic Dichroism

XAS provides interesting possibilities for studying rare-earth magnetic systems [170]. In the presence of a magnetic field, the $(2J + 1)$-fold degenerate ground state $4f^m(J)$ splits into its $2M_J + 1$ sub-levels and only the level $-J$ is populated at $T = 0$ K. For an incident radiation polarized in the magnetic field direction, in the 3d range, the dipole transitions take place from the M_J level of the ground state to the $M_{J'}$ levels of the final configuration $3d^9 4f^{m+1}(J, J \pm 1)$. The magnetic splitting of the final levels is generally small with respect to the experimental resolution and cannot be observed. In contrast, intensity dependence on the polarization and the temperature is observable. In the case where the transitions to the final levels, characterized by $J + 1, J, J - 1$, can be separated the spectral analysis supplies a direct measurement of the weighted average of M^2, i.e. of the square of the local magnetic moment of the rare-earth atom, as a function of the temperature.[1] It was shown in the particular case of dysprosium that the $J + 1$, J, $J - 1$ components of the transition $4f^9$–$3d_{5/2}^{-1}4f^{10}$ were clearly separated in a non-magnetic system. The dysprosium spectrum in magnetically ordered materials was predicted to depend on the polarization and the temperature. This effect occurs in most of the rare-earths. Above the magnetic ordering temperature, the polarization dependence disappears. The polarization of the rare-earth 3d absorption spectra depends on the magnetic structure of the ground state $4f^m(J)$, and therefore on the unequal population of the M_J sub-levels in the presence of a magnetic field. Then, XAS can be used to determine the local magnetic moment of the rare-earth atoms in a magnetically ordered material as well as the orientation of the local moment with respect to the total magnetization direction and its dependence on the field and the temperature. Experiments were made in the cerium L and M absorption ranges for its inter-metallic compounds [171, 172].

Magnetic dichroism can be measured by using the partial fluorescence yield given by RIXS. This method was used in the cerium L_{III} range for the $CeFe_2$ ferromagnetic intermetallic compound [173, 174]. A magnetic field of about 0.3 T was applied to the polycrystalline sample. The field direction was aligned parallel or antiparallel to the incident circularly polarized beam while the angle between sample surface and radiation was chosen in order to make negligible the magnetic anisotropy. The sample was cooled to about 20 K. Double structures were observed but owing to the low intensity of the experimental signals no numerical data could be obtained. These observations have revealed the presence of different final con-figurations, in agreement with that which is calculated for an α-cerium type com-pound such as $CeFe_2$.

4.3.1.3   BIS, IPS and EELS

The BIS spectra observed for the γ and α phases in cerium metal have clearly shown the difference between the two phases. A sharp peak is present above the Fermi level in the α phase while no such peak is observed in the γ phase [175]. This peak is characteristic of the $4f^1$ final configuration. A similar feature was observed in the BIS spectra of α-type compounds, $CeIr_2$ and $CeRu_2$, in contrast with the spectrum of γ-type compound CeAl where this peak was absent [5]. In addition, a $4f^1$ peak was observed in $CeRh_3$ and $CePd_3$ and not in $CeIn_3$, $CeSn_3$ and $CePb_3$ [176]. These results are compatible with the prediction that the 4f electron hybridizes stronger with the nd distributions of the transition metals than with the sp metal distributions and that mixed-valence compounds often contain transition elements. For $CePt_5$, a $4f^1$ peak is present and is clearly narrower than the $4f^2$ multiplet characteristic of the Ce^{3+} initial configuration. The same observation was made for $CeNi_5$ while strong

---

[1]The local magnetic moment is defined as $|M| = -g\mu_B J B_J$ where g is the Landé factor, μ_B the Bohr magneton and B_J the Brillouin function.

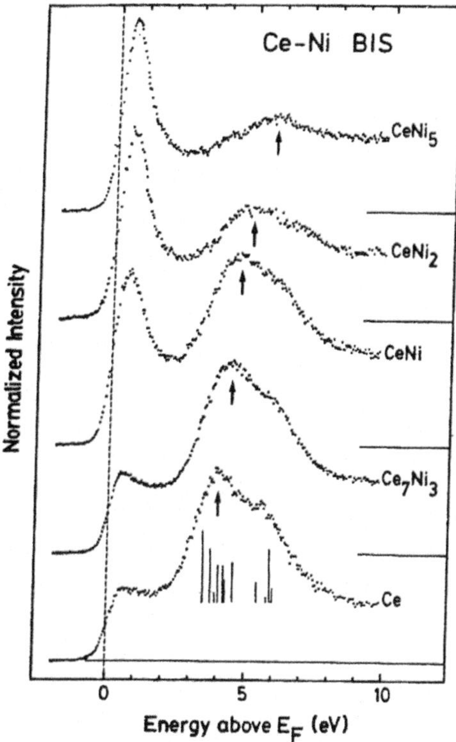

Fig. 4.18 BIS of Ce–Ni compounds [177]

increase of the $4f^2$ multiplet was observed with increase of the cerium concentration in a series of Ce–Ni alloys (Fig. 4.18) [177]. Thus, in Ce_7Ni_3, the $4f^1$ peak practically disappears while the $4f^2$ multiplet completely dominates the spectrum. On the other hand, for $PrPt_5$, $NdPt_5$ and $SmPt_5$, only $4f^m$ final state multiplets with $m = 3$, 4 and 6, are observed. These multiplets are shifted to higher energies as compared to those observed for the metals. No peak characteristic of the $(m - 1)$ configuration was observed for these three compounds.

A prominent peak was also observed by BIS in the band gap of CeO_2 and attributed to the occupation of initially empty 4f levels (Fig. 4.19) [178]. Its full width at half maximum is 1.2 eV. This peak was attributed to a $4f_{5/2,7/2}$ spin–orbit doublet broadened by the lifetime and the instrumental function. Its presence was considered as representative of the cerium $4f^0$ configuration of the ground state.

From IPS, the presence of the $4f^{12}$ and $4f^{13}$ thulium configurations has been confirmed in TmSe. Agreement was obtained between the IPS measurement and the previous determinations of the valence, which is estimated to be 2.6 ± 0.08 [179]. As underlined by the authors, the presence of the change $4f^{13}$–$4f^{14}$ shows the stability of 4f inner potential well, which may traps one 4f electron more than in divalent atomic ground state.

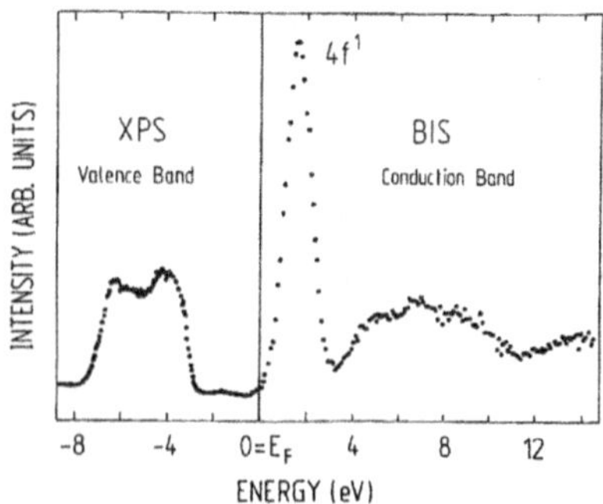

Fig. 4.19 XPS and BIS of CeO$_2$ [178]

The 3d EELS spectra are characteristic of the 4f shell occupancy in the excited ion and oxidation states of cerium and protactinium in compounds were determined by this method, which is very sensitive when the second derivate of the spectrum is used [180]. On the other hand, damage of CeO$_2$ by an electron beam of high energy (100 keV) was observed in a transmission electron microscope. It is well known that a reduction of the oxides is expected under bombardment of electrons of high energy. Indeed, the analysis of the cerium 3d absorption spectrum by EELS showed the presence of both Ce^{4+} and Ce^{3+} ions [181].

4.3.2   Density of Occupied States

4.3.2.1   Photoemission and Resonant Auger

Valence photoemission of the isostructural γ-α phase transition in cerium metal has been widely studied. Two peaks are present in each phase. In the γ phase, their binding energies were 0.2 and 2.0 eV below E_F. In the α phase, one was near E_F and the other at 2.1 eV below E_F (Fig. 4.20a) [182]. In the γ-phase, with the energy of incident photons E_{hv} varying from 40 to 60 eV, the peak at -2 eV increased more rapidly than the peak at -0.2 eV while both peaks increased approximately in the same manner in the α-phase (Fig. 4.20b). The strong intensity increase of the peaks at -2 eV with E_{hv} confirmed their attribution to the 4f photoemission. The variation of the lower energy peaks showed that the contribution of the 4f electrons is clearly higher in the α-phase than in γ-cerium. However, the total number of 4f electrons appeared to remain the same in the two phases. The peaks near E_F presented a difference of shape. In the γ phase this peak had a shoulder near E_F whereas in the α phase it was shifted closer to E_F and exhibited a sharp cutoff. The contribution of the 4f electrons was thus important at the Fermi level in the α-phase where they were predicted to hybridize with the 5d6s valence electrons. However, the phase transition did not show a sudden change. Different interpretations of the two peak structure of metal cerium were proposed [120, 183–185]. The peak at -2 eV can be considered to correspond to the excitation of a 4f electron while the peak "near E_F" could be due to

Fig. 4.20 Valence
photoemission of cerium:
a γ- and α-cerium for incident
photons of 55 eV normalized
to −1.0 eV; **b** α-cerium for
incident photons of 40–60 eV.
The intensity is normalized at
the energy of −1.0 eV [182]

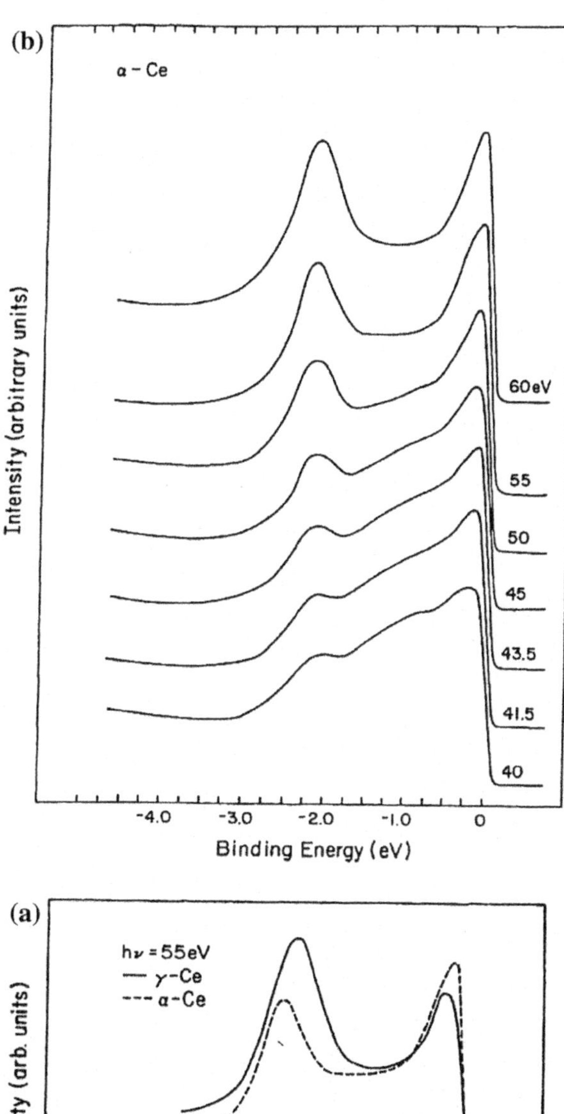

the excitation of a d electron, partially hybridized with a 4f electron in α-cerium. This model explained the difference in peak shapes of the two phases. By taking into account the strong variation of the 4f cross section with the energy $h\nu$ of the incident photons, it has been possible to explain the difference of emissions observed at $h\nu$ equal 21.2 and 40.8 eV. Then, two very weak peaks of equal intensity appeared at E_F and at -0.28 eV for γ-cerium with increasing $h\nu$. They were attributed to a very small contribution of the $4f_{5/2}$ and $4f_{7/2}$ levels, respectively. For α-cerium, the shape of the spectrum changed. The intensity of the $4f_{7/2}$ peak at -0.28 eV strongly decreased and this indicated a loss of the atomic character of the 4f levels at the phase change [186, 187].

From the 3d photoemission of the γ and α phases it was shown that multiplet effects due to the coupling of the 3d hole with the open 4f sub shell were observable in the two phases. But a $3d^9 4f^0$ component was present in α-cerium and absent in γ-cerium. These observations confirmed the presence of a phase-dependent change in the electronic structure of cerium metal [175]. Thus, the 4f conduction electron hybridization increased through the phase transition [188] while the number of electrons occupying the 4f levels remained integral. The 4f delocalization took place in the denser phase, i.e. in the α phase, and it was mainly responsible for the phase transition. A 4f quasi-band was formed. However, the 4f levels do not form a band in the usual sense but retain a partially localized character. The term *strongly correlated* electrons is very often used to designate the 4f electrons. The 3d and 4f photoemissions and also BIS of γ and α cerium were calculated from a generalized Anderson Hamiltonian by using local spin-density approximation in the density functional formalism (Fig. 4.21) [189]. Agreement was found with the experimental results when the valence change at the surface was taken into account.

In fact, a change in the 4f electronic structure of α-cerium towards that of γ-cerium was observed at the metal surface. The distributions of the 4f levels in γ- and α-cerium were obtained experimentally from resonant 4d and 3d Auger spectra. This experiment was based on the different surface sensitivity of the two types of spectra and on the strong contribution of the 4f emissions in these processes. The contribution of the 5d6s valence band emission was eliminated by subtracting the part of the spectrum observed at energy smaller than the cerium 4d threshold. The 4f distributions of the two γ- and α-cerium phases are compared in Fig. 4.22 [190]. The surface sensitivity increases in the 4d range and a substantial surface dependent change of the 4f electronic distribution of α-cerium metal towards the γ-like distribution was demonstrated. It is manifested by an intensity decrease of the hybridized 4f peak close to the Fermi level. The same type de variation was observed for $CeIr_2$ [191].

It is interesting to compare the above valence photoemission of cerium with that of praseodymium metal observed with photons of energy E_0 between 32 and 80 eV [61]. Two peaks were also observed for praseodymium. At $E_0 = 32$ eV, a broad one, located between -0.3 and -0.8 eV, is observed to be the stronger one. Its intensity decreases when E_0 increases. At $E_0 = 80$ eV, it is replaced by a large band with a

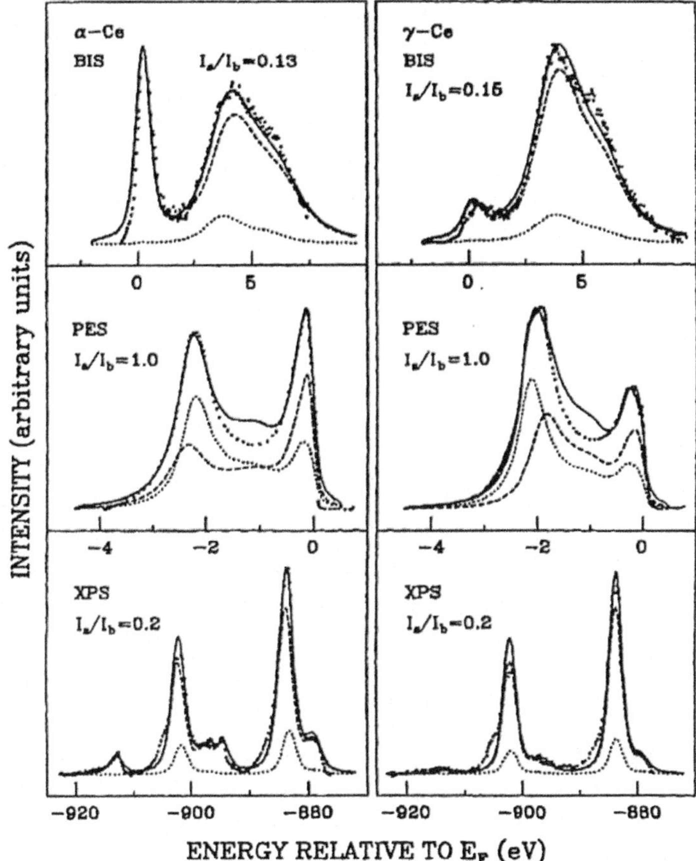

Fig. 4.21 3d photoemission, valence photoemission and 4f BIS of γ- and α-cerium: comparison of calculated (*solid lines*) and experimental (*dotted lines*) spectra. Surface (*dotted lines*) and bulk contributions (*dashed lines*) are indicated. The surface to bulk emission ratio I_S/I_B is indicated for each spectrum [189]

weak peak located at about -0.8 eV. The second peak is broad; it is located at 3.6 eV and its intensity increases considerably when E_0 increases. It is due to the 4f electrons. The band is essentially 6s5d character. The weak peak, which appeared at 0.8 eV, contains a small 4f contribution, which was attributed to a 4f–5d hybridization [114, 192]. The double-peak structure observed for the light rare-earths decreases substantially in intensity and moves away from E_F as one goes from cerium to praseodymium to neodymium. Only one peak is observed for samarium.

In the intermetallic compounds of light rare-earths of the type $(RE)Ru_2$, (RE) Pd_2, $(RE)Ir_2$ and $(RE)Rh_2$ [78, 193], the 4f distribution was characterized by a double-peaked structure with roughly constant separation between the peaks. These distributions were obtained by resonant Auger from the $4d^9 4f^{m+1}$ excited configuration. The peaks move together toward larger energy in going from cerium to

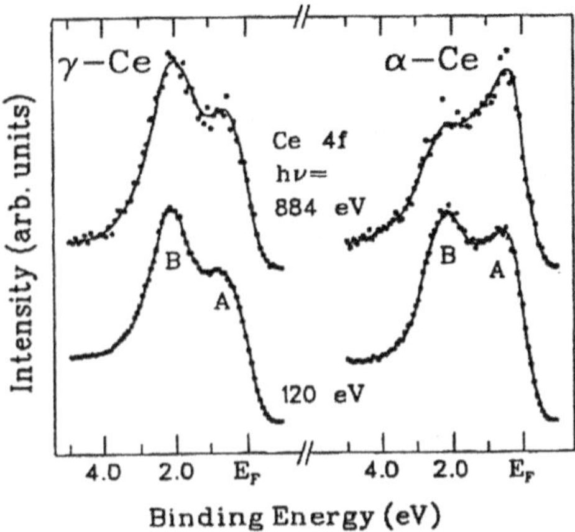

Fig. 4.22 Resonant Auger of γ- and α-cerium observed at the 4d–4f (120 eV) and 3d–4f (884 eV) resonances [190]

praseodymium and to neodymium but their energy separation remains approximately of the order of -2.5 eV. For cerium compounds, the low energy peak is rather narrow and very near E_F. For the other rare-earth compounds, this peak is broadened and its intensity decreases. Near E_F the 4f contribution is, therefore, larger in the cerium compounds than in the other systems. However, a difference can exist between cerium compounds of analogous chemical formulas. Thus, from 4d resonant Auger, a peak was observed at E_F in CeRhSb while no peak but a gap was present at E_F in CeRhAs. These results can be easily understood since the first case is a semimetal and the second a semiconductor [194]. Temperature dependence of the valence photoemission was also observed. For CeNiSn, the intensity near E_F decreased when the temperature became inferior to 50 K while for LaNiSn, no significant change with temperature was observed. The intensity decrease near E_F provided evidence for the presence of a pseudo gap at about 50 K. In these experiments, very low energy photons were used because, in this energy range, the measurements are bulk sensitive. Independently, from comparison of the 4d and 3d resonant Auger spectra of $CeRu_2$ and $CeRu_2Si_2$, it was shown that the 4f electrons are more delocalized in $CeRu_2$, i.e. when cerium is bound only to a transition metal [93]. Difference of the photoemission observed at hv equal to 21.2 and 40.8 eV in $CeRu_2$ showed the presence of the very weak $4f_{5/2}$ and $4f_{7/2}$ peaks [195]. A small part of localized 4f electrons is predicted to be mixed with the itinerant 4f electrons. However this result has not been confirmed [196]. Further studies using high-quality single crystals are needed in order to obtain a description of the electronic structure.

It is important to underline that a change of the electronic structure was observed for $CeIr_2$, $CePd_3$ and $CeRh_3$ by varying the surface sensitivity part of the experimental analysis. This change is analogous to the one seen for the α-cerium. Indeed, from simultaneously valence and 3d photoemissions, the electronic structure of these α-like cerium compounds was observed to be of the γ-like cerium type at the surface, thus to be more localized [197]. This surface valence change can play an important role in the physico-chemical properties of these materials, as for example their surface reactivity.

Numerous other compounds, in which cerium is associated with aluminium, silicon or transition elements, were analyzed by photoemission or resonant Auger. Two peaks are observed, of which one is present near E_F. This peak was generally interpreted by the single impurity Anderson model as being a *Kondo resonance* of f-electron character [119, 198]. Among those numerous compounds $CeSi_2$ was widely investigated. Two photoemission peaks were observed, one near -2 eV, characteristic of the $4f^0$ configuration, the other near E_F, characteristic of the $4f^1$ configuration, [199]. The spectra were recorded under 21.2 eV (HeI) and 40.8 eV (HeII) irradiations, and the difference between them was used to single out the contribution of the 4f electrons. The variation with the temperature of the low energy peak was observed (Fig. 4.23) [200]. The natural width of the incident radiations is about 2 meV and a spectrometer with a spectral resolution of about 5 meV was used. The resolution was better than 20 meV. The photoemission observed at 300 K shows two very weak peaks, one near E_F, the other at -0.28 eV. This corresponds to the $4f_{5/2}$–$4f_{7/2}$ splitting. At 12 K, the shape of the lower energy peak changed considerably and a narrow maximum appeared at E_F. It was interpreted in the Kondo model. However, contrary to what was expected from this model, it was suggested that the dependence of this feature on the temperature can be accounted for by

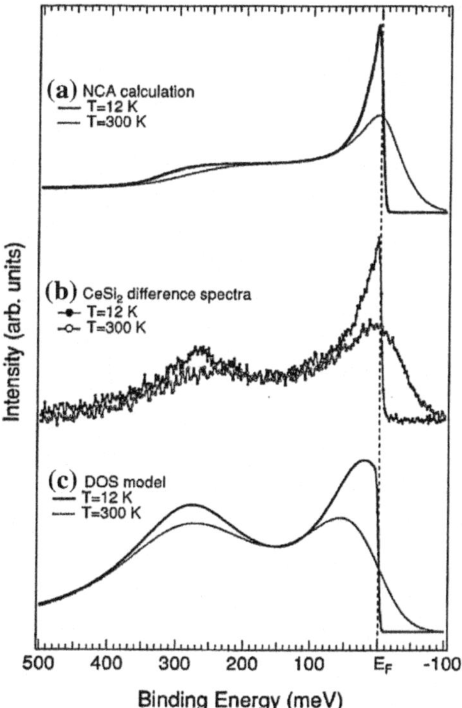

Fig. 4.23 Difference between $CeSi_2$ photoemission excited by He II and He I and observed at 12 and 300 K (**b**) compared to two different theoretical models (**a**) and (**c**) calculated at the two temperatures

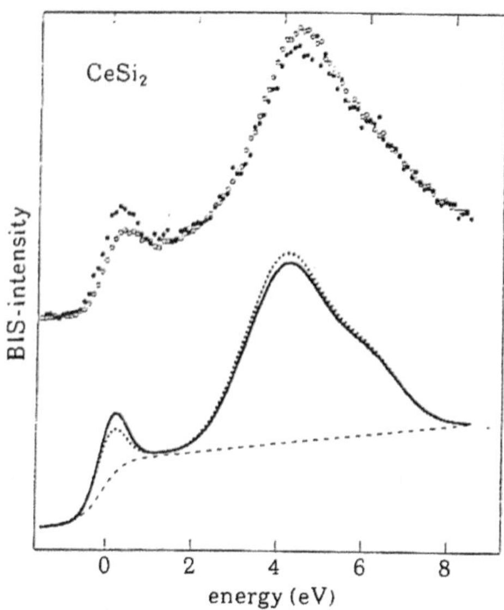

Fig. 4.24 Experimental BIS of CeSi$_2$ at 15 K (*full circles*) and 300 K (*open circles*) compared to the calculated spectra at 15 K (*full line*) and 300 K (*dotted line*) [202]

phonon broadening alone [201]. BIS experiments have then been made. Peaks characteristic of the f^0 and f^1 initial configurations, respectively, were observed at 0.5 eV and about 5 eV (Fig. 4.24) [202]. An intensity decrease of the peak characteristic of the 4f^0 configuration with a transfer of intensity from this peak to the peak characteristic of 4f^1 was observed when the temperature varies from 15 to 300 K. This transfer was attributed to a change of the electron distribution. This is in agreement with the Kondo model. Nevertheless, spectroscopic parameters such as transition probabilities are not included in the interpretations and an instrumental resolution approaching kT_K, where T_K is the Kondo temperature, was considered necessary in further observations. Moreover, phonon effect cannot be discarded.

The compounds of type (RE)(M)$_2$(Si, Ge)$_2$ where RE is cerium or ytterbium, M represents a transition or noble metal ion, were studied extensively. Let us mention the study of the compounds with Si and M = Cu, Ag, Au, Pd, by valence photoemission and 4d resonant Auger [203]. The cerium 4f distributions were deduced from the difference between the spectra observed at $hv = 112$ and 122 eV, i.e. by the difference between the spectra of photoemission and resonant Auger. For Cu, Ag, Au compounds, two peaks were observed at about −2.5 and −0.6 eV, all with analogous shapes except for the lower energy peak of the Cu compound. The lower energy peak is weak but it increases from Cu to Au. For the Pd compound, the two peaks are at about −2 and −0.5 eV and the intensity of the lower energy peak is clearly higher than for the other compounds. The presence of these two features was often attributed to the two different screening modes of the 4f hole in the final state of the emission. The higher binding energy feature corresponds to screening of the

4f hole by an itinerant electron and the other by a localized electron. However, the screening is mostly caused by itinerant electrons. Moreover, due to such screening, a 4f feature is predicted to be blurred and pulled down below the Fermi level, in contrast with the observations. In fact, the two features could be due to the difference in the character of the 4f electron, which may be either localized or hybridized-like. The final configurations involved in the emission are then either f^0 or f^1.

Large variation of valence exists in ytterbium intermetallics. Thus, valence is around 2.9 in $YbCu_{5-x}Al_x$ and $YbRh_2Si_2$ while it is of the order of 2.7 in $YbInCu_4$ and $YbAl_3$. Electronic structures of β-$YbAlB_4$ and its polymorph α-$YbAlB_4$ were particularly investigated because β-$YbAlB_4$ was the first superconductor discovered among the ytterbium compounds. Its transition temperature is 80 mK. Yb^{3+} and Yb^{2+} components were observed in the $3d_{3/2}$ and $3d_{5/2}$ photoemission spectra of each of these two polymorphic compounds [204]. This is a direct evidence of the coexistence of the two configurations $4f^{13}$ and $4f^{14}$. All ytterbium sites are crystallographically equivalent in α- and β-$YbAlB_4$ and the possibility of a spatial separation of divalent and trivalent ions is thus excluded. From the ratio of the peak areas, valence of 2.73 ± 0.02 and 2.75 ± 0.02 is obtained for α- and β-$YbAlB_4$, respectively. These values deviate significantly from 3. Consequently strong hybridization exists between the 4f and valence electrons in β-$YbAlB_4$, despite the presence of a local magnetic moment along the c axis. All the same, the valence band of β-$YbAlB_4$ clearly shows the presence of Yb^{2+} 4f peak at the Fermi level. These observations are in disagreement with the conventional schemes used to describe the superconductivity and show the need to generalize the measurements at low temperature.

Anomalous physical properties of $CeOs_4Sb_{12}$ have also attracted attention. The electric resistivity shows a typical metallic character above about 50 K but increases rapidly below this temperature. The electronic specific heat at low temperature clearly increases. A localized f-electron model was suggested from the magnetic susceptibility measurements while it was excluded by neutron scattering. Study of the electronic structure in the vicinity of the Fermi level was then undertaken [205]. In order to raise the contribution of the 4f levels in the photoemission, incident photons of energy equal to the maximum of the $3d_{5/2}$ absorption line, 882 eV, were used. A peak characteristic of the $4f^1$ final configuration was observed around E_F, accompanied by a shoulder at -0.25 eV due to spin–orbit splitting. A structure characteristic of the $4f^0$ final configuration was observed at about -3.2 eV. In contrast, the spectrum obtained with incident photons of energy lower than 882 eV was mainly due to the osmium 5d density of states, in agreement with the calculated electronic distribution. From these observations, the cerium 4f levels dominate the electronic structure at E_F. With decreasing temperature, a pseudo-gap of about 50 meV is formed while another sharp feature is observed in the gap several meV above E_F. The pseudo-gap was attributed to a valence-4f hybridization gap. Its dependence on the temperature shows a change of the hybridization. The feature just above E_F becomes sharper with decreasing temperature and results in the anomalous coexistence of hybridization-gap semiconducting and heavy fermion thermal states. These anomalies were attributed to a dual character of the 4f electrons.

Other studies were made at very low temperature. Let us notice the analysis of $CeNi_2Ge_2$ below 20 K by UV angle-resolved photoemission spectroscopy (ARPES) and its comparison with augmented spherical wave band calculations by the DFT-LDA method [206]. Around -2 eV, the levels of nickel 3d character were interpreted to be predominant. Their contribution to the spectrum decreases with decreasing binding energy and almost vanishes at E_F while cerium 4f levels are present from about -0.4 eV and their contribution increases rapidly in approaching E_F. A high density of unoccupied cerium 4f levels is also present above E_F. In this interpretation, only the extended levels were discussed. Dependence of the hybridization between extended and 4f levels as a function of the symmetry of the crystal field is also an important theme of research.

Studies by ARPES were made for other $Ce(M)_2Ge_2$ compounds by using soft X-ray radiation in order to decrease the surface contribution [207]. From the same experimental method with soft X-ray radiation and relativistic band calculation based on LDA it was shown that ytterbium is nearly divalent in $YbCu_2Ge_2$. The Yb 4f levels cross E_F and weak valence fluctuations are expected [208].

From cerium 3d or 4d photoemission of a mixed-valence compound, two sets of photoemission peaks, separated by approximately ten electronvolts, were observed. Each set is characteristic of one of the valences. In the 3d range, the spin–orbit separation is of about 18 eV and the two series of $3d_{3/2}$ and $3d_{5/2}$ peaks are well resolved. In the 4d range, the spin–orbit splitting is only of the order of a few eV and the different peaks can be difficult to resolve and interpret. In the mixed-valence cerium compounds, 3d peaks characteristic of the $4f^1$ (Ce^{3+}) and $4f^0$ (Ce^{4+}) initial configurations are well observed. A feature characteristic of the well-screened $4f^2$ configuration is also present. This feature is expected to be more intense in the mixed-valence compounds because of the presence of a 4f–5d6s hybridization. The valence can be estimated from the intensity ratio of the peaks attributed to the $4f^1$ and $4f^0$ initial configurations. From this estimate, it was confirmed that cerium is trivalent in $CeAl_2$, $CeCu_2Si_2$ and has mixed valence in $CePd_3$ [209], in $CePt_3$ and in the Ce–Ni and Ce–Co alloys [50]. But the valence of the mixed-valence compounds deduced from 3d photoemission was generally too low. Let us remark that the $3d_{3/2}$ and $3d_{5/2}$ peaks characteristic of the Ce^{4+} configuration are narrow and similar to the lanthanum $^1P_1$ and $^3D_1$ peaks. The transition probabilities towards these levels are different from the probabilities towards partially hybridized levels and it has to be taken into account in the determination of the valence.

In summary, even a small change in the characteristics of the 4f electrons, like their partial delocalization, has a very significant influence on the properties of the intermetallic materials and is at the origin of those special properties.

Let us consider now photoemission involving the valence electrons of chemical compounds. In an insulator compound with a large band gap like CeO_2, the situation is different from that of the intermetallic materials. It was shown, from self-consistent band calculations using the muffin-tin approximation, that in CeO_2, of energy gap 6.9 eV, each cerium ion was associated with two oxygen ions by covalent bonds [210]. In this model, the oxygen 2p band was filled, cerium was tetravalent and a charge equivalent to that of about two electrons was present in the interstitial region between the rare-earth and oxygen ions. Some delocalized-like 4f electrons were mixed with the valence ones and unoccupied 4f levels were expected in the band gap. Consequently, despite the fact that the bonding is mainly between cerium 5d and oxygen 2p band, the 4f orbital is predicted to participate in the chemical bonding [211]. Valence photoemission and BIS were found in agreement with these predictions (Fig. 4.19) [178]. In photoemission, there was no indication of the presence of localized 4f electrons in the ground state while a prominent peak observed by BIS in the band gap was attributed to the filling of initially empty levels by localized 4f electrons. This result was in agreement with other results obtained by L_{III} absorption and discussed in the previous paragraph. The same electronic distribution was predicted in CeF_4. Indeed, from comparison between CeF_3 and CeF_4 valence photoemission, a peak due to a localized cerium 4f level in CeF_3 was observed just above the valence band, while this peak was absent in CeF_4 [142]. This result confirmed that CeF_4 is a tetravalent compound. In PrO_2, the Pr 4f–O2p overlap is predicted to decrease because the localization of the 4f electrons increases with the atomic number. However, PrO_2, as well as CeO_2, are calculated to be tetravalent compounds [210]. When PrO_2 is in the ground state, 4f levels situated in the band gap are expected to contain one localized electron.

Controversy exists whether or not it is possible to differentiate covalence and mixed valence by high energy spectroscopy. It has been claimed that core electron photoemission, as well as L absorption spectra, could not distinguish between the two properties [142, 212–214]. As already underlined, when an atom is ionized to a core hole, the extended levels are weakly affected while the localized levels undergo an important increase of their energy. The creation of a core hole leads to a renormalization of all the levels, including the 4f levels, and can cause the appearance of a $4f^{m+1}$ configuration. When 4f levels are mixed with a filled valence band, it was shown that interaction exists between these extended 4f levels and the other 4f levels [114, 120]. Consequently, in CeO_2, extended covalent orbitals with f-symmetry and localized 4f orbitals associated with a deep hole are present at the same time and it is necessary to take them into account together each time a core hole is suddenly created in cerium. A strongly relaxed configuration is created simultaneously with the normal configuration and a second peak is associated with them. Thus, the presence of double peaks does not imply the existence of two different valences in the ground state of the material.

In 3d photoemission, the double peak intensity depends on the interactions between 4f electrons and the core hole and the relative intensities of the peaks are not usually predicted but only their energies. In each $3d_{5/2}$, or $3d_{3/2}$, photoemission of CeO_2, an intense peak corresponding to $3d^9 4f^1$ atomic-like final configuration was observed along with the main peak associated with the $3d^9 4f^0$ final configuration [178]. These two peaks are separated by about 15 eV. This large energy separation is characteristic of the large energy change which accompanies the transfer of one localized 4f electron to a more extended orbital. A satellite associated with the $3d^9 4f^1$ peak is also present. The intense peak associated with $3d^9 4f^0$

indicates the presence of extended covalent orbitals with f symmetry while the other peaks correspond to configurations in which localized 4f orbitals are populated in the presence of the core hole. These peaks were predicted by a many-body calculation taking into account the coupling between extended f-symmetry orbitals and the different final populations. It was concluded that discrete final levels in core level spectroscopy do not necessary reflect the presence of an initial mixed-valence state. Let us mention that in the two Ce_2O_3 and CeO_2 insulator oxides, no shake-down satellite is present. Indeed, the $3d_{5/2}^{-1}4f^1$ configuration is observed at about 884 eV in the two compounds.

Similar spectrum was observed for CeF_4 but the separation between $3d^94f^1$ and $3d^94f^0$ has, in this case, dropped down to 13.5 eV and the separation between $3d^94f^1$ and $3d^94f^2$ is also reduced [142]. Moreover, the relative intensity of the $3d^94f^1$ and $3d^94f^2$ peaks is lower than for CeO_2 because the covalence is weaker in the fluoride. It was concluded that the compound CeF_4 is tetravalent, as is CeO_2.

In contrast with the valence photoemission of chalcogenides that exhibits a single 4f peak of binding energy 2.4–2.6 eV, two peaks are present in the valence photoemission of the cerium pnictides. From UV photoemission and 4d autoionization, the two 4f peaks of CeP, CeAs and CeBi, are observed at about -0.6 eV and about -3.0 eV, respectively in the three compounds, but with different intensities [215]. The attribution of these peaks to the presence of 4f electrons was verified by increasing the energy of the incident photons. The peak at -3 eV was found characteristic of the $4f^1$ configuration while the peak at -0.6 eV indicated the presence of a hybridization of the cerium 4f and the pnictide p orbitals, due to the covalent character of the bonding. The hole present at the final state of the photoemission is partially delocalized on pnictide p levels. The lower energy peak is the most intense in CeP. Its intensity decreases in the CeP, CeAs, CeBi sequence, that is with the size of the ligand, and it becomes very low for CeBi.

For CeN, a narrow peak superimposed on a broader band was observed at E_F in X-ray photoemission from the valence band. This peak was attributed to an electron of cerium fluctuating between f and d orbitals [76, 77]. Hardly distinguishable for CeP, this peak remained unresolved in the other pnictides. Large valence band was observed about -2 to -3 eV below E_F for all these compounds. A structure was present on the high-binding energy side of this band at about -3 to -4 eV from E_F. In CeN, in contrast with the other pnictides, a 4f level is very near E_F and energetically mixed with the valence band in this compound.

In the 3d photoemission, only two main $3d_{3/2}$ and $3d_{5/2}$ peaks corresponding to trivalent cerium and two weak shake-down peaks were observed for all the pnictides except for CeN. For this compound, two additional peaks were observed, each one shifted to higher energies of its corresponding main peak by 11 eV. Additional peaks were also observed in the 4d photoemission of CeN and cerium metal at low temperature (Fig. 4.25) [76, 77]. They were attributed to the tetravalent $3d^94f^0$ configuration, which is predicted to be very well separated from the $3d^94f^1$ one. The presence of these Ce^{4+} $3d_{3/2}$ and $3d_{5/2}$ peaks in the CeN spectrum and their absence from the CeP, CeAs, CeBi spectra as well as in the CeSe spectrum, had already

Fig. 4.25 Cerium 4d photoemission in CeN and metal: **a** at liquid-nitrogen temperature; **b** at standard temperature [76]

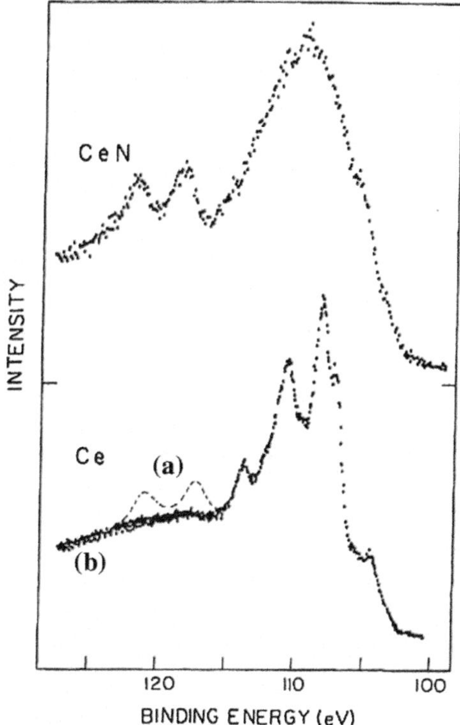

shown that CeN occupied a particular position among the pnictides because cerium has a mixed valence between 3 and 4 in this compound.

From self-consistent energy band calculations, the theoretical lattice constant was derived and compared with its measured value [216]. It was deduced that the 4f electrons are delocalized-like in CeN while they are localized and non-bonding in the heavier pnictides. Several recent calculations using the LDA/GGA + DMFT approach [80] or more simply using the LDA + U method [217] have confirmed that the lattice constants and physical properties of the pnictides were correctly described by a model of delocalized 4f electrons for CeN and localized ones for the heavier pnictides.

Typical examples of valence fluctuating compounds are present among the samarium, thulium and ytterbium compounds as, for example, thulium monochalcogenides. At standard pressure, TmS is a trivalent metal and TmSe is a mixed-valence compound. TmTe is a divalent magnetic semiconductor up to a pressure of 2 GPa; between 2 and 6 GPa it shows a mixed valence and becomes a trivalent metal above 6 GPa. The anomalous magnetic properties of TmSe were believed to be caused by fluctuations between divalent and trivalent states of the Tm ion. Both these valence states were clearly shown by X-ray photoelectron spectroscopy [218]. Features observed in the photoemission spectra of these materials are not reproduced from calculations of the electronic structure using the approximations LSDA and LDA + U [219, 220] because the atomic multiplet effects are not taken into account in these calculations. In contrast, agreement between theory and experiment was obtained when the approach described in reference [221] was adopted. In this approach, the atomic multiplet effects enter naturally in the one-particle Green function and the model is adapted to treat localized electron systems such the 4f electrons in these compounds. Spectroscopic data were obtained, and are reported in references [222, 223]. In TmS and TmSe, the thulium ions have localized $4f^{12}$ electrons and a narrow f band is present near E_F while TmTe has localized $4f^{13}$ electrons. Comparison between experiment and theory was made in reference [224]. An isolated $4f^n$ ion was embedded in the solid described by the Hubbard approximation. The presence of this ion was taken into account by introducing its atomic self-energy in the crystal Hamiltonian. The observed multiplet structures were well described theoretically by considering the ground state as an almost pure f^{12} configuration in TmS, a mixture $f^{12} + f^{13}$ configuration in TmSe and a pure f^{13} configuration in TmTe.

Valence fluctuations were also observed in the 4f and core level photoemission of samarium monochalcogenides [225]. In SmS, fluctuations between divalent and trivalent configurations exist. The valence fluctuations between them are less than 3 % and the ground configuration was found very close to $4f^6$. The reduction of the number of the 4f electrons from 6 to 5 becomes more noticeable above 6 kbar. In SmSe and SmTe, samarium is divalent. No chemical shift or splitting of the core levels of samarium in the three divalent chalcogenides was observed. In contrast, in $Sm_{0.82}Gd_{0.18}S$, the presence of the gadolinium ions can be considered as equivalent to an external pressure and divalent and trivalent samarium ions coexist. Distribution of 4f levels associated with the Sm^{2+} ion in this compound is identical to that of the divalent compounds. Owing to the presence of the Gd^{3+} ion, multiplets associated with the Sm^{3+} ion have different energies from those associated with trivalent samarium in SmSb for example. These energy variations are connected with an electronic semiconductor-metal transition without a structural change. A delocalization of a 4f electron to the 5d, 6s conduction band at standard pressure is associated with this transition. Recently, partial 4f delocalization induced by the pressure was observed in samarium and ytterbium monochalcogenides [226].

Among rare-earth borides, SmB_6 and YbB_{12} are typical examples of valence fluctuating semiconductors. SmB_6 was the first studied mixed-valence compound. The presence of the two configurations $4f^5$ and $4f^6$ was rapidly confirmed by photoemission [227]. With 150 eV incident photons, i.e. in surface sensitive conditions, photoemission demonstrated that the SmB_6 valence changes from uniformly mixed in the bulk to non-uniformly mixed on the surface while in samarium metal the changes are from single valence in the bulk to inhomogeneously mixed valence on the surface [70]. It was verified that these observations were not due to some superficial contamination. These experiments used incident photons in the energy range of the 4d $\rightarrow$ 4f photoabsorption and took advantage of the large intensity of the 4f electron emission in this range: indeed, they were resonant Auger experiments. Clearly resolved multiplets associated with the $4f^6$ and $4f^5$ initial configurations were observed in high-resolution angle-resolved photoemission at the 4d–4f resonant energy of 140 eV [228]. From this experiment the lowest Sm^{2+} $4f^6$ multiplet structure is observed at -15 meV below E_F. Actually, SmB_6 is the subject of great interest because it has a very low temperature residual conductivity that is not well understood. It was suggested that its unusual transport properties are related to the presence of 4f-character surface states in the band gap and to their temperature dependence. Consequently, surface properties of SmB_6 have been widely investigated. Recently, the study of cleaved single crystals at low temperature was made by X-ray photoemission [229]. Valence around 2.6 was obtained, in agreement with the previous values from Mossbauer spectroscopy or X-ray absorption spectroscopy [150]. This photoemission analysis showed the complexity of the SmB_6 surface. Stoichiometric deviation and formation of a sub-oxide with increasing temperature have been observed. But difficulties in the interpretation of surface sensitive measurements have been underlined. This is the case, for example, in the recent analysis of 4f-character surface states by angle-resolved photoemission. In summary, various physical properties of SmB_6 vary with the temperature and these variations follow the variation of the samarium valence. This shows the importance of valence in understanding the physical properties of SmB_6 and, in general, its decisive role in the physical properties of the materials.

4.3.2.2   X-ray Emission

Cerium 3d emission and absorption of $CeNi_2$ have been observed under the same experimental conditions (Fig. 4.26) [157]. In absorption, a comparison between the experimental curve and the theoretical curve calculated as described in Sect. 3.2.4 shows the presence of the two final configurations, $4f^1$ and $4f^2$. The valence is found equal to 3.22, in agreement with the value predicted from the L_{III} absorption. An additional structure, observed at about 5.5 eV above the multiplets of the $4f^1$ final configuration, is characteristic of the transitions towards the extended conduction

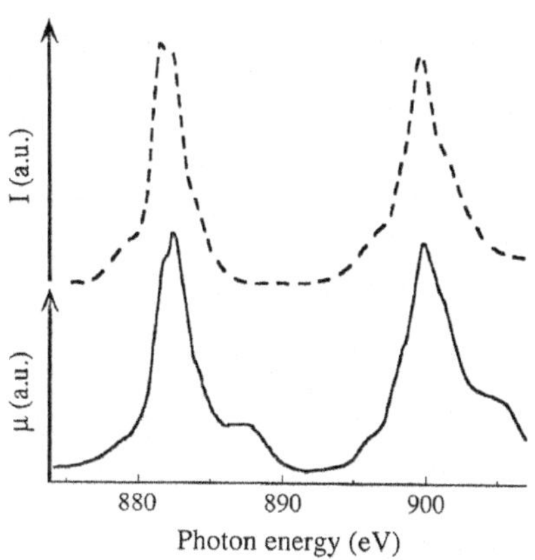

Fig. 4.26 3d absorption (*full line*) and emission (*dotted line*) of $CeNi_2$ [157]

levels above E_F. In emission spectrum, lines in coincidence with the absorption lines are observed. They are resonance emissions, whose observation confirms the presence of the excited configurations $4f^1$ and $4f^2$ and, consequently, the existence in this compound of ions with the configurations $4f^0$ and $4f^1$ in the ground state, respectively. An unresolved feature is observed towards the lower energies of the resonance lines. It is characteristic of the extended valence states and its presence makes the shape of the emission spectrum not rigorously identical to that of the absorption spectrum. From the intensity of the emissions characteristic of the configurations $4f^0$ and $4f^1$, the valence is only 3.2. These results underline again the many possibilities offered by the resonant emissions, in conjunction with the corresponding absorptions, to determine the number of the localized electrons and the electron configurations of the rare-earths. Spectral resolution is mediocre but sufficient to supply data compatible with those obtained by other more recently used methods.

The rare-earth nitrides that can be used as high polarization materials in spintronics have been studied extensively. Among them, the most promising are GdN, which is expected to become a future magnetic sensor, and EuN. The electronic structure of these compounds must be well known in order to make easier their understanding. Concerning GdN, absorption and emission have been observed at the NK edge. Absorption was measured in the total fluorescent yield mode and the spectral distribution was found in good agreement with the theoretical densities of states [230]. Thus, energies of the two Gd t_{2g} and e_g levels, which correspond to the Gd 5d orbitals split by the crystal field, are observed, and agree with their theoretical predicted energies. Agreement between experiment and theory was also observed for the N 2p valence band, whose the maximum is at about -2 eV below the absorption threshold. The Gd 4f levels are predicted at -8 eV. From Gd 3d absorption, a single $3d^9 4f^{m+1}$ excited configuration was observed, which corresponds to the ground configuration $4f^7$. The experimental conditions employed were not adapted to obtain information on the occupied 4f distributions and the proposed attribution for the Gd Mα and Mβ emissions cannot be accepted. However, it may be concluded that, for the considered epitaxial GdN film, no interaction exists between the Gd 4f and N 2p levels, the 4f electrons are strongly localized and no 4f level is present near the zero energy. In contrast, low density of valence states was predicted with a band gap of a few meV at the zero energy [231]. These valence states are thought to be at the origin of the half-metallic electronic structure of this compound.

In EuN, the presence of ions Eu^{3+} ($J = 0$) was predicted. However, from X-ray magnetic dichroism of EuN thin films, it was shown that few percents of Eu^{2+} are also present [232]. These Eu^{2+} ions are in the origin of the magnetic properties and strongly localized levels close to E_F are predicted to be associated with them. The presence of such levels could have a large influence on the conduction properties. Consequently, theoretical and experimental studies of the electronic structure of this material have been developed. Concerning nitrogen, analogy was found between the N 1s absorption and emission of EuN and those previously observed for GdN [232]. Thus, the energy separation between the RE 5d t_{2g} and e_g levels is approximately the same in the two compounds. The valence level distributions are thus similar in the two materials. From Eu 3d absorption, Eu^{3+} was found to be prevalent but the presence of a small amount of Eu^{2+} was confirmed. From DMFT (LSDA+U) calculations, EuN appears as a semiconductor with a minimum indirect gap of 0.3 eV. X-ray photoemission shows that the 4f levels associated with Eu^{3+} are about -7.5 eV below the valence band. On the other hand, the presence of a Eu^{2+} f^7 ^8S level close to the top of the valence band has been demonstrated. Valence band and atomic-like 4f levels of Eu^{2+} are energetically mixed and a possible interaction between the valence levels and a small concentration of 4f levels associated with the magnetic Eu^{2+} ions could explain the complex properties of this material.

4.3.2.3 Resonant Inelastic X-ray Scattering

RIXS in the 2p–3d range was used to probe the γ-α transition in cerium metal under pressure [233]. The number of localized 4f electrons was found to decrease from 0.97 in the γ phase at 1.5 kbar to 0.81 in the α phase at 20 kbar. This decrease had already been observed but here the change is larger than expected. Mixing of different electron configuration was also expected in ytterbium intermetallics due to the possible presence of the Yb^{3+} ($4f^{13}$) and Yb^{2+} ($4f^{14}$). RIXS study as a function of the temperature was made for $YbAgCu_4$ and $YbInCu_4$ [155] in the range from 15 to 300 K. In this range, the RIXS signal decreases continuously for $YbAgCu_4$. The average number of 4f holes varies continuously from 0.87 at 15 K to 0.93 at 300 K. In contrast, for $YbInCu_4$, a sharp valence transition was observed at 42 K with a sudden variation from 0.83 to 0.96 of the number of 4f holes.

The γ-α phase transition of Ce–Th and Ce–Sc solid solutions was also observed by RIXS in the 2p–3d range [234]. The structural transition was obtained as a function of temperature by substituting thorium or scandium for cerium, i.e. by using *chemical pressure*. Indeed, volume contraction of about 12.3 % was obtained in the $Ce_{0.93}Sc_{0.07}$ solid solution. The main component of the cerium 2p–3d spectrum corresponds to the $4f^1$ configuration. A weak "pre-edge" structure, observed towards the lower energies of the main peak, at about 877 eV, was associated with a hybridization process of the 4f orbitals in the presence of a core hole. It corresponds to the peak already discussed in 2p–3d RIXS of CeF_3. It can be identified with the low energy photoemission peak observed in cerium metal and intermetallics. To the γ-α phase transition, an increase of the low energy peak, associated with an increase of the 4f-valence hybridization, was deduced from the measurement of the ratio of the observed peaks.

RIXS in the 3d–4f range was applied to CeO_2 [235] and PrO_2, [236]. The two oxides were treated as intermediate valence compounds while CeO_2 is generally considered as the typical compound with a 4f quasi-delocalized electron. The cerium 3d absorption has been described in Sect. 4.3.1.1 (Fig. 4.17). When the incident photon energy is that of the main peak, i.e. 884.1 eV, which is characteristic of the $3d^9 4f^1$ configuration (Fig. 4.27), the fluorescence spectrum comprises

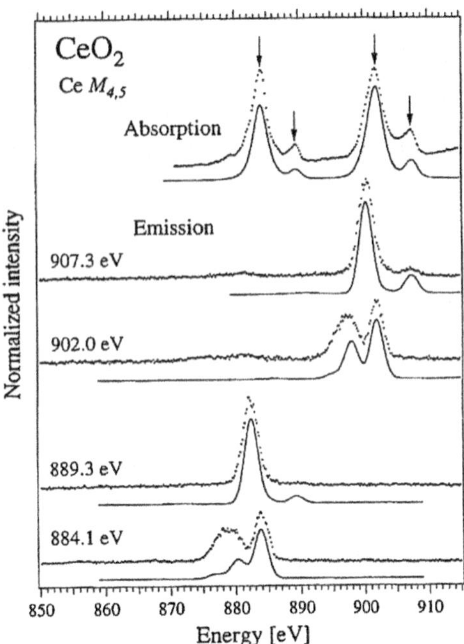

Fig. 4.27 Cerium 3d absorption and RIXS in CeO$_2$: experimental (*dots*) and calculated (*solid lines*). The arrows on the absorption indicate the excitation energies used for the RIXS [235]

the resonant emission, $3d^9 4f^1 - 3d^{10} 4f^0$, at the energy of the absorption line, accompanied by a peak characteristic of the hybridized 4f-valence electrons, $3d^9 4f^1 - 3d^{10} 4f^1 V^{-1}$ towards the lower energies. When the incident photon energy is that of the structure located at 889.3 eV, the fluorescence consists of a single peak that can be designated as the emission in the presence of a hole in the valence band. Indeed, the secondary absorption peak can be considered as reflecting the presence of the extended 4f states mixed with the conduction states.

4.4   Conclusion

A large variety of unusual phenomena such as the coexistence of a semiconducting behaviour and a significantly high specific heat at low temperature, as well as the presence of unconventional superconductivity, can be encountered in numerous rare-earth compounds. These phenomena are generally attributed to a competition between the localized and itinerant character of the 4f electrons, associated with a strongly anisotropic 4f spatial distribution related to the shape of the atomic potential. Changes in the 4f localization associated with effects of temperature, pressure or alloying are observed in the light rare-earths. These changes are more important for cerium than for praseodymium and remains weak for neodymium. Characteristics of the 4f distributions are governed by the interactions among different sub shells and a spectral analysis of the 4f electron distributions is essential. Concerning the heavier rare-earths, one characteristic of their compounds is the phenomenon of intermediate or mixed valence, where two different 4f electron occupancies coexist. The concept of mixed valence is suitable to describe compounds that have a narrow 4f level as well as a broad 5d band at the Fermi level. The 4f electrons do not participate in bonding and no delocalized 4f electrons are present in these compounds. The intermediate valence compounds have interesting physical properties directly related with the non-integer number of the 4f electrons.

The strong surface reactivity of the rare-earths perturbs the observations. Superficial oxidation of the metals can be accompanied by a change of the valence, as in the case of the tetravalent cerium oxide. Intrinsic differences of valence between surface and bulk materials have also been observed. This factor has to be considered in the choice of the analysis techniques.

Photoemission spectroscopy is actually the most used technique. Large improvement of the resolution to a few meV and development of experiments at low temperature have achieved big progress in X-ray photoemission spectroscopy in recent years. Its handicap is its high sensitivity at the surface under usual conditions. Incident radiation of the order of 5–8 keV has been used to increase the escape depth of the photoelectrons thus making it possible to probe the bulk electronic structure. The presence of a hole at the final state can induce secondary effects in the photoemission from the core levels and the method is used actually to determine the valence distributions.

Along with the advantageous intense synchrotron radiation, RIXS and related analyzes have been developed but their interpretations are difficult. The most important advance concerns the use of these spectroscopies to determine magnetic structures [170].

Among the methods using incident X-ray radiation, XAS is the only one giving a direct determination of the number of localized nf electrons in a solid. Because the excited electron screens the core hole, the perturbation due to the presence of the hole is predicted to be negligible during the excitation process. Then, the configuration of the ground state may be obtained directly provided that the broadening due to the width of the core level is sufficiently small. Considering the important role of the d–f dipolar transition probabilities, the nd absorptions can determinate directly the number of the localized nf electrons present in a solid.

EXES describes the densities of states of a selected orbital moment and atomic number. The excited configurations can be analyzed leading to the direct measurement of the number of localized 4f electrons present in any rare-earth compound. These experiments are the most suitable in the 3d range. In the 4d range the presence of a giant resonance and the low intensity of the other peaks make the precise determination of the number of localized electrons more difficult. Well chosen experimental conditions made the method suitable for the bulk. EXES has been used only for the metals and several oxides. Generalization of this method is in need to be developed. The same needs to be done for its non-radiative counterpart, the electron-excited resonant Auger emission. To make these methods more efficient, experiments at low temperature are necessary.

Observations of transitions from excited configurations play a very important role in the determination of the electron configurations in the 4f element compounds. In these materials, the excitation is a *single-site process* and the excited states can be described with the help of an atomic model. It has been shown that solid-state effects have only a small influence on these configurations. The number of localized electrons determines the magnetic ordering, which appears as a second order process. In contrast, hybridization of 4f levels with the other valence levels plays an essential role in the electronic properties of some intermetallic compounds. Electron distributions and spectral transitions are generally explained by a single impurity model. But, the electron distributions must be considered together with the local surrounding, crystal field and temperature. For example, a study of the relaxation dynamics of the photoexcited heavy electron compounds as a function of temperature can be mentioned. It was shown that the relaxation rate decreased strongly with the temperature and dropped by more than two orders of magnitude at liquid helium temperature [237]. Two interpretations were considered: –the electron-phonon thermalization inducing the slowing down of the relaxation; –the presence of a narrow gap in the density of states near the Fermi level. The slowing down of the relaxation at low temperature can be explained by a phonon "bottleneck" mechanism. This anomalous slowing down of the decay dynamics upon cooling was initially studied by low energy spectroscopy but the same observations are expected in other energy ranges.

Rare-earth compounds with their unusual and surprising properties form one of the challenging classes of materials in modern condensed matter physics. Considering the interplay between different models that takes into account simultaneously their principal physical parameters is essential in order obtain an insight into all the exotic properties of these recently synthesized compounds.

References

1. M.F. Lopez, C. Laubschat, G. Kaindl, Europhys. Lett. **23**, 538 (1993)
2. M.R. Norman, A.J. Freeman, Phys. Rev. B **33**, 8896 (1986)
3. C. Bonnelle, unpublished data
4. G. Wendin, *Structure and Bonding*, vol. 45, (Springer, Berlin, 1981)
5. J.W. Allen, S.-J. Oh, M.B. Maple, M.S. Torikachvili, Phys. Rev. B **28**, 5347 (1983)
6. A. Kotani, Y. Toyozawa, J. Phys. Soc. Jpn. **37**, 912 (1974)
7. M.O. Krause, J.H. Oliver, J. Phys. Chem. Ref. Data **8**, 329 (1979)
8. J.M. Mariot, R.C. Karnatak, Solid St. Comm. **16**, 611 (1975)
9. K. Hämäläinen, D.P. Siddons, J.B. Hastings, L.E. Berman, Phys. Rev. Lett. **67**, 2850 (1991)
10. D.K. Wohlleben, in *Valence Fluctuation in Solids*, eds. by L.M. Falicov, W. Hanke, M.B. Maple (North-Holland, Amsterdam, 1981), p. 1
11. J.M. Tarascon, Y. Isikawa, B. Chevalier, J. Etourneau, P. Hagenmuller, M. Kasaya, J. Phys. France **41**, 1135 (1980)
12. P.A. Lee, E.A. Stewardson, J.E. Wilson, Proc. Phys. Soc. A **65**, 668 (1952)

13. E.A. Stewardson, J.E. Wilson, Proc. Phys. Soc. London **A69**, 93 (1956)
14. C. Bonnelle, R.C. Karnatak, J. Phys. France **32**, C4-230 (1971)
15. C. Bonnelle, R.C. Karnatak, J. Sugar, Phys. Rev. A. **9**, 1920 (1974)
16. V.F. Demekhin, Sov. Phys. Solid State. **16**, 659 (1974)
17. S.L.M. Schroeder, Solid State Com. **98**, 405 (1996)
18. A. Owens, S.C. Bayliss, G.W. Fraser, S.J. Gurman, Nucl. Instr. Meth. **A385**, 556 (1997)
19. B.T. Thole, G. van der Laan, J.C. Fuggle, G.A. Sawatzky, R.C. Karnatak, J.M. Esteva, Phys. Rev. B **32**, 5107 (1985)
20. J. Sugar, Phys. Rev. A **6**, 1764 (1972)
21. T.M. Zimkina, V.A. Fomichev, S.A. Gribovskii, I.I. Zhukova, Sov. Phys. Solid State **9**, 1128 (1967)
22. V.A. Fomichev, T.M. Zimkina, S.A. Gribovskii, I.I. Zhukova, Sov. Phys. Solid State **9**, 1163 (1967)
23. R. Haensel, P. Rabe, B. Sonntag, Solid State Commun. **8**, 1845 (1970)
24. O. Aita, K. Ichikawa, M. Kamada, M. Okusawa, H. Nakamura, K. Tsutsumi, J. Phys. Soc. Japan **56**, 649 (1987)
25. K.C. Williams, Proc. Phys. Soc. **87**, 983 (1966)
26. J.L. Dehmer, A.F. Starace, Phys. Rev. B **5**, 1792 (1972)
27. H.W. Wolff, R. Bruhn, K. Radler, B. Sonntag, Phys. Lett. **59A**, 67 (1976)
28. E.R. Radtke, J. Phys. B. Atom. Molec. Phys. **12**, L77 (1979)
29. M. Richter, M. Meyer, M. Pahler, T. Prescher, E.V. Raven, B. Sonntag, H.E. Wetzel, Phys. Rev. A **39**, 5666 (1989)
30. C.G. Olson, D.W. Lynch, J. Opt. Soc. Am. **72**, 88 (1982)
31. H. Ogasawara, A. Kotani, J. Synchrotron Rad. **8**, 220 (2001)
32. P. Trebbia, C. Colliex, Phys. Stat. Sol. (b) **58**, 523 (1973)
33. M. Cukier, B. Gauthe, C. Wehenkel, J. Physique **41**, 603 (1980)
34. C. Colliex, M. Gasgnier, P. Trebbia, J. Physique **37**, 397 (1976)
35. K.T. Moore, B.W. Chung, S.A. Morton, A.J. Schwartz, J.G. Tobin, S. Lazar, F.D. Tichelaar, H.W. Zandbergen, P. Söderlind, G. van der Laan, Phys. Rev. B **69**, 193104 (2004)
36. I.B. Borovskii, E. YaKomarov, O. L'vov, Phys. Met. Metallogr. **44**, 57 (1979)
37. J.A.D. Matthew, G. Strasser, F.P. Netzer, Phys. Rev. B **27**, 5839 (1983)
38. G. Strasser, G. Rosina, J.A.D. Matthew, F.P. Netzer, J. Phys. F. Met. Phys. **15**, 739 (1985)
39. J. Bloch, N. Shamir, M.H. Mintz, U. Atzmony, Phys. Rev. **30**, 2462 (1984)
40. J.A. Bradley, K.T. Moore, G. vander Laan, J.P. Bradley, R.A. Gordon, Phys. Rev. B **84**, 205105 (2011)
41. J.K. Lang, Y. Baer, P.A. Cox, Phys. Rev. Lett. **42**, 74 (1979)
42. J.K. Lang, Y. Baer, P.A. Cox, J. Phys. F: Metal Phys. **11**, 113 (1981)
43. J.K. Lang, Y. Baer, P.A. Cox, J. Phys. F: Metal Phys. **11**, 121 (1981)
44. F. Gerken, J. Barth, C. Kunz, X-ray and atomic inner-shell physics, X-82. AIP Conf. Proc **94**, 602 (1982)
45. P. Sakellaridis, J. Physique Radium **16**, 422 (1955)
46. C. Bonnelle, R.C. Karnatak, C.R. Seances, Acad. Sci. Ser. A **268**, 494 (1969)
47. C. Bonnelle, in *Advances in X-ray spectroscopy*, eds. C. Bonnelle, C. Mande (Pergamon, New York, 1982), p. 104
48. A.S. Shulakov, A.P. Braiko, T.M. Zimkina, Sov. Phys. Solid State **23**, 1081 (1981)
49. C. Bonnelle, P. Motais, Phys. Rev. A **73**, 042504 (2006)
50. J.C. Fuggle, F.U. Hillebrecht, Z. Zolnierek, R. Lässer, Ch. Freiburg, O. Gunnarsson, K. Schönhammer, Phys. Rev. B **27**, 7330 (1983)
51. R.E. LaVilla, Phys. Rev. A **9**, 1801 (1974)
52. S.A. Gribovskii, V.A. Fomichev, T.M. Zimkina, Sov. Phys. Solid State **15**, 894 (1973)
53. T.M. Zimkina, V.A. Fomichev, S.A. Gribovskii, Sov. Phys. Solid State **15**, 1097 (1973)
54. V.A. Fomichev, S.A. Gribovskii, T.M. Zimkina, Sov. Phys. Solid State **15**, 1880 (1974)
55. T.M. Zimkina, V.A. Fomichev, S.A. Gribovskii, Sov. Phys. Solid State **15**, 1786 (1974)
56. K. Ichikawa, A. Nisawa, K. Tsutsumi, Phys. Rev. B **34**, 6690 (1986)

57. J.G. Tobin, S.W. Yu, B.W. Chung, G.D. Waddill, L. Duda, J. Nordgren, Phys. Rev. B **83**, 085104 (2011)
58. S. Tanaka, A. Kotani, J. Phys. Soc. Jpn. **61**, 4212 (1992)
59. P.O. Hedén, H. Löfgren, S.B.M. Hagström, Phys. Rev. Lett. **26**, 432 (1971)
60. D.E. Eastman, M. Kuznietz, Phys. Rev. Lett. **26**, 846 (1971)
61. D.M. Wieliczka, C.G. Olson, D.W. Lynch, Phys. Rev. Lett. **52**, 2180 (1984)
62. G. Rossi, A. Barski, Solid State Comm. **57**, 277 (1986)
63. G.K. Wertheim, A. Rosencwaig, R.L. Cohen, H.J. Guggenheim, Phys. Rev. Lett. **27**, 505 (1971)
64. P.A. Cox, Y. Baer, C.K. Jorgensen, Chem. Phys. Lett. **22**, 433 (1973)
65. F. Gerken, J. Phys. **F13**, 703 (1983)
66. S. Lebègue, A. Svane, M.I. Katsnelson, A.I. Lichtenstein, O. Eriksson, Phys. Rev. B **74**, 045114 (2006)
67. S. Lebègue, A. Svane, M.I. Katsnelson, A.I. Lichtenstein, O. Eriksson, J. Physics, Condensed Matter **18**, 6329 (2006)
68. F. Gerken, A.S. Flodström, J. Barth, L.I. Johansson, C. Kunz, Phys. Scr. **32**, 43 (1985)
69. W.D. Schneider, C. Laubschat, G. Kalkowski, J. Haase, Z. Puschmann, Phys. Rev. B **28**, 2017 (1983)
70. J.W. Allen, L.I. Johansson, I. Lindau, S.B. Hagstrom, Phys. Rev. B **21**, 1335 (1980)
71. G.K. Wertheim, G. Crecelius, Phys. Rev. Lett. **40**, 813 (1978)
72. A.F. Orchard, G. Thornton, J. Electron Spectr. Rel. Phenom **10**, 1 (1977)
73. B.D. Padalia, W.C. Lang, P.R. Norris, L.M. Watson, D.J. Fabian, Proc. Roy. Soc. Lond. A **354**, 269 (1977)
74. P.A. Cox, Struct. Bonding (Berlin) **24**, 79 (1975)
75. S. Sato, J. Phys. Soc. Jpn **41**, 913 (1976)
76. Y. Baer, Ch. Zürcher, Phys. Rev. Lett. **39**, 956 (1977)
77. Y. Baer, R. Hauger, Ch. Zürcher, M. Campagna, G.K. Wertheim, Phys. Rev. B **18**, 4433 (1978)
78. R.D. Parks, S. Raaen, M.L. den Boer, Y.S. Chang, G.P. Williams, Phys. Rev. Lett. **52**, 2176 (1984)
79. M.R. Norman, D.D. Koelling, A.J. Freeman, Phys. Rev. B **32**, 7748 (1985)
80. M.S. Litsarev, I. DiMarco, P. Thunström, O. Eriksson, Phys. Rev. B **86**, 115116 (2012)
81. H. Yamada, T. Fukawa, T. Muro, Y. Tanaka, S. Imada, S. Suga, D.-X. Li, T. Suzuki, J. Phys. Soc. Jpn **65**, 1000 (1996)
82. F. Leuenberger, A. Parge, W. Felsch, K. Fauth, M. Hessler, Phys. Rev. B **72**, 014427 (2005)
83. S.S. Stoyko, K.K. Ramachandran, C.S. Mullen, A. Mar, Inorg. Chem. **52**, 1040 (2013)
84. J.F. Herbst, R.E. Watson, J.W. Wilkins, Phys. Rev. B **13**, 1439 (1978)
85. J.F. Herbst, R.E. Watson, J.W. Wilkins, Phys. Rev. B **17**, 3089 (1978)
86. R.C. Karnatak, J.M. Esteva, H. Dexpert, M. Gasgnier, P.E. Caro, L. Albert, Phys. Rev. B **36**, 1745 (1987)
87. P. Lagarde, A.-M. Flank, H. Ogasawara, A. Kotani, J. Electron. Spectrosc. Relat Phenom **128**, 193 (2003)
88. G. Dufour, C. Bonnelle, J. Physique Lett. **35**, L255 (1974)
89. G. Dufour, R.C. Karnatak, J.-M. Mariot, C. Bonnelle, J. Microsc. Spectrosc. Electron. **1**, 105 (1976)
90. M. Richter, T. Prescher, M. Meyer, E.V. Raven, B. Sonntag, H.E. Wetzel, S. Aksela, Phys. Rev. B **38**, 1763 (1988)
91. J.C. Riviere, F.P. Netzer, G. Rosina, G. Strasser, J.A.D. Matthew, J. Electron Spectrosc. Relat Phenom **36**, 331 (1985)
92. A. Sekiyama, T. Iwasaki, K. Matsuda, Y. Saitoh, Y. Onuki, S. Suga, Nature **403**, 396 (2000)
93. F. Gerken, J. Barth, K.L.I. Kobayashi, C. Kunz, Solid State Commun. **35**, 179 (1980)
94. F. Gerken, J. Barth, C. Kunz, Phys. Rev. Lett. **47**, 993 (1981)
95. M. Croft, A. Franciosi, J.H. Weaver, A. Jayaraman, Phys. Rev. B **24**, 544 (1981)
96. M. Aono, T.-C. Chiang, J.A. Knapp, T. Tanaka, D.E. Eastman, Phys. Rev. B **21**, 2661 (1980)

97. J.-J. Gallet, J.-M. Mariot, C.F. Hague, F. Sirotti, M. Nakazawa, H. Ogasawara, A. Kotani, Phys. Rev. B **54**, R14238 (1996)
98. A. Moewes, T. Eskildsen, D.L. Ederer, J. Wang, J. McGuire, T.A. Callcott, Phys. Rev. B **57**, R8059 (1998)
99. A. Moewes, S. Stadler, R.P. Winarski, D.L. Ederer, T.A. Callcott, J. Electron Spectr. Relat. Phenom **110–111**, 189 (2000)
100. A. Moewes, M.M. Grush, T.A. Callcott, D.L. Ederer, Phys. Rev. B **60**, 15728 (1999)
101. A. Moewes, D.L. Ederer, M.M. Grush, T.A. Callcott, Phys. Rev. B **59**, 5452 (1999)
102. M.H. Krisch, C.C. Kao, F. Sette, W.A. Caliche, K. Hämälainen, J.B. Hastings, Phys. Rev. Lett. **74**, 4931 (1995)
103. J.-J. Gallet, J.-M. Mariot, L. Journel, C.F. Hague, A. Rogalev, H. Ogasawara, A. Kotani, M. Sacchi, Phys. Rev. B **60**, 14128 (1999)
104. C.F. Hague, J.-M. Mariot, L. Journel, J.-J. Gallet, A. Rogalev, G. Krill, J.P. Kappler, J. Elect. Spectrosc. Relat. Phenom. **110–111**, 179 (2000)
105. P. Carra, M. Fabrizio, B.T. Thole, Phys. Rev. Lett. **74**, 3700 (1995)
106. M. van Veenendaal, P. Carra, B.T. Thole, Phys. Rev. B **54**, 16010 (1996)
107. L. Journel, J.-M. Mariot, J.-P. Rueff, C.F. Hague, G. Krill, M. Nakazawa, A. Kotani, A. Rogalev, F. Wilhelm, J.P. Kappler, G. Schmerber, Phys. Rev. B **66**, 045106 (2002)
108. N. Spector, C. Bonnelle, G. Dufour, C.K. Jorgensen, H. Berthou, Chem. Phys. Lett. **41**, 199 (1976)
109. G. Crecelius, G.K. Wertheim, D.N.E. Buchanan, Phys. Rev. B **18**, 6519 (1978)
110. J.C. Fuggle, M. Campagna, Z. Zolnierek, R. Lässer, A. Platau, Phys. Rev. Lett. **45**, 1597 (1980)
111. J.C. Fuggle, O. Gunnarsson, G.A. Sawatzky, K. Schönhammer, Phys. Rev. B **37**, 1103 (1988)
112. H. Berthou, C.K. Jorgensen, C. Bonnelle, Chem. Phys. Lett. **38**, 199 (1976)
113. G. Dufour, R.C. Karnatak, J.M. Mariot, C. Bonnelle, Chem. Phys. Lett. **42**, 433 (1976)
114. W.-D. Schneider, B. Delley, E. Wuilloud, J.-M. Imer, Y. Baer, Phys. Rev. B **32**, 6819 (1985)
115. K.-H. Park, S.-J. Oh, Phys. Rev. B **48**, 14833 (1993)
116. S. Suzuki, T. Ishii, T. Sagawa, J. Phys. Soc. Jpn **37**, 1334 (1974)
117. C.K. Jorgensen, *Modern Aspects of Ligand Field Theory* (North-Holland, Amsterdam, 1971)
118. A. Kotani, H. Ogasawara, J. Electron Spectrosc. Relat. Phenom **60**, 257 (1992)
119. O. Gunnarsson, K. Schönhammer, Phys. Rev. Lett. **50**, 604 (1983)
120. O. Gunnarsson, K. Schönhammer, Phys. Rev. B **28**, 4315 (1983)
121. C. Suzuki, J. Kawai, M. Takahashi, A.-M. Vlaicu, H. Adachi, T. Mukoyama, Chem. Phys. **253**, 27 (2000)
122. S.P. Kowalczyk, N. Edelstein, F.R. McFeely, L. Ley, D.A. Shirley, Chem. Phys. Lett. **29**, 491 (1974)
123. W.C. Lang, B.D. Padalia, L.M. Watson, D.J. Fabian, P.R. Norris, Faraday Dis. Chem. Soc. **60**, 37 (1975)
124. A.F. Orchard, G. Thornton, J. Electron Spectrosc. Relat. Phenom **13**, 27 (1978)
125. W. Lenth, F. Lutz, J. Barth, G. Kalkoffen, C. Kunz, Phys. Rev. Lett. **41**, 1185 (1978)
126. C.M. Varma, Rev. Mod. Phys. **48**, 219 (1976)
127. P. Wachter, *Handbook on the Physics and Chemistry of Rare earths*, vol. 19 (Elsevier Science, Amsterdam, 1994)
128. B. Lengeler, G. Materlik, J.E. Müller, Phys. Rev. B **28**, 2276 (1983)
129. K. Buschow, Rep. Prog. Phys. **42**, 1373 (1979)
130. L. Falicov, W. Hanke, M.B. Maple (eds.), *Valence Fluctuations in Solids*, (North-Holland, Amsterdam, 1981)
131. G. Krill, J.P. Kappler, A. Meyer, L. Abadli, M.F. Ravet, J. Phys. F: Metal Phys. **11**, 713 (1981)
132. D. Gignoux, F. Givord, R. Lemaire, H. Launois, F. Sayetat, J. Physique (France) **43**, 173 (1982)
133. M. Croft, R. Neifeld, C.U. Segre, S. Raaen, R.D. Parks, Phys. Rev. B **30**, 4164 (1984)

134. R.A. Neifeld, M. Croft, T. Mihalisin, C.U. Segre, M. Madigan, M.S. Torikachvili, M.B. Maple, L.E. Delong, Phys. Rev. B **32**, 6928 (1985)

135. F. Lu, M. Croft, E.G. Spencer, Phys. Rev. B **33**, 5950 (1986)

136. D. Malterre, G. Krill, J. Durand, G. Marchal, Phys. Rev. B **38**, 3766 (1988)

137. J.G. Sereni, G. Nieva, J.P. Kappler, P. Haen, J. Phys. Fr **1**, 1499 (1991)

138. M. Daniel, S.-W. Han, C.H. Booth, A.L. Cornelius, P.G. Pagliuso, J.L. Sarrao, J.D. Thompson, Phys. Rev. B **71**, 054417 (2005)

139. P. Burroughs, A. Hamnett, A.F. Orchard, G. Thornton, J. Chem. Soc. Dalton Trans. **17**, 1686 (1976)

140. A. Bianconi, A. Marcelli, H. Dexpert, R. Karnatak, A. Kotani, T. Jo, J. Petiau, Phys. Rev. B **35**, 806 (1987)

141. H. Dexpert, R.C. Karnatak, J.M. Esteva, J.P. Connerade, M. Gasgnier, P.E. Caro, L. Albert, Phys. Rev. B **36**, 1750 (1987)

142. G. Kaindl, G.K. Wertheim, G. Schmiester, E.V. Sampathkumaran, Phys. Rev. Lett. **58**, 606 (1987)

143. G. Kaindl, G. Schmiester, E.V. Sampathkumaran, P. Wachter, Phys. Rev. B **38**, 10174 (1988)

144. G. Kalkowski, G. Kaindl, G. Wortmann, D. Lentz, S. Krause, Phys. Rev. B **37**, 1376 (1988)

145. D. Ravot, C. Godart, J.C. Achard, P. Lagarde, in *Valence Fluctuations in Solids*, eds. by L.M. Falicov, W. Hanke, B. Maple (North Holland Publishing Company, Amsterdam, 1981), p. 423

146. D. Malterre, C. Brouder, G. Krill, E. Beaurepaire, B. Carrière, D. Chandesris, Europhys. Lett. **15**, 687 (1991)

147. J.P. Kappler, E. Beaurepaire, G. Krill, J. Serenis, C. Godart, G. Olcese, J. Physique I (France) **1**, 1381 (1991)

148. J.M. Tarascon, Y. Isikawa, B. Chevalier, J. Etourneau, P. Hagenmuller, M. Kasaya, J. Physique (France) **41**, 1141 (1980)

149. E. Beaurepaire, J.P. Kappler, G. Krill, Solid State Commun. **57**, 145 (1986)

150. M. Mizumaki, S. Tsutsumi, F. Iga, J. Phys. Conf. Ser **176**, 012034 (2009)

151. R.M. Martin, J.B. Boyce, J.W. Allen, F. Holtzberg, Phys. Rev. Lett. **44**, 1275 (1980)

152. H. Launois, M. Rawiso, E. Holland-Moritz, R. Pott, D. Wohlleben, Phys. Rev. Lett. **44**, 1271 (1980)

153. E.E. Vainshtein, S.M. Blokhin, Y.B. Paderno, Sov. Phys. Solid State **6**, 2318 (1965)

154. J. Jarrige, H. Ishii, Y.Q. Cai, J.-P. Rueff, C. Bonnelle, T. Matsumura, S.R. Shieh, Phys. Rev. B **72**, 075122 (2005)

155. C. Dallera, M. Grioni, A. Shukla, G. Vanko, J.L. Sarrao, J.P. Rueff, D.L. Cox, Phys. Rev. Lett. **88**, 196403 (2002)

156. D. Ottewell, E.A. Stewardson, J.E. Wilson, J. Phys. B: Atom. Molec. Phys. **6**, 2184 (1973)

157. A. Sonder, Thèse, University Paris VI, 1988

158. A. Sonder, E. Belin, C. Bonnelle, J. Physique (France) C9-1021, (1987)

159. A. Fujimori, Phys. Rev. B **28**, 2281 (1983)

160. A. Fujimori, Phys. Rev. B **28**, 4489 (1983)

161. J.W. Allen, J. Mag. Mag. Mat. **47–48**, 168 (1985)

162. G. Kaindl, W.D. Brewer, G. Kalkowski, F. Holtzberg, Phys. Rev. Lett. **51**, 2056 (1983)

163. K.R. Bauchspiess, W. Boksch, E. Holland-Moritz, H. Launois, R. Pott, D. Wollheben, in *Valence Fluctuations in Solids*, ed. by L.M. Falicov, W. Hanke, M.B. Maple (North-Holland, 1981), p. 417

164. J.C. Fuggle, F.U. Hillebrecht, J.-M. Esteva, R.C. Karnatak, O. Gunnarsson, K. Schönhammer, Phys. Rev. B **27**, 4637 (1983)

165. G. Kaindl, G. Kalkowski, W.D. Brewer, B. Perscheid, F. Holtzberg, J. Appl. Phys. **55**, 1910 (1984)

166. P. Hansmann, A. Severing, Z. Hu, M.W. Haverkort, C.F. Chang, S. Klein, A. Tanaka, H.H. Hsieh, H.-J. Lin, C.T. Chen, B. Fak, P. Lejay, L.H. Tjeng, Phys. Rev. Lett. **100**, 066405 (2008)

167. D. Wieliczka, J.H. Weaver, D.W. Lynch, C.G. Olson, Phys. Rev. B **26**, 7056 (1982)
168. T. Hanyu, H. Ishii, M. Yanagihara, T. Kamada, T. Miyahara, H. Kato, K. Naito, S. Suzuki, T. Ishii, State Commun. **56**, 381 (1985)
169. G. Kalkowski, C. Laubschat, W.D. Brewer, E.V. Sampathkumaran, M. Domke, G. Kaindl, Phys. Rev. B **32**, 2717 (1985)
170. B.T. Thole, G. vander Laan, G.A. Sawatzky, Phys. Rev. Lett. **55**, 2086 (1985)
171. C. Giorgetti et al., Phys. Rev. B **48**, 12732 (1993)
172. A. Delobbe, A.-M. Dias, M. Finazzi, L. Stichauer, J.P. Kappler, G. Krill, Europhys. Lett. **43**, 320 (1998)
173. A. Kotani, Eur. Phys. J. B **47**, 3 (2005)
174. N. Jaouen, S.G. Chiuzbaian, C.F. Hague, R. Delaunay, C. Baumier, J. Lüning, A. Rogalev, G. Schmerber, J.-P. Kappler, Phys. Rev. B **81**, 180404(R) (2010)
175. E. Wuilloud, H.R. Moser, W.-D. Schneider, Y. Baer, Phys. Rev. B **28**, 7354 (1983)
176. C. Laubschat, W. Grentz, G. Kaindl, Phys. Rev. B **36**, 8233 (1987)
177. F.U. Hillebrecht, J.C. Fuggle, G.A. Sawatzky, M. Campagna, O. Gunnarsson, K. Schönhammer, Phys. Rev. B **30**, 1777 (1984)
178. E. Wuilloud, B. Delley, W.D. Schneider, Y. Baer, Phys. Rev. Lett. **53**, 202 (1984)
179. S.-J. Oh, J.W. Allen, Phys. Rev. B **29**, 589 (1984)
180. J.A. Fortner, E.C. Buck, Appl. Phys. Lett. **68**, 3817 (1996)
181. L.A.J. Garvie, P.R. Buseck, J. Phys. Chem. Solids **60**, 1943 (1999)
182. D. Wieliczka, C.G. Olson, D.W. Lynch, Phys. Rev. B **29**, 3028 (1984)
183. S.H. Liu, K.-M. Ho, Phys. Rev. B **26**, 7052 (1982)
184. B.I. Min, H.J.F. Jansen, T. Oguchi, A.J. Freeman, Phys. Rev. B **33**, 8005 (1986)
185. H. Eckardt, L. Fritsche, J. Phys. F: Met. Phys. **16**, 1731 (1986)
186. F. Patthey, B. Delley, W.-D. Schneider, Y. Baer, Phys. Rev. Lett. **55**, 1518 (1985)
187. F. Patthey, J.-M. Imer, W.-D. Schneider, H. Beck, Y. Baer, B. Delley, Phys. Rev. B **42**, 8864 (1990)
188. J.W. Allen, R.M. Martin, Phys. Rev. Lett. **49**, 1106 (1982)
189. L.Z. Liu, J.W. Allen, O. Gunnarsson, N.E. Christensen, O.K. Andersen, Phys. Rev. B **45**, 8934 (1992)
190. E. Weschke, C. Laubschat, T. Simmons, M. Domke, O. Strebel, G. Kaindl, Phys. Rev. B **44**, 8304 (1991)
191. C. Laubschat, E. Weschke, M. Domke, C.T. Simmons, G. Kaindl, Surf Sci **269/270**, 605 (1992)
192. M.D. Nunez-Regueiro, M. Avignon, Phys. Rev. Lett. **55**, 615 (1985)
193. G. Kalkowski, E.V. Sampathkumaran, C. Laubschat, M. Domke G. Kaindl, Solid State. Commun. **55**, 977 (1985)
194. M. Taniguchi, J. Alloy Compd **362**, 133 (2004)
195. S.-H. Yang, H. Kumigashira, T. Yokoya, A. Chainani, T. Takahashi, H. Takeya, K. Kadowaki, Phys. Rev. B **53**, R11946 (1996)
196. J.-S. Kang, C.G. Olson, M. Hedo, Y. Inada, E. Yamamoto, Y. Haga, Y. Onuki, S.K. Kwon, B.I. Min, Phys. Rev. B **60**, 5348 (1999)
197. C. Laubschat, E. Weschke, C. Holtz, M. Domke, O. Strebel, G. Kaindl, Phys. Rev. Lett. **65**, 1639 (1990)
198. J.W. Allen, S.J. Oh, O. Gunnarsson, K. Schönhammer, M.B. Maple, M.S. Torikachvill, I. Lindau, Adv. Phys. **35**, 275 (1986)
199. F. Patthey, W.-D. Schneider, Y. Baer, B. Delley, Phys. Rev. Lett. **58**, 2810 (1987)
200. M. Garnier, K. Breuer, D. Purdie, M. Hengsberger, Y. Baer, B. Delley, Phys. Rev. Lett. **78**, 4127 (1997)
201. J.J. Joyce, A.J. Arko, J. Lawrence, P.C. Canfield, Z. Fisk, R.J. Bartlett, J.D. Thompson, Phys. Rev. Lett. **68**, 236 (1992)
202. D. Malterre, M. Grioni, P. Weibel, B. Dardel, Y. Baer, Europhys. Lett. **20**, 445 (1992)
203. R.D. Parks, B. Reihl, N. Martensson, F. Steglich, Phys. Rev. B **27**, 6052 (1983)

204. M. Okawa, M. Matsunami, K. Ishizaka, R. Eguchi, M. Taguchi, A. Chainani, Y. Takata, M. Yabashi, K. Tamasaku, Y. Nishino, T. Ishikawa, K. Kuga, N. Horie, S. Nakatsuji, S. Shin, Phys. Rev. Lett. **104**, 247201 (2010)
205. M. Matsunami, R. Eguchi, T. Kiss, K. Horiba, A. Chainani, M. Taguchi, K. Yamamoto, T. Togashi, S. Watanabe, X.Y. Wang, C.T. Chen, Y. Senba, H. Ohashi, H. Sugawara, H. Sato, H. Harima, S. Shin, Phys. Rev. Lett. **102**, 036403 (2009)
206. D. Ehm, F. Reinert, G. Nicolay, S. Schmidt, S. Hüfner, R. Claessen, V. Eyert, C. Geibel, Phys. Rev. B **64**, 235104 (2001)
207. T. Okane, T. Ohkochi, Y. Takeda, S.-I. Fujimori, A. Yasui, Y. Saitoh, H. Yamagami, A. Fujimori, Y. Matsumoto, M. Sugi, N. Kimura, T. Komatsubara, H. Aoki, Phys. Rev. Lett. **102**, 216401 (2009)
208. A. Yasui, S.-I. Fujimori, I. Kawasaki, T. Okane, Y. Takeda, Y. Saitoh, H. Yamagami, A. Sekiyama, R. Settai, T.D. Matsuda, Y. Haga, Y. Onuki, Phys. Rev. B **84**, 195121 (2011)
209. R. Lässer, J.C. Fuggle, M. Beyss, M. Campagna, F. Steglich, F. Hulliger, Physica **102B**, 360 (1980)
210. D.D. Koelling, A.M. Boring, J.H. Wood, Solid State. Commun. **47**, 227 (1983)
211. M.V. Ryzhkov, V.A. Gubanov, YuA Teterin, A.S. Baev, Z. Phys. B: Condensed Matter **59**, 1 (1985)
212. A. Fujimori, Phys. Rev. Lett. **53**, 2518 (1984)
213. E. Wuilloud, B. Delley, W.-D. Schneider, Y. Baer, Phys. Rev. Lett. **53**, 2519 (1984)
214. A. Bianconi, T. Miyahara, A. Kotani, Y. Kitajima, T. Yokoyama, H. Kuroda, M. Funabashi, H. Arai, T. Ohta, Phys. Rev. B **39**, 3380 (1989)
215. A. Franciosi, J.H. Weaver, N. Martensson, M. Croft, Phys. Rev. B **24**, 3651 (1981)
216. M.S.S. Brooks, J. Magn. Magn. Mater. **47–48**, 260 (1985)
217. H. Ai-Min, B. Jing, Chin. Phys. B **22**, 107102 (2013)
218. M. Campagna, E. Bucher, G.K. Wertheim, D.N.E. Buchanan, L.D. Longinotti, Phys. Rev. Lett. **32**, 885 (1974)
219. H.J.F. Jansen, A.J. Freeman, R. Monnier, J. Magn. Mater **47–48**, 459 (1985)
220. V.N. Antonov, B.N. Harmon, A.N. Yaresko, Phys. Rev. B **63**, 205112 (2001)
221. A.I. Lichtenstein, M.I. Katsnelson, Phys. Rev. B **57**, 6884 (1998)
222. Y. Ufuktepe, S. Kimura, T. Kinoshita, K.G. Nath, H. Kumigashira, T. Takahashi, T. Matsumura, T. Suzuki, H. Ogasawara, A. Kotani, J. Phys. Soc. Jpn **67**, 2018 (1998)
223. P. Wachter, S. Kamba, M. Grioni, Phys. B **252**, 178 (1998)
224. S. Lebègue, G. Santi, A. Svane, O. Bengone, M.I. Katsnelson, A.I. Lichtenstein, O. Eriksson, Phys. Rev. B **72**, 245102 (2005)
225. M. Campagna, E. Bucher, G.K. Wertheim, L.D. Longinotti, Phys. Rev. Lett. **33**, 165 (1974)
226. J. Jarrige, H. Yamaoka, J.-P. Rueff, J.-F. Lin, M. Taguchi, N. Hiraoka, H. Ishii, K.D. Tsuei, K. Imura, T. Matsumura, A. Ochiai, H.S. Suzuki, A. Kotani, Phys. Rev. B **87**, 115107 (2013)
227. J.-N. Chazalviel, M. Campagna, G.K. Wertheim, P.H. Schmidt, Phys. Rev. B **14**, 4586 (1976)
228. J.D. Denlinger, J.W. Allen, J.-S. Kang, K. Sun, B. Min, D.-J. Kim, Z. Fisk, JPS. Conf. Proc **3**, 017038 (2014)
229. N. Heming, U. Treske, M. Knupfer, B. Büchner, D.S. Inosov, N.Y. Shitsevalova, V.B. Filipov, S. Krause, A. Koitzsch, Phys. Rev. B **90**, 195128 (2014)
230. A.R.H. Preston, B.J. Ruck, W.R.L. Lambrecht, L.F.J. Piper, J.E. Downes, K.E. Smith, H.J. Trodahl, Appl Phys. Lett. **96**, 032101 (2010)
231. Z.C. Gernhart, J.A.C. Santana, L. Wang, W.N. Mei, C.L. Cheung, Symposium M: Materials and Technology for non volatile Memories
232. J.H. Richter, B.J. Ruck, M. Simpson, F. Natali, N.O.V. Plank, M. Azeem, H.J. Trodahl, A.R.H. Preston, B. Chen, J. McNulty, K.E. Smith, A. Tadich, B. Cowie, A. Svane, M. van Schilfgaarde, W.R.L. Lambrecht, Phys. Rev. B **84**, 235120 (2011)
233. J.-P. Rueff, J.-P. Itié, M. Taguchi, C.F. Hague, J.-M. Mariot, R. Delaunay, J.P. Kappler, N. Jaouen, Phys. Rev. Lett. **403** (2006)

234. J.-P. Rueff, C.F. Hague, J.-M. Mariot, L. Journel, R. Delaunay, J.-P. Kappler, G. Schmerber, A. Derory, N. Jaouen, G. Krill, Phys. Rev. Lett. **93**, 067402 (2004)
235. S.M. Butorin, D.C. Mancini, J.-H. Guo, N. Wassdahl, J. Nordgren, M. Nakazawa, S. Tanaka, T. Uozumi, A. Kotani, Y. Ma, K.E. Myano, B.A. Karlin, D.K. Shuh, Phys. Rev. Lett. **77**, 574 (1996)
236. S.M. Butorin, L.-C. Duda, J.-H. Guo, N. Wassdahl, J. Nordgren, M. Nakazawa, A. Kotani, J. Phys/Condens. Matter **9**, 8155 (1997)
237. J. Demsar, J.L. Sarrao, A.J. Taylor, J. Phys.: Condens. Matter **18**, R281 (2006)

Chapter 5
Actinide Spectroscopy

Abstract Differences and similarities between 5f and 4f electrons are pointed out. Changes of the characteristics of the 5f electrons along the series are described. Spectroscopic analysis of actinide compounds throwing light on the localized or delocalized character of the 5f electrons are discussed by analogy with the observations made for the rare-earths.

Keywords Localized 5f electrons · Actinide metal · Actinide inter-metallics · Actinide compounds

5.1   Introduction

Experimental studies of the actinides beyond neptunium are difficult because they are highly radioactive. The observations are limited by the severe safety precautions that are required. Other constraints are due to the high surface reactivity of the actinides, principally connected with oxidation processes. Actinide metals oxidize rapidly when exposed to air and moisture. One of the major technical difficulties in actinide spectroscopy research is the fast diffusion of oxygen, which leads to a surface degradation and is manifested by the presence of the oxygen spectrum. The studies are often restricted to the range in which the diffusion processes are slowed down by low temperature. The surface oxide layer can be removed in situ by ion sputtering. After several repeated sputtering cycles, oxidation of the surface is generally not detectable. But, the cleaning of bulk samples can prove very difficult.

The presence of each new radioactive element has always been confirmed by the observation of some of its intense X-ray lines. The first identification using this method was made as early as 1939 when Cauchois and Hulubei, the discoverers of astatine ($Z = 85$), proved unequivocally their discovery by their observation of its three emission lines $K\alpha_1$, $L\alpha_1$ and $L\beta_1$ [1, 2]. Concerning the actinides, one can mention as an example the identification of 102 nobelium from the observation of

© Springer Science+Business Media Dordrecht 2015

C. Bonnelle and N. Spector, *Rare-Earths and Actinides in High Energy Spectroscopy*, Progress in Theoretical Chemistry and Physics 28, DOI 10.1007/978-90-481-2879-2_5

its K emissions [3]. Electron probe microanalysis was widely employed to characterize actinide compounds in single crystal or polycrystal form [4], to control the quality of newly synthesized compounds, to detect impurities and composition variations [5].

X-ray absorption spectra were observed in various energy ranges, L_{III}, M_{III}, M_{IV-V}, N_{IV-V}, initially in order to determine the energy of the levels, then to obtain the densities of unoccupied states, in relation with the theoretical predictions. X-ray emission spectra stimulated by electrons were observed in the same energy ranges with particular attention to the nd–$5f$ resonant emissions. However, the number of the elements studied in a systematic manner is small.

Photoemission has been one of the mostly used methods for the experimental determination of the actinide electronic structure. The valence states were observed preferably with the He II line at 40.8 eV as incident radiation. But at this energy, the photoionisation cross sections of 5f and 6d sub shells are nearly equal and it is difficult to distinguish between their contributions. However, the 5f photoionisation cross section is considerably larger at 40.8 eV than at 21.2 eV (He I line) and comparison between the results obtained using these two different excitation energies brings out the contribution of the 5f electrons to the spectrum. XPS was used to measure the binding energies of the core levels in metals and compounds and to observe the valence bands. Particular attention was given to the changes undergone by the 5f levels with a variation of the valence electron distribution. Auger electrons emitted in resonant conditions were also observed. As already underlined, the photoemission is extremely sensitive to the surface contaminants among which oxygen is the most abundant, easily detectable by its O2p emission. Because of the fast oxygen diffusion and degradation of the solid samples, thin layers deposited on an ad hoc substrate have often to be used.

Studies using synchrotron radiation have been developed in order to observe the inelastic X-ray scattering and fluorescence emission in the vicinity of an absorption threshold. Experiments were made in the region of the L_{III}, $M_{IV,V}$ and $N_{IV,V}$ thresholds but are not numerous.

These experimental studies have as essential aim to determine the presence and the number of 5f localized electrons in solid actinides. Among the parameters studied, the intensity of the spectral characters associated with the $5f_{5/2}$ or $5f_{7/2}$ sub shells can be considered. These parameters depend on the filling of the 5f sub shells, therefore on the coupling in the 5f sub shells. For non-localized 5f electrons the two $5f_{5/2}$ or $5f_{7/2}$ sub shells are filled simultaneously and the coupling is of the LS type. In contrast, when localized electrons are present, the $5f_{5/2}$ sub shell is filled preferentially and for the elements near the half-shell j–j coupling prevails. This is the case of americium with six 5f electrons and no observed magnetic moment. For the other actinides, the problem is complex, particularly for plutonium because its number of 5f electrons is considered [6] to be 5.2 and the metal is not magnetic.

5.2 Density of Unoccupied States

5.2.1 X-ray Absorption

The L_{III} ($2p_{3/2}$) absorption energies in thorium[7], uranium [8] and plutonium [9] have been known for a long time. This absorption takes place in the Fermi level and, consequently, its energy was considered as equal to the energy of the L_{III} level and used as reference. The other levels were deduced by combining the energy of the L_{III} absorption discontinuity with those of the L emission lines [10]. The values obtained for metal uranium are given in Table 5.1. The levels of plutonium were determined from the L_{III} absorption of the metal [9] and the L emissions emitted by fluorescence from a sample of PuO_2 [11, 12]. They are given also in Table 5.1. The accuracy was approximately 1 eV. These values are in agreement with those obtained later from photoelectron spectroscopy [13–16] except for the more exterior unresolved levels.

Table 5.1 Energy levels of Uranium and Plutonium: (1) [10]; (2) [12]; (3) [14]

	Plutonium		Uranium	
	(2)	(3)	(1)	(3)
L_I	23,099	23098.9	21,753	21757.4
L_{II}	22,259	22262.4	20,943	20947.6
L_{III}	18,050	18053.2	17,163	17166.3
M_I	5928	5929.3	5548	5548.0
M_{II}	5546	5542.9	5181	5182.2
M_{III}	4562	4558.4	4302	4303.4
M_{IV}	3968	3969.0	3725	3727.6
M_V	3774	3774.6	3550	3551.7
N_I	1556	1555.0	1439	1440.8
N_{II}	1379	1373.8	1272	1272.6
N_{III}	1119	1116.5	1042	1044.9
N_{IV}	846	845.1	780	780.4
N_V	797	797.9	738	737.7
$N_{VI,VII}$	420	418.6	380	391.3
				380.9
O_I	345	348.2	325	323.7
O_{II}	282	275.7	256	259.3
O_{III}	212	208.2	198	195.1
O_{IV}	102	112.3	97	105.0
O_V	102	101.6	97	96.3
P_I			72	70.7
$P_{II,III}$			28	42.3
				32.3

In the L_{III} spectral range of the actinides, 16.3–19.9 keV from thorium to curium, the energy resolution is low and the L_{III} level is between 7 and 10 eV wide. The dipole absorption transitions take place towards the distribution of unoccupied 6d levels. The transitions to 5f are of quadrupolar nature and thus much weaker. Despite the low energy resolution, numerous experiments were made in this spectral range. Experiments are easier because the protective safety screen absorbs well the particles but lets pass the radiation without a marked intensity loss. Consequently, it is possible to work safely around the sample and still get a significant X-ray signal.

L_{III} absorption of uranium glasses and oxides with different oxygen concentrations have been analyzed [17]. A broad main peak, more than 10 eV large, is present at the threshold. It corresponds to transitions from the U $2p_{3/2}$ level to the lowest d, s-like unoccupied band, consisting of the U 6d, 7s orbitals. Additional structures were observed between 10 and 15 eV above the main peak. They are characteristic of the atomic distribution around the emitting atoms. Information on the oxygen–uranium bonds as a function of the inter-atomic distances and the oxygen concentration was deduced from these observations. An analogous spectral shape was observed for the M_{III} absorption. However, the M_{III} main peak is narrower and the resolution of the spectra is better because the width of the inner level is smaller. For the neptunium compounds, the L_{III} absorption edge was observed to shift to higher energies when the valence increases [18]. The presence of a second peak towards the higher energies of the main peak was attributed to the existence of different configurations in the final state, due to a different number of 5f electrons. However, this second peak is located about 15 eV above the threshold and, in this energy range, such a peak could also depend on the atomic arrangement.

In the L_{III} range, the resolution is strongly limited by the lifetime of the $2p_{3/2}$ core hole. In order to improve the energy resolution, partial fluorescence yield mode was used [19]. The spectrometer is tuned to the $L\alpha_1$ ($2p_{3/2}$–$3d_{5/2}$) fluorescence line and the absorption is recorded by monitoring the $L\alpha_1$ intensity as a function of the incident energy. The spectral width is then limited by the width of the final state, i.e. the width of the $3d_{5/2}$ core hole, of the order of 4 eV. However, as underlined in Chap. 3, absorption and fluorescence spectra can be different because relaxation of the configuration with a core hole can take place prior to the fluorescence emission while such a process is absent in absorption. Indeed, an additional peak was observed towards the lower energies of the main peak. It was attributed to the 2p–5f quadrupolar transition. But as already underlined in Chap. 4, the probability of the p–f transition is very small and another possibility has been considered for the rare-earths. The main peak is due to the U $2p_{3/2}$–6d transition. It is observed at lower and slightly higher energy, respectively, in U^{5+} and U^{6+} compounds with respect to the U^{4+} in UO_2 oxide. This unexpected variation between U^{4+} on the one hand and U^{5+} and U^{6+} on the other hand was explained by a very short U–O bond in the U^{5+} and U^{6+} compounds, which has as consequence a strong mixing of the U6d and O2p orbitals and an increase of the covalence character of the bond. The peak located

15 eV above the main peak was attributed to multiple scattering of the U $2p_{3/2}$ photoelectrons along the bond axes. Its energy depends on the chemical interactions, i.e. on the metal–ligand distances and varies according to the compounds.

The L_{III} absorption spectra have recently been used in order to determine what *multiconfigurational states* take part in the absorption process for uranium, plutonium and their inter-metallic compounds. Indeed, the screening of the core hole is different for each electronic configuration and consequently a different energy is associated with each of them in the absorption spectrum. Since the peaks corresponding to each configuration cannot be resolved, the determination of the different configurations is deduced from the analysis of the shape and the width of the total absorption curve. It was deduced from these experiments that, in inter-metallic compounds, uranium could have the f^3, f^2 and f^1 configurations while plutonium could have the f^6, f^5 and f^4 configurations [20].

L_{III} absorption of metal americium was observed simultaneously with RIXS for values of the pressure between the standard value and 23 GPa [21]. No change of the electron configuration was observed except for a slight shift of the L_{III} edge towards the higher energies and a small increase of the 6d bandwidth. Although this spectral range is not adapted to the analysis of the 5f electrons, these results suggest that the stability of the $5f^6$ configuration should be sufficient to explain the absence of an observable change of the 5f electron number under pressure.

The M_{III} ($3p_{3/2}$) absorption corresponds to the transition of a $3p_{3/2}$ electron to the 6d7s distributions and its inflexion point gives the position of the Fermi level in the metal. This absorption was observed together with M_{IV-V} ($3d_{3/2}$–$3d_{5/2}$) absorptions for uranium and plutonium metal and oxide [22, 23]. By comparing the energies of the M_{IV} and M_V absorption lines and the energies of the M_{IV} and M_V levels given in Table 5.1, it is possible to deduce the energy interval between the $5f^{n+1}$ orbitals and the Fermi level E_F in the metal. However, for uranium and plutonium, inaccuracy on the level energies is of the order of the energy differences between E_F and the absorption lines. Then, it is possible to conclude only that the $5f^{n+1}$ orbitals are very near E_F.

As discussed for the rare-earths, the nd photoabsorption gives directly the unoccupied nf level distribution. The variation of the photoabsorption coefficient with photon energy was observed near the M_V ($3d_{5/2}$) and M_{IV} ($3d_{3/2}$) thresholds of Th and ThO_2 [24]. One slightly asymmetric absorption line is seen at each threshold. These lines show a close analogy with those observed in lanthanum. They are characteristic of the presence of localized unoccupied 5f levels energetically mixed with the conduction levels.

M_V ($3d_{5/2}$) and M_{IV} ($3d_{3/2}$) absorptions of uranium in metal and dioxide have been analyzed [25]. Each spectrum presents one absorption line, slightly asymmetric towards the higher energy in the metal, quasi-symmetric and narrower in the oxide. The thickness of the absorber was chosen to have a negligible effect on the shape of the absorption lines (cf. Chap. 3). Indeed, the absorption coefficient varies strongly in this range and the thickness effect is very large. It should lead to a change in the shape of the lines, which should appear strongly asymmetric towards the higher energies in the presence of such effect. Consequently, the experimental shape of the absorption lines can be determined only if the thickness of the absorbent is known. The lines in the oxide are shifted with respect to the metal towards the higher energies by 1.6 eV for M_V and 1.2 eV for M_{IV}. This shift is characteristic of an increase of the oxidation state of the uranium. The absence of absorption jump in suitably chosen experimental conditions shows that the unoccupied 5f levels are highly localized on each ion in the two materials. The N_{IV} ($4d_{3/2}$) and N_V ($4d_{5/2}$) absorptions are analogous to $3d_{3/2}$ and $3d_{5/2}$ absorptions. They present also narrow lines.

In the $5d_{3/2}$–$5d_{5/2}$ edge range, a high absorption probability forms the so-called "giant" resonance [26]. The 5d absorptions of thorium and uranium were observed by XAS and EELS and compared to theoretical predictions deduced from an independent particle model using Hartree–Fock wave functions [27]. Qualitative agreement obtained between experiment and theory confirmed the interpretation of this resonance as the 5d–εf transitions to the continuum states. Independently, photoexcitation cross sections were calculated for metal uranium of configuration $(5f_{5/2})^3$ $(7s6d)^3$ in a Dirac-Fock model while the relativistic photoionization cross sections were calculated in the electric dipole approximation [28]. By comparison with the previous experimental 5d spectrum of metal uranium, the peak at 98 eV was attributed to the $5d_{5/2} \rightarrow 5f_{7/2}$ transitions and the unresolved peak located towards the higher energies to the $5d_{3/2} \rightarrow 5f_{5/2}$ transitions while the broad maximum centred at 112 eV was attributed to the photoionization. The agreement between experimental and theoretical results confirmed that in its metallic state uranium has the configuration $(5f_{5/2})^3$. Another calculation in the random phase approximation has correctly predicted the giant dipole excitation [29]. On the other hand, the configuration in the gas is known as being $(5f_{5/2})^4 7s^2 (J = 0)$. Excitation from 5d can take place only to the $J = 1$ level. This explains why no multiplet splitting exists in the excited state and only a single high peak is seen in the corresponding excitation spectrum [30].

For U_3O_8, the $5d_{3/2,5/2}$ absorption was found in agreement with the spectrum calculated for the $5f^2$ configuration [31]. These results are discussed together with the emission at Sect. 5.3.1. It must be underlined that the very large variation of the absorption coefficient in the 5d region requires the use of very thin absorber screen, difficult to achieve in solids. The thicknesses of the absorbers used are generally too large and this makes the spectra unusable.

The distribution of the unoccupied part of the oxygen np orbitals in UO_2 can be deduced from the O 1s absorption. Comparison between the O 1s absorptions of UO_2 and CeO_2 revealed a close analogy between the two spectra [32]. The two oxides crystallize in the fluorite structure and belong to the same space group. Their unit cell parameters are very close, 5.47 Å for UO_2 and 5.40 Å for CeO_2 and the arrangement of the anions around each cation is the same. One expects the energy distribution of the valence orbitals to be also the same. In CeO_2, no localized 4f electron is present (Chap. 4). All the same, four valence electrons of uranium including one 5f electron participate in the chemical bond. Their orbitals are hybridized with the O2p ones and the ensemble forms the valence band. The two

Fig. 5.1 The $3d_{5/2}$ absorptions of α and δ plutonium [35]

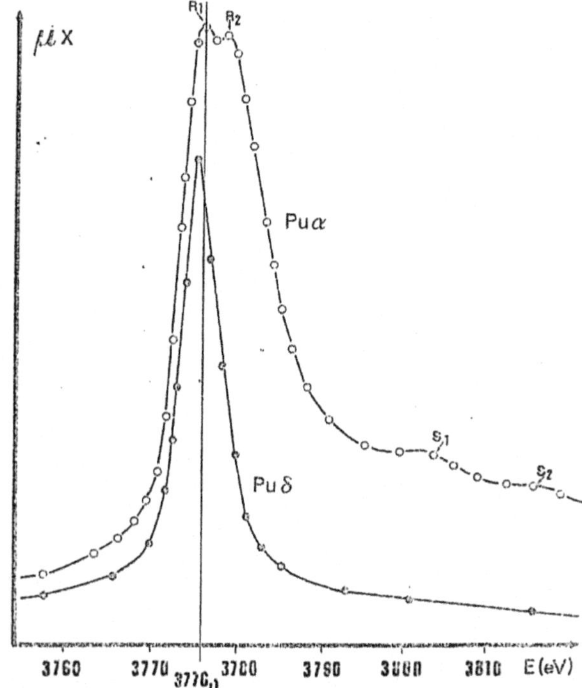

remaining 5f electrons are localized and do not interact with those in the band. The O 1s absorptions of UO_2, U_3O_8 and UO_3 have been compared [33]. Their threshold energies differ by only a few tenths of eV, showing that the transitions take place to unoccupied levels of the same symmetry. The shapes of the O 1s absorption edges observed for UO_2 and U_3O_8 differ only slightly. For UO_3, the absorption edge is clearly less pronounced. This may be explained by an increase of the density of empty delocalized 5f orbitals, situated at the bottom of the conduction band and to which no transition can take place. Beyond 4 eV, the absorption structures differ because they depend on different atomic arrangements.

The 3d photoabsorption of plutonium in its α and δ phases is presented in Fig. 5.1 [34]. The same sample was observed at standard temperature (phase α) and at 400 °C (phase δ) [35]. Marked difference between the absorption spectrum of the two phases near the M_V ($3d_{5/2}$) and M_{IV} ($3d_{3/2}$) thresholds was observed. One quasi-symmetric line is observed for δ-plutonium; it is shifted by about −0.8 eV with respect to the first line of α-plutonium. When the electron localization increases, the screening of the nuclear charge is enhanced. This can explain the above shift to the lower energies. Moreover, no absorption jump is observed for δ-plutonium, showing that the empty 5f levels are not mixed with the conduction states. The absorption shape is similar to that observed for the rare-earths and describes the transitions to unoccupied 5f levels of $3d_{3/2}$ or $3d_{5/2}$ electrons, which are highly localized on each ion. In contrast, for the α-plutonium phase, an absorption jump is observed along with two structures, S_1 and S_2, denoting the presence of transitions to 5f levels, that are hybridized with continuum levels. However, one resonance line is observed in the emission spectrum in coincidence with the R_2 peak. The presence of this line shows that part of the empty 5f levels in α-plutonium is localized; these are the levels towards which a 3d electron can be excited. Consequently, the spectral changes observed between the α and δ phases can be interpreted as due to an increase of the 5f level localization in the high-temperature δ-phase.

5.2.2 Electron Energy Loss Spectroscopy (EELS)

Numerous actinide compounds were investigated by EELS in a transmission electron microscope. When the spectra are produced by a beam of electrons at high accelerating voltages, they are equivalent to X-ray absorption spectra. From observations made at standard temperature and at liquid N_2 temperature, it was shown that UO_2 is stable under the electron beam. However, some materials showed a reduction under electron beam. Thus, reduction of the uranium was seen in $SrCa_2UO_6$ at standard temperature [36]. It was shown that the reduction by the electron beam decreases at low temperature.

As seen previously for the rare-earths, the observations by EELS confirmed a correlation between the intensity ratios of the M_4 and M_5 absorption lines and the number of the 5f electrons. The ratio of the areas beneath the M_5 and M_4 peaks was calculated and it was shown to increase along the series. It increases also significantly with the actinide valence. Empirical relation was then searched between the branching ratio $M_5/(M_4 + M_5)$ and the number of 5f electrons. It was suggested that the branching ratio depended on the spin–orbit coupling of the electrons in the valence band and on the electrostatic interactions between the core hole and valence electrons. It may also depend on solid state effects such as the bonding, but these effects have a minor influence in the actinides. The measurement of the branching ratio associated with M_4 and M_5 absorption lines was then used to determine the valence of uranium compounds and observe its quasi-linear variation with the number of the 5f electrons. An advantage of this method of determination of the oxidation states is that different parts of the same sample can be probed with a high spatial resolution in an electronic microscope.

Analogous experiments were conducted in the $N_{4,5}$ range [37]. The branching ratio $N_5/(N_4 + N_5)$ was obtained [38]. For uranium and plutonium, it is higher for the dioxide than for the metal and the lines of the oxide are narrower. However, it is concluded from these measurements that the valence remains the same in the two materials. Change of the branching ratio between α-Pu, δ-Pu and aged δ-Pu was observed. This change results from the increase of the localization of the 5f electrons due to self-irradiation damage with time. The δ-Pu structure is not centrosymmetric as expected because the electronic structure is dominated by 5f electrons. Consequently, the bonds between each atom of plutonium and its 12 nearest neighbours have largely varying strengths and that makes the crystal highly anisotropic compared to other fcc metals [38].

The branching ratio associated with the transition of a core 4d electron to the unoccupied 5f levels depends on the 5f spin–orbit interaction and the available 5f levels. When the splitting due to the 5f spin–orbit interaction can be neglected in comparison with the other broadenings, the branching ratio depends only on the statistical weights $(2J + 1)$ of the $4d_{5/2}$ and $4d_{3/2}$ levels and is equal to 3/5. In a solid, if the 5f electrons are delocalized into the valence band, their spin–orbit interaction decreases and the branching ratio has the statistical value 3/5. If the electrons are localized, the system is well described in the intermediate coupling scheme. The strength of the spin–orbit interaction was estimated for the actinides from the partial occupancy of valence levels at a given site. The X-ray absorption branching ratio from 4d to 5f levels was determined in order to follow its progress along the series [39]. According to these authors a connection exists between the angular part of the spin–orbit contribution and the degree of 5f localization. Indeed, they have obtained a good agreement between the parameters deduced from the measurement of the branching ratios and those obtained from LDA–DMFT calculations. Comparison was made for the metals and the oxides from uranium to curium.

The $N_{4,5}$ EELS spectra of thorium, uranium, neptunium, plutonium, americium and curium metals observed in a transmission electron microscope are shown in Fig. 5.2 [40]. The intensity variation of the $4d_{3/2}$ peak confirms that the 5f electrons occupy preferentially the $5f_{5/2}$ level up to americium, which has the $5f_{5/2}$ almost full half sub shell with six 5f electrons while only a small number of electrons are present in the $5f_{7/2}$ sub shell, resulting in J–j coupling for this metal. In contrast the $4d_{3/2}$ peak increases for curium, due to intermediate coupling. All the same, in the 5d range, agreement exists between the electron energy loss spectra and the direct photoabsorption peaks. This was observed, for example, for the uranium 5d photoabsorption peaks derived from energy loss data. It was shown that the electron energy loss spectra exhibit similar shapes and positions in the fluorite UF_4 and in the metal [41]. More recently, observation was made in this energy range for thorium, uranium, neptunium, americium and curium metals (Fig. 5.3) [42].

A giant resonance was observed for all the studied metals. It corresponds to the ensemble of the 5d–5f dipole transitions. Indeed, the contribution of the 5d spin–orbit interaction is smaller than that of the 5d–5f electrostatic interactions and all the dipole-allowed transitions are considered as contained within the giant resonance, except for the first actinides, thorium, uranium, neptunium. Structures of low energies, named "prepeak", are observed for these metals. They are due to transitions with $\Delta S = 1$, allowed as a consequence of the influence of the 5d spin–orbit interaction. This has been widely discussed for the transition f^0–d^9f^1. These transitions appear energetically resolved from the giant peak for the lighter actinides. Beyond the americium they become absorbed in the giant peak and therefore energetically unresolved from it. On the other hand, the width of the giant

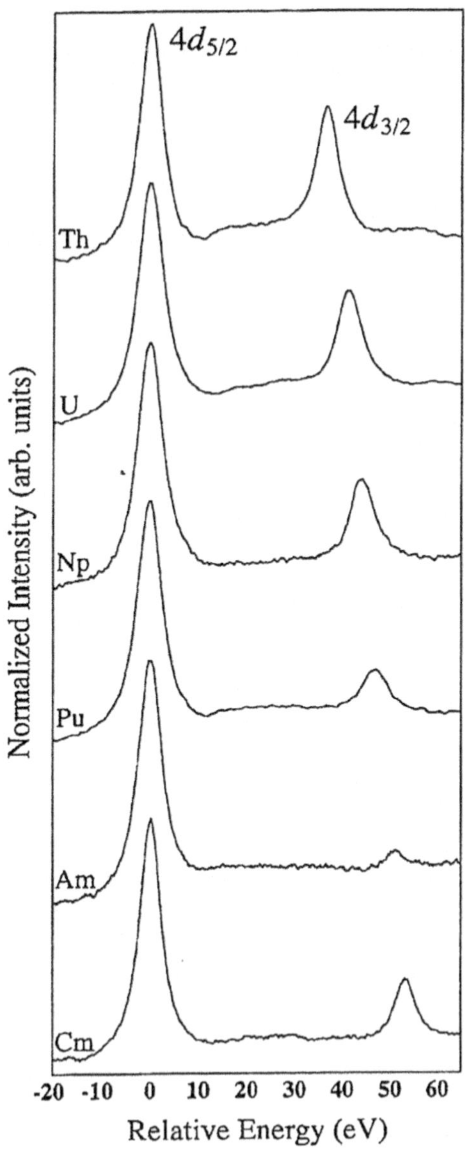

Fig. 5.2 The 4d EELS spectra for Th, U, Np, Pu, Am and Cm metals, normalized to the $4d_{5/2}$ peak height [40]

resonance is strongly reduced between plutonium and americium. This is because for americium the $5f_{5/2}$ half sub shell is filled.

In EELS, a core hole is present in the final state and this hole is not screened, in contrast with photoabsorption, where the photoelectron remains in the vicinity of the core hole and screens it. The presence of the core hole could contribute to partially increase the localization of the 5f electrons.

Fig. 5.3 The 5d EELS spectra for the α-phases of Th, U, Np, Pu, Am and Cm metals [42]

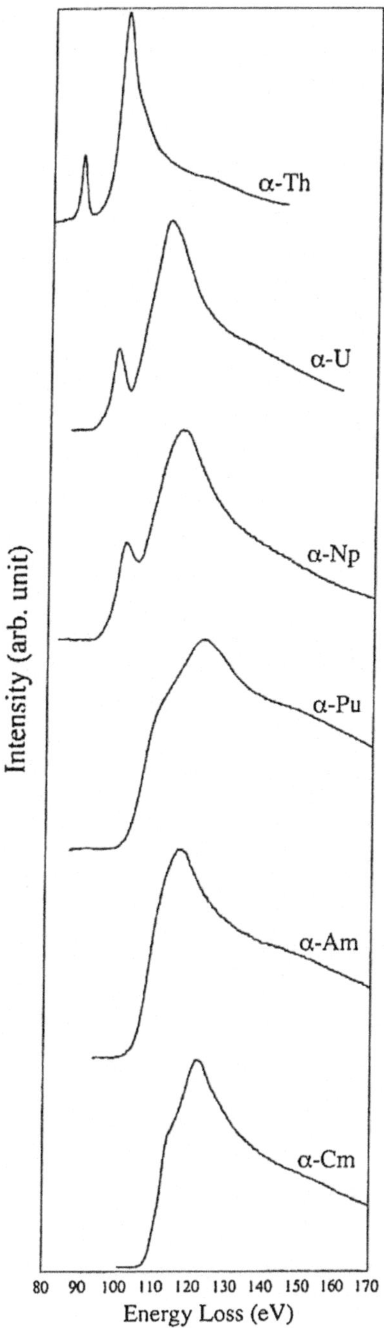

5.2.3   BIS and IPES

No core hole is created by these two experimental methods and this explains the analogy between the spectra obtained by BIS and by IPES despite differences in the energy of the analyzed radiation and, consequently, in the analyzed depth and the spectral resolution.

The BIS spectra of thorium and α-uranium metals show narrow bands characteristic of the 5f unoccupied level distributions in the absence of core hole [43]. The shape of the spectra is rather similar in the two metals but for uranium a band is "pinned" to E_F and the spectrum spreads up to 3 eV while for thorium it is present between 3 and 6 eV above E_F and no 5f level is present at E_F.

The spectra obtained for UO_2 by BIS [44, 45] and by IPES [46] were similar. A broad peak centred at about 5 eV above the bottom of the conduction band was attributed to a distribution of initially unoccupied 5f levels. This interpretation was confirmed by an intensity diminution of the peak in the case of an oxidation of the sample, thus of a decrease in the number of 5f electrons. A shoulder observed at about 2.5 eV was attributed to the presence of unoccupied 6d levels. That confirmed the results obtained from reflectivity measurements of UO_2 single crystals attributing the observed energy gap of 2–3 eV to the transition 5f–6de$_g$ [47, 48]. From these experimental data, the excitation from the valence levels of higher energy to the unoccupied levels of lowest energy was suggested to be of the 5f–6d type and not 5f–5f, as expected for an f–f Mott–Hubbard insulator. More recently, spectra observed by BIS and quasi-identical to that cited previously, were differently interpreted [49]: it suggested the mixing of unoccupied U6d–O2p levels located beyond +6 eV from the conduction band minimum. However, this suggestion appears unrealistic.

The BIS spectrum of UC obtained with MgKα radiation was also analyzed [50]. A sharp main peak was observed at +2.4 eV above E_F with a shoulder located at +1 eV. The two structures were identified as a spin–orbit split pair of the U 5f level and they are followed by a long tail that extends up to +12 eV into the uranium 6d energy region. However, because of the large U 5f cross section at the MgKα energy, contribution of the U 5f levels is also expected in this energy range.

5.3   Density of Valence States

5.3.1   X-ray Emission

The M_V ($3d_{5/2}$) and M_{IV} ($3d_{3/2}$) emissions of thorium in metal were analyzed in the vicinity of the $3d_{5/2}$ and $3d_{3/2}$ thresholds [34]. The core hole was created by electronic bombardment at energy near the threshold. The emission spectrum was compared to the absorption spectrum observed under the same experimental conditions (Fig. 5.4). As in the emission spectrum of lanthanum, $3d_{5/2}$ and $3d_{3/2}$

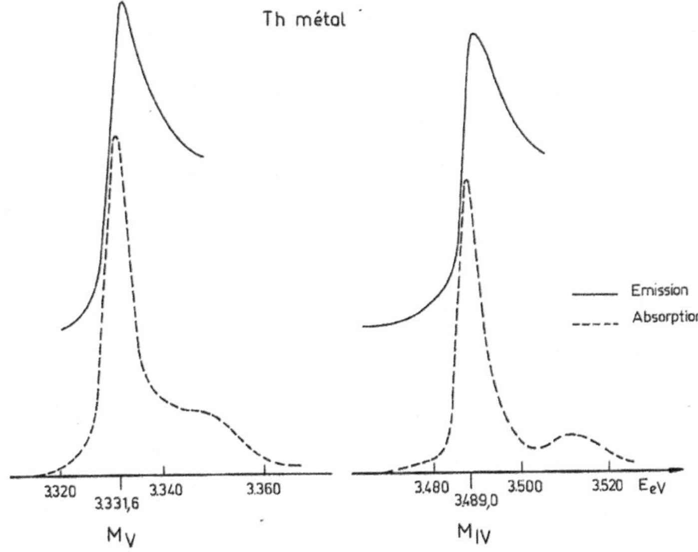

Fig. 5.4 The 3d absorption and emission spectra of thorium metal

resonance lines were recorded in coincidence with the corresponding absorption lines. The appearance of resonance lines shows the strongly localized character of the 5f electron of the $3d^{-1}5f^1$ excited configuration increasing its recombination probability during the lifetime of the 3d hole. The intensity ratio of the $3d_{5/2}$ and $3d_{3/2}$ lines is the same in emission and absorption. However, their shape is different. The resonance emission lines are highly asymmetrical towards higher energies and wider than the corresponding absorption lines. These characteristics reveal the existence of an interaction between the excited 5f levels and the other unoccupied ones. This interaction is absent for lanthanum, whose emission and absorption lines have similar shapes. The thorium 5f excited levels are 1.7 eV above the Fermi level and no 5f electron is initially present in the metal. Weak features were observed towards the low energies of the resonance lines in the oxide ThO_2 (Fig. 5.5). They are characteristic of the presence of valence electrons. Indeed, no 4f electrons are present in the La_2O_3 3d emissions and strong analogy is observed between the ThO_2 5f–5d and La_2O_3 4f–4d resonant emissions [51].

The M_V and M_{IV} emissions and absorptions of uranium in metal are represented in Fig. 5.6 [25]. Large variation of the emission intensity as a function of the energy E_0 of the incident electrons was observed around the $3d_{5/2}$ threshold. The energy of the $3d_{5/2}$ level is equal to 3.55 keV. For E_0 equal to 4.3 keV, the emission spectrum includes the M_{IV}–O_{III} normal line and two intense resonance lines in coincidence with the two M_{IV} and M_V absorption lines. The energy of the M_{III} threshold is close to 4.3 keV. For E_0 equal to 5 keV, the ionization can take place from the M_{III} level. The M_{III}–N_{IV} and M_{III}–N_V ($M\gamma$) lines with energy near the $3d_{5/2}$ threshold were

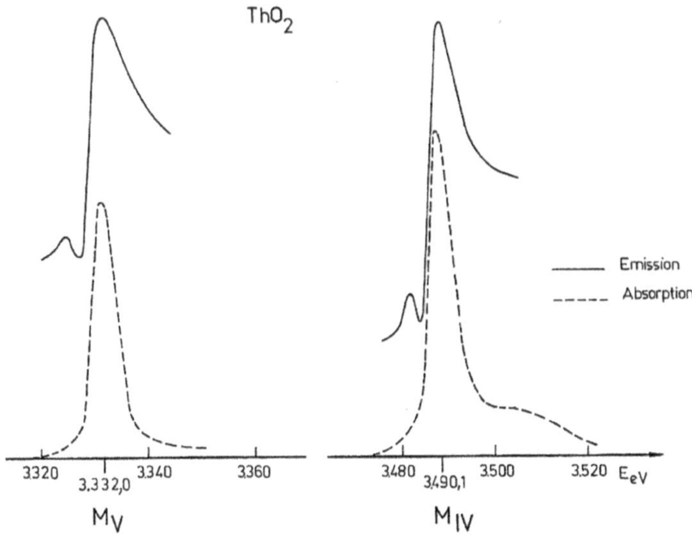

Fig. 5.5 The 3d absorption and emission spectra of thorium oxide ThO_2

observed. An additional emission was observed towards the low energies of the M_V resonance line. It was interpreted as the transition of an itinerant 5f electron towards the $3d_{5/2}$ hole. This transition can be considered as a 5f band emission and reveals the presence of itinerant 5f electrons. It is predicted towards the lower energies of the resonance line, in contrast to the rare-earth 4f–3d normal emissions of the $4f^{n-1}$ final configuration, which correspond to the transition of a localized electron and are located towards the higher energies of the resonance lines. As expected, when E_0 increases, the intensity of the resonance lines decreases while that of the normal lines and the valence band increases strongly. Consequently, at E_0 equal to 8 keV, the M_V resonance line was not resolved from the 5f band and especially from the very close and intense Mγ line. From these observations, it was deduced that localized 5f electrons, of the type analogous to the 4f electrons of the rare-earths, were present in the uranium metal simultaneously with itinerant 5f electrons. No transition of 5f delocalized electrons to $3d_{3/2}$ core hole was observed and the $3d_{3/2}/3d_{5/2}$ intensity ratio of the resonance lines is higher than the ratio of the absorption lines. Differences between the M_{IV} and M_V emission spectra suggest that the 5f electrons are localized mainly in $J = 5/2$ levels. The transition probabilities of localized electrons are higher than those of delocalized electrons (cf. Chap. 3) and therefore the relative number of the two types of electrons cannot be deduced from the intensities of the emissions. The M_{III} absorption threshold was measured in the same sample in order to determine the energy of the Fermi level. Based on the energy differences between the M_{III}, M_{IV} and M_V levels, the itinerant 5f electrons were centred around -2 eV under the Fermi level and the excited states less than 1 eV above [52].

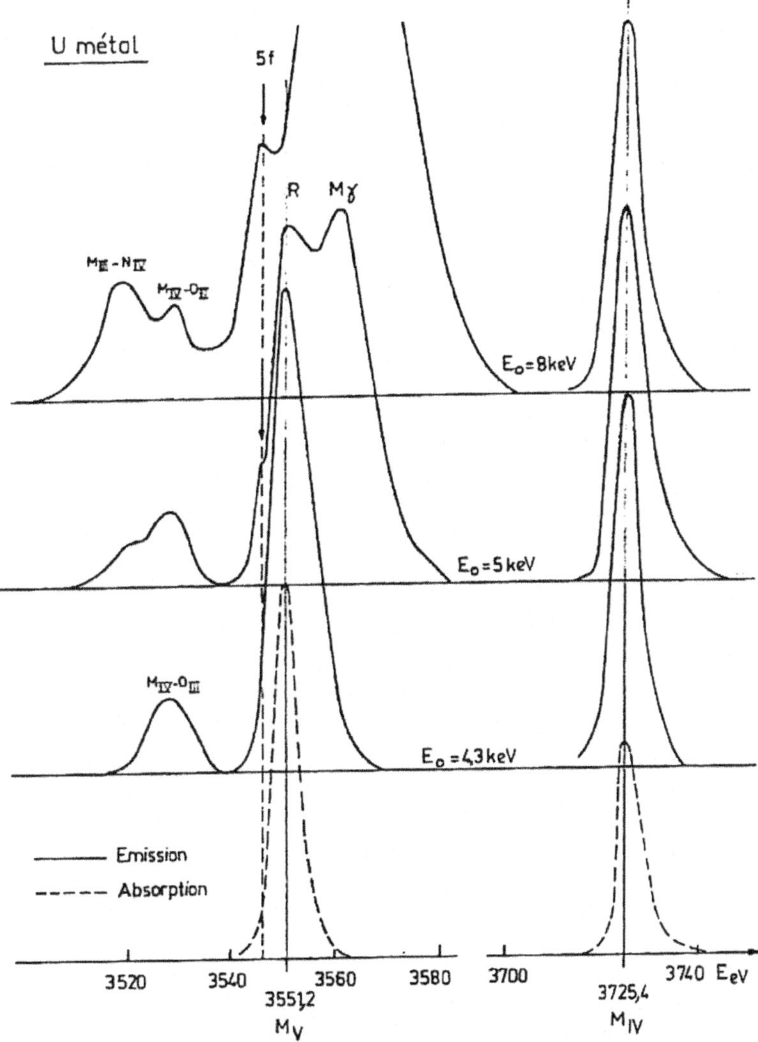

Fig. 5.6 The 3d absorption and emission spectra of α-uranium: the emissions are observed under irradiation by electrons of 4.3, 5 and 8 keV [25]

The α-plutonium emission spectrum near the M_V absorption is plotted in Fig. 5.7 [34, 53]. At 5.3 keV, the self-absorption effect remains weak and so does the intensity of the normal line Mγ. The resonance line R is the strongest of the emission spectrum. It coincides with the R_2 absorption peak while no emission is observed at the energy of the R_1 absorption peak. At higher incident energies, the intensity of R decreases while that of Mγ increases strongly. Mγ becomes highly asymmetric towards the higher energy masking the emission due to 5f itinerant

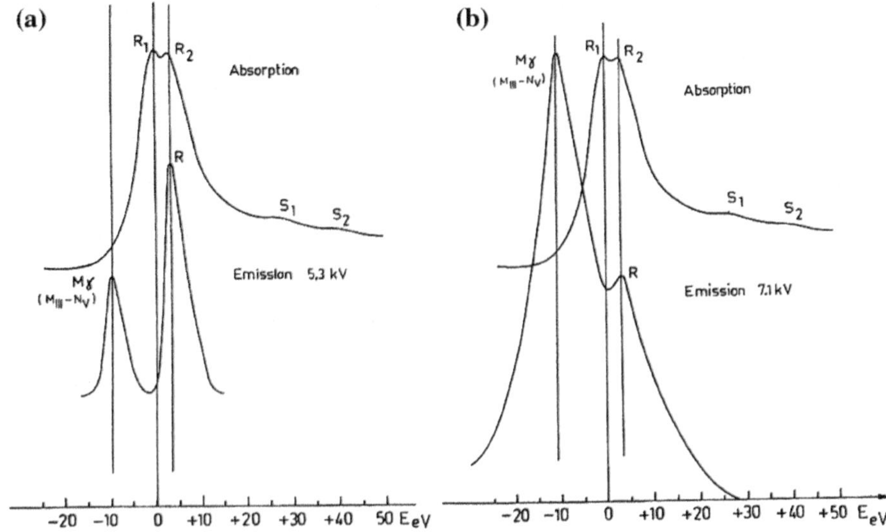

Fig. 5.7 The $3d_{5/2}$ absorption and emission spectra of α-plutonium: the emission is observed under irradiation by electrons **a** of 5.3 keV; **b** of 7.1 keV [34]

electrons. In the M_{IV} range, only an absorption peak is observed, no resonance line is present and emission is observed towards the lower energies of the absorption peak. From these observations, the configuration of α-plutonium is a mixture of delocalized and localized 5f electrons. Since the $5f_{5/2}$ discrete levels are preferentially occupied, the $3d_{3/2}$–5f excitation is very weak and no resonance line is observed in this range (Fig. 5.8) [34]. The $3d_{3/2}$ emission located towards the small energies of the absorption is characteristic of the presence of a 5f band. From the energy of this emission, the average energy of the occupied $5f^n$ distribution is at about 2 eV below the Fermi level.

Changes of the 3d spectra were seen in UO_2 (Fig. 5.9) [25]. They were shifted towards the higher energies by more than 1 eV with respect to the metal spectra. The resonance lines were stronger, narrower and quasi-symmetrical in the oxide. They were still observed at 8 keV. In contrast, the emission due to 5f itinerant electrons was weaker. The 5f electrons are clearly more localized in the oxide than in the metal. The emission from the 5f itinerant electrons was shifted by 3.5 eV towards the higher binding energies with respect to the emission in the metal (Fig. 5.10) [25]. The 5f itinerant electrons in the oxide are then approximately at 6 eV below the Fermi level of the metal and are energetically mixed with the 2p states of the oxygen. Energy level diagrams were deduced from these experimental results [52].

Uranium 4d emission spectrum in UO_2 was also analyzed [54]. The X-ray emission lines taking place between core sub shells are well described by an atomic model in the solid. Let us consider the $4d_{5/2}$–$5p_{3/2}$ emission. In the absence of open

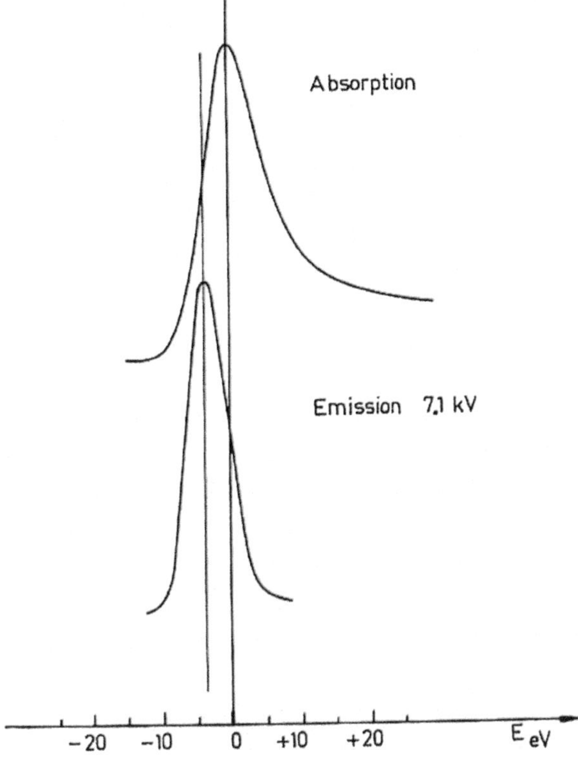

Fig. 5.8 The $3d_{3/2}$ absorption and emission spectra of α-plutonium: the emission is observed under irradiation by electrons of 7.1 keV [34]

sub shell, one expects it to be a discrete line of Lorentzian shape and of width equal to the sum of the widths of the two levels, $4d_{5/2}$ and $5p_{3/2}$. In fact, the observed emission has a complex shape and is extended over about 6 eV. Such a shape suggests that the electrons of the initial and final configurations interact with localized non-coupled electrons of uranium. The $4d_{5/2}(5f^2)$–$5p_{3/2}(5f^2)$ emission was calculated in the ion $5f^2$ with the help of the multiconfigurational Dirac–Fock method (cf. Chap. 3). Each line has a natural Lorentzian shape. This natural shape has little influence on the final shape, which is composed of a large number of transitions between the numerous levels of the initial and final configurations. The above ab initio atomic calculation reproduces well the shape of the emission. The error in the absolute value of the energy is only about 2 %.

Another confirmation of the presence of the $5f^2$ configuration of uranium in UO_2 was obtained from the observation of the $4d_{5/2}^5\ 5f^3$–$5f^2$ resonant emission and $4d_{3/2}^3\ 5f^3$–$6p_{1/2}^1 5f^3$ emission in the presence of an additional excited 5f electron forming a "spectator electron" situation [54]. Each of these two emissions consists of a large number of lines, whose observed mean energies are 743 and 762.5 eV, respectively. The calculated energies of the $4d_{5/2}$–5f transitions in the ion and the excited atom differ only by a few eV's. This difference is of the order of or smaller than the difference between the calculated and experimental values. The same is

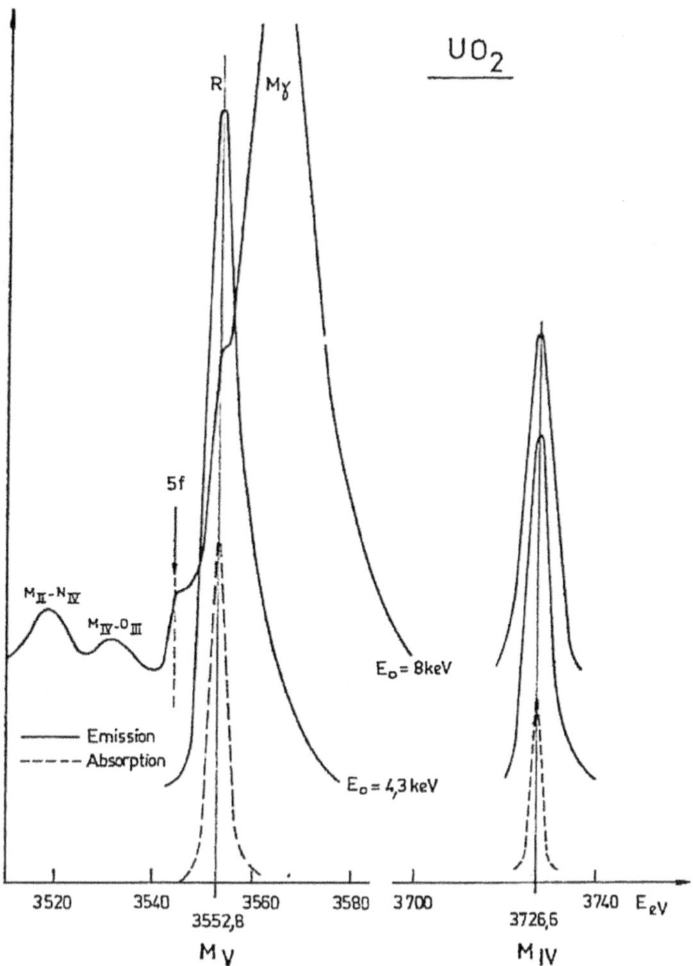

Fig. 5.9 The 3d absorption and emission spectra of UO_2: the emissions are observed under irradiation by electrons of 4.3 and 8 keV [25]

true for the $4d_{3/2}$–6p transition, making it impossible to base the interpretations on calculated energies. In the absence of photoabsorption spectrum, the identification of the resonant emissions was based on the measurement of their relative intensities at different incident electron energies E. In an energy range of several eV around the threshold, excitation is much more probable than ionization. A value of 75 was evaluated for the ratio of the photoexcitation to photoionization cross sections at the U $4d_{5/2}$ threshold energy E_S. Conditions for the observation of resonant spemissions are favourable with incident electrons of energies from E_S up to 1.5 time E_S. In this range, the more intense emissions are from the excited atom and the transitions in the ion are observed as structures of low intensity. However, 4d–5f emission in the

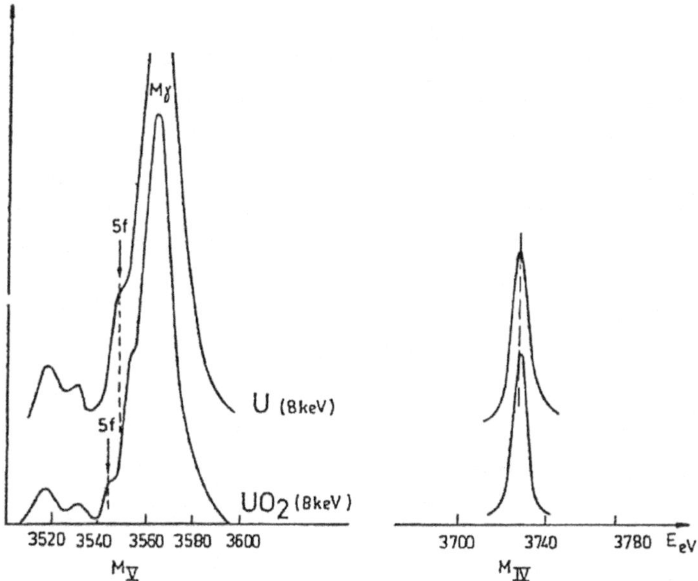

Fig. 5.10 The 3d emission spectra of α-uranium and UO$_2$: the emissions are observed under irradiation by electrons of 8 keV [25]

$4d_{5/2}^{-1}$ ion seems to have been observed [49]. It was accompanied by a line, attributed to a satellite, but, taking into account the energies of the levels as measured in Ref. [55] and the spectra observed in Ref. [54], this line can be attributed to the $4d_{3/2}$–$6p_{1/2}$ emission whose intensity is known to be low. The line in the ion has thus a low intensity.

In the 5d range, both absorption and emission were observed for U_3O_8 [31]. The general shape of the spectra depends on the experimental conditions (Fig. 5.11) [54]. The absorber thickness was too large. In emission, the incident electron energies used were equal to $4E_S$, $12.5E_S$, $33E_S$, leading to strong self-absorption effects. These experimental constraints change the relative heights of the peaks and introduce broadenings that modify strongly their shapes. However, it appears clearly that absorption and emission maxima coincide. The experimental data obtained for U_3O_8, which is not purely $5f^2$, were compared to calculated spectra for the configuration f^2. Despite the above constraints, a general good agreement was obtained between experimental and calculated spectra. As an example, the structure observed at 100 eV in absorption and emission appears also in the calculated spectra and the peak observed in emission at 80 eV is identified as a transition in the presence of a spectator electron. These results confirm the localized character of the $5f^2$ electrons in uranium oxide.

It should be noted that oxygen Kα emission of the three uranium oxides has been observed by RIXS at the oxygen 1s edge [33]. Excitation energy dependence was observed for UO_3 but not for UO_2 and U_3O_8. These results were explained by

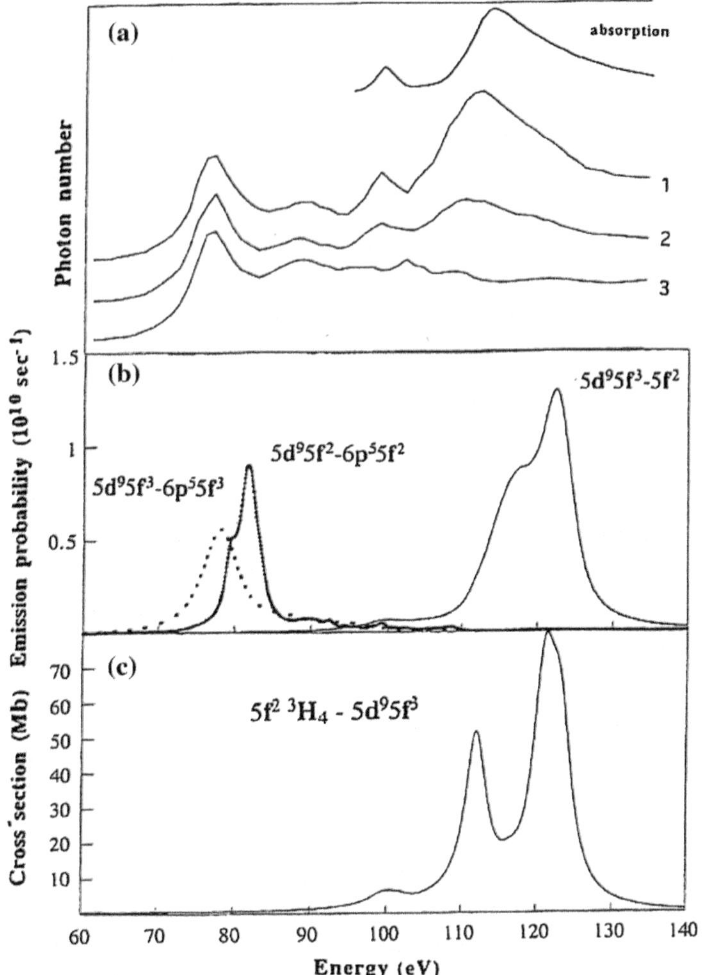

Fig. 5.11 The 5d spectra of UO_2: **a** observed emission at E_0 equal to (1) 500 eV, (2) 1500 eV, (3) 4000 eV, and observed absorption [51]. **b** Calculated emission probability (10^{10} s^{-1}) of U^{4+} resonance emission, U^{4+} $5d^9 5f^3$–$6p^5 5f^3$ and U^{5+} $5d^9 5f^2$–$6p^5 5f^2$. **c** Calculated photoexcitation cross section from U^{4+} $5f^2$ $^3H_4$ [54]

structural differences. In the case of UO_3, the oxygen–uranium hybridization varies with the oxygen site and the contribution of non-equivalent oxygen sites depends on the excitation energy.

In conclusion, as in the 3d spectra of the rare-earths, in the 3d and 4d emission spectra of the actinides transitions to excited and ionized states are simultaneously present. Resonant emissions as well as emissions in the presence of an excited spectator electron are characteristic of configurations with localized electrons and take place in a time scale similar to that of the core-hole lifetime, i.e. of the order of

10^{-14}–10^{-16} s. From the observation of such transitions, one can obtain information on the dynamics of the electrons in the excited configurations of high energy that throw light on the presence and the number of strongly localized electrons. Excitation and decay are two independent processes. Consequently, no interference exists between them and this makes easy the interpretations. This approach can be generalized to complex compounds because of the inner shell aspect of the transitions involved. From the energies of resonance and normal lines, it is possible to deduce the respective energy levels of the localized and itinerant 5f electrons and therefore the energy level diagram of the material.

5.3.2   Photoemission

The valence band of the metal actinides was observed by UPS and XPS [43, 55–57]. Comparison between the photoemission of thorium and uranium metals had revealed the characteristics of the 5f electrons in uranium (Fig. 5.12). An intense

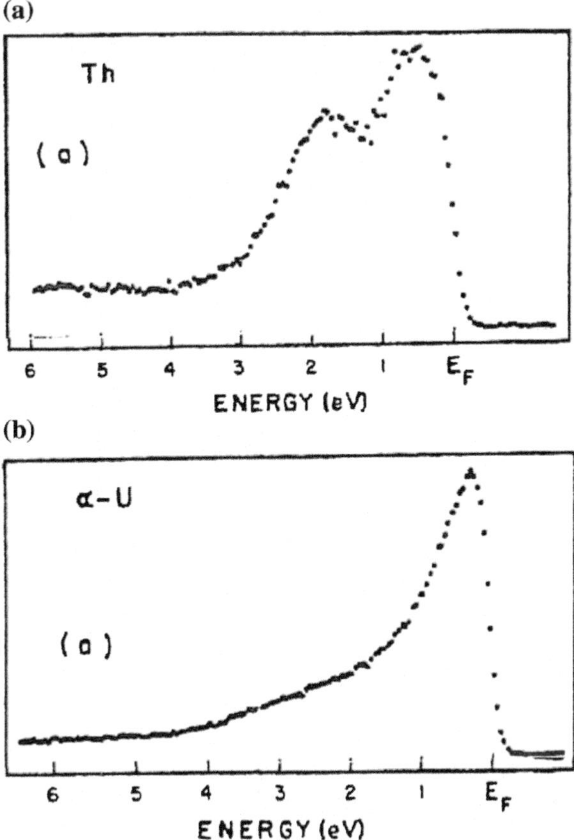

Fig. 5.12 X-ray valence photoemission of **a** thorium; **b** α-uranium [43]

narrow peak characteristic of the U 5f electrons was observed just below the Fermi level in the photoemission of uranium. This peak is strongly asymmetric to the higher binding energies and thus reveals a partial mixing of 5f electrons with the distribution of the 6d7s valence electrons centred some tenth of an eV under E_F. In contrast, for thorium, only the 6d7s valence distributions was observed.

The same situation exists for neptunium and for plutonium in the α-phase as in the δ-phase [58–61]. The photoemission of the two phases of plutonium extends through 4 eV and has a similar shape except near the Fermi energy. In α-plutonium, a broad peak is present and reaches maximum intensity at about 100 meV below E_F. In δ-plutonium, a much narrower intense peak is present just at E_F, in agreement with the theoretical predictions; it is followed by a small peak at about 1 eV below E_F. A very weak structure is observed between these two peaks. The shape of the spectrum observed for the δ-phase is consistent with the increased tendency towards localization. Indeed, the spectrum of a partially filled 5f shell with a localized character presents a multiplet structure. The δ-phase spectrum, calculated assuming four localized 5f electrons out of the five 5f ones, reproduced rather well the observed photoelectron spectrum (Fig. 5.13) [62, 63]. The four localized 5f electrons created a stable singlet state, δ-Pu $5f^4$, located under E_F. For α-Pu, a large region of high binding energies reveals that 5f electrons strongly correlated with the valence electrons are also present. In summary, from uranium to plutonium, the valence band extends towards the higher binding energies. The maximum, which is

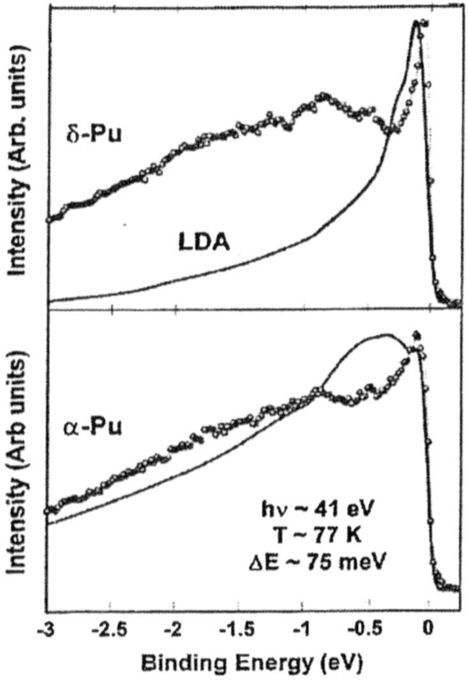

Fig. 5.13 He II valence photoemission of α- and δ-plutonium (*points*) compared to LDA calculations (*solid curves*) [62]

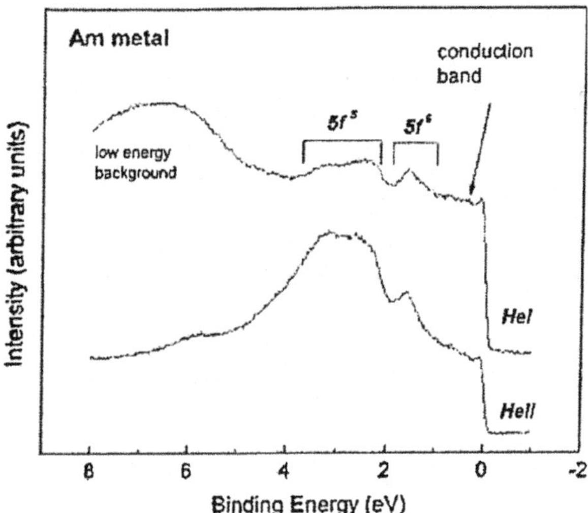

Fig. 5.14 He I and He II valence photoemission of high purity americium metal film [65]

pinned at E_F, becomes narrower and more pronounced with increasing atomic number. At this energy, the observed intensity was attributed to the presence of itinerant 5f electrons. On the other hand, the multiplets observed in the structures towards the greater binding energy of this maximum are clearly more accentuated in plutonium and indicated the presence of an increasing number of localized 5f electrons.

For americium metal, there is a density of occupied states at E_F. However, in the photoemission obtained with He II, the intensity observed at E_F is low compared to the high intensity observed between -5 and -1 eV below E_F [64, 65]. Indeed, in this energy range the observed emission includes a broad peak centred round E_F-3 eV and a secondary peak located at about E_F-1.8 eV. Its shape is almost the same as that of the emission excited by incident photons of low energy (He I). Only the relative intensities of the peaks are different (Fig. 5.14) [65]. This intensity difference is due to the important role played by the photoionisation cross sections in photoemission spectra. The excitation cross section of the 5f electrons increases with the energy of the photons more rapidly than that of the other electrons. The same is verified in the X-ray range, where the probability of the 5f electron excitation is an order of magnitude higher than that of the 6d and 7s electron. From these comparisons it follows that, unlike early actinides, in americium the 5f levels are not centred at E_F but are shifted to higher binding energies. Only the 6d7s valence electrons are present at E_F while the peaks observed below E_F in the He II spectrum are attributed to the 5f electrons. These observations provide the experimental evidence that a change from itinerant to localized 5f electrons occurs between metallic plutonium and americium, in agreement with the difference of their crystalline structures. According to these data, all the 5f electrons of americium would be localized. Owing to the absence of magnetic order in its ground state, americium metal is considered as trivalent, with the ground configuration

$5f^6$ ($J = 0$). This configuration is stable in the actinides. The strong 5f spin–orbit interaction favours the filling of the $5f_{5/2}$ sub shell. The photoemission should, then, give the description of a final state of the configuration $5f^5$. However, the peak observed at E_F-1.8 eV was considered as due to the $5f^6$ final configuration and only the peak at E_F-3 eV to the $5f^5$ final configuration. No interaction exists between the delocalized valence distribution and the localized electrons but an energy mixing is present and that could make possible the localization of an initially valence electron in the 5f levels and explain the stability of the $5f^6$ final configuration. Another interpretation of the two peaks was also suggested, based on the presence of the strong 5f spin–orbit interaction, which could introduce a splitting of almost 1 eV between final levels associated with two different spin components. It must be underlined that the theoretical models used are not adapted to describe localized electrons and no satisfactory fit is obtained between the theoretical predictions and the observed structures. On the other hand, in a research undertaken in order to determine the changes of the electronic structure under pressure [66], partial delocalization of the 5f electrons has been predicted theoretically and experimental work is in progress in this subject.

As indicated in Chap. 1, the localization of the valence electrons increases with the decreasing size of the sample. From UPS and XPS observations of plutonium ultrathin layers, it was suggested that the degree of localization of the 5f electrons also depends on the layer thickness. For a single monolayer, the spectrum is dominated by a broad asymmetrical peak centred on 1.6 eV under E_F (Fig. 5.15) [59]. This non-structured peak was attributed to the presence of localized 5f elec-trons. However, it must be underlined that such a large peak is also present in the oxide spectrum since the oxidation of an ultrathin layer is easy. With increase of the number of layers, this peak is gradually wiped out and its intensity is spread out between 0 and 1 eV below E_F. The shape of the emissions also depends on the

Fig. 5.15 He II valence photoemission of plutonium recorded after ion etching at various temperatures [59]

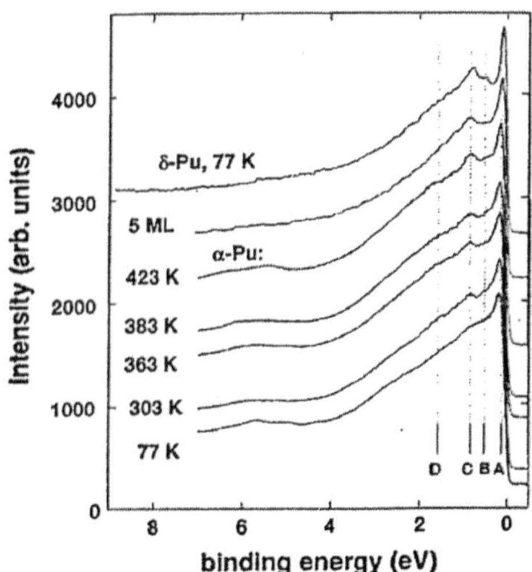

surface characteristics. The 5f electrons present on the surface are calculated to have a more localized character than generally observed because of the decreased coordination number and the presence of defects. These modifications are not reproduced by the calculations.

As already underlined, the oxidation of the actinide surface is rapid. The strong dependence of the photoemission on the surface permits the detection of any contaminant of even a fraction of a monolayer. The photoemission spectra are therefore well adapted to probe a sample of small thickness and analyze its chemical composition. The He II line is mainly used as incident radiation. At this energy, the cross sections of the plutonium 5f, 6d and oxygen 2p sub shells are comparable and the photoelectrons are emitted almost entirely from the first five monolayers. Let us recall that the bond is not ionic in the oxides, contrary to what is often claimed [67]. Significant covalent character is present [68] and increases along the series and with the metal valence. From protactinium to plutonium, each element has several oxidation states. This appears as typical of the partial localization of the 5f electrons. In the dioxides, four electrons of the actinide contribute to the chemical bond. From the XPS spectra of UO_2, the O2p–U5f6d7s valence band is located between E_F-4 and E_F-10 eV. The presence of the localized 5f electrons is detected around −2 eV under the Fermi level. The attribution of this structure to 5f electrons is confirmed by its absence in the ThO_2 photoemission (Fig. 5.16a) [69]. No electron is present at E_F in agreement with the insulating properties and with recent calculations [67]. The same results were obtained by UPS using He II radiation [70]. It was shown

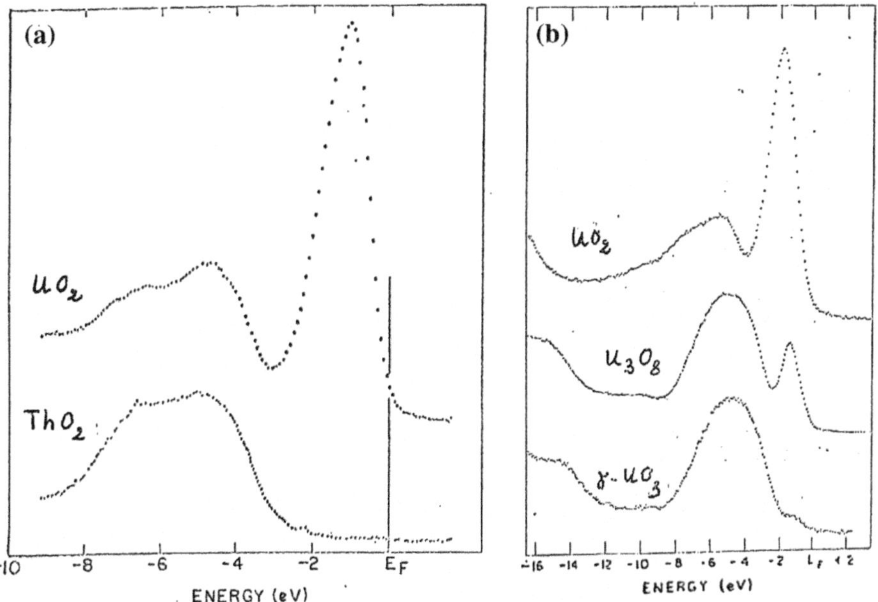

Fig. 5.16 X-ray valence photoemission of a ThO_2 and UO_2. b UO_2, U_3O_8 and γ-UO_3 [69]

that assuming the presence of only one 5f electron in the valence band is sufficient to account for the experimental binding energies [71]. The intensity of the peak at E_F-2 eV decreases with an increased oxidation, namely with a decrease of the non-hybridized 5f electron number. It vanishes for UO_3 (Fig. 5.16b). This confirms the interpretation proposed for this peak. Indeed, in this oxide, all the uranium valence electrons are present in the valence band and no quasi-pure 5f orbital is occupied. The same observation is made in all the hexavalent uranium compounds [72]. For NpO_2, as for the UO_2, the presence of the 5f electrons is characterized by a single narrow line.

Photoemission was observed for UO_2 with photon energies near the U $5d_{IV,V}$ absorption thresholds [73]. The spectra were interpreted as taking place from the $5f^2$ ground state to $5d^9 5f^3$ excited configuration. As for the metal, the transitions to the levels of higher spin were found to correspond to weaker peaks located at 98–100 eV. The transitions to the multiplets with S equal to 1/2 or 1 could be responsible for the strong features at about 108–110 eV. The latter partially overlap the transitions to the continuum and this explains why the excited configurations have low probability to decay by direct recombination.

The experimental densities of states of Pu_2O_3 and PuO_2 are compared in Fig. 5.17 [74] to densities of states obtained from a hybrid DFT-type calculation without spin–orbit splitting. The photoionization cross section of the 2p valence electrons of oxygen is 1.5 bigger than those of the 5f and 6d electrons of plutonium, which are almost equal. However, it was not taken into account in the theoretical curves. Pu_2O_3 is characterized by two peaks at 1.6 and 5.5 eV and PuO_2 by two peaks at 2.5 and 4.6 eV below E_F. The two oxides are predicted to be insulators, in agreement with the experiment. The peaks of lower energies correspond to 5f

Fig. 5.17 Valence photoemission of Pu_2O_3 at room temperature and PuO_2 at 77 K (*points*) compared to calculated densities of states (*line*) [74]

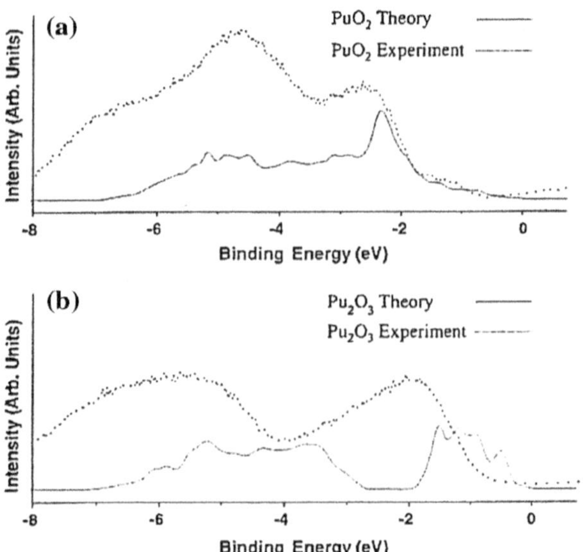

plutonium electrons whereas the peaks of higher binding energies correspond to valence 2p O electrons mixed with the valence electrons of the metal. No fine structure is observed in the spectra. The 5f levels of PuO_2 at E_F-2.5 eV are more bound than those of Pu_2O_3 at E_F-1.6 eV because the valence and the ion charge are higher. The separation energy between the plutonium 5f and valence orbitals decreases with the stabilization of the 5f orbitals. This energy decrease in PuO_2 with respect to Pu_2O_3 is explained by the difference in their crystalline arrangement. In PuO_2, eight oxygen neighbours are present at 2.34 Å whereas in Pu_2O_3 only four neighbours at 2.36 Å and three neighbours at 2.62 Å are present. These changes induce an increase of the Pu–O bond strength and of the covalence in PuO_2.

Actinide dioxides are often non-stoichiometric. Their surfaces can be easily perturbed during the measurements. Thus, UO_2 can oxidize rapidly. It was shown that a short exposure of UO_2 to atomic oxygen at room temperature oxidizes the surface up to UO_3. The formation of this high oxide is manifested by the disappearance of the U 5f emission peaks in the photoemission spectrum. The case of plutonium has been particularly studied because of its technological interest. Thin plutonium films can be provided as reasonably clean surfaces but the surface of the bulk metal oxidizes at room temperature and the thickness of the oxide layer depends strongly on the environment. In the presence of molecular oxygen, the sesquioxide (Pu(III)) is formed, then the dioxide (Pu(IV)). In contrast to UO_2, atomic oxygen does not incorporate into the bulk PuO_2, which is only covered by chemisorbed oxygen that is released at 200 °C. The presence of a higher oxide PuO_{2+x} had been considered but most recent results show it is not stable [75] while the stability of the Pu $5f^4$ configuration is higher than that of U $5f^2$.

Study of a clean plutonium surface and its oxidation was carried out on gallium-stabilized δ-plutonium [76]. Under a residual pressure of some 10^{-11} Torr at 77 K, the metal surface initially cleaned by laser ablation undergoes a superficial contamination from residual gases, which can take place over a time scale of several hours. This vacuum environment favours the initial formation of Pu_2O_3. Then, PuO_2 grows on the Pu_2O_3 layer. At 150 °C, Pu_2O_3 forms within minutes. Both oxides coexist over a wide range of thickness. However, after an extended period of time under the same pressure but at room temperature, the only oxide present at the surface was Pu_2O_3.

As in the other actinide oxides, in Am_2O_3, the photoemission intensity is zero at the Fermi level in agreement with the non-metallic character of the oxide. The O 2p–Am 6d7s valence band is located between E_F-6 to −3 eV and the peak towards the lower energies at about E_F-2 eV is attributed to americium 5f levels. This interpretation is confirmed by the intensity increase of this last peak with the increase of the incident photon energy (Fig. 5.18) [65]. The energy of the 5f peak is not shifted with respect to that in the metal. This is characteristic of the presence of localized 5f electrons, whose number remains the same in the two materials. Americium has three valence electrons in the metal and in its trivalent oxide. The final state of the photoemission is predicted to have the $5f^5$ configuration. However, the observed photoemission has no structure. It was suggested that no interaction

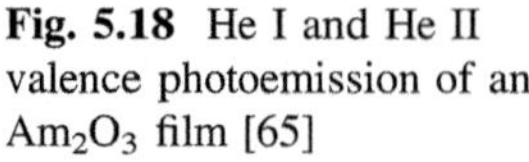

Fig. 5.18 He I and He II valence photoemission of an Am$_2$O$_3$ film [65]

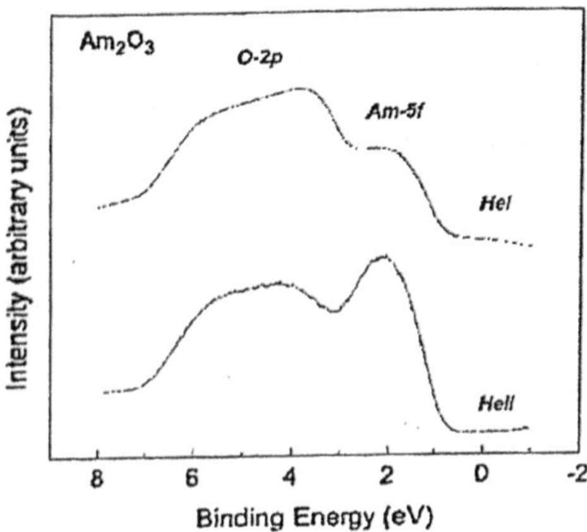

exists between the 5f^5 final configuration and the valence electron distribution in the oxide. The 5f^5 configuration is calculated to be very strongly localized and stable and only its ground state is predicted to be observed. Further studies are needed to confirm this interpretation.

The actinides are divalent in the monochalcogenides. US is known as having the most stable ferromagnetic state among these compounds. From VUV photoemission, the valence state distribution of US has revealed the presence of two narrow peaks, one located around E_F, the other in the range from -1 to -0.5 eV below it, and of a broad weak structure around -5 eV (Fig. 5.19) [77]. The more intense part of the spectrum corresponds to the 5f electrons. The peak at E_F is attributed to itinerant 5f electrons while the feature near -1 eV was suggested to be due to the 5f^3 configuration. The broad band centred around E_F-5 eV was the S3p–U6d7s valence electron band. These interpretations were based on the dual character of the 5f electrons. More recently, XPS has been observed from single-crystalline samples of US at 20 K. A single asymmetrical peak was observed near E_F. The additional structure detected by UPS was attributed to the presence of surface states and it was concluded that the U 5f electrons were itinerant in US [77]. However, the energy resolution with which the asymmetrical peak is observed by XPS could be reduced by the possible presence of secondary electrons and this could explain the difference between the two experimental results.

For USe and UTe, the XUV photoemission peaks have energies similar to those of US. Thus, a broader feature is present near -1 eV and a narrow peak is seen at the Fermi level. But the relative intensities are different. Here the more intense part of the spectrum is the wide emission located between about E_F-1.5 and E_F-0.2 eV, which corresponds to the 5f electrons [78]. The lattice constants are 5.489 and 6.155 Å for US and UTe, showing an increase of the distance between the uranium atoms passing from US to UTe, followed by an increase of the localization of the 5f electrons, as expected from the intensity increase of the emission characteristic of the localized 5f electrons.

Fig. 5.19 Valence
photoemission of US at low
temperature and various
incident photon energies. The
spectra are normalized to the
S 3p peak height [77]

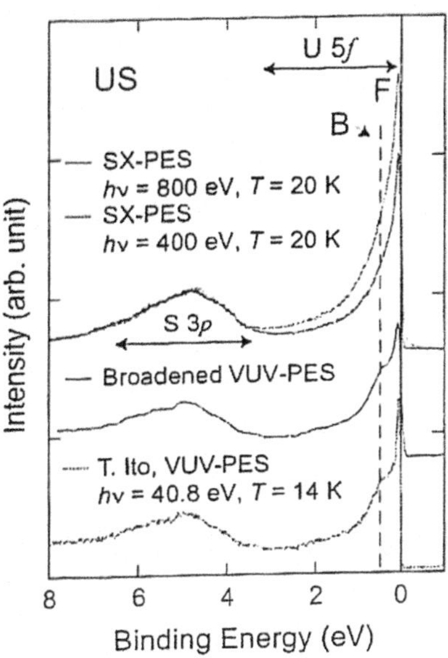

The plutonium divalent compounds are weakly paramagnetic. It was suggested
that the configuration of plutonium in these compounds was close to $5f^6$. As for the
uranium divalent compounds, the VUV photoemission observed for PuSe [79]
reveals the presence of several peaks. A large structure is present between E_F-2.5
and E_F-1.5 eV. A non-negligible part of the spectral intensity is concentrated in a
narrow peak near E_F. A peak is also present at about E_F-1 eV and a much smaller
feature exists between these two peaks. All the three peaks have already been
observed in the Pu δ-phase. They represent a characteristic feature of the plutonium
5f electron distribution, which is largely independent on a particular crystal environment. From LSDA and LSDA-SIC as well as LDA + U calculations, a peak is
predicted in the vicinity of the Fermi level but the structures round E_F-1 eV are not
well represented by these theoretical models [80, 81]. In contrast, the DMFT calculations give a picture similar to the experimental spectra for PuSe and PuTe [82].
However, the agreement is better for PuTe than for PuSe, probably because the
PuTe photoemission was obtained from a single crystal while for the PuSe spectra
thin films were used. This theoretical model is less satisfactory for UTe. The above
comparisons confirm that in order to describe the electronic structure of actinide
compounds dynamic correlations must be taken into account due to the partially
localized character of the 5f electrons in these materials.

The difference between the 5f distributions of the divalent compounds of uranium (UX) and plutonium (PuX) results from the difference in the number of their
localized 5f electrons. In UX, a large part of the 5f electron spectral intensity is
located between E_F-1 and E_F-0.5 eV and a narrow peak is present at E_F. In PuX, the
5f distribution is much more spread out and structured. A broad maximum is
centred at about E_F-2 eV and the three peaks already seen in δ-Pu between E_F-1 and

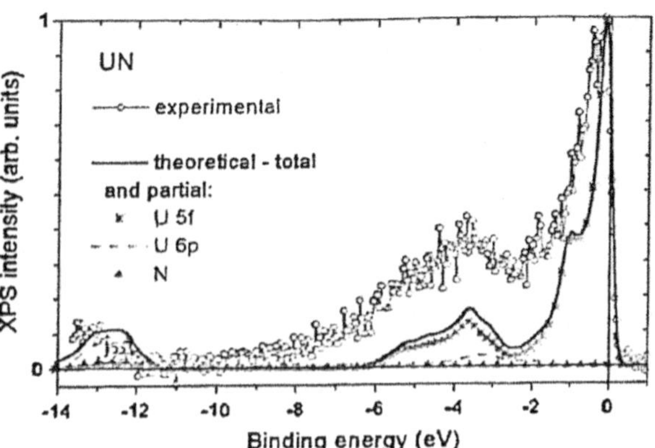

Fig. 5.20 Valence photoemission of UN compared with the calculated total valence band [83]

E_F are present. The maximum seen at E_F-2.0 eV in PuSe was attributed to an unresolved final-state $5f^4$ multiplet. This feature is not noticed for bulk metal plutonium but develops gradually when the thickness of plutonium layers is reduced down to a few monolayers. Its relative intensity can be taken as a measure of the 5f localization. By analogy, the maximum centred at E_F-1 eV in USe was attributed to a multiplet, which could belong to the $5f^2$ final configuration.

Among the monopnictides, UN was well studied. At T_N = 53 K it is antiferromagnetically ordered. The U–U distance is 3.46 Å. This distance is smaller than that of the other UX compounds. The N 2p band has a lower binding energy than the O 2p band. It is closer to the 5f levels. These different parameters favour the presence of a partially itinerant 5f character in UN. The U 5f levels were predicted theoretically to be present in a large energy range from E_F down to −6 eV. A relatively intense narrow peak, slightly mixed with U 6d band, was expected at E_F as well as a broad band combining U 6d band and N 2p levels between −3 and −6 eV [83]. The single crystal photoemission spectrum is in agreement with these predictions (Fig. 5.20) [83]. The electronic distribution spreads over a wide energy range. Towards the higher energies, it is characteristic of the valence states and of the presence of a covalent bonding. In this energy range, the spectrum has a shape similar to that observed for ThN [84] but it is more intense because 5f electrons are present and are hybridized with U 6d and N 2p electrons. Simultaneously, the 5f electrons appear in an asymmetrical peak near E_F extending up to −2 eV and the 6d density vanishes in this range. UN seems to be the compound with the widest energy distribution of 5f electrons. The 5f peak near to E_F could reveal the presence of quasi-localized 5f electrons. The 5f wave functions extend outside core regions and partially overlap the N 2p orbitals. There is no contamination by oxygen. Consequently, all the observed features are characteristic of UN.

USb is an antiferromagnet with T_N = 214 K. Various experimental results, such as a large magnetic moment and a small electronic specific heat coefficient, indicate a partially localized model as adequate to describe the 5f electrons in this compound. This is in agreement with the fact that the lattice parameters of USb are larger than the UN ones. Two peaks are present below E_F between about -0.6 and -0.06 eV [85]. They are predominantly 5f in character. These two peaks were assigned to the $5f^2$ final-state multiplet. They indicate the presence of localized U 5f electrons. The intensity of the multiplet structure is higher in USb than in UN. A band located at about 1 eV is ascribed to the U 6d electrons and a deeper and wider band to Sb 5p electrons.

Electronic structure calculations in mononitrides from thorium to plutonium resulted in theoretical values for the lattice parameters that agree with their experimental ones. This seems to demonstrate the presence of delocalized 5f electrons while a rapid lattice increase for AmN indicates the presence of 5f localization. However, the possibility to have intermediate localization as is the case in compounds of PuX type must be considered. The plutonium–nitrogen phase diagram exhibits only one known phase, PuN. This compound is antiferromagnetically ordered at $T_N = 13$ K. The emission intensity due to plutonium 5f electrons is concentrated around -2 eV while a large peak due to the N 2p–Pu 6d7s valence band is present between -2 and -5 eV. Comparison between the spectra obtained with He I and He II radiation revealed clearly the contribution of the 5f electrons to the photoemission (Fig. 5.21) [86]. A sharp peak is seen at E_F and another one at -0.85 eV as well as a weak peak at -0.5 eV. Contrary to the observations made for the monochalcogenides, the maximum at -2 eV is absent in PuN and the PuN 5f features are found similar to that observed for the metal δ-Pu. Consequently, the

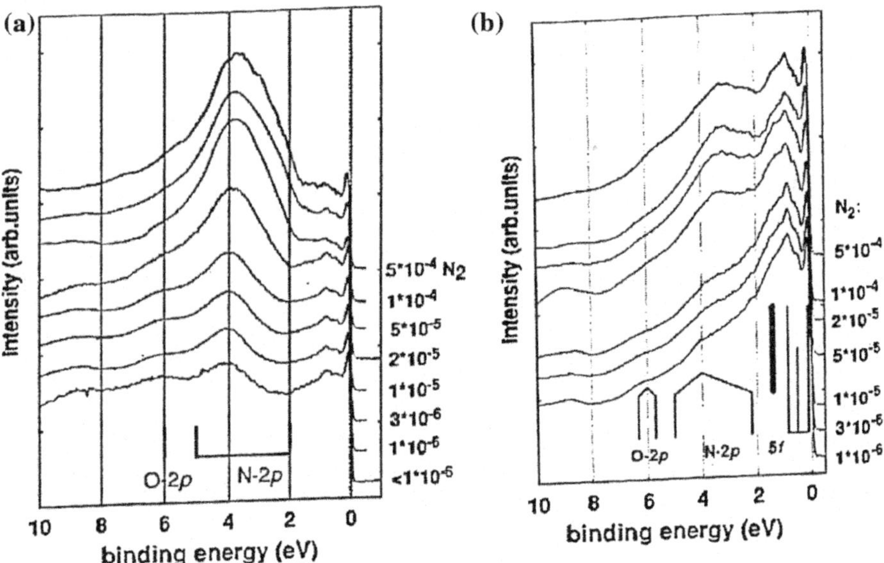

Fig. 5.21 Valence photoemission of PuN prepared by sputter deposition under varying nitrogen partial pressure with **a** He I, **b** He II incident radiation [86]

Fig. 5.22 Valence photoemission of PuSb and PuTe at He II [78]

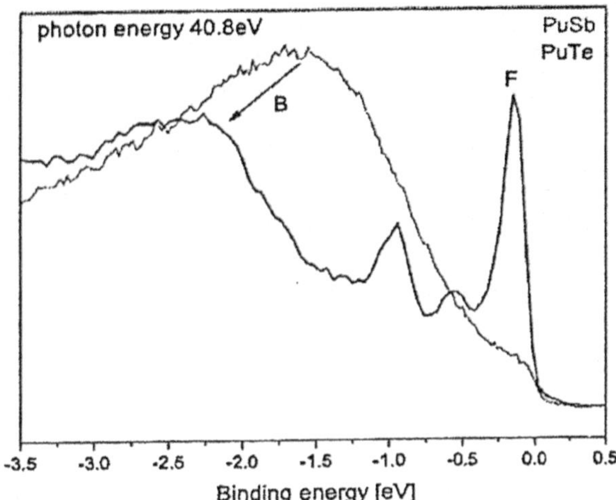

degree of 5f localization in PuN does not differ much from that of the plutonium metal. These same three peaks are also observed in the carbides [87].

PuSb crystallizes in the NaCl-type structure. It is ordered magnetically below $T_N \approx 85$ K with a magnetic moment of $0.75\mu_B$. These characteristics suggested that the 5f electrons were mostly localized and only a weak hybridization with conduction electron was present [88]. The density of states at E_F is very low. The intensity is present in a non-structured broad maximum centred at about -1.7 eV (Fig. 5.22) [77]. This structure is not observed in bulk plutonium phases but develops when the thickness is reduced down to few plutonium monolayers. Its intensity increases for energy incident photons jumping from 21.2 to 40.8 eV. This was attributed to an unresolved final-state $5f^4$ multiplet. A similar broad structure is observed in other trivalent compounds.

NaCl-type metallic UC was observed by XPS with the help of Mg $K\alpha$ radiation [50]. At this energy, the photoionisation cross section of U 5f levels is approximately 20 times greater than that of U 6d levels and 40 times greater than that of U 7s and C2s levels. The cross section of C2p is negligible. A single narrow peak was observed at -0.6 eV and ascribed to the U 5f electrons. A shoulder was seen between -2 and -4 eV and ascribed to U 5d levels mixed with U 5f levels. C2s band is seen around -9 eV. Differences between experimental results and band calculations suggest the presence of correlation effects between the 5f electrons.

In most of the plutonium metallic systems as well as in a broad range of its compounds three equally sharp peaks are observed within 1 eV below E_F. Only a few details are modified from one compound to another. For example, the sharp peak located just at E_F can undergo a slight shift towards higher binding energies or an additional very weak peak can appear. The spectral intensities can also vary considerably between different compounds. These peaks are expected to have a common origin. Their dependence on the incident photon energy reveals their 5f character. It reveals also that they are not related to the surface. A certain 6d admixture could be present. The observations of these peaks at the same energies in very different materials show that they cannot be related to any particular band structure and are quasi-independent of the surrounding. They are characteristic of the presence of localized 5f electrons. It was suggested that they were associated with the $5f^5$ multiplet while the structure present between 1 and 2 eV could be related to the $5f^4$ multiplet.

Study of the inter-metallic compounds is actually in progress. Their valence states photoemission is observed in order to follow the electron distribution changes responsible for their various macroscopic properties [89–91]. In UB_2, the U–U distance is only 3.123 Å and the specific heat coefficient is 10 mJ/mol K^2. The photoemission shows an intense sharp peak just below E_F, characteristic of the 5f electrons (Fig. 5.23) [91]. In $UFeGa_5$ and $UPtGa_5$, the U–U distances are, respectively, 4.258 and 4.341 Å. The spectral distribution reveals also the presence of an intense peak at E_F, particularly for $UFeGa_5$, which is paramagnetic as UB_2. The same intense peak is seen in UGe_2, UCoGe and URhGe, which are super-conductors under 0.8 K. In other compounds, such as UPt_3 a peak is present at E_F but it is weaker than the higher energy structures. Contrary to the previous compounds, in UPd_3 no peak is seen just below E_F, where the density of states is very low. The maximum of the valence states distribution is located at −3 eV and the contribution of U 5f levels was predicted to be noticeable in the −1 to −0.5 eV energy range. This compound has a specific heat coefficient of only 7.6 mJ/mol K^2 while it is equal to 420 mJ/mol K^2 for UPt_3, which has a high density of states near

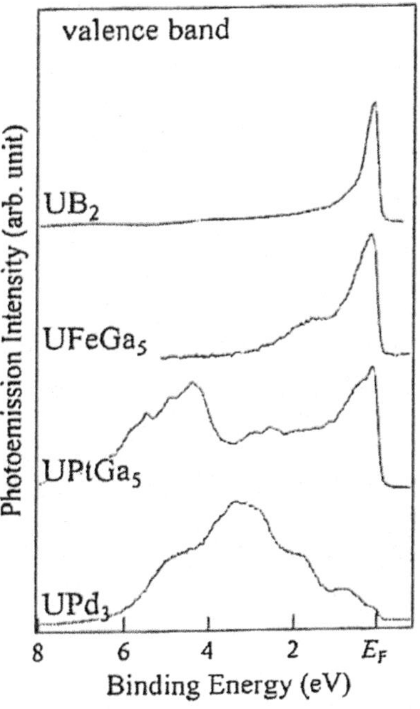

Fig. 5.23 Valence photoemission of UB_2, $UFeGa_5$, $UPtGa_5$ and UPd_3 at 800 eV [91]

E_F. From the intensity variation of the narrow peak near E_F, the number of the localized 5f electrons appears to decrease from UB_2 to UPd_3, where uranium is tetravalent [92]. However, it should be underlined that the density of the valence states of palladium is much higher than that of boron. This makes the comparison difficult. Analysis of the core levels was made simultaneously from the same samples in order to evaluate the number of the localized 5f electrons and, consequently, to deduce the relative importance of the hybridization between uranium 5f and ligand electrons in these materials (cf. Sect. 4.3).

Inter-metallic uranium compounds with other valence states have been observed and present analogous characteristics, namely an intense narrow peak just below E_F, a broad feature between about -1 and -0.5 eV and complex multiple peaks on the high binding energy side of this peak. For example, such distribution was observed for UNi_2Al_3 and UPd_2Al_3. For URu_2Si_2, analysis of the spectrum was made by varying the energy of the incident photons [93]. The narrow peak observed at E_F was attributed to U 5f electrons, while the intense feature centred around -2 eV was attributed to Ru 4d valence states. In a general point of view, it was considered that for a large majority of compounds the spectra are characteristic of a material with one of the three U 5f electrons being itinerant and contributing to the valence band while the remaining two 5f electrons are localized.

The isostructural $PuCoGa_5$, $PuRhGa_5$ and $PuCoIn_5$ superconductors, called Pu-115 series, have critical temperatures an order of magnitude higher than their cerium counterparts. Their superconducting temperatures are between those of the Ce-based and d-electron superconductors. In contrast, $UCoGa_5$ shows no evidence of strongly correlated electron characteristics. The photoemission of Pu-115 compounds presents a strong peak close to E_F and a structure located around 0.5 eV [94]. The peak and the structure were interpreted as revealing the presence of itinerant and localized 5f electrons, respectively [95].

5.4 Core Levels

5.4.1 *X-ray Photoelectron Spectroscopy (XPS)*

Core-level binding energies were determined experimentally from the X-ray absorptions and emissions [9, 10] and from the photoemission peaks [14]. The values obtained for thorium, uranium and plutonium are given in Table 5.2 and compared to the energies of the N, O and P levels measured for thorium and uranium by XPS [55, 96, 97]. The core-level energies obtained from X-ray spectroscopy are in agreement with those obtained by photoelectron spectroscopy except for the $P_{II,III}$ (6p) levels [98]. The proximity between the 6p levels and the ligand levels had induced confusion in the primary measurements, which have been corrected.

Table 5.2 Energy levels of Thorium and Uranium: (1) [10]; (2) [55]; (3) [98]

| | Energy levels | | | | | |
| | Thorium | | | Uranium | | |
	(1)	(2)	(3)	(1)	(2)	(3)
L_I	20,460			21,753		
L_{II}	19,688			20,943		
L_{III}	16,296			17,163		
M_I	5181			5548		
M_{II}	4821			5181		
M_{III}	4038			4302		
M_{IV}	3488			3725		
M_V	3330			3550		
N_I	1323			1439		
N_{II}	1160			1272		
N_{III}	959	966.4		1042	1043.0	
N_{IV}	711	712.4	713.0	780	778.3	780.1
N_V	676	675.2	675.7	738	736.2	737.7
N_{VI}	{335	342.4	343.9	{380	388.2	390.5
N_{VII}	{335	333.1	334.6	{380	377.4	379.6
O_I	290			325		
O_{II}	224	233.8		256	258.4	
O_{III}	173	177.2	178.9	198	194.8	197.2
O_{IV}	{87	92.5		{97	102.8	
O_V	{87	85.4	87.0	{97	94.2	96.0
P_I	60	41.4		72	43.9	
P_{II}	{41	24.5	25.2	{28	26.8	28.1
P_{III}	{41	16.6	17.2	{28	16.8	17.2

The spectra of the $n = 5$ core levels present secondary peaks, whose number increases along the series. This splitting of the peaks results from the strong interaction of the 5l hole present at the final state with the localized electrons of the incomplete 5f sub shell. The splitting is particularly important for the 5d levels. The $5d_{3/2}$ and $5d_{5/2}$ spin–orbit components are not recognized in the spectra of the heavier elements. Atomic-like calculations are necessary in this aim but it is difficult to identify the electronic structure of the investigated material from the 5d photoemission.

In contrast to the 5d core levels, the 4f core levels were observed extensively. The chemical shift of these levels was measured and the valence was deduced from these measurements. As an example, the binding energy of U $4f_{7/2}$ peak increases by 1.7–1.8 eV as the oxidation state of uranium increases from U(IV) in UO_2 to U(VI) in UO_3 [99]. From these measurements, it was shown that U(IV) and U(VI) ions, analogous to those found in UO_2 and UO_3, were mixed in U_3O_8 and U_2O_5 while no U(V) was present in these oxides [100]. Similar results were obtained by the same authors for the uranium fluorides.

Shake-up satellites are known to be located towards the higher energies of the main peak. They are attributed to the excitation of an electron from an occupied orbital to an empty or partially filled electronic level, which takes place simultaneously with the creation of the core hole. In the compounds, they correspond generally to an inter-atomic shake-up process with transfer of a ligand electron to an unoccupied metal level. In the actinides, no satellites associated with the 4d or 5d peaks were found. On the contrary, satellites were observed to the high binding energy side of the 4f peaks in numerous thorium and uranium compounds [101]. These satellites are very sensitive to the local chemical environment and their analysis gives information on the electronic structure and on the valence and conduction levels in the solid.

The observation of the shake-up satellites associated with the actinide 4f levels was developed at first for the uranium compounds [102, 103]. Energetically separated satellites were recorded for UO_2 and UO_3 and used to identify the presence of the U(IV) or U(VI). Initially, two satellites were predicted for each uranium oxide [103]. Independently, the 4f core-level spectra of the dioxides were recorded from thorium to californium (Fig. 5.24) [98] and a satellite was associated with each 4f peak of the light actinides. This satellite is attributed to the excitation of a valence electron to the 5f levels, often unsuitably named "charge transfer". Its position depends on the energy necessary to transfer an O2p valence electron to the empty actinide 5f orbitals and its intensity is a function of the hybridization strength between the O2p and actinide 5f orbitals. Its presence was confirmed in several experiments (Fig. 5.25) [70]. The intensity of these satellites increases from thorium to plutonium and decreases for americium and for curium; their shapes remain similar up to curium. For berkelium and californium, strong shape distortions are observed and indicate the existence of additional exchange interactions, which are present when the number of localized 5f electrons is superior to six.

The 4f core-level spectrum of UN is complex (Fig. 5.26) [83]. Three satellites are seen at higher energies of the $4f_{7/2}$ peak. Only two are observed for $4f_{5/2}$ because the presence of the N 1s peak perturbs the observation of the third one. The satellite observed at +7 eV of the $4f_{7/2}$ peak is analogous to others seen in numerous compounds of uranium and attributed to an inter-atomic shake-up. The two other unresolved satellites are at +3 and +1 eV of the main peak. It was suggested that this unresolved strongly asymmetric main peak corresponds to a mixing of the $5f^3$ and $5f^2$ configurations, the $5f^3$ configuration being connected with the peak of the lowest binding energy.

Inter-metallic compounds have been studied recently (Fig. 5.27) [91]. The U $4f_{7/2}$ peak of UB_2 is very close to that observed for the metal. Its slightly asymmetric shape is typical of the shape expected for a simple metallic material and no satellite is present. The asymmetry increases for UGe_2, $UFeGa_5$. A satellite of higher energy may also be present, for example for UNi_2Al_3 and UPd_2Al_3. For UPd_3 and UPt_3, a shoulder was observed approximately at the binding energy of the main peak observed for the previous materials. This shoulder can be attributed to a

Fig. 5.24 4f photoemission of oxides from thorium to californium [98]

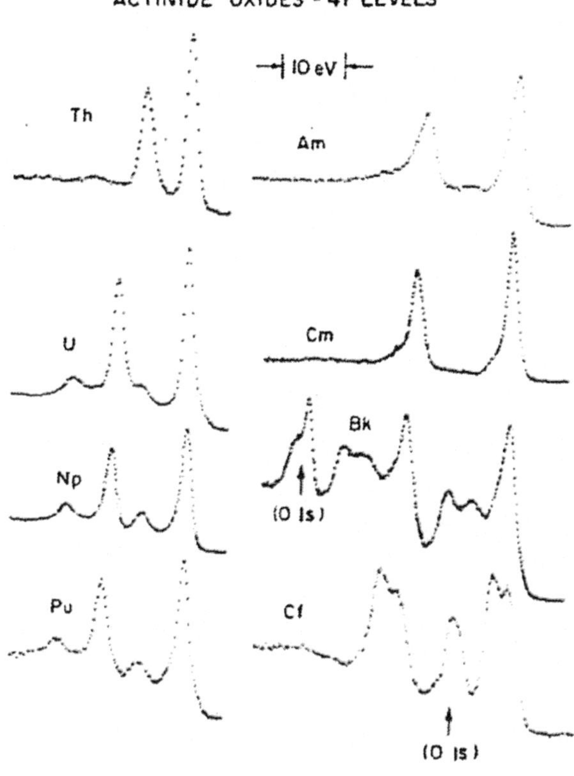

Fig. 5.25 Uranium 4f photoemission **a** of UO_2; **b** of $UO_{2.2}$ for $MgK\alpha$ incident radiation [67]

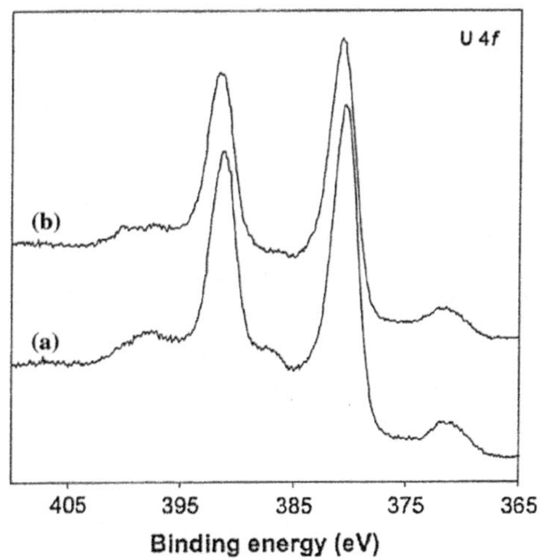

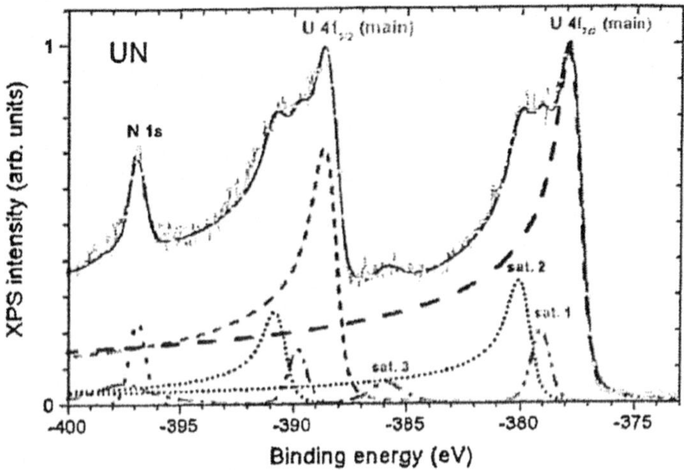

Fig. 5.26 Uranium 4f photoemission of UN, decomposed into the *main lines* and their satellites and N 1s photoemission [83]

Fig. 5.27 Uranium 4f photoemission of **a** UB₂, **b** UFeGa₅, **c** UPtGa₅ and **d** UPd₃ at 800 eV [91]

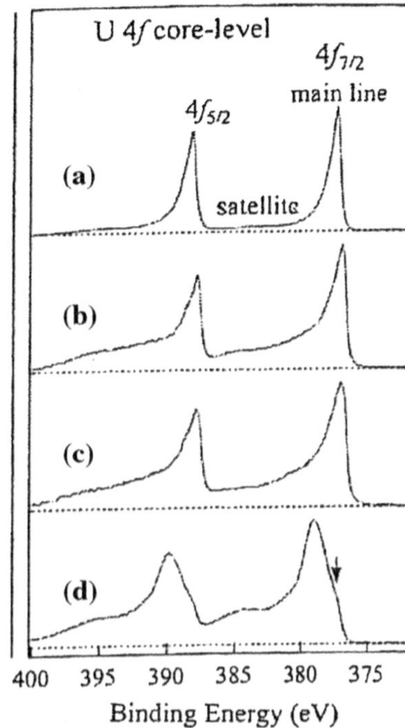

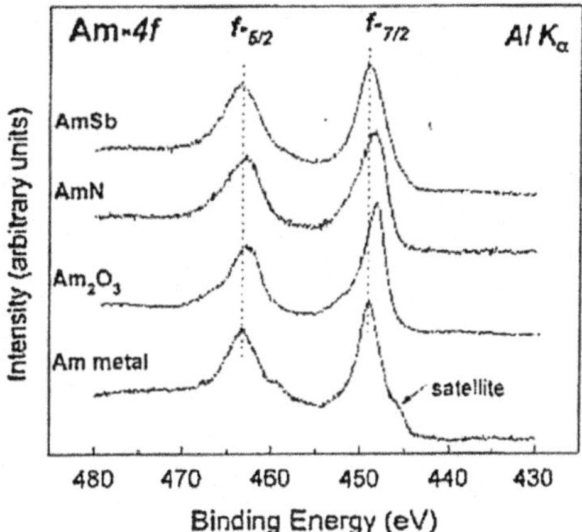

Fig. 5.28 Americium 4f photoemission of metal, Am₂O₃, AmN and AmSb. The low energy satellite is observed only in the metal [65]

well-screened peak and thus associated with the $5f^{n+1}$ configuration. The main peak, shifted by about 1–2 eV towards the higher energies, is thus associated with the $5f^n$ configuration. From this energy shift, the number of localized 5f electrons is smaller in UPd₃ and UPt₃ than in the other materials. It was suggested that this number was close to two while it could be close to three in all the other uranium inter-metallic compounds.

Few changes of the 4f core-level spectra were observed between americium in metal and trivalent compounds (Fig. 5.28) [65]. A satellite was seen towards the low binding energies in the metal and not in the compounds. This satellite was designated as a well-screened peak. By analogy with the 3d photoemission of the light rare-earth metals (cf. Chap. 4), this peak can be attributed to a shake-down satellite corresponding to the excited configuration $4f^{-1}5f^{n+1}$. The presence of this configuration can be related to the localized character of the 5f electrons in americium.

As for the rare-earth metals, the excited $(nl)^{-1}5f^{n+1}$ configuration can be created simultaneously with the formation of a core hole in a conductor but not in insulator compounds because of the presence of the energy gap. The energy of this "well-screened" configuration is lower than that of the $(nl)^{-1}5f^n$ ionized configuration, which corresponds to the main peak, often designed as "poorly screened" peak. As already underlined, the presence of a shake-down satellite depends on the energy interval between the top of the valence band and the empty 5f orbital. Its intensity depends on the coupling energy between valence band and localized $5f^n$ levels. This coupling is weak making the shake-down satellite weak. Its intensity decreases with an increase of the 5f orbital localization and it can fade out. These satellites are expected only for the metals and inter-metallic compounds of light actinides. For the insulator compounds, the empty 5f orbitals are generally pulled down into the forbidden band, leaving no observable shake-down satellite.

The energy of the main photoemission peak is equal to the energy of the core level as determined from X-ray emission and absorption or EELS and comparison between these various data can be used to identify the main peak. This was done for the rare-earths and resulted in the elimination of several ambiguities (cf. Chaps. 3 and 4). For the actinides, the levels concerned are generally the 4f levels. No comparison between the energies of the levels determined by photoemission and by other methods is made and no absolute energy scale is given in the figures. Consequently, numerous doubts exist about the designation of the poorly- and well-screened peaks. The peak of lower energy was often considered as a well-screened peak. The main peak was then either practically absent in americium [64] or very weak in thorium [55] and plutonium compounds [59, 86], in contradiction with the descriptions reported above.

Indeed, interpretations were based on a *new* model, different from the one used to describe the rare-earth spectra and initially developed to interpret the valence band spectra of the actinides. In this new model, the final configuration associated with the low binding energy peak is characterized by a strong mixing of the 5f orbitals with the valence band. The high binding energy peak is, then, associated with the presence of localized 5f electrons. Consequently, the low energy peak is dominant in the hybridized systems while the high energy peak is intense in localized systems. This description should not to be confused with the one previously developed in this text, from which the low energy peak, designated as well screened, corresponds to the formation of the $(nl)^{-1}n'f^{m+1}$ excited configuration and has a low or negligible intensity.

Theoretical analysis of the 4f photoemission spectra for dioxides was carried out using the impurity Anderson model and including the exchange interaction among 5f electrons [104]. It was shown that the energy Δ needed to transfer a O2p valence electrons to an actinide empty 5f orbital decreases from thorium to curium, then jumps up for berkelium. Good agreement was obtained between experimental and calculated spectra. The effect due to exchange interaction remains weak up to neptunium but becomes pronounced beyond. The valence was found to have almost integral values except for neptunium, plutonium and berkelium. It was about 4.6 in PuO_2 and 7.5 in BkO_2. Consequently, these two oxides were considered as mixed-valence materials. The correlation energy U is inferior to Δ in the beginning of the series. Neptunium on, Δ becomes smaller than U and the oxides are not of the Mott–Hubbard type. They have the characteristics of the oxygen–actinide charge transfer, which are responsible for the insulator properties of the material.

It should be remarked that some 4f core-level spectra of actinides were not correctly interpreted: the main peak is considered as having at the final state a 4f hole, well screened by the 5f electrons; the satellite is then interpreted as being the poorly-screened main peak. This interpretation is equivalent to the identification of the main peak with a shake-down satellite while these satellites are known as having a weak intensity. Let us recall that the energies of the core transitions can be determined from the level energies as deduced from X-ray spectra and, consequently, no ambiguity can exist in the interpretations.

5.4.2   Resonant Auger Emission

As for the rare-earths, resonant Auger taking place from an excited configuration towards a singly ionized one was observed also in the actinide spectra but initially identified with a shake-up satellite. This is the case for α-uranium metal [105]. A peak corresponding to the resonant Auger transition $(5d_{5/2})^{-1}5f^{n+1}$–$5f^{n-1}$ was observed at -2.3 eV below E_F when the incident photons have energy near the $5d_{5/2}$ threshold, i.e. when the excitation probability is larger than the probabilities of the other processes. The observation of this peak has revealed the presence of localized uranium $5f^{n+1}$ electrons with n integer.

Auger process was observed at the 3d threshold in UO_2 [106]. From the excited configuration $3d^9 4f^{14} 5f^3$, different decay processes are possible, like the autoionization to $3d^{10} 4f^{13} 5f^2 \varepsilon g$ and Auger decay in the presence of the spectator 5f electron to $3d^{10} 4f^{12} 5f^3 \varepsilon g$. The multiplet structure of this Auger process was observed near the $4d_{3/2}$ photoemission peak and was difficult to resolve (Fig. 5.29) [106]. However, it was clearly identified. This observation showed the presence of the excited configuration $3d^9 4f^{14} 5f^3$ in UO_2.

Auger decay processes of the super Coster–Kronig type, $5d^9 5f^{n+1} \rightarrow 5d^{10} 5f^{n-1} +$ e^-, were observed in single crystals of $U_x Th_{1-x} Sb$, UTe, UPd_3 and UO_2 [107]. The observation of these transitions involves the decay of the core excited $5d^9 5f^{n+1}$ configuration and indicates that localized 5f electrons are present in these materials. Indeed, the spectra of these four compounds are dominated by the transitions involving localized 5f electrons. From the same excited configuration, the Coster–Kronig transition, $5d^9 5f^{n+1} \rightarrow 5d^{10} 5f^n 6d^{-1} + e^-$, can also take place. At the resonance, i.e. for incident photons of 98 eV, an asymmetric peak was observed near E_F in α-U [108]. It was attributed to the 6d valence band mixed with 5f states towards the higher energies. Asymmetry towards the higher energies slowly increases in the presence of gallium, which is known to stabilize the fcc phase with respect to the less symmetrical phases. That favoured the localization of the 5f electrons and explained the extension of the band towards the higher energies up to -3 eV. In UO_2, no peak is observed near E_F. A quasi-symmetrical peak is observed at -2 eV, at the energy of the asymmetrical extension observed in the metal and attributed to the 5f localized electrons. A large band attributed to the 2p oxygen orbitals mixed with the valence electrons of uranium is present between -5 and -8 eV. This description is in agreement with the one obtained by valence photoemission.

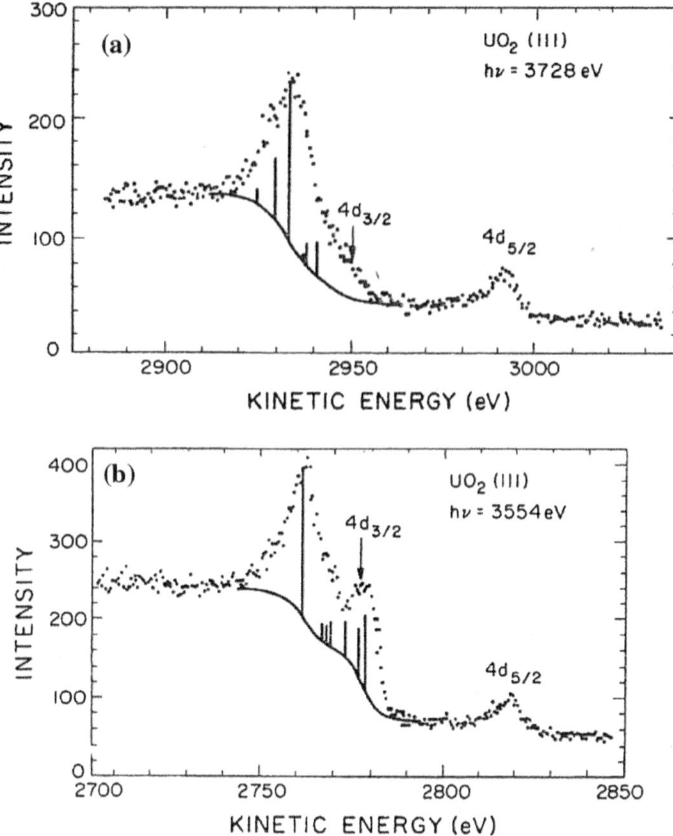

Fig. 5.29 Uranium 4d photoemission of UO$_2$ and MN$_{6,7}$N$_{6,7}$ Auger transitions at **a** 3d$_{3/2}$ and **b** 3d$_{5/2}$ resonant photon energies. The *vertical bars* indicate calculated major components of the multiplets [106]

5.5 Conclusion

The physical properties of actinides and their compounds are governed by the presence of the 5f electrons. In the metallic materials, the 5f electrons were initially considered as forming narrow bands. However, while the density of states of thorium was correctly described by theoretical calculations, it was difficult to treat in the band formalism the nearly localized 5f electrons of uranium. It was then shown that the 5f electrons may either be localized or itinerant and it was difficult to treat theoretically the intermediate stage between localization and itinerancy. With the exception of cerium, there is a difference between the actinides and the lanthanides. Whereas the 4f electrons always keep their localized character in all the trivalent rare-earths, in the actinides the 5f electrons behave with a more complex manner. In the heavy actinides, the 5f electrons are localized. In the light actinides, they can be in narrow bands or be partially localized and partially hybridized with the valence electrons. Delicate balance between 5f bonding and localization appears to be the valid solution to be retained. The 5f electrons have a dual character. Thus, hybridization of 5f orbitals with the 6d, 7s orbitals exists for the light actinides but it is only partial. Two-fluid models, localized atomic-like and itinerant band-like, were suggested to treat the dual character of the 5f electrons in the light actinides and in a number of their compounds. It was, moreover, suggested that the impact of the 5f localization could depend on the other atoms present in the compounds of light actinides and on the close overlap between the 5f orbitals and the ligand orbitals.

For the heavy actinides, from americium, the lattice constant suddenly increases and the above overlap is reduced. Moreover, orbital contraction accompanies the increase of the nuclear charge. Both these effects lead to a localization of the 5f electrons in metal and compounds. However, this localization is often considered as fragile so that a pressure increase might be enough to make the 5f electrons lose their localized character.

The fundamental question is, then, to know the localization regime of the 5f electrons. Valence changes and fluctuations already observed in rare-earth materials have a high probability to be present also in the actinide compounds and the debate concerning the characteristics of the 5f orbitals seems yet incompletely resolved. This is because very few experimental techniques yield direct evidence of the number of the localized 5f electrons or, conversely, of the number of the orbitals participating in bond formation. Moreover, the spectra give an instantaneous description of the electronic configuration but eventual valence fluctuations are unobservable. The shape of the transitions involving a localized 5f electron is not very sensitive to the configurations concerned. Only their energy position and their intensity vary and these parameters have rarely been measured with precision. That explains why a method such as photoemission has been privileged: indeed, the strong variation of the peak shape from one compound to another could be a useful parameter in the interpretation of the photoemission spectra.

However, to determine the *number of localized* 5f *electrons*, the direct method is spectral analysis of the X-ray emissions stimulated by electrons including excitations from the nd core levels to 5f levels. Indeed, the observed nd–5f resonance lines demonstrate the localization of 5f electrons on the atomic sites having an nd core hole and give information on their dynamics. As already underlined, resonance lines are only observed if the interactions between the excited and continuum levels are weak, that is to say if the excited electron remains localized on the same atom with the same spin orientation during the lifetime of the core hole. It appears that only the nf electrons undergo sufficiently small relaxation effects for the observation of resonance lines to be possible while resonance emissions are not observed in the spectra of the transition elements and their compounds. This can be explained by the difference between the characteristics of the d and f wave functions.

The normal nd–5f emissions vary also strongly according the character of the 5f electrons. If itinerant 5f electrons are present, normal emission is expected towards the lower energies of the absorption transitions. If the 5f electrons are localized, normal emissions in the ions with an 3d or 4d hole are predicted towards the higher energies of the resonance lines as observed in the rare-earth spectra. The big advantage of this type of experiment is the possible comparison between the experimental and calculated 5f emissions, which enables the determination of the number of the localized 5f electrons. Electron stimulated X-ray emission is thus an excellent method to obtain precise information on the localization or itinerancy of the 5f electrons in the actinides. Owing to the remarkable characteristics of plutonium, observation of the spectra of its six crystallographic phases in order to determine its 5f electron spatial distribution could turn out very revealing.

In summary, for the light actinides, as for cerium, the number of localized nf electrons decreases with an increase of the oxidation state while beyond americium the oxidation state remains constant. The most studied typical case is that of uranium, for which the valence varies between 3 and 6. To this variation is associated a decrease of the number of localized 5f electrons from 3 to 0 while, in contrast, the number of the 5f electrons in the valence band increases but is not an integer. Particular electronic properties and complex crystallographic and physical characteristics are inherently connected with the simultaneous presence of localized and itinerant 5f electrons. This makes the study of the electronic structure of actinide compounds a very important challenge.

References

1. H. Hulubei, Y. Cauchois, C.R. Séances, Acad. Sci. Ser. C **209**, 39 (1939)
2. B.F. Thornton, S.C. Burdette, Bull. Hist. Chem. **35**(2), 86 (2010)
3. P.F. Dittner, C.E. Bemis, D.C. Hensley, R.J. Silva, C.D. Goodman, Phys. Rev. Lett. **26**, 1037 (1971)
4. K. Lassmann, C. O'Carroll, J. Laar, C.T. Walker, J. Nucl. Mater. **208**, 201 (1994)
5. F. Wastin, C.T. Walker, A.D. Stalios, J. Rebizant, in *Proceeding of Actinide 97*, Baden-Baden, Germany, Sept 1997
6. J.H. Shim, K. Haule, G. Kotliar, Nature **446**, 513 (2007)
7. Y. Cauchois, M.L. Allais, Philo. Trans. Roy. Soc. London A **1**, 44 (1940)
8. Y. Cauchois, I. Manescu, Disq. Mat. Phys. **1**, 117 (1940)
9. Y. Cauchois, I. Manescu, L. de Bersuder, C. R. Acad. Paris **253**, 1042 (1961)
10. Y. Cauchois, J. Phys. Radium **13**, 113 (1952)
11. Y. Cauchois, I. Manescu, L. Le Berquier, C. R. Acad. Paris **239**, 1780 (1954)
12. Y. Cauchois, in *Proceedings of Xth Colloquium Spectroscopicum Internationale 1962*, Spartan Book
13. S. Hagström, C. Nordling, K. Siegbahn, *Table of Electron Binding Energies, Alpha, Beta and Gamma Ray Spectroscopy*, ed. by K. Siegbahn, North-Holland (1964)
14. J.A. Bearden, A.F. Burr, Rev. Mod. Phys. **39**, 125 (1967)
15. M.O. Krause, C.W. Nestor, Phys. Scr. **16**, 285 (1977)
16. F.T. Porter, M.S. Freedman, J. Phys. Chem. Ref. Data **7**, 4 (1978)
17. J. Petiau, G. Calas, D. Petitmaire, A. Bianconi, M. Benfatto, A. Marcelli, Phys. Rev. B **34**, 7350 (1986)
18. S. Bertram, G. Kaindl, J. Jové, M. Pagès, J. Gal, Phys. Rev. Lett. **63**, 2680 (1989)
19. T. Vitova, K.O. Kvashnina, G. Nocton, G. Sukharina, M.A. Denecke, S.M. Butorin, M. Mazzanti, R. Caciuffo, A. Soldatov, T. Behrends, H. Geckeis, Phys. Rev. B **82**, 235118 (2010)

20. C.H. Booth, Y. Jiang, D.L. Wang, J.N. Mitchell, P.H. Tobash, E.D. Bauer, M.A. Wall, P.G. Allen, D. Sokaras, D. Nordlung, T.-C. Weng, M.A. Torrez, J.L. Sarrao, Proc. National Acad. Sciences **109**, 10205 (2012)
21. S. Heathman, J.-P. Rueff, L. Simonelly, M.A. Denecke, J.-C. Griveau, R. Caciuffo, G.H. Lander, Phys. Rev. B **82**, 201103 (2010)
22. Y. Cauchois, C. Bonnelle, L. de Bersuder, C. R. Acad. Sc. Paris **257**, 2980 (1963)
23. Y. Cauchois, C. Bonnelle, I. Manescu, C. R. Acad. Sc. Paris **267**, 817 (1968)
24. C. Bonnelle, *Theoretical Chemistry and Physics of Heavy and Superheavy Elements*, ed. by U. Kaldor, S. Wilson, (Kluwer Academic Publication, 2003) pp. 115–170
25. C. Bonnelle, G. Lachère, J. de Physique **35**, 295 (1974)
26. J.L. Dehmer, A.F. Starace, U. Fano, J. Sugar, J.W. Cooper, Phys. Rev. Lett. **26**, 1521 (1971)
27. M. Cukier, P. Dhez, B. Gauthé, P. Jeaglé, Cl Wehenkel, F. Combet-Farnoux, J. Physique Lettres **39**, L315 (1978)
28. I.M. Band, M.B. Trzhaskovskaya, J. Moscow Phys. Soc. **1**, 359 (1991)
29. G. Wendin, Phys. Rev. Lett. **53**, 724 (1984)
30. M. Pantelouris, J.P. Connerade, J. Phys. B At. Mol. Phys. **16**, L23 (1983)
31. T.M. Zimkina, A.S. Shulakov, A.P. Braiko, in *Proceeding of the joint UK-USSR Seminars*, Daresbury Laboratory, Nov 1982
32. F. Jollet, T. Petit, S. Gota, N. Thromat, M. Gautier-Soyer, A. Pasturel, J. Phys. Condens. Matter. **9**, 9393 (1997)
33. M. Magnuson, S.M. Butorin, L. Werme, J. Nordgren, K.E. Ivanov, J.-H. Guo, D.K. Shuh, Appl. Surf. Sci. **252**, 5615 (2006)
34. C. Bonnelle, Struct. Bond. **31**, 23 (1976)
35. C. Bonnelle, A. Courtois, D. Calais, 5th International Conference on Plutonium and Other Actinides, North-Holland, Amsterdam, 1976
36. M. Colella, G.R. Lumpkin, Z. Zhang, E.C. Buck, K.L. Smith, Phys. Chem. Minerals **32**, 52 (2005)
37. K.T. Moore, G. Vander Laan, R.G. Haire, M.A. Wall, A.J. Schwartz, Phys. Rev. B **73**, 033109 (2006)
38. K.T. Moore, P. Söderlind, A.J. Schwartz, D.E. Laughlin, Phys. Rev. Lett. **96**, 206402 (2006)
39. J.H. Shim, K. Haule, G. Kotliar, EPL **85**, 17007 (2009)
40. K.T. Moore, G. vander Laan, M.A. Wall, A.J. Schwartz, R.G. Haire, Phys. Rev. B **76**, 073105 (2007)
41. M. Cukier, B. Gauthe, C. Wehenkel, J. Physique **41**, 603 (1980)
42. M.T. Butterfield, K.T. Moore, G. vander Laan, M.A. Wall, R.G. Haire, Phys. Rev. B **77**, 113109 (2008)
43. Y. Baer, J.K. Lang, Phys. Rev. B **21**, 2060 (1980)
44. Y. Baer, J. Schoenes, Sol. State Commun. **33**, 885 (1980)
45. Y. Baer, J. Schoenes, Sol. State Commun. **43**, 783 (1982)
46. P. Roussel, P. Morrall, S.J. Tull, J. Nuclear Mat. **385**, 53 (2009)
47. J. Schoenes, J. Appl. Phys. **49**, 1463 (1978)
48. J. Schoenes, J. Magn. Magn. Mat. **9**, 57 (1978)
49. J.G. Tobin, S.-W. Yu, Phys. Rev. Lett. **107**, 167406 (2011)
50. T. Ejima, K. Murata, S. Suzuki, T. Takahashi, S. Sato, T. Kasuya, Y. Onuki, H. Yamagami, A. Hasegawa, T. Ishii, Phys. B **186–188**, 77 (1993)
51. T.M. Zimkina, A.S. Shulakov, A.P. Braiko, Sov. Phys. Solid State **23**, 1171 (1981)
52. G. Lachéré, C. Bonnelle, J. de Physique, 41, C5-15, 1980
53. C. Bonnelle, G. Lachère, 5th International Conference on Plutonium and Other Actinides, North-Holland, Amsterdam, 1976
54. C. Bonnelle, P. Jonnard, C. Barré, G. Giorgi, J. Bruneau, Phys. Rev. A 3422, 1997
55. J.C. Fuggle, A.F. Burr, L.M. Watson, D.J. Fabian, W. Lang, J. Phys. F Metal Phys. **4**, 335 (1974)
56. B.W. Veal, D.J. Lam, Phys. Rev. B **10**, 4902 (1974)
57. J.K. Lang, Y. Baer, P.A. Cox, Phys. Rev. Lett. **42**, 74 (1979)

58. D.T. Larson, J. Vac. Sci. Technol. **17**, 55 (1980)
59. T. Gouder, L. Havela, F. Wastin, J. Rebizant, Europhys. Lett. **55**, 705 (2001)
60. S.Y. Savrasov, G. Kotliar, E. Abrahams, Nature **410**, 793 (2001)
61. J.M. Wills, O. Eriksson, A. Delin, P.H. Andersson, J.J. Joyce, T. Durakiewicz, M.T. Butterfield, A.J. Arko, D.P. Moore, L.A. Morales, J. Electron Spectr. Relat. Phenom. **1**(35), 163 (2004)
62. A.J. Arko, J.J. Joyce, L. Morales, J. Wills, J. Lashley, F. Wastin, J. Rebizant, Phys. Rev. B **62**, 1773 (2000)
63. A.B. Shick, V. Drchal, L. Havela, Europhys. Lett. **69**, 588 (2005)
64. J.R. Naegele, L. Manes, J.C. Spirlet, W. Müller, Phys. Rev. Lett. **52**, 1834 (1984)
65. T. Gouder, P.M. Oppeneer, F. Huber, F. Wastin, J. Rebizant, Phys. Rev. B **72**, 115122 (2005)
66. P. Söderling, K.T. Moore, A. Landa, B. Sadigh, J.A. Bradley, Phys. Rev. B **84**, 075138 (2011)
67. J.J. Joyce, T. Durakiewicz, K.S. Graham, E.D. Bauer, D.P. Moore, J.N. Mitchell, J.A. Kennison, R.L. Martin, L.E. Roy, G.E. Scuseria, J. Phys. Conf. Series **273**, 012023 (2011)
68. V.A. Gubanov, A. Rosen, D.E. Ellis, J. Phys. Chem. Solids **40**, 17 (1979)
69. B.W. Veal, D.J. Lam, Phys. Lett. **49A**, 466 (1974)
70. T. Gouder, C. Colmenares, J.R. Naegele, J. Verbist, Surf. Sci. **235**, 280 (1990)
71. L.E. Cox, J. Electron Spectros. Relat. Phenom. **26**, 167 (1982)
72. B.W. Veal, D.J. Lam, W.T. Carnall, H.R. Hoekstra, Phys. Rev. B **12**, 5651 (1975)
73. L.E. Cox, W.P. Ellis, R.D. Cowan, J.W. Allen, S.-J. Oh, I. Lindau, B.B. Pate, A.J. Arko, Phys. Rev. B **35**, 5761 (1987)
74. M.T. Butterfield, T. Durakiewicz, I.D. Prodan, G.E. Scuseria, E. Guziewicz, J.A. Sordo, K.N. Kudin, R.L. Martin, J.J. Joyce, A.J. Arko, K.S. Graham, D.P. Moore, L.A. Morales, Surf. Sci. **571**, 74 (2004)
75. T. Gouder, A. Seibert, L. Havela, J. Rebizant, Surf. Sci. **601**, L77 (2007)
76. M.T. Butterfield, T. Durakiewicz, E. Guziewicz, J.J. Joyce, A.J. Arko, K.S. Graham, D.P. Moore, L.A. Morales, Surf. Sci. **571**, 74 (2004)
77. Y. Takeda, T. Okane, T. Ohkochi, Y. Yamagami, A. Fujimori, A. Ochiai, Phys. Rev. B **80**, 161101(R) (2009)
78. T. Durakiewicz, J.J. Joyce, G.H. Lander, C.G. Olson, M.T. Butterfield, E. Guziewicz, A.J. Arko, L. Morales, J. Rebizant, K. Mattenberger, O. Vogt, Phys. Rev. B **70**, 205103 (2004)
79. T. Gouder, F. Wastin, J. Rebizant, L. Havela, Phys. Rev. Lett. **84**, 3378 (2000)
80. P.M. Oppener, T. Kraft, M.S.S. Brook, Phys. Rev. B **61**, 12825 (2000)
81. L. Petit, A. Svane, W.M. Temmerman, Z. Szotek, Eur. Phys. J. B **25**, 139 (2002)
82. L.V. Pourovskii, M.I. Katsnelson, A.I. Lichtenstein, Phys. Rev. B **72**, 115106 (2005)
83. M. Samsel-Czekala, E. Talik, P. de V. Du Plessis, R. Troc, H. Misiorek, C. Sulkowski, Phys. Rev. B 76, 144426, 2007
84. P.R. Norton, R.L. Tapping, D.K. Creber, W.J.L. Buyers, Phys. Rev. B **21**, 2572 (1980)
85. T. Ito, H. Kumigashira, S. Souma, T. Takahashi, T. Suzuki, J. Magn. Magn. Mat. **226–230**, 68 (2001)
86. L. Havela, F. Wastin, J. Rebizant, T. Gouder, Phys. Rev. B **68**, 085101 (2003)
87. T. Gouder, L. Havela, A.B. Shick, F. Huber, F. Wastin, J. Rebizant, J. Phys. Condens. Matter **19**, 476201 (2007)
88. G.J. Hua, B.R. Cooper, Phys. Rev. B **48**, 12743 (1993)
89. T. Ohkochi, S. Fujimori, H. Yamagami, T. Okane, Y. Saitoh, A. Fujimori, Y. Haga, E. Yamamoto, Y. Onuki, Phys. Rev. B **78**, 165110 (2008)
90. T. Ito, H. Kumigashira, S. Souma, T. Takahashi, Y. Haga, Y. Tokiwa, Y. Onuki, Phys. Rev. B **66**, 245110 (2002)

91. S. Fujimori, T. Ohkochi, I. Kawasaki, A. Yasui, Y. Takeda, T. Okane, Y. Saitoh, A. Fujimori, H. Yamagami, Y. Haga, E. Yamamoto, Y. Tokiwa, S. Ikeda, T. Sugai, H. Ohkuni, N. Kimura, Y. Ōnuki, J. Phys. Soc. Jpn. **81**, 014703 (2012)
92. K.A. McEwen, H.C. Walker, P. Boulet, E. Colineau, F. Wastin, S.B. Wilkins, D. Fort, J. Phys. Soc. Jpn. Suppl. **75**, 20 (2006)
93. I. Kawasaki, S. Fujimori, Y. Takeda, T. Okane, A. Yasui, Y. Saitoh, H. Yamagami, Y. Haga, E. Yamamoto, Y. Onuki, J. Phys. Conf. Series **273**, 012039 (2011)
94. J.J. Joyce, T. Durakiewicz, K.S. Graham, E.D. Bauer, D.P. Moore, J.N. Mitchell, J.A. Kennison, R.L. Martin, L.E. Roy, G.E. Scuseria, J. Phys. Conf. Ser. **273**, 012023 (2011)
95. M.E. Pezzoli, K. Haule, G. Kotliar, Phys. Rev. Lett. **106**, 016403 (2011)
96. C. Nordling, S. Hagström, Ark. Fys. **15**, 431 (1959)
97. C. Nordling, S. Hagström, Z. Phys. **178**, 418 (1964)
98. B.W. Veal, D.J. Lam, H. Diamond, H.R. Hoekstra, Phys. Rev. B **15**, 2929 (1977)
99. D. Chadwick, Chem. Phys. Lett. **21**, 291 (1973)
100. J. Verbist, J. Riga, J.J. Pireaux, R. Caudano, J. Electron Spectr. Relat. Phenom. **5**, 193 (1974)
101. G.C. Allen, P.M. Tucker, Chem. Phys. Lett. **43**, 254 (1976)
102. C. Keller, C.K. Jorgensen, Chem. Phys. Lett. **32**, 397 (1975)
103. J.J. Pireaux, J. Riga, E. Thibaut, C. Tenret-Noel, R. Caudano, J.J. Verbist, Chem. Phys. **22**, 113 (1977)
104. A. Kotani, T. Yamazaki, Prog. Theor. Phys. Suppl. **108**, 117 (1992)
105. M. Iwan, E.E. Koch, F.-J. Himpsel, no published
106. L.E. Cox, W.P. Ellis, R.D. Cowan, J.W. Allen, S.-J. Oh, Phys. Rev. B **31**, 2467 (1985)
107. B. Reihl, N. Martensson, D.E. Eastman, A.J. Arko, O. Vogt, Phys. Rev. B **26**, 1842 (1982)
108. B. Reihl, M. Domke, G. Kaindl, G. Kalkowski, C. Laubschat, F. Hulliger, W.D. Schneider, Phys. Rev. B **32**, 3530 (1985)

Subject Index

© Springer Science+Business Media Dordrecht 2015
C. Bonnelle and N. Spector, *Rare-Earths and Actinides in High Energy Spectroscopy*, Progress in Theoretical Chemistry and Physics 28, DOI 10.1007/978-90-481-2879-2

Chemical Index

© Springer Science+Business Media Dordrecht 2015

C. Bonnelle and N. Spector, *Rare-Earths and Actinides in High Energy Spectroscopy*, Progress in Theoretical Chemistry and Physics 28, DOI 10.1007/978-90-481-2879-2